AFFINITY CHROMATOGRAPHY

Edited by **Sameh Magdeldin**

Affinity Chromatography
Edited by Sameh Magdeldin

Published by InTech
Janeza Trdine 9, 51000 Rijeka, Croatia

Publishing Process Manager Bojan Rafaj

Technical Editor Teodora Smiljanic

Cover Designer InTech Design Team

Additional hard copies can be obtained from orders@intechopen.com

Affinity Chromatography, Edited by Sameh Magdeldin
p. cm.
ISBN 978-953-51-0325-7

For EU product safety concerns:
IN TECH d.o.o., Prolaz Marije Krucifikse Kozulić 3, 51000 Rijeka, Croatia,
info@intechopen.com or visit our website at intechopen.com.

Meet the editor

Sameh Magdeldin is senior post doctoral researcher in medical school, Niigata University, Japan and academic associate professor (Lecturer) in Physiology dept. SCU, Egypt. He received his M.V.Sc and Ph.D in Physiology and expecting his 2nd Ph.D in proteomics by July, 2012. He is expertise in shot-gun proteomics analysis, reversed- phase chromatography and label- free comparative proteomic approaches. Dr. Magdeldin created the outstanding "All and None" methodology for analyzing large throughput proteomic data published in a highly respected proteomics journal. Beside his published articles, he currently serves as a guest editor, associate editor and a peer-reviewer in several international journals. Dr. Magdeldin received several grants and awards such as 8th HUPO congress young investigator award, JSN award, Grant- in – aid for young scientist, and young researcher overseas grant from Japan society for the promotion of science JSPS.

Contents

Preface

With the huge number of books and publications that utilizes different aspects of affinity chromatography, it is still hard to find a freely accessed book that gathers a solid and concise understanding of affinity principles together with its applications in a single volume. The vision of this book is not more than an attempt to provide an open source single volume demonstrating the concept of affinity chromatography with some of its applications that meets the current throughput screening demands of scientists and researchers. This book "*Affinity chromatography*" starts with introductory chapter that passes through the basic principles and snapshot applications of this analytical technique, followed by a wide ranged practical research chapters utilizing affinity chromatography, written by leading experts worldwide. It is worthy to say that the scope of the information contained in this book is still limited to be covered in a single volume.

Affinity chromatography is aimed mainly at those interested in different analytical and separation techniques, particularly, biochemists, biologists, pharmacists, advanced graduate students and post graduate researchers.

Finally, I am grateful to the all experts who showed positive and specific "*affinity*"to participate and share in this book with their valuable experience. Indeed, without their participation, this book won't come to light.

Sameh Magdeldin, Ph.D
Senior Post Doc Researcher and Proteomics Team Leader
Medical School, Niigata University,
Japan
Ass. Prof. (Lecturer), Physiology Dept.
Suez Canal University,
Egypt

Part 1

Principles and Applications of Affinity Chromatography

1

Affinity Chromatography: Principles and Applications

Sameh Magdeldin[1,2] and Annette Moser[3]
[1]*Department of Structural Pathology, Institute of Nephrology, Graduate School of Medical and Dental Sciences, Niigata University,*
[2]*Department of Physiology, Faculty of Veterinary Medicine, Suez Canal University, Ismailia,*
[3]*Department of Chemistry, University of Nebraska at Kearney, Kearney, NE,*
[1]*Japan*
[2]*Egypt*
[3]*USA*

1. Introduction

Since the inception of affinity chromatography 50 years ago (Cuatrecasas et al, 1968), traditional purification techniques based on pH, ionic strength, or temperature have been replaced by this sophisticated approach. It has been stated that over 60% of all purification techniques involve affinity chromatography (Lowe, 1996). The wide applicability of this method is based on the fact that any given biomolecule that one wishes to purify usually has an inherent recognition site through which it can be bound by a natural or artificial molecule. Thus, we can say that affinity chromatography is principally based on the molecular recognition of a target molecule by a molecule bound to a column.

Affinity purification involves 3 main steps:

a. Incubation of a crude sample with the affinity support to allow the target molecule in the sample to bind to the immobilized ligand.
b. Washing away non-bound sample components from the support.
c. Elution (dissociation and recovery) of the target molecule from the immobilized ligand by altering the buffer conditions so that the binding interaction no longer occurs.

Since the beginning of this technique, the term affinity chromatography has raised many controversies among researchers. Some say it would be more accurate if termed bioaffinity chromatography (O'Carra et al, 1974) or hydrophobic affinity (Shaltiel, 1974). Nonetheless, the term affinity chromatography has been expanded to describe a potential method of separating biomolecule mixtures on the basis of specific biological interactions. Recently, a modern form of liquid chromatography referred to as "flash chromatography" was introduced.

2. History of affinity chromatography

In 1910, the German scientist, Emil Starkenstein published an article which described the concept of resolving macromolecule complexes via their interactions with an immobilized substrate. This manuscript discussed the influence of chloride on the enzymatic activity of liver α-amylase and opened the door for the early beginnings of this approach by several researchers (Arsenis & McCormick, 1966; Bautz & Hall, 1962; Campbell et al, 1951; Sander et al, 1966). Later on, the term affinity chromatography introduced in 1968 by Pedro Cuatecasas, Chris Anfinsen and Meir Wilchek in an article that briefly described the technique of enzyme purification via immobilized substrates and inhibitors (Cuatrecasas et al, 1968). Other early articles described the activation of a Sepharose matrix using a cyanogen bromide (CNBr) reaction (Axen et al, 1967) and the use of a spacer arm to alleviate steric hindrance (Cuatrecasas et al, 1968).

Affinity chromatography is still developing. It has played a central role in many "Omics" technologies, such as genomics, proteomics and metabolomics. The breakthrough development of affinity liquid chromatography has enabled researchers to explore fields such as protein-protein interactions, post translational modifications and protein degradation that were not possible to be examined previously. Finally, the coupling of reversed phase affinity chromatography with mass spectrometry has ultimately aided in discovery of protein biomarkers.

3. Fundamental principles of affinity chromatography

Separation of a desired protein using affinity chromatography relies on the reversible interactions between the protein to be purified and the affinity ligand coupled to chromatographic matrix. As stated earlier, most of the proteins have an inherent recognition site that can be used to select the appropriate affinity ligand. The binding between the protein of interest and the chosen ligand must be both specific and reversible.

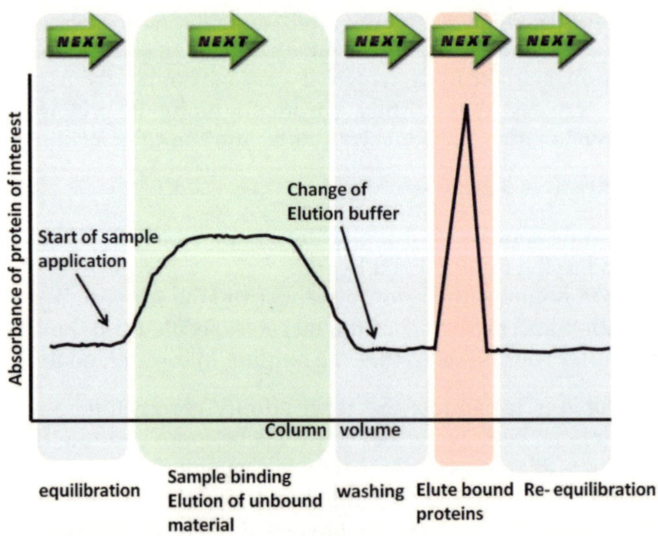

Fig. 1. Typical affinity chromatography purification

A typical affinity purification is shown in Figure 1 and involves several steps. First, samples are applied under conditions that favor maximum binding with the affinity ligand. After sample application, a washing step is applied to remove unbound sustances, leaving the desired (bound) molecule still attached to the affinity support. To release and elute the bound molecules, a desorption step is usually performed either 1) specifically using a competitive ligand or 2) non-specifically by changing the media atmosphere (e.g. changing the ionic strength, pH or polarity) (Zachariou, 2008). As the elution is perfomed, the purified protein can be collected in a concentrated form.

3.1 Biomolecules purified by affinity chromatography

Antibodies were first purified using affinity chromatography in 1951 when Campbell et al. used affinity chromatography to isolate rabbit anti-bovine serum albumin antibodies (Campbell et al, 1951). For their purification, bovine serum albumin was used as the affinity ligand on a cellulose support. Two years later, this technique was expanded to purify mushroom tyrosinase using an immobilized inhibitor of the enzyme (azophenol) (Lerman, 1953). Since then, affinity chromatography is commonly used to purify biomolecules such as enzymes, recombinant proteins, antibodies, and other biomolecules.

Affinity chromatography is often chosen to purify biomolecules due to its excellent specificity, ease of operation, yield and throughput. In addition, affinity chromatography has the ability to remove pathogens, which is necessary if the purified biomolecules are to be used in clinical applications. The purity and recovery of target biomolecules is controlled by the specificity and binding constant of the affinity ligand. In general, the association constants of affinity ligands used for biomolecule purification range from 10^3 - 10^8 M^{-1} (Janson, J-C, 1984). A common affinity ligand used in these purifications is an antibody, but other affinity ligands such as biomimetic dye-ligands, DNA, proteins and small peptides have been used as well. Figure 2 shows a wide variety of molecules that can be purified by affinity chromatography based on their polarity and volatility.

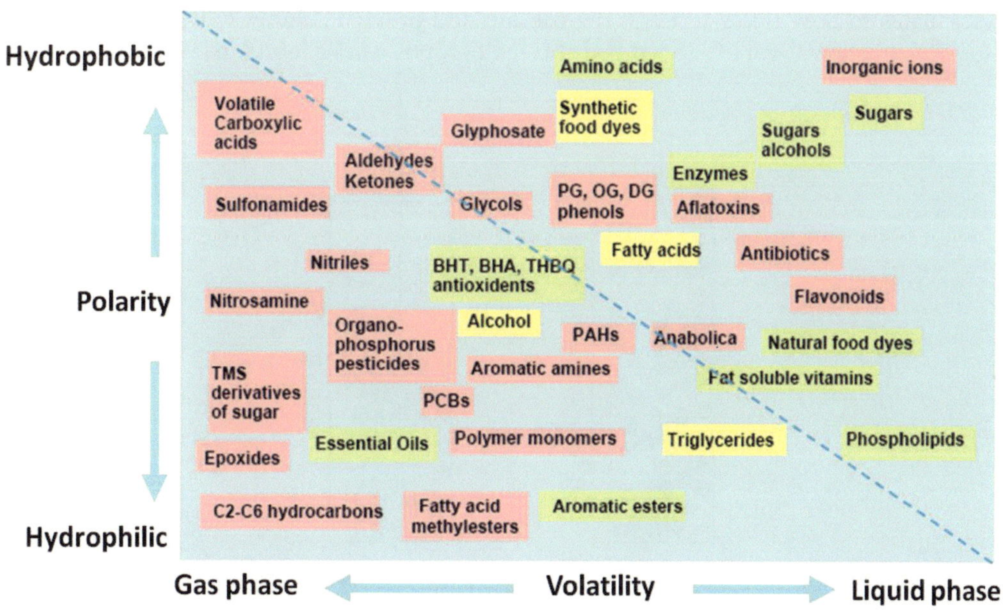

Fig. 2. Illustration showing different molecules that can be purified using affinity chromatography.

3.2 Components of affinity medium

When affinity chromatography is used for the purification and separation of large biomolecules from complex mixtures, the support (matrix), spacer arms, and ligand must be considered.

3.2.1 Affinity supports (matrix)

Traditionally, affinity chromatography support materials have consisted of porous support materials such as agarose, polymethacrylate, polyacrylamide, cellulose, and silica. All of these support materials are commercially available and come in a range of particle and pore sizes. Some supports may be available with common affinity ligands already immobilized (e.g. protein A, Cibacron Blue, heparin). Other types of support materials are being developed including nonporous supports, membranes, flow-through beads (perfusion media), monolithic supports, and expanded-bed adsorbents.

Nonporous support materials consist of nonporous beads with diameters of 1- 3 µm. These supports allow for fast purifications, but suffer from low surface areas when compared to traditional porous supports. Membranes used in affinity chromatography also lack diffusion pores which limits surface area, but like the nonporous beads allow for fast separations. Flow-through beads or perfusion media (originally developed for ion-exchange chromatography) have both small and large pores present. The addition of the large flow-through pores allows substances to be directly transported to the interior of the particle which means only short distances are required for diffusion. Monolithic supports are based on the same principle as perfusion media – they contain both large flow-through pores and small diffusion pores. Expanded-bed adsorbents were designed to prevent column clogging and utilize a reverse in flow to allow for the expansion of the column bed which allows for particulates to flow freely through the column and prevent column fouling. See Figure 3. More information about expanded-bed chromatography can be found in (Mattiasson, 1999).

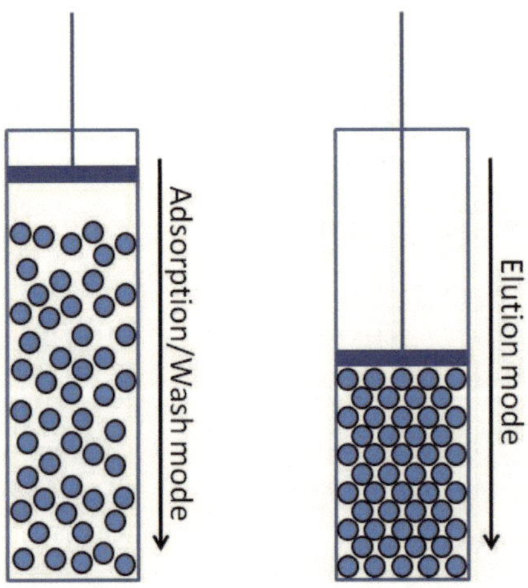

Fig. 3. Expanded-bed chromatography. In this type of chromatography, elution is performed in a normal packed-bed, but during the adsorption-wash step, the flow is reversed and the column bed expanded. This allows for particulate contaminates to pass freely through the column and prevent column clogging.

Regardless of the type of support used in the affinity purification, several factors must be considered when choosing a support material. These include chemical inertness, chemical stability, mechanical stability, pore size, and particle size.

Chemical inertness of the support material requires that the affinity support bind only the molecule of interest and have little or no nonspecific binding. While the specificity is related to the affinity ligand immobilized onto the support, the properties of the support must be chosen to limit the nonspecific binding of other molecules. Supports which have little or no nonspecific binding mimic the properties of the aqueous mobile phase. Therefore, chemically inert support materials are hydrophilic. In addition, most separations are performed in low ionic strength media. As a result, the number of charges on the support should be minimized to prevent nonspecific ionic interactions.

In addition, a support material must be chemically stable under normal operating conditions. This includes resistance to degradation by all enzymes and microbes, elution buffers, regenerating solvents, and cleaning agents that will be used within the column. These stability considerations must also be expanded to the stability of the affinity ligand-matrix linkage. Agarose-based support materials meet all of these requirements as they can be used between pH 3 and 12, are not attacked by enzymes, and are not affected by most aqueous eluants. However, ligand attachment in agarose support materials is often not as stable, depending on the type of linkage used.

Mechanical stability is another consideration when choosing a chromatographic support material for affinity chromatography. Support materials must be able to withstand the backpressures encountered during normal separations without compressing. While most commercial packing materials meet this requirement, the build-up of particulate contaminants may restrict column flow and lead to high backpressures. Under these pressures, soft porous gel supports such as agarose beads will compress and increase the pressure even further causing collapse of the support structure. More mechanically stable supports (e.g. silica and heavily cross-linked polymers) are able to withstand these high pressures, but the build-up of particulate contaminants should be avoided if at all possible.

Particle size is an additional consideration when choosing a support material. Ideally, small particle sizes are desired to limit mass transfer effects and limit band broadening. In addition, smaller particle sizes tend to offer greater surface area of the support material and allow for a larger number of affinity ligands to be immobilized on the surface of the support. Unfortunately, as particle size is decreased, backpressures are increased. In addition, when using smaller particles, the potential for the build-up of particulate contaminants and column fouling is increased. For this reason, in preparative applications large particles (30 – 100 μm) are often used. An alternative method to avoid the potential build-up of particulates is to use an expanded-bed support material as discussed earlier and seen in Figure 3.

Pore size is another item that must be considered when using affinity chromatography since the biomolecules of interest must be able to not only pass through the column but also be able to fully interact with the affinity ligand. Based on the Renkin equation which allows the estimation of the effective diffusion coefficient (Renkin, 1954), the pore diameter should be at least 5 times the diameter of the biomolecule being purified (Gustavsson & Larsson, 2006). Therefore, a typical protein with a 60 Å diameter would need a support with at least a 300 Å pore size. Often, the optimal pore size takes into account the ability of the affinity ligand to interact with the biomolecule as well as the surface area of the column since increasing pore size leads to a decrease in the surface area which limits the number of affinity ligands which can be immobilized to the support material.

3.2.2 Spacer arms

Due to the fact that binding sites of the target molecule are sometimes deeply located and difficult to access due to steric hindrance, a spacer arm is often incorporated between the matrix and ligand to facilitate efficient binding and create a more effective and better binding environment. See Figure 4.

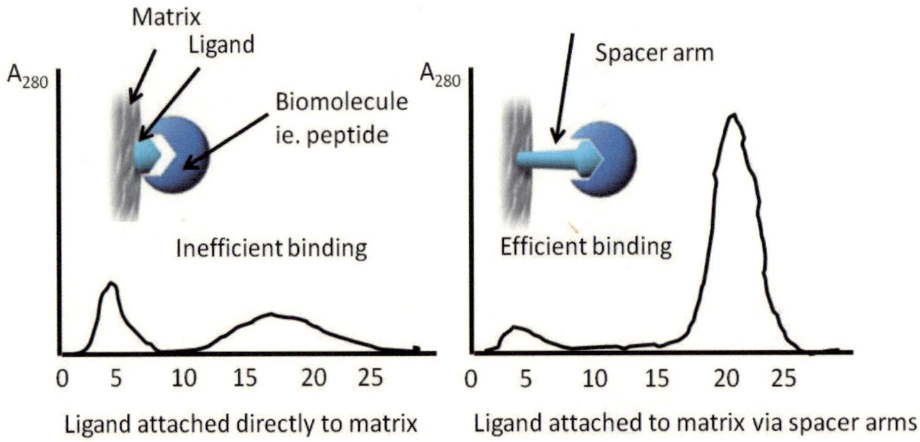

Fig. 4. Chromatogram showing better ligation and elution when spacer arms are introduced between the ligand and matrix

The length of these spacer arms is critical. Too short or too long arms may lead to failure of binding or even non-specific binding. In general, the spacer arms are used when coupling molecules less than 1000 Da.

3.2.3 Ligands used in affinity chromatography

Antibodies have several advantages including their high specificity and relatively large binding constants. Antibodies or immunoglobulins are a type of glycoprotein produced when a body's immune system responds to a foreign agent or antigen. Due to the variability of the amino acid sequence in the antibody binding sites (F_{ab} regions shown in Figure 5), it has been estimated that antibodies can be produced for millions or even billions of different foreign agents.

Antibodies which are produced by separate cell lines are referred to as polyclonal antibodies. Monoclonal antibodies are produced when a single antibody producing cell is combined with a carcinoma cell to create a hybridoma which can be grown in a cell culture. Monoclonal antibodies are often more desirable than polyclonal antibodies in affinity chromatography due to their lack of variability which allows for the creation of a more uniform affinity support.

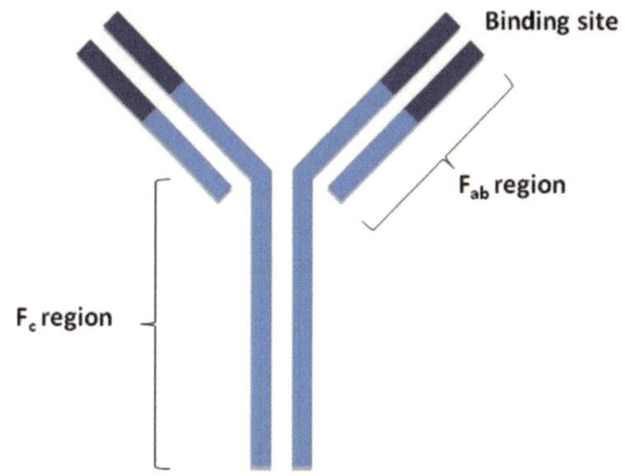

Fig. 5. Typical structure of an antibody. The amino acids in the F_c region generally have the same sequence, whereas the amino acids in the F_{ab} region have variable amino acid sequences which allows for the specificity of the binding interaction against a wide range of antigens.

Another type of affinity ligand which can be used to purify biomolecules from complex mixtures is a dye-ligand. Dye ligand chromatography originated in 1968 when Haeckel et al. were purifying pyruvate kinase using gel filtration chromatography and found that Blue Dextran (a small dye molecule) co-eluted with the protein (Haeckel et al, 1968). After further investigation, it was determined that binding between the dye and enzyme caused this co-elution. The dye-enzyme binding was later utilized in the purification of pyruvate kinase using a Blue Dextran column in 1971 (Staal et al, 1971).

Biomimetic dye-ligand chromatography takes dye-ligand chromatography one step further and utilizes modified dyes which mimic the natural receptor of the target protein. In addition to offering better binding affinities, these modified dyes were initially developed as a result of the concerns over purity, leakage, and toxicity of the original commercial dyes (Lowe et al, 1992). Cibacron Blue 3GA is one of the most common modified triazine dyes that has been used for protein purification. Its structure can be seen in Figure 6. Covalent attachment of the dye can be achieved through nucleophilic displacement of the dye's chlorine atom by hydroxyl groups on the support's surface (Labrou, 2000; Labrou et al, 1995; Labrou, 2002).

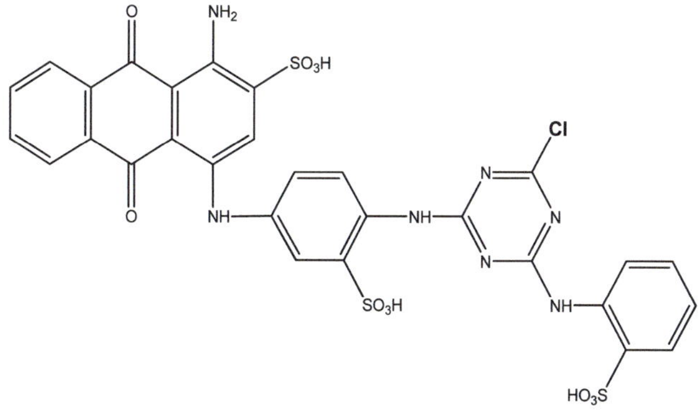

Fig. 6. Chemical structure of blue sepharose dye-ligand (Cibacron Blue 3GA) commonly used for purification of albumin as well as enzymes (NAD^+ and $NADP^+$).

Chlorotriazine polysulfonated aromatic molecules (triazine dyes) have been used for the purification of albumin, oxidoreductases, decarboxylases, glycolytic enzymes, nucleases, hydroloases, lyases, synthetases, and transferases (Labrou, 2000; Labrou et al, 1995). The main advantages of using dye-ligands and biomimetic dye-ligands are their low cost and resistance to chemical and biological degradation. The main disadvantage of these synthetic ligands is that the selection process for a particular biomolecule is empirical and requires extensive screening processes during method development. More information on biomimetic dyes can be found in reference (Clonis et al, 2000).

DNA can also be used as an affinity ligand. It can be used to purify DNA-binding proteins, DNA repair proteins, primases, helicases, polymerases, and restriction enzymes. The scope of biomolecules which can be purified using DNA is expanded when aptamers are utilized. Aptamers are single-stranded oligonucleotides which have a high affinity for a target molecule. SELEX (Systematic Evolution of Ligands by Exponential Enrichment) allows for the isolation of these oligonucleotide sequences and allows for a wide range of potential targets including biomolecules which typically have no affinity for DNA or RNA. The SELEX process for DNA is shown in Figure 7.

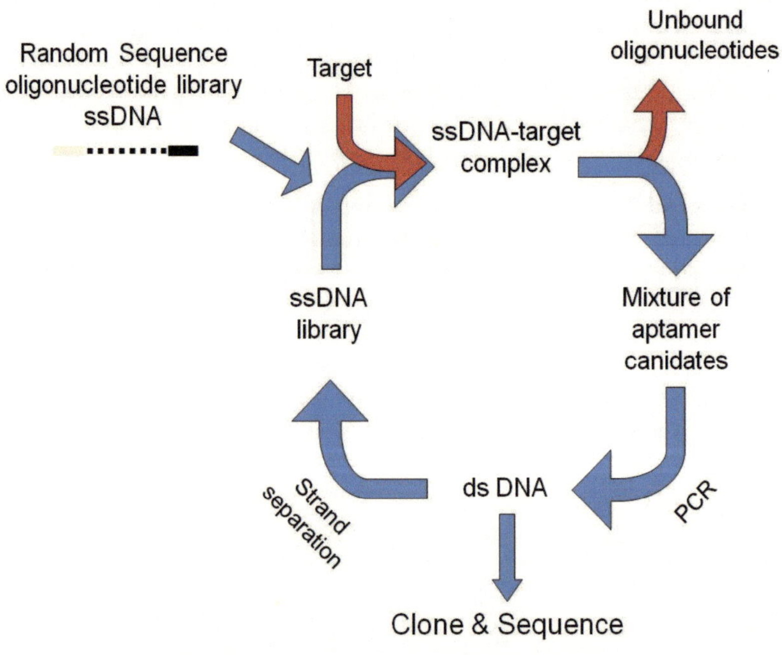

Fig. 7. Diagram depicting the SELEX process for the selection of aptamers against a target. First a random short stranded ssDNA (or RNA) library (10^{14} sequences) is exposed to the target compound and allowed to bind. The unbound oligonucleotides are then separated from the ssDNA-target complexes and removed. The remaining complexes are then disrupted leaving a mixture of aptamer candidates and the target compound. The ssDNA aptamer candidates are then amplified using PCR, the strands separated and the cycle repeated. After multiple cycles (typically 5 – 15), the initial DNA library will have been condensed down to a few sequences which tightly bind the target. These candidates can then be cloned, sequenced and used for affinity chromatography.

A similar process can be used to develop RNA affinity ligands. Once a potential aptamer sequence is identified, it can synthesized *in vitro* and used as the affinity ligand on a chromatographic support. An example of aptamers usage as in purification of L-selectin (Romig et al, 1999) and RNA binding proteins (Dangerfield et al, 2006; Windbichler & Schroeder, 2006).

Peptide affinity chromatography is another method which can be used for purifying biomolecules. Peptide affinity ligands are typically identified using one of two techniques (Wang et al, 2004); biological combinatorial peptide libraries (e.g phage-displayed libraries) (Cwirla et al, 1990; Devlin et al, 1990; Smith & Scott, 1993) or solid-phase combinatorial libraries (e.g. one-bead-one-peptide libraries) (Lam et al, 1991). Since then, peptide sequences have been isolated for a wide range of targets (Casey et al, 2008) and have been used to purify staphylococcal enterotoxin B (Wang et al, 2004), β-tryptase (Schaschke et al, 2005), and α-cobratoxin (Byeon & Weisblum, 2004). The main advantages of using peptides as affinity ligands are their low cost and stability.

Other ligands can be used in affinity chromatography for biomolecule purification. For more information on all types of affinity ligands see references (Clonis, 2006; Hage, 2006).

3.2.4 Immobilization of affinity ligands

Immobilization of the affinity ligand is also very important when designing an affinity chromatography method for biomolecule purification. When immobilizing an affinity ligand, care must be taken to ensure that the affinity ligand can actively bind the desired target after the immobilization procedure. Activity of the affinity ligand can be affected by multi-site attachment, orientation of the affinity ligand, and steric hindrance. See Figure 8.

Multi-site attachment occurs when an affinity ligand is attached through more than one functional group on a single ligand molecule. If these multiple attachment sites cause the affinity ligand to become denatured or distorted, multisite attachment can lead to reduced binding affinity. However, in some instances, the additional attachment sites can result in more stable ligand attachment. In general, it is best to try for site-specific attachment of the affinity ligands to limit the potential for multi-site attachment. For example, when immobilizing antibodies, covalent attachment is often directed toward the carbohydrate moieties within the F_c region of the antibody. Not only does this limit the number of attachment sites, but it can also help direct the binding and, thus, help orientate the antibody so the binding regions (F_{ab}) are exposed. Another way to prevent multi-site attachment is to use a support that has a limited number of reactive sites. By limiting the reactive sites, the potential for multiple attachments from a single affinity ligand is greatly reduced. A general rule of thumb is the larger the affinity ligand to be immobilized, the fewer number of reactive sites on the support needed.

Obviously, when performing affinity purifications, it is important to ensure the affinity ligands are immobilized so that the binding regions are exposed and free to interact and bind with the target molecule(s). Ideally, immobilization methods which specifically avoid attaching the affinity ligand via functional groups within the binding site(s) are used. One way to achieve this when immobilizing proteins is to use site-directed mutagenesis to introduce a single cysteine residue at a site known to be far away from the binding site(s) (Huang et al, 1997).

Once the cysteine residue is introduced, the protein can be immobilized using a cysteine specific coupling reagent such as N-γ-maleimidobutyryl-oxysuccimide ester.

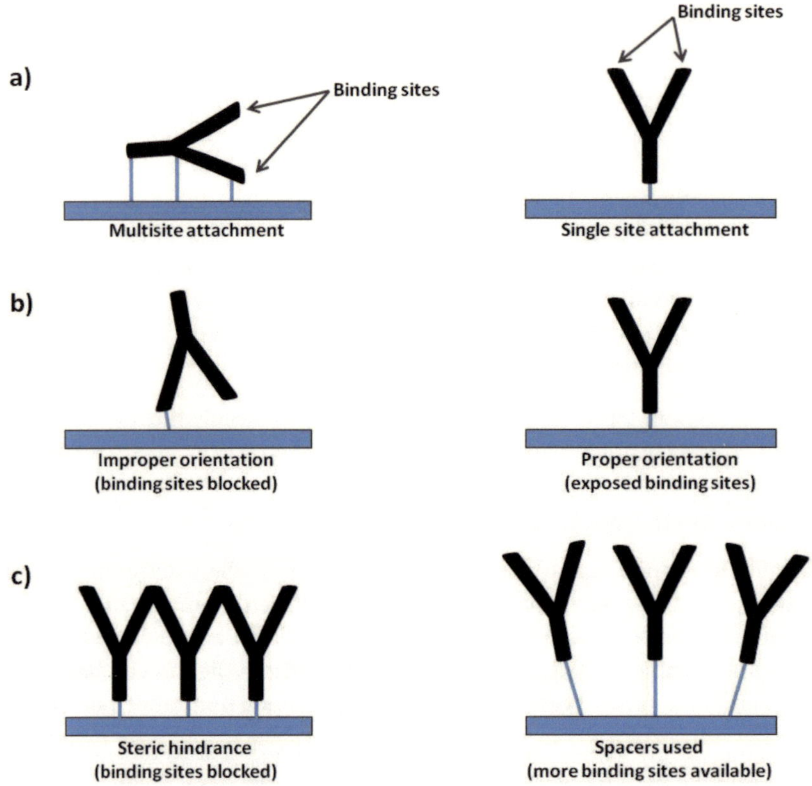

Fig. 8. Potential immobilization problems which can affect affinity ligand activity by a) multi-site attachment, (b) improper orientation, and (c) steric hindrance.

Affinity ligands can be covalently immobilized, adsorbed onto a surface via nonspecific or biospecific interactions, entrapped within a pore, or coordinated with a metal ion as in metal-ion affinity chromatography (IMAC). Each of these methods has advantages and disadvantages and is briefly discussed below.

Covalent immobilization is one of the most common ways of attaching an affinity ligand to a solid support material. There is a wide range of coupling chemistries available when considering covalent immobilization methods. Amine, sulfhydryl, hydroxyl, aldehyde, and carboxyl groups have been used to link affinity ligands onto support materials. More information about these specific reactions can be found in reference (Kim & Hage, 2006). Although covalent attachment methods are more selective than other immobilization methods, they generally require more steps and chemical reagents. While this may lead to a greater initial cost of preparation, the stability of these supports typically is greater and the support does not need to be periodically regenerated with additional affinity ligands as is typically the case when using adsorption techniques. As a result, covalent immobilization may be more economical in the long-term for the immobilization of costly affinity ligands.

Adsorption of affinity ligands may also be used to immobilize affinity ligands onto support materials. The adsorption can be either nonspecific or specific. In nonspecific adsorption the affinity ligand simply adsorbs to the surface of the support material and is a result of Coulombic interactions, hydrogen bonding, and/or hydrophobic interactions. Biospecific adsorption is commonly performed by using avidin or streptavidin for the adsorption of biotin containing affinity ligands or protein A or protein G for the adsorption of antibodies. Both of these immobilization methods allow for site-specific attachment of the affinity ligand which minimizes binding site blockages. When biospecific adsorption is used for immobilization, the primary ligand (i.e. avidin, streptavidin, protein A or protein G) must first be immobilized onto the support material. Avidin, streptavidin, protein A and protein G can be immobilized using amine-reactive methods. Avidin is glycosylated and can also be immobilized through its carbohydrate residues.

Entrapment of affinity ligands was demonstrated by Jackson et al. when human serum albumin (HSA) was entrapped using hydrazide-activated supports and oxidized glycogen as a capping agent (Jackson et al, 2010). Their method can be used with other affinity ligands ranging from 5.8 to 150 kDa. This type of immobilization method is generally less harsh than other immobilization methods and does not require the use of recombinant proteins. In addition, no linkage exists between the affinity ligand and the support which eliminates the potential immobilization problems seen in Figure 8.

Sol-gel entrapment is another method of encapsulation of affinity ligands (Avnir et al, 2006; Jin & Brennan, 2002; Pierre, 2004). The sol-gel entrapment process is as follows: First, the sol is formed from a silica precursor (e.g. alkoxysilane or glycerated silane). Once the sol has been formed, the buffered protein solution is added and the gelation reaction initiated. This is followed by an aging process in which the sol-gel is dried and further crosslinking of the silica occurs leaving the protein physically trapped within the cross-linked silica gel.

4. Current techniques involving affinity chromatography

Affinity chromatography is currently being used for a wide variety of applications ranging from the study of drug-protein binding interactions to the depletion of high abundance proteins to enhance the detection/quantification of dilute proteins.

Affinity chromatography can be used to study drug-protein binding interactions. Frontal analysis, zonal elution, and the Hummel-Dreyer method can be used to measure drug-protein binding constants, to quantify kinetic properties of the various interactions, to quantify allosteric interactions, and to identify drug binding sites. More information about the measurement of drug-protein binding constants can be found in two review articles (Hage, 2002; Hage et al, 2011). Information on quantifying kinetic properties of drug-protein interactions can be found in a review by (Schiel & Hage, 2009). A discussion on the quantification of allosteric interactions by affinity chromatography can be seen in an article by (Chen & Hage, 2004). Additional information on the identification of drug-binding sites can be found in a review article (Hage & Austin, 2000).

When trying to analyze low abundance proteins, it is often necessary to remove high abundance proteins prior to analysis. This removal effectively enriches low abundance proteins and allows more of them to be identified and quantified. Removal of the top 7 or top 14 high-abundance proteins has been shown to result in a 25% increase in identified proteins (Tu et al, 2010). Moreover, affinity chromatography is widely used in many 'omics' studies (e.g. proteomics, metabolomics and genomics) and is currently used in tandem with other methods to develop high-throughput screening methods for potential drugs.

5. Biokinetics of affinity chromatography

The reaction between the ligand (L) and target compound (T) in an affinity atmosphere (either adsorption or desorption) is represented in Figure 9.

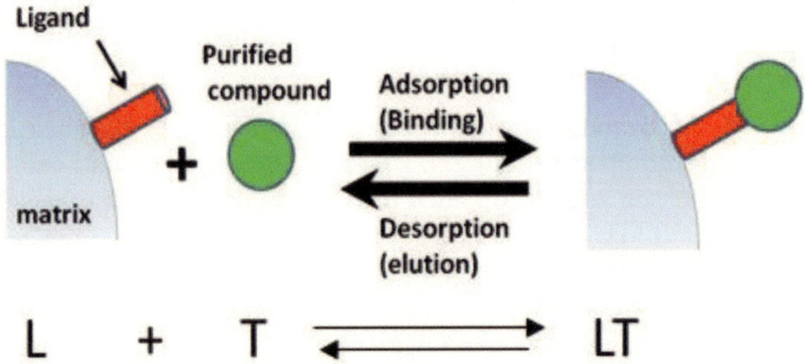

Fig. 9. Basic reaction between compound to be purified and ligand.

The standard definition of the term equilibrium dissociation constant [K_D] can be expressed in equation 1,

$$K_D = \frac{[L]*[T]}{[LT]} \tag{1}$$

where [L] is the free ligand, [T] is the target compound, and [LT] is the ligand-target complex.

According to the postulation of (Graves & Wu, 1974), the bound target-total target ratio can by represented as shown in equation 2,

$$\frac{\text{Bound target}}{\text{Total bound}} \approx \frac{L_0}{K_D} + L_0 \tag{2}$$

where L_0 is the concentration of the ligand (usually 10^{-4} - 10^{-2} M). To achieve successful binding, the ratio of bound to total target must be near 1. Therefore, K_D should be small compared to ligand concentration. K_D can be greatly affected by changing in pH, ionic strength, and temperature. Thus changing these parameters can be used to control the binding and elution efficiency of the reaction and can be expressed as seen in equations 3 and 4,

$$L + T \leftrightharpoons LT \tag{3}$$

$$LT \leftrightharpoons L + T \tag{4}$$

In equation 3, the K_D range is between 10^{-6} - 10^{-4} M which means there is more binding and less elution. In equation 4, the K_D is decreased to 10^{-1} - 10^{-2} M by the elution conditions which results in less binding and more elution of the target compound.

The interactions described in equations 3 and 4 apply under non-selective (noncompetitive) elution conditions. In case of selective elution or competitive elution, the interaction can be represent as shown in Figure 10.

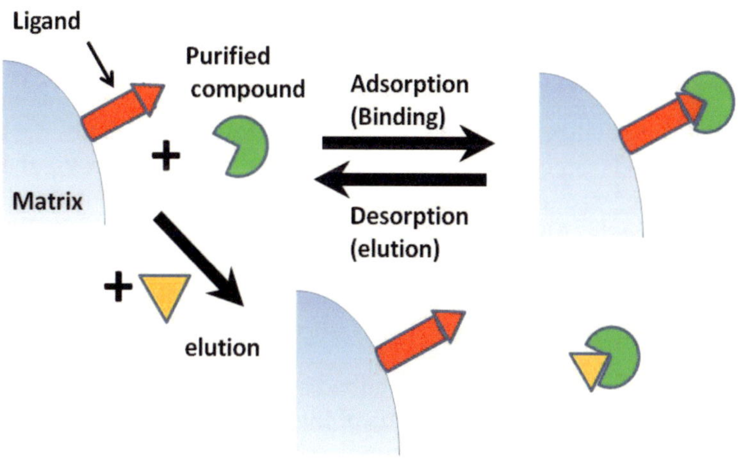

Fig. 10. Competitive elution of the target by adding a competitive free ligand (triangle).

When adding a competing binding substance or a free ligand (C) that binds to the purified compound of interest during elution, the interaction can be represented as shown in Equation 5.

$$C + T \leftrightharpoons CT \tag{5}$$

The equilibrium constant, K_D, is calculated according to equation 6

$$K_D = \frac{[C]*[T]}{[CT]} \tag{6}$$

where [C] and [T] are the concentration of the free competing ligand and target, respectively and [CT] is the concentration of the competing ligand-target complex.

(Graves & Wu, 1974) have shown that the eluted target to total target compound ratio can be represented by equation 7

$$\frac{\text{Eluted target}}{\text{Total bound target}} \approx \left({}^{p}\!/\!{}_{p+1} \right)^{*} \left[\frac{pC_0}{pC_0 + \dfrac{K_{DComp} * L0}{K_D}} \right] \tag{7}$$

where, p is the ratio between the volume of competitor added and the pore volume of the gel, K_D is the dissociation constant for coupled ligand, K_{Dcomp} is the dissociation constant for the free competing ligand, C_0 is the concentration of the competing ligand (usually 10^{-2} – 10^{-1} M) and L_0 is the concentration of the coupled ligand, usually 10^{-4} – 10^{-2} M.

If both K_D and K_{Dcomp} are similar, then the concentrations of the competing and coupled ligand must be similar to achieve an efficient elution. On the other hand, if K_{Dcomp} is equal to $5*K_D$ we would expect that the concentration of the competing ligand will need to be 5x higher to achieve successful elution.

6. Applications and uses of affinity chromatography

6.1 Immunoglobulin purification (antibody immobilization)

Antibodies can be immobilized by both covalent and adsorption methods. Random covalent immobilization methods generally link antibodies to the solid support via their free amine groups using cyanogen bromide, N-hydroxysuccinimide, N,N'-carbonyldiimidazole, tresyl chloride, or tosyl chloride. Alternatively, free amine groups can react with aldehyde or free epoxy groups on an activated support. As these are random immobilization methods, the antibody binding sites may be blocked due to improper orientation, multi-site attachment or steric hindrance as shown in Figure 8.

Site-specific covalent immobilization of antibodies can be achieved by converting the carbohydrate residues located in the F_c region of the antibody to produce aldehyde residues which can react with amine or hydrazide supports (Ruhn et al, 1994). Another site-specific immobilization of antibodies can be accomplished by utilizing the free sulfhydryl groups of F_{ab} fragments. These groups can be used to couple the antibody fragments to an affinity support using a variety of established methods including epoxy, divinylsulfone, iodoacetyl, bromoacetyl, thiol, maleimide, TNB-thiol, tresyl chloride, or tosyl chloride methods (Hermanson et al, 1992).

Antibodies can also be immobilized by adsorbing them onto secondary ligands. For example, if an antibody is reacted with hydrazide biotin, the hydrazide can react with oxidized carbohydrate residues on the F_c region of the antibody. The resultant biotinylated antibody can then be adsorbed onto an avidin or streptavidin affinity support. This type of biotin immobilization allows for site-specific immobilization of the antibody and can be performed using commercially available biotinylation kits.

Alternatively, antibodies can be directly adsorbed onto a protein A or protein G support due to the specific interaction of antibodies with protein A and G. Immobilized antibodies on the protein A or G support can easily be replaced by using a strong eluent, regenerating the protein A/G, and re-applying fresh antibodies. Generally, this method is used when a high capacity/high activity support is needed. If a more permanent immobilization is desired, the adsorbed antibodies may be cross-linked to the support material using carbodiimide (Phillips et al, 1985) or dimethyl pimelimidate (Schneider et al, 1982; Sisson & Castor, 1990).

6.2 Recombinant tagged proteins

Purification of proteins can be easier and simpler if the protein of interest is tagged with a known sequence commonly referred to as a tag. This tag can range from a short sequence of amino acids to entire domains or even whole proteins. Tags can act both as a marker for protein expression and to help facilitate protein purification.

The properties of fusion tags allow tagged proteins to be easily manipulated in the laboratory. Most significantly, the well-characterized tag-ligand chemistry enables single-step affinity purification of tagged molecules using immobilized versions of their corresponding affinity ligands. In addition, antibodies to fusion tags are also available and can be used for "universal" purification and detection of tagged proteins (i.e., without having to obtain or develop a probe for each specific recombinant protein).

In general, the most commonly used tags are glutathione-S-transferase (GST), histidine fusion (His or polyHis tag) and protein A fusion tags. Other types of fusion tags are also available including maltose-binding protein (di Guan et al, 1988), thioredoxin (LaVallie et al, 1993), NusA (Whetstone et al, 2004), GB1 domain for protein G (Davis et al, 1999), and others (Balbas, 2001; Thorn et al, 2000). The decision to use any of these tagging methods depends mainly on the needs of of the researcher. Table 1 compares GST and $(His)_6$ tags and may help when deciding which tag is appropriate for a particular purification.

GST tag	$(His)_6$ tag
Can be used in any expression system	
Purification procedure gives high yields of pure product	
Selection of purification products available for any scale	
Site-specific proteases enable cleavage of tag if required	Site-specific proteases enable cleavage of tag if required. N.B. enterokinase sites that enable tag cleavage without leaving behind extra amino acids are preferable
GST tag easily detected using an enzyme assay or an immunoassay	$(His)_6$ tag easily detected using an immunoassay
Simple purification. Very mild elution conditions minimize risk of damage to functionality and antigenicity of target proteins	Simple purification, but elution conditions are not as mild as for GST fusion proteins. Purification can be performed under denaturing conditions if required. Neutral pH but imidazole may cause precipitation. Desalting to remove imidazole may be necessary
GST tag can help stabilize folding of recombinant proteins	$(His)_6$ -dihydrofolate reductase tag stabilizes small peptides during expression
Fusion proteins form dimers	Small tag is less likely to interfere with structure and function of fusion partner
	Mass determination by mass spectrometry not always accurate for some $(His)_6$ fusion proteins

Table 1. Comparison of GST and His tags for protein purification. The information from this table is summarized fromthe Amersham recombinant protein handbook and (Geoghegan et al, 1999).

In the following sections, the most commonly used purification techniques and methods utilizing affinity chromatography will be discussed.

6.2.1 GST tagged purification

Glutathione S-transferase (GST) is a 26 kDa protein (211 amino acids) located in cytosole or mitochondria and present both in eukaryotes and prokaryotes (e.g. *Schistosoma japonicum*). The enzymes have various sources both native and recombinantly expressed by fusion to the N-terminus of target proteins (Allocati et al, 2009; Allocati et al, 2011; Sheehan et al, 2001; Udomsinprasert et al, 2005). GST-fusion proteins can also be produced in *Escherichia coli* as recombinant proteins. Separation and purifcation of GST-tagged proteins is possible since the GST tag is capable of binding its substrate, glutathione (tripeptide, Glu-Cys-Gly).

When glutathione is reduced (GSH), it can be immobilized onto a solid support through its sulfhydryl group. This property can be used to crosslink glutathione with agarose beads and, thus, can be used to capture pure GST or GST-tagged proteins via the enzyme-substrate binding reaction (Beckett & Hayes, 1993; Douglas, 1987). Binding is most efficient near neutral physiological conditions (pH 7.5) using Tris saline buffer and mild conditions to preserve the structure and enzymatic function of GST. As a result of the potential for permanent denaturation, denaturing elution conditions are not compatible with GST purification. In addition, upon denaturation or reduction, the structure of the GST fusion tag often degrades.

Following a washing step to remove unbound samples, the bound GST-fusion protein can be recovered by the addition of excess reduced glutathione since the affinity of GST for free glutathione is higher than the affinity for immobilized glutathione. The free glutathione replaces the immobilized glutathione and releases the GST-tagged protein from the matrix allowing its elution from the column.

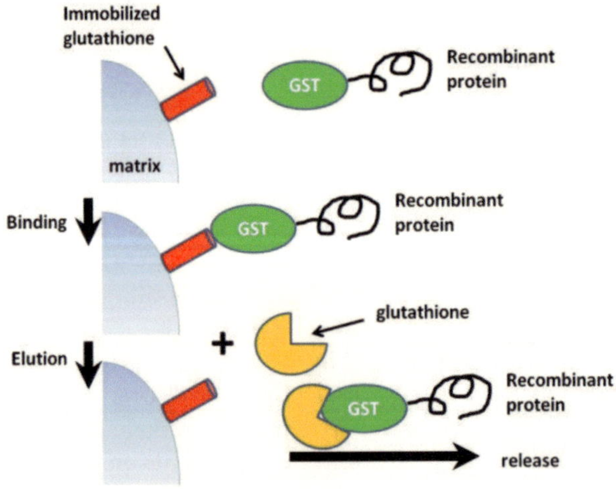

Fig. 11. GST- tagged protein immobilization.

6.2.2 His-tagged protein purification

Recombinant proteins which have histidine tags can be purified using immobilized metal ion chromatography (IMAC). The His-tag can be placed on either the N- or C-terminus. Optimal binding and, therefore, purification efficiency is achieved when the His-tag is freely accessible to metal ion support (Dong et al, 2010).

Histidine tags have strong affinity for metal ions (e.g. Co^{2+}, Ni^{2+}, Cu^{2+}, and Zn^{2+}). One of the first support materials used immobilized iminodiacetic acid which can bind metal ions and allow for the coordination complex with the His-tagged protein. One difficulty with iminodiacetic acid supports is the potential for metal ion leaching leading to a decreased protein yield. Modern support materials, including nickel-nitrilotriacetic acid (Ni-NTA) and cobalt-carboxymethylasparate (Co-CMA), show limited leaching and, therefore, result in more efficient protein purifications. The coordination of a His-tag with a Ni-NTA support can be seen in Figure 12. Once the tagged protein is bound by the immobilized chelating agent, it can be eluted by introducing a competing agent for the chelating group (imidazole) or an additional metal chelating agent (EDTA).

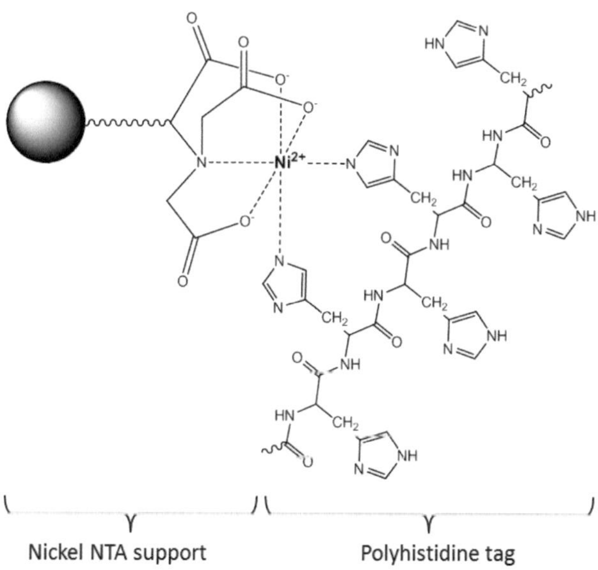

Fig. 12. showing the complex formed between the poly-histidine tag and a nickel NTA support.

One advantage of using His-tags for protein purification includes the small size of the affinity ligand. Due to the small size, it has minimal effects on the folding of the protein. In addition, if the His-tag is placed on the N-terminal end of the protein, it can easily be removed using an endoprotease. Another advantage of using His-tag purification methods is that polyhistidine tags can bind proteins under both native and denaturing conditions. The use of denaturing conditions becomes important when proteins are found in inclusion bodies and must be denatured so they can be solubilized.

Disadvantages of using His-tag protein purification include potential degradation of the His-tag, dimer and tetramer formation, and coelution of other histidine-containing proteins. First, when a few histidine residues are proteolytically degraded, the affinity of the tagged protein is greatly reduced leading to a decrease in the protein yield. Second, once a protein has a His-tag added to its structure, it has the potential to form dimers and tetramers in the presence of metal ions. While this is often not a large problem, it can lead to inaccurate molecular mass estimates of the tagged protein. A third disadvantage of protein purification using His-tags is coelution of proteins that naturally have two or more adjacent histidine residues.

6.3 Protein A, G, and L purification

Proteins A, G, and L are native or recombinant proteins of microbial origin which bind specifically to immunoglobulins including immunoglobulin G (IgG). IgG represents 80% of serum immunoglobulins. Native and recombinant protein A can be cloned in *Staphylococcus aureus*. Recombinant protein G (cell surface protein) is cloned in *Streptococcus* while recombinant protein L is cloned from *Peptostreptococcus magnus*. Both protein A and G specifically bind the F_c region of IgG while protein L binds to the kappa light chains of IgG.

The most popular matrixes or supports for affinity applications which utilize protein A, G, or L is beaded agarose (e.g. Sepharose CL-4B; agarose crosslinked with 2,3-dibromopropanol and desulphated by alkaline hydrolysis under reductive conditions), polyacrylamide, and magnetic beads (Grodzki & Berenstein, 2010; Hober et al, 2007; Katoh et al, 2007; Tyutyulkova & Paul, 1995).

All three proteins bind extensively with the IgG subclass. In general, protein A is more suitable for cat, dog, rabbit and pig IgG whereas protein G is generally more preferable when purifying mouse or human IgG. A combination of protein A and G is also applicable for purifying a wide range of mammalian IgG samples. Since protein L binds to the kappa light chain of immunoglobulins and these light chains exist in other immunoglobulins (i.e IgG, IgM, IgA, and IgE), protein L is suitable for the purification of different classes of antibodies. The binding characteristics of antibody binding proteins (A, G and L) to a variety immunoglobulin species is summarized in Table 2. IgGs from most species bind to protein A and G near physiological pH and ionic strength. To elute purified immunoglobulins from protein G sepharose, the pH should be less than 2.7.

Species	Protein A	Protein G	Protein L*
Human	Strong	Strong	Strong
Mouse	Strong	Strong	Strong
Rat	Weak	Medium	Strong
Cow	Weak	Medium	Strong
Goat	Weak	Strong	No binding
Sheep	Weak	Strong	No binding
Horse	Weak	Strong	Unknown
Rabbit	Strong	Strong	Weak
Guinea pig	Strong	Weak	Unknown
Pig	Strong	Weak	Strong
Dog	Strong	Weak	Unknown
Cat	Strong	Weak	Unknown
Chicken	Unknown	Unknown	Unknown

*Binding affinity based on total IgG binding, L proteins binds to Kappa light chains while Proteins A and G bind to F_c region.

Table 2. Binding affinity for proteins A, G, and L with a variety of immunoglobulin species.

6.4 Biotin and biotinylated molecules purification

If a biotin tag can be incorporated into a biomolecule, it can be used to purify the biomolecule using a streptavidin or avidin affinity support. One way is to insert a biotinylation sequence into a recombinant protein. Biotin protein ligase can then be used to add biotin in a post-translational modification step (Cronan & Reed, 2000). Biotin, also known as vitamin H or vitamin B_7, is a relatively small cofactor present in cells. In affinity chromatography it is often used an affinity tag due to its very strong interactions with avidin and streptavidin. One advantage of using biotin as an affinity tag is that it has a minimal effect on the activity of a large biomolecule due to its small size (244 Da).

Streptavidin is a large protein (60 kDa) that can be obtained from *Streptomyces avidinii* and bind biotin with an affinity constant of 10^{13} M^{-1}. Avidin is a slightly larger glycoprotein (66 kDa) with slightly stronger binding to biotin (10^{15} M^{-1}). Both avidin and streptavidin have four subunits that can each bind one biotin molecule. To purify biotinylated biomolecules, streptavidin is immobilized onto a support material and used to extract the biotinylated molecules out of solution. Both avidin and streptavidin may be immobilized using amine-reactive coupling chemistries. In addition, avidin can also be immobilized via its carbohydrate residues.

Due to the strong interaction between biotin and (strept)avidin, harsh elution conditions are required to disrupt the binding. For example, 6 M guanidine hydrochloride at pH 1.5 is commonly used to elute the bound biotinylated biomolecule. This prevents the recovery of most proteins in their active form. To overcome this difficulty, modified (strept)avidin or modified biotin may be used to create a lower affinity interaction. In one study chemically modified avidin had relatively strong binding (>10^9 M^{-1}), but was also able to completely release biotinylated molecules at pH 10 (Morag et al, 1996). In addition, at any pH between 4 and 10, a 0.6 mM biotin solution could be used to displace and elute the biotinylated molecules.

Biotin is also used in isotopically coded affinity tags (ICATs) which can be used to compare the protein content in two different samples (Bottari et al, 2004). The ICAT consists of two labels, one which contains deuterium (heavy) and one which contains only hydrogen (light). The two labels (light and heavy) are added separately to the cell lysates being compared. Since the reagent contains a thiol-specific reactive group, it will covalently bind free cysteines on proteins. The labeled lysates are combined, digested with trypsin, and then isolated on a streptavidin column. After a second separation step, the labeled proteins are analyzed using mass spectrometry. The change in protein expression between the two cell lysates can then be quantified and related to the different conditions applied to the two sets of cell lysates.

6.5 Affinity purification of albumin and macroglobulin contamination

Affinity purification is a helpful tool for cleaning up and removing excess albumin and α_2-macroglobulin contamination from samples since these components can mask or interfere with subsequent steps of analysis (e.g. mass spectrometry and immunoprecipitation). One purification method which can be used to remove these contaminants either before or after other purification steps is Blue sepharose affinity chromatography. In this method, the dye ligand is covalently coupled to sepharose via a chlorotriazine ring. Albumin binds in a non-specific manner by electrostatic and/or hydrophobic interactions with the aromatic anionic ligand (Antoni et al, 1978; Peters et al, 1973; Travis & Pannell, 1973; Young & Webb, 1978). The most commonly used dye is Cibacron blue F-3-GA which can be immobilized onto sepharose to create an affinity column. See Figure 6. This dye is capable of removing over 90% of albumin in the sample (Travis et al, 1976).

6.6 Lectin affinity chromatography

Lectin affinity chromatography is one of the most powerful techniques for studying glycosylation as a protein post translational modification (Hirabayashi et al, 2002; Spiro, 2002). Lectins are carbohydrate binding proteins that contain two or more carbohydrate binding sites and can be classified into five groups according to their specificity to the monosaccharide. They exhibit the highest affinity for: mannose, galactose/N-acetylgalactosamine, N-acetylglucosamine, fucose, and N-acetylneuraminic acid (Sharon, 1998). In this affinity technique, protein is bound to an immobilized lectin through its sugar moeities (N-linked or O-linked). Once the glycosylated protein is bound to the affinity support, the unbound contaminants are washed away, and the purifed protein eluted.

Currently, many lectins are commercially available in an immobilized form. Among them, Concanavaline A (Con A) Sepharose and wheat germ agglutinin (WGA) are the most popular for glycoprotein purification. As shown in Table 3, several different types of lectin may be used in affinity chromatography.

Acronym, Organism and source	Metal ions required	Sugar specificity	Elution conditions	Useful for binding
Con A (*Canavalia ensiformis;* jack bean seeds)	Ca^{2+}, Mn^{2+}	α-Man > α-Glc	0.1–0.5 M α-MeMan	High-Man, hybrid, and biantennary N-linked chains
LCA or LCH (*Lens culinarus;* lentil seeds)	Ca^{2+}, Mn^{2+}	α-Man > α-Glc	0.1–0.5 M α-MeMan	Bi- and triantennary N-linked chains with Fuc α1-6 in core region
PSA (*Pisum sativum;*peas)	Ca^{2+}, Mn^{2+}	α-Man	0.1–0.5 M α-MeMan	Similar to LCA/LCH
WGA (*Triticum vulgaris;* wheat germ)	Ca^{2+}, Mn^{2+}	ß-GlcNAc	0.1–0.5 M GlcNAc	GlcNAc- and Sia- terminated chains, or clusters of *O*-GlcNAc; succinylated form selectively binds GlcNAc>Sia
HPA (*Helix promatia;* albumin gland of edible snail)	-	α-GalNAc	0.1–0.5 M GalNAc	Proteins with terminal α-GalNAc or GalNAcα-*O*-Ser/Thr (Tn antigen)
UEA-I (*Ulex europaeus;* furze gorse seeds)	-	α-L-Fuc	0.1–0.5 M L-Fuc or methyl-α-L-Fuc	Sugar chains with terminal α-Fuc, especially in α1-2 linkage, but much less with α1-3 or α1-6 linkages
LBA (*Phaseolus lunatus;* lima bean)	Mn^{2+}, Ca^{2+}	Terminal α-GalNAc	0.1–0.5 M GalNAc	Proteins with blood group A structure GalNAcα1-3(Fucα1-2)Gal–

Table 3. Some examples of lectins used for glycoprotein purification modified from current protocols in protein science.

Lectin affinity columns can be prepared by immobilizing lectins with different specificities toward oligosaccharides to a variety of matrices, including agarose (West & Goldring, 2004), silica (Geng et al, 2001), monolithic stationary phases (Okanda & El Rassi, 2006) and cellulose (Aniulyte et al, 2006). These immobilized lectins are invaluable tools for isolating and separating glycoproteins, glycolipids, polysaccharides, subcellular particles and cells. In addition, lectin affinity columns can be used to purify detergent-solubilized cell membrane components. They also are useful for assessing changes in levels or composition of surface glycoproteins during cell development and in malignant or virally transformed variants. In subsequent chapters, more detailed examples of lectin affinity purification can be found.

6.7 Reversed phase chromatography

Reversed phase chromatography is a kind of affinity interaction between a biomolecule dissolved in a solvent (mobile phase) that has some hydrophobicity (e.g. proteins, peptides, and nucleic acids) and an immobilized hydrophobic ligand (stationary phase) (Dorsey & Cooper, 1994). Reversed phase chromatography is generally more suitable for separating non-volatile molecules. The term "reversed phase" was adopted because the binding occurs between a hydrophobic ligand (octadecyl; C18) and molecules in a polar aqueous phase which is reversed from normal phase chromatography [where a hydrophilic polar ligand binds to molecules in a hydrophobic nonpolar mobile phase].

In general, the macromolecules (e.g. protein or peptides) are adsorbed onto the hydrophobic surface of the column. Elution is achieved using a mobile phase which is usually a combination of water and organic solvents (such as acetonitrile or methanol) applied to the column as a gradient (e.g. starting with 95:5 aqueous:organic and gradually increasing the organic phase until the elution buffer is 5:95 aqueuos:organic). The macromolecules bind the hydrophobic surface of the column and remain until the concentration of the organic phase is high enough to elute the macromolecules from the hydrophobic surface.

When using reversed phase chromatography, the most polar macromolecules are eluted first and the most nonpolar macromolecules are eluted last: the more polar (hydrophilic) a solute is, the faster the elution and vice versa. In summary, separations in reversed phase chromatography depend on the reversible adsorption/desorption of solute molecules with varying degrees of hydrophobicity to a hydrophobic stationary phase.

As illustrated in Figure 13, the initial step of reversed phase separation involves equilibration of the column under suitable conditions (pH, ionic strength and polarity). The polarity of the solvent can be modified by adding a solvent such as methanol or acetonitrile and an ion pairing agent such as formic acid or trifluoroacetic acid may be added. Next, sample is applied and bound to the immobilized matrix. Following this step, desorption and elution of the biomolecules is achieved by decreasing the polarity of the mobile phase (by increasing the percentage of organic modifier in the mobile phase). At the end of the separation, the mobile phase should be nearly 100% organic to ensure complete removal of all bound substances. Once everything has eluted from the column, the initial mobile phase is reapplied to the column to reequilibrate the column for a subsequent sample application.

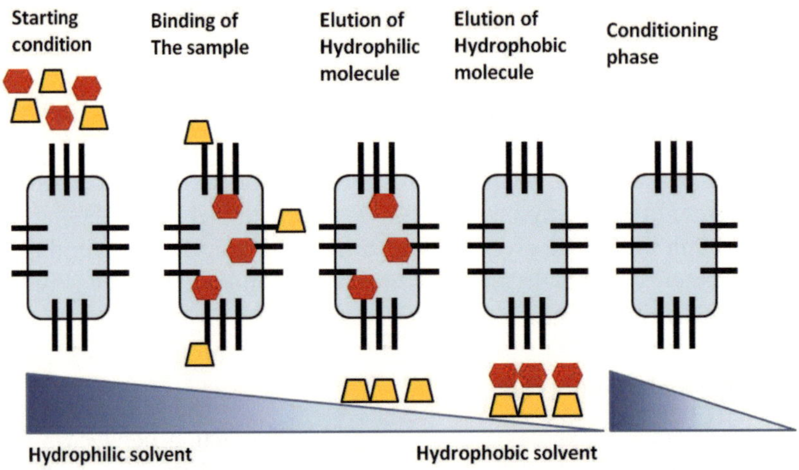

Fig. 13. Steps of a of reversed phase chromatography separation.

7. Funds

This work was supported by JSPS (Japan Society for Promotion of Science) Grant-in-Aid for scientific research (B) to Dr. Sameh Magdeldin (23790933) from Ministry of Education, Culture, Sports, Science and Technology of Japan. The funders had no role in the decision to publish or in the preparation of this chapter.

8. References

Allocati N, Federici L, Masulli M, Di Ilio C (2009) Glutathione transferases in bacteria. FEBS J 276: 58-75

Allocati N, Federici L, Masulli M, Di Ilio C (2011) Distribution of glutathione transferases in Gram-positive bacteria and Archaea. Biochimie

Aniulyte J, Liesiene J, Niemeyer B (2006) Evaluation of cellulose-based biospecific adsorbents as a stationary phase for lectin affinity chromatography. J Chromatogr B Analyt Technol Biomed Life Sci 831: 24-30

Antoni G, Botti R, Casagli MC, Neri P (1978) Affinity chromatography on blue dextran-sepharose of serum albumin from different animal species. Boll Soc Ital Biol Sper 54: 1913-1919

Arsenis C, McCormick DB (1966) Purification of flavin mononucleotide-dependent enzymes by column chromatography on flavin phosphate cellulose compounds. J Biol Chem 241: 330-334

Avnir D, Coradin T, Livage J (2006) Recent bio-applications of sol-gel materials. J Chem Mater 16: 1013-1030

Axen R, Porath J, Ernback S (1967) Chemical coupling of peptides and proteins to polysaccharides by means of cyanogen halides. Nature 214: 1302-1304

Balbas P (2001) Understanding the art of producing protein and nonprotein molecules in Escherichia coli. Mol Biotechnol 19: 251-267

Bautz EK, Hall BD (1962) The isolation of T4-specific RNA on a DNA-cellulose column. Proc Natl Acad Sci U S A 48: 400-408

Beckett GJ, Hayes JD (1993) Glutathione S-transferases: biomedical applications. Adv Clin Chem 30: 281-380

Bottari P, Aebersold R, Turecek F, Gelb MH (2004) Design and synthesis of visible isotope-coded affinity tags for the absolute quantification of specific proteins in complex mixtures. Bioconjug Chem 15: 380-388

Byeon WH, Weisblum B (2004) Affinity adsorbent based on combinatorial phage display peptides that bind alpha-cobratoxin. J Chromatogr B Analyt Technol Biomed Life Sci 805: 361-363

Campbell DH, Luescher E, Lerman LS (1951) Immunologic Adsorbents: I. Isolation of Antibody by Means of a Cellulose-Protein Antigen. Proc Natl Acad Sci U S A 37: 575-578

Casey JL, Coley AM, Foley M (2008) Phage display of peptides in ligand selection for use in affinity chromatography. Methods Mol Biol 421: 111-124

Chen J, Hage DS (2004) Quantitative analysis of allosteric drug-protein binding by biointeraction chromatography. Nat Biotechnol 22: 1445-1448

Clonis YD (2006) Affinity chromatography matures as bioinformatic and combinatorial tools develop. J Chromatogr A 1101: 1-24

Clonis YD, Labrou NE, Kotsira VP, Mazitsos C, Melissis S, Gogolas G (2000) Biomimetic dyes as affinity chromatography tools in enzyme purification. J Chromatogr A 891: 33-44

Cronan JE, Jr., Reed KE (2000) Biotinylation of proteins in vivo: a useful posttranslational modification for protein analysis. Methods Enzymol 326: 440-458

Cuatrecasas P, Wilchek M, Anfinsen CB (1968) Selective enzyme purification by affinity chromatography. Proc Natl Acad Sci U S A 61: 636-643

Cwirla SE, Peters EA, Barrett RW, Dower WJ (1990) Peptides on phage: a vast library of peptides for identifying ligands. Proc Natl Acad Sci U S A 87: 6378-6382

Dangerfield JA, Windbichler N, Salmons B, Gunzburg WH, Schroder R (2006) Enhancement of the StreptoTag method for isolation of endogenously expressed proteins with complex RNA binding targets. Electrophoresis 27: 1874-1877

Davis GD, Elisee C, Newham DM, Harrison RG (1999) New fusion protein systems designed to give soluble expression in Escherichia coli. Biotechnol Bioeng 65: 382-388

Devlin JJ, Panganiban LC, Devlin PE (1990) Random peptide libraries: a source of specific protein binding molecules. Science 249: 404-406

di Guan C, Li P, Riggs PD, Inouye H (1988) Vectors that facilitate the expression and purification of foreign peptides in Escherichia coli by fusion to maltose-binding protein. Gene 67: 21-30

Dong XY, Feng XD, Sun Y (2010) His-tagged protein purification by metal-chelate affinity extraction with nickel-chelate reverse micelles. Biotechnol Prog 26: 1088-1094

Dorsey JG, Cooper WT (1994) Retention mechanisms of bonded-phase liquid chromatography. Anal Chem 66: 857A-867A

Douglas KT (1987) Mechanism of action of glutathione-dependent enzymes. Adv Enzymol Relat Areas Mol Biol 59: 103-167

Geng M, Zhang X, Bina M, Regnier F (2001) Proteomics of glycoproteins based on affinity selection of glycopeptides from tryptic digests. J Chromatogr B Biomed Sci Appl 752: 293-306

Geoghegan KF, Dixon HB, Rosner PJ, Hoth LR, Lanzetti AJ, Borzilleri KA, Marr ES, Pezzullo LH, Martin LB, LeMotte PK, McColl AS, Kamath AV, Stroh JG (1999) Spontaneous alpha-N-6-phosphogluconoylation of a "His tag" in Escherichia coli: the cause of extra mass of 258 or 178 Da in fusion proteins. Anal Biochem 267: 169-184

Graves DJ, Wu YT (1974) On predicting the results of affinity procedures. Methods Enzymol 34: 140-163

Grodzki AC, Berenstein E (2010) Antibody purification: affinity chromatography - protein A and protein G Sepharose. Methods Mol Biol 588: 33-41

Gustavsson, P-E, Larsson, P-O (2006) Support Materials for Affinity Chromatography, In: *Handbook of Affinity Chromatography*, Hage, DS, pp 15-33, Taylor & Francis, Boca Raton, FL.

Haeckel R, Hess B, Lauterborn W, Wuster KH (1968) Purification and allosteric properties of yeast pyruvate kinase. Hoppe Seylers Z Physiol Chem 349: 699-714

Hage DS (2002) High-performance affinity chromatography: a powerful tool for studying serum protein binding. J Chromatogr B Analyt Technol Biomed Life Sci 768: 3-30

Hage DS (2006) Handbook of Affinity Chromatography: Taylor & Francis, Boca Raton, FL.

Hage DS, Anguizola J, Barnaby O, Jackson A, Yoo MJ, Papastavros E, Pfaunmiller E, Sobansky M, Tong Z (2011) Characterization of drug interactions with serum proteins by using high-performance affinity chromatography. Curr Drug Metab 12: 313-328

Hage DS, Austin J (2000) High-performance affinity chromatography and immobilized serum albumin as probes for drug- and hormone-protein binding. J Chromatogr B Biomed Sci Appl 739: 39-54

Hermanson GT, Mallia AK, Smith PK (1992) Immobilized Affinity Ligand Techniques, New York: Academic Press.

Hirabayashi J, Hashidate T, Kasai K (2002) Glyco-catch method: A lectin affinity technique for glycoproteomics. J Biomol Tech 13: 205-218

Hober S, Nord K, Linhult M (2007) Protein A chromatography for antibody purification. J Chromatogr B Analyt Technol Biomed Life Sci 848: 40-47

Huang W, Wang J, Bhattacharyya D, Bachas LG (1997) Improving the activity of immobilized subtilisin by site-specific attachment to surfaces. Anal Chem 69: 4601-4607

Jackson AJ, Xuan H, Hage DS (2010) Entrapment of proteins in glycogen-capped and hydrazide-activated supports. Anal Biochem 404: 106-108

Janson, J-C. (1984) Large-scale affinity purification — state of the art and future prospects. Trends in Biotechnology 2: 31-38

Jin W, Brennan JD (2002) Properties and applications of proteins encapsulated within sol–gel derived materials. Analytica Chimica Acta 461: 1-36

Katoh S, Imada M, Takeda N, Katsuda T, Miyahara H, Inoue M, Nakamura S (2007) Optimization of silica-based media for antibody purification by protein A affinity chromatography. J Chromatogr A 1161: 36-40

Kim, HS, Hage, DS, Immobilization methods for affinity chromatography, In: *Handbook of Affinity Chromatography*, Hage, DS, pp 35-78, Taylor & Francis, Boca Raton, FL.

Labrou NE (2000) Dye-ligand affinity chromatography for protein separation and purification. Methods Mol Biol 147: 129-139

Labrou NE, Karagouni A, Clonis YD (1995) Biomimetic-dye affinity adsorbents for enzyme purification: application to the one-step purification of Candida boidinii formate dehydrogenase. Biotechnol Bioeng 48: 278-288

Labrou NE, Clonis Y.D (2002) Immobilized synthetic dyes in affinity chromatography, In *Theory and Practice of Biochromatography,* Vijayalakshimi MA, pp 335-351, Taylor & Francis, London

Lam KS, Salmon SE, Hersh EM, Hruby VJ, Kazmierski WM, Knapp RJ (1991) A new type of synthetic peptide library for identifying ligand-binding activity. Nature 354: 82-84

LaVallie ER, DiBlasio EA, Kovacic S, Grant KL, Schendel PF, McCoy JM (1993) A thioredoxin gene fusion expression system that circumvents inclusion body formation in the E. coli cytoplasm. Biotechnology (N Y) 11: 187-193

Lerman LS (1953) A Biochemically Specific Method for Enzyme Isolation. Proc Natl Acad Sci U S A 39: 232-236

Lowe CR (1996) Analytical biotechnology. Curr Opin Biotechnol 7: 1-3

Lowe CR, Burton SJ, Burton NP, Alderton WK, Pitts JM, Thomas JA (1992) Designer dyes: 'biomimetic' ligands for the purification of pharmaceutical proteins by affinity chromatography. Trends Biotechnol 10: 442-448

Mattiasson B, (1999) Expanded bed chromatography (special issue), 8, 1-271.

Morag E, Bayer EA, Wilchek M (1996) Reversibility of biotin-binding by selective modification of tyrosine in avidin. Biochem J 316 (Pt 1): 193-199

O'Carra P, Barry S, Griffin T (1974) Interfering and complicating adsorption effects in bioaffinity chromatography. Methods Enzymol 34: 108-126

Okanda FM, El Rassi Z (2006) Affinity monolithic capillary columns for glycomics/proteomics: 1. Polymethacrylate monoliths with immobilized lectins for glycoprotein separation by affinity capillary electrochromatography and affinity nano-liquid chromatography in either a single column or columns coupled in series. Electrophoresis 27: 1020-1030

Peters T, Jr., Taniuchi H, Anfinsen CB, Jr. (1973) Affinity chromatography of serum albumin with fatty acids immobilized on agarose. J Biol Chem 248: 2447-2451

Phillips TM, Queen WD, More NS, Thompson AM (1985) Protein A-coated glass beads. Universal support medium for high-performance immunoaffinity chromatography. J Chromatogr 327: 213-219

Pierre AC (2004) The sol-gel encapsulation of enzymes. Biocatal Biotransform 22: 145-170

Renkin EM (1954) Filtration, diffusion, and molecular sieving through porous cellulose membranes. J Gen Physiol 38: 225-243

Romig TS, Bell C, Drolet DW (1999) Aptamer affinity chromatography: combinatorial chemistry applied to protein purification. J Chromatogr B Biomed Sci Appl 731: 275-284

Ruhn PF, Garver S, Hage DS (1994) Development of dihydrazide-activated silica supports for high-performance affinity chromatography. J Chromatogr A 669: 9-19

Sander EG, McCormick DB, Wright LD (1966) Column chromatography of nucleotides over thymidylate-cellulose. J Chromatogr 21: 419-423

Schaschke N, Gabrijelcic-Geiger D, Dominik A, Sommerhoff CP (2005) Affinity chromatography of tryptases: design, synthesis and characterization of a novel matrix-bound bivalent inhibitor. Chembiochem 6: 95-103

Schiel JE, Hage DS (2009) Kinetic studies of biological interactions by affinity chromatography. J Sep Sci 32: 1507-1522

Schneider C, Newman RA, Sutherland DR, Asser U, Greaves MF (1982) A one-step purification of membrane proteins using a high efficiency immunomatrix. J Biol Chem 257: 10766-10769

Shaltiel S (1974) Hydrophobic chromatography. Methods Enzymol 34: 126-140

Sharon N (1998) Lectins: from obscurity into the limelight. Protein Sci 7: 2042-2048

Sheehan D, Meade G, Foley VM, Dowd CA (2001) Structure, function and evolution of glutathione transferases: implications for classification of non-mammalian members of an ancient enzyme superfamily. Biochem J 360: 1-16

Sisson TH, Castor CW (1990) An improved method for immobilizing IgG antibodies on protein A-agarose. J Immunol Methods 127: 215-220

Smith GP, Scott JK (1993) Libraries of peptides and proteins displayed on filamentous phage. Methods Enzymol 217: 228-257

Spiro RG (2002) Protein glycosylation: nature, distribution, enzymatic formation, and disease implications of glycopeptide bonds. Glycobiology 12: 43R-56R

Staal GE, Koster JF, Kamp H, van Milligen-Boersma L, Veeger C (1971) Human erythrocyte pyruvate kinase. Its purification and some properties. Biochim Biophys Acta 227: 86-96

Thorn KS, Naber N, Matuska M, Vale RD, Cooke R (2000) A novel method of affinity-purifying proteins using a bis-arsenical fluorescein. Protein Sci 9: 213-217

Travis J, Bowen J, Tewksbury D, Johnson D, Pannell R (1976) Isolation of albumin from whole human plasma and fractionation of albumin-depleted plasma. Biochem J 157: 301-306

Travis J, Pannell R (1973) Selective removal of albumin from plasma by affinity chromatography. Clin Chim Acta 49: 49-52

Tu C, Rudnick PA, Martinez MY, Cheek KL, Stein SE, Slebos RJ, Liebler DC (2010) Depletion of abundant plasma proteins and limitations of plasma proteomics. J Proteome Res 9: 4982-4991

Tyutyulkova S, Paul S (1995) Purification of antibody light chains by metal affinity and protein L chromatography. Methods Mol Biol 51: 395-401

Udomsinprasert R, Pongjaroenkit S, Wongsantichon J, Oakley AJ, Prapanthadara LA, Wilce MC, Ketterman AJ (2005) Identification, characterization and structure of a new Delta class glutathione transferase isoenzyme. Biochem J 388: 763-771

Wang G, De J, Schoeniger JS, Roe DC, Carbonell RG (2004) A hexamer peptide ligand that binds selectively to staphylococcal enterotoxin B: isolation from a solid phase combinatorial library. J Pept Res 64: 51-64

West I, Goldring O (2004) Lectin affinity chromatography. Methods Mol Biol 244: 159-166

Whetstone PA, Butlin NG, Corneillie TM, Meares CF (2004) Element-coded affinity tags for peptides and proteins. Bioconjug Chem 15: 3-6

Windbichler N, Schroeder R (2006) Isolation of specific RNA-binding proteins using the streptomycin-binding RNA aptamer. Nat Protoc 1: 637-640

Young JL, Webb BA (1978) Two methods for the separation of human alpha-fetoprotein and albumin. (A) Affinity chromatography using Blue Sepharose CL-6B and (B) ampholyte displacement chromatography. Anal Biochem 88: 619-623

Zachariou M (2008) Affinity chromatography: methods and protocols. Preface. Methods Mol Biol 421: vii-viii

2

Affinity Interactions as a Tool for Protein Immobilization

Eva Benešová and Blanka Králová
The Institute of Chemical Technology, Prague
Czech Republic

1. Introduction

Proteins, biopolymers composed of proteinogenic amino acids, are molecules with irreplaceable functions in human organism. They have among others a role as structural components, they are involved in motion processes, they appear as significant parts of immune response and different types of protection events, they play important roles in transport and storage processes and they also occur in signalling processes. One of the most important protein functions is their role as natural biocatalysts - enzymes, because these compounds, increasing the rate of metabolic reactions, are necessary for almost all reactions in human body. The potential of enzyme application in biotechnological processes was discovered many years ago. The rapid development of molecular biology and protein engineering, enabling targeted designing of proteins with suitable features and their production in recombinant form, contributed in decisive way to the final anchoring of proteins in biotechnological practice including such areas as food industry and medicine. Also different methods of protein immobilization represent a way enabling their common and easy application in biotechnology (Demain & Vaishnav, 2009; García-Junceda et al., 2004; Murray et al., 2002; Vodrážka, 1999). A part of this improvement, focused on utilization of affinity interactions in immobilization processes, will be the major part of this chapter.

First part of this chapter will be aimed to the explanation of the importance of enzyme application in biotechnological processes and the improvement of their usage caused by immobilization. Advantages and limitations of immobilization processes will be described and some examples of their practical application in biotechnology and pharmaceutical industry will be mentioned. This part will also contain a brief summary of common immobilization techniques.

Second part of this chapter will explain the reasons for recombinant proteins preparation and the advantage of proteins modifications by techniques of molecular biology. A simple strategy for recombinant protein preparation in the simplest expression system of *E. coli* will be described for a better comprehensibility. Different methods of exploitation of affinity interactions for the protein immobilization will be referred in the last part of the chapter. The importance of achieved results for the biotechnological practice will be summarized.

2. Immobilization: What, why, how?

At the beginning of the topic describing immobilization techniques and their advantages we should explain the meaning of the word "immobilization" in this text. A molecule or a cell is referred to be immobilized, if its mobility in the reaction space is artificially restricted. Many various immobilization protocols were evolved as we will show in the next chapter part. It is also necessary to stress at this point that although we have decided to focus on enzyme immobilization, also other molecules (antibodies, DNA etc.) and various cells and cellular organelles may be immobilized. In some cases also systems containing more than one immobilized enzyme were prepared. Such systems may gradually catalyse subsequent reactions in biochemical process (Aehle, 2007; Brena & Batista-Viera, 2006; Costa et al., 2005; García-Junceda et al., 2004; Guisan, 2006; Hernandez & Fernandez-Lafuente, 2011; Krajewska, 2004; Rao et al., 1998; Tischer & Wedekind, 1999).

Generally it is possible to say that enzymes are excellent biocatalysts working under mild reaction conditions (temperature, pressure, pH) and evincing high substrate and reaction specificity i.e. biotechnologically important characteristics, which result in production of desired end-product without by-products contamination. For these reasons enzymes found their place in a wide variety of biotechnological areas including among others food industry, medicine and pharmaceutical industry, analytical applications, cosmetics or e.g. textile and paper industry and new and new applications are constantly announced (Aehle, 2007; Brena & Batista-Viera, 2006; Costa et al., 2005; Cowan & Fernandez-Lafuente, 2011; Guisan, 2006; Krajewska, 2004; Rodrigues et al., 2011; Sheldon, 2007).

As one example for all we can mention β-D-galactosidase representing an enzyme very popular in food industry applications. Its importance for the milk processing lies in its ability to hydrolyse lactose, as this ability offers the possibility of lactose free milk preparation, an important product for lactose-intolerant people. The ability of some of these enzymes to catalyze transglycosylation reactions is often utilized for health beneficial galactooligosaccharides production. Crystallization prevention, cheeses ripen improving or whey lactose hydrolysis are other examples of possible biotechnological β-D-galactosidase application. Several methods were successfully used for immobilization of this enzyme, e.g. physical adsorption, gel entrapment or covalent binding. Also techniques utilizing affinity interactions with fusion β-D-galactosidases were evolved (Aehle, 2007; Cowan & Fernandez-Lafuente, 2011; Krajewska, 2004; Panesar et al., 2006).

In order to make enzyme application in biotechnological processes more favourable, different methods for the cost decrease are implemented, immobilization techniques being one of them. Moreover, enzyme application suffers from various other limitations resulting e.g. from low stability, sensitivity to process conditions or from tendency to be inhibited by high concentrations of reaction components of some of these biocatalysts. Improvement of these characteristics is still a challenge for modern biotechnological research and some of these problems found their solution in precisely designed immobilization processes (Aehle, 2007; Cao, 2005; Costa et al., 2005; Guisan, 2006; Hernandez & Fernandez-Lafuente, 2011; Krajewska, 2004).

Enzyme immobilization enables primarily the re-use or continuous use of the biocatalysts and it also substantially simplifies the manipulation with the biocatalyst and the control of the reaction process. Also the separation of the enzyme from the reaction mixture is significantly easier and protein contamination of final product is minimized. Moreover, immobilization is next to molecular biology and protein engineering an alternative method for improving natural features of enzymes as e.g. stability, activity, specificity or selectivity. Anyway, various combinations of above mentioned approaches are possible, i.e. immobilization of suitably modified enzymes or, reversely, modification (physical or chemical) of already immobilized enzyme. Also the unwanted enzyme inhibition caused by reaction components, by aggregation, adsorption, by dissociation into subunits or by autolysis or proteolysis can be positively influenced by precise design of immobilization process. In special cases (therapeutic application etc.) also additional advantages as e.g. prolonged blood circulation lifetime or lower immunogenicity may be observed. As a conclusion it is possible to summarize that enzyme immobilization increases the productivity of these biocatalysts and improves their features, which make them more attractive for various applications. However, in some cases immobilization can cause a lowering of enzyme activity or changes of natural enzyme features in undesirable way. These situations must be prevented. Other complications, which have to be solved during immobilization process designing, represent mass transfer limitations (Aehle, 2007; Cao, 2005; Chern & Chao, 2005; Costa et al., 2005; Cowan & Fernandez-Lafuente, 2011; García-Junceda et al., 2004; Guisan, 2006; Hernandez & Fernandez-Lafuente, 2011; Krajewska, 2004; Liu & Scouten, 1996; Mateo et al., 2007; Panesar et al., 2006; Rodrigues et al., 2011; Sheldon, 2007; Tischer & Wedekind, 1999; Turková, 1999).

2.1 Immobilization techniques

The main goal of this part is to introduce some basic information about immobilization techniques to the reader. In fact there are few basic protocols used during immobilization processes, all of them having many variations. The description of all details about these methods and their modifications go far beyond the extent of our topic and for this reason we recommend to find details in referred publications.

Immobilization methods are classified differently in various publications. We have chosen for immobilization techniques classification system of three major classes 1) binding to a carrier, 2) entrapment and 3) enzyme molecules cross-linking. However, another ways of sorting e.g. according to the reversibility of the process are also frequently used. In fact all classifications suffer from the fact that many newly evolved immobilization procedures exceed the border of simple sorting (Aehle, 2007; Brenda & Batista-Viera, 2006; Cao, 2005; Costa et al., 2005; Krajewska, 2004; Sheldon, 2007; Tischer & Wedekind, 1999).

2.1.1 Binding to a carrier

These methods take the advantage of the fact that proteins contain amino acids with different features. Functional groups in side chains of these amino acids can be involved in binding to the support by various types of linkages and interactions. Many types of carriers with diverse properties were evolved for immobilization processes; however, the suitability of their application in individual cases needs a careful consideration and very often also laborious method adaptation. Preparing a suitable carrier may be seen as another problem and may incur additional costs for the whole immobilization procedure (Aehle, 2007; Brenda & Batista-Viera, 2006; Sheldon, 2007).

a. Covalent binding

For covalent binding of the protein to the carrier amino acid residues not involved in the reaction mechanism may be used. Generally used functional groups are amino groups of lysines and arginines, sulphydryl groups of cysteins, carboxyl groups of aspartic and glutamic acids and hydroxyl groups of tyrosines, serines and threonines. Strength of the linkage not allowing enzyme release from the support during the reaction process and frequent enzyme stability increase counterbalance limitations caused by the possibility of unwanted changes in active structure of the enzyme and thus possible decrease of enzyme activity caused by strong enzyme - carrier interactions (Aehle, 2007; Brenda & Batista-Viera, 2006; Costa et al., 2005; Ho et al., 2004; Panesar et al., 2006; Rao et al., 1998; Sheldon, 2007).

b. Physical adsorption

In this method, typical for its simple performance and little effect on biocatalysts conformation, several different types of noncovalent interactions (e.g. hydrogen bonds, hydrophobic interactions and van der Waals forces) are involved in the immobilization process. The weakness of support - enzyme interactions, which can be easily influenced by reaction conditions, causing enzyme desorption represents its major limitation (Aehle, 2007; Brenda & Batista-Viera, 2006; Costa et al., 2005; Ho et al., 2004; Panesar et al., 2006; Sheldon, 2007).

c. Ionic binding

Ion-ion interactions, providing a stronger binding of immobilized molecules than physical adsorption, are utilized during immobilization process. This method also doesn´t change enzyme conformation in a substantial extent. Disadvantages of this relatively simple process may lie in the usage of highly charged supports, which can interact with charged substrates or products and in laborious finding of suitable conditions providing sufficiently strong interaction and preserving the activity of the enzyme (Aehle, 2007; Brenda & Batista-Viera, 2006; Costa et al., 2005).

d. Affinity binding

This method is based on the principle of complementary biomolecules interactions, which represents its biggest advantage i.e. high selectivity (Brenda & Batista-Viera, 2006; Costa, 2005; García-Junceda et al., 2004). Detailed description of this method will be given in the part 2.2.

Principles of above mentioned methods are for illustration presented in the figure 1.

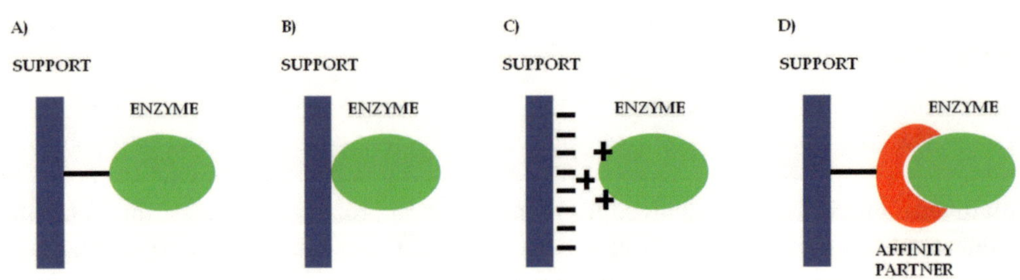

Fig. 1. Immobilization methods exploiting binding to a carrier. A) covalent binding, B) physical adsorption, C) ionic binding, D) affinity binding

2.1.2 Enzyme molecules cross-linking

In these methods bi- or multifunctional compounds are used for cross-linking of desired enzyme molecules (figure 2). Lysines amino groups are usually involved in covalent bonds formation; however, other amino acids functional groups may be used, too. Since several drawbacks accompany this method, e.g. low activity yields, poor reproducibility or manipulation difficulties, new improvements including cross-linking of enzyme crystals or enzyme aggregates, co-cross-linking with inert materials or cross-linking on solid support or in gels were evolved (Aehle, 2007; Costa et al., 2005; Sheldon, 2007; Tischer & Wedekind, 1999).

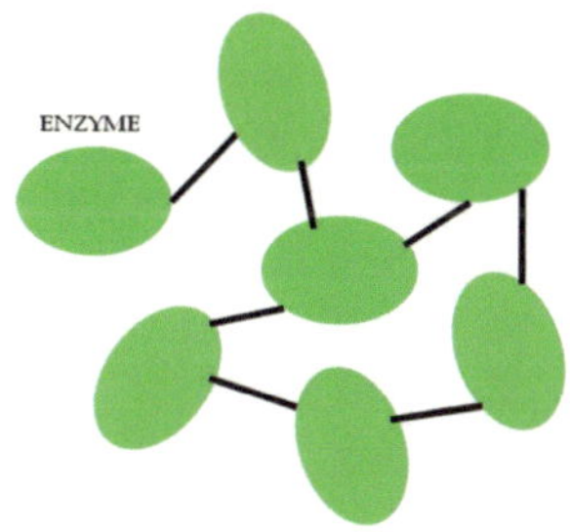

Fig. 2. Cross-linking of enzyme molecules

2.1.3 Entrapment

The basis of this method is the inclusion of the biocatalyst within a polymeric network of different types (figure 3). These protocols comprise among others entrapment into gel matrices, microencapsulation or fiber entrapping. Its major advantages consist in simplicity of performance, in the possibility to use similar procedures for different enzymes or even in their simultaneous immobilization and in elimination of inhibition by proteases and inhibitors of high molecular weight. On the other hand, diffusion constraints and the possibility of enzyme leakage belong to the major method limitations (Aehle, 2007; Brenda & Batista-Viera, 2006; Cao, 2005; Costa et al., 2005; Ho et al., 2004; Panesar et al., 2006; Sheldon, 2007).

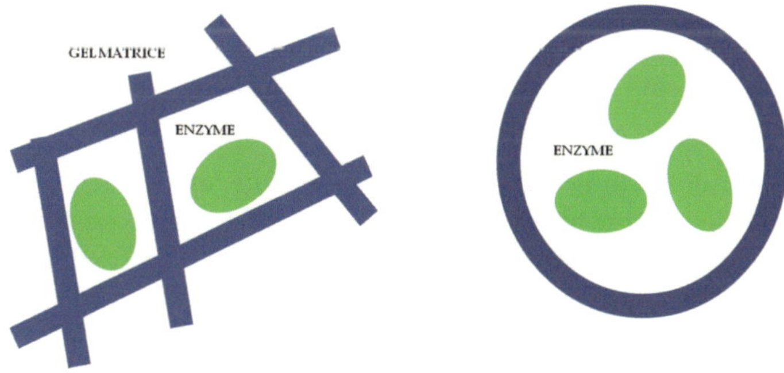

Fig. 3. Enzyme immobilization by entrapment. A) entrapment into gel matrices, B) microencapsulation

Another aspect, which can be used for immobilization methods classification and which should be considered during the choice of immobilization process, is the influence of immobilization on enzyme orientation. Standard immobilization protocols without a control of binding mode are usually considered as random immobilization. Such immobilization processes result in heterogeneity in binding nature. This random immobilization may suffer from various drawbacks as e.g. involving of catalytic amino acids residues in the binding process with the subsequent loss of enzymatic activity or restricted accessibility of the active site for substrate molecule. These limitations were overcome in the method of site-specific enzyme immobilization exploiting the attachment of the enzyme due to defined binding sites. As a nice example of different immobilization outcomes a comparative study measured for β-galactosidase by Vishwanath and co-workers may serve. A conjugate of β-galactosidase with a polypeptide tail and unmodified β-galactosidase were immobilized. While a rapid drop of enzymatic activity was observed for the random immobilization of unmodified enzyme (only 1.8 % of original activity detected), tagged β-galactosidase preserved 87.7 % of original activity, which suggests the attachment due to the polypeptide tail and a positive influence of oriented immobilization. Immobilization techniques utilizing affinity interactions, which will be described in the following part of the chapter, are representatives of oriented immobilization methods (Cao, 2005; Hernandez & Fernandez-Lafuente, 2011; Kumada et al., 2010; Liu & Scouten, 1996; Rao et al., 1998; Turková, 1999; Vishwanath et al., 1995).

Although various methods for protein immobilization were evolved, precise designing of immobilization process is still not a routine task. For different enzymes different supporting materials and different immobilization techniques are suitable. Thus in every case of immobilization protocol designing three major things are important to be considered - the enzyme of interest, suitable carrier type and used immobilization method. From the biotechnological point of view also the economical suitability has to be taken into account. Nowadays, the effort to evolve rationally designed and specialized immobilization processes, ideally realized by simple methodologies, is accompanied and facilitated by modern biochemical methods (Aehle, 2007; Brena & Batista-Viera, 2006; Cao, 2005; Guisan, 2006; Krajewska, 2004; Mateo, 2007; Sheldon, 2007).

2.2 Methods exploiting affinity interactions for enzymes immobilization

Following part of the chapter will be focused on the possibility to take an advantage of affinity interactions for protein immobilization. Detailed description will be devoted to the usage of affinity tags and also some other immobilization methods utilizing affinity interactions will be presented at the end of the chapter.

Affinity immobilization techniques exploit the selectivity of specific interactions, which occur in almost all important biological processes in living organisms. Various methods, based on the ability of different affinity partners to bind selectively together (antibodies and antigens or haptens, lectins and free saccharidic chains or glycosylated macromolecules, nucleic acids and nucleic acid-binding proteins, hormones and their receptors, avidin and biotin, polyhistidine tag and metal ions etc.), were discovered for utilization in affinity chromatography or in affinity immobilization methodologies (Carlsson et al., 1998; Nelson & Cox, 2005; Nilsson et al., 1997; Roy & Gupta, 2006; Saleemuddin, 1999).

As mentioned earlier the major advantage of utilization of affinity interactions for enzyme immobilization lies in the selectivity of the method. Also the possibility to control the orientation of immobilized enzyme and minimal conformational changes caused by this type of binding resulting in high retention of the immobilized molecule activity represent an important benefit. The reversibility of the methods, enabling the support reloading, and the possibility of direct enzyme immobilization from crude cell lysate without additional purification steps contribute to the better applicability of these methods in practice. Mild reaction conditions and relative simplicity of these immobilization processes should also be emphasized (Andreescu et al., 2006; Brena & Batista-Viera, 2006; Bucur et al., 2004; Clare et al., 2001; Costa et al., 2005; Daunert et al., 2007; Kumada et al., 2010; Saleemuddin, 1999).

2.2.1 Fusion protein affinity tags utilized for protein immobilization

One way how to use affinity binding in the immobilization process is the approach exploiting possibilities of molecular biology generating recombinant fusion proteins (chimeric proteins), i.e. proteins of interest containing specific parts suitable for affinity interactions. These parts, called affinity tags, are usually amino acids sequences, ranging in size from a few amino acids to whole proteins. A wide variety of these fusion partners was prepared, some of them with the ability to bind small ligands, others interacting with a suitable protein partner. Not negligible advantage of use of fusion protein approach consists in the possibility to attach one type of fusion partner to different proteins of interest. This fact enables the usage of one support type, preparation of which can sometime be very expensive or complicated, for immobilization of various enzymes (Arnau et al., 2006; Chern & Chao, 2005; Costa et al., 2005; Daunert et al., 2007; Nilsson et al., 1997; Sørensen & Mortensen, 2005; Rao et al., 1998; Saleemuddin, 1999; Terpe, 2003).

It is important to mention that immobilization is not the only reason for recombinant fusion proteins production, although in their ability to enable simple purification (affinity chromatography) and immobilization processes lies their most important advantage. In some cases proteins are produced with suitable fusion partners also for increase of solubility (fusion with maltose-binding protein, thioredoxin, glutathione S-transferase etc.) and stability of original enzyme (maltose-binding protein, thioredoxin etc.), for an improvement of the correct folding of the protein (maltose-binding protein, ubiquitin-based tags etc.) or for an increase of the synthesis of proteins, which are translated only poorly under normal conditions (green fluorescent protein) etc. Fusion partners can be used as specific folding and expression reporters, too (green fluorescent protein, rainbow tags) (Arnau et al., 2006; Altenbuchner & Mattes, 2005; Arechaga et al., 2003; Baneyx, 1999; Jacquet et al., 1999; Nilsson et al., 1997; Rao et al., 1998; Sørensen & Mortensen, 2005; Waldo et al., 1999).

Description of various ways of immobilizations enabled by affinity interactions mediated by affinity tags and particular examples will be described in the part 2.2.3. For a better comprehensibility of the recombinant proteins preparation next part of the chapter will be devoted to a brief description of one of the simplest ways for recombinant protein production.

2.2.2 Recombinant proteins preparation

Increasing usage of enzymes in broad spectra of biotechnological applications led to a demand on cost-effective processes producing sufficient amounts of desired proteins. The isolation from original sources was very often not suitable (low production in original organism, high costs of cultivation process, pathogenicity of the original organisms etc.), but new techniques of molecular biology opened the way for the production of recombinant proteins in host organisms, thus solving many of above mentioned complications (Demain & Vaishnav, 2009; García-Junceda et al., 2004; Sørensen & Mortensen, 2005).

Many different expression systems were evolved, including Gram-positive and Gram-negative bacteria, yeasts, fungi and plant, insect or mammalian cells. The reason for different expression systems preparation is the fact, that every system has its advantages and drawbacks and no expression system could be suitable for all demanded proteins. Generally it is possible to say that the choice of correct expression system depends on several factors as e.g. cost, size of the protein, demanded posttranslational modifications, final yield etc. (Arnau et al., 2006; Demain & Vaishnav, 2009; Gellissen et al., 2005; Primrose et al., 2001). As it is beyond the theme of this chapter, it is not possible to explain here all details related to all individual expression systems. For illustration we will briefly describe the usage of one mostly used expression system using the best known and described bacteria (*E. coli*) and we will also outline one of possible strategies for recombinant proteins production.

E. coli is the microorganism serving for a long decades for scientific purposes. For its convenience it is nowadays one of the most often used organisms for research and also for industrial purposes in the recombinant enzymes production. But even this expression system has its limitations. The inability to produce glycosylated proteins can be named as one example. Therefore, research based on genetic engineering focuses aims to overcome some shortcomings and thus increase the possibility of application of these microorganisms in many new industrial processes (Baneyx, 1999; Demain & Vaishnav, 2009; Altenbuchner & Mattes, 2005; Sørensen & Mortensen, 2005).

Every strategy for preparation of recombinant protein in a host cell consists of four basic steps (Primrose et al., 2001).

1. In the first step it is necessary to have the possibility to prepare DNA fragment containing the gene encoding the desired protein.
2. Prepared DNA is inserted into a chosen vector. Plasmids, nowadays mostly commercially prepared, are often used as suitable DNA vectors.
3. Prepared expression plasmid is introduced into host cells.
4. Host cells after plasmid DNA introduction are cultivated on suitable media and positive colonies containing expression plasmid for recombinant protein production are detected (Altenbuchner & Mattes, 2005; Primrose et al., 2001; Sørensen & Mortensen, 2005; Vodrážka, 1999).

A scheme of the process of recombinant protein production is for a better comprehensibility shown in the figure 4.

Another advantage offered by recombinant proteins production is the possibility of modification of the original nucleotide sequence encoding the protein of interest. For example suitable alterations in the original structure of the protein may result in desirable changes of its stability, activity or specificity. However, for the topic of this chapter the fact that on both termini of the gene special DNA sequences may be ligated, which are responsible for a production of original polypeptide chain with additional polypeptide sequences, which can be used as affinity tag for immobilization processes, is important. These positions of fusion tags usually don´t represent any obstacle for the catalytic centre and thus for a correct enzymatic function of the protein (Andreescu et al., 2006; Arnau et al., 2006; Bucur et al., 2004; Demain & Vaishnav, 2009; Rao et al., 1998).

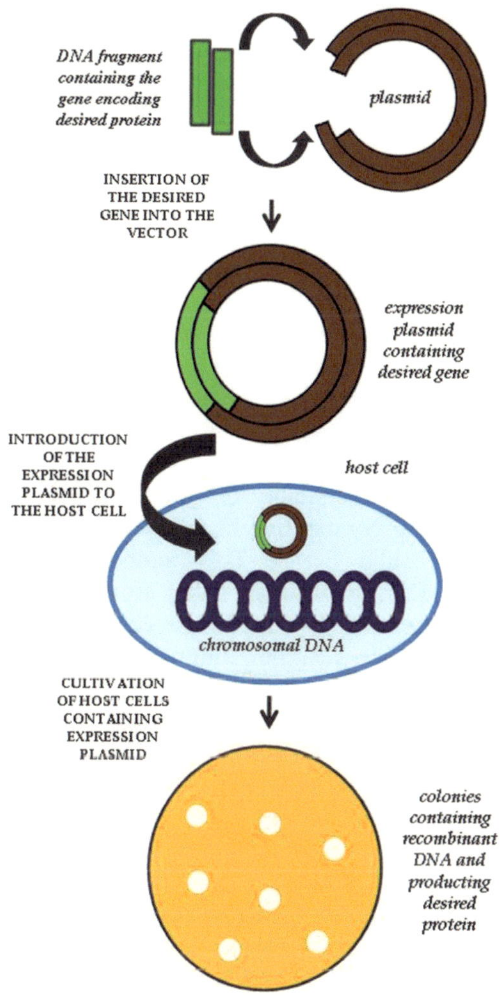

Fig. 4. Recombinant protein preparation

It should be once more emphasized that only one of the simplest strategies for recombinant protein production was described and for more precise description of cloning methods, gene manipulations and expression system advantages the reader is recommended to search in mentioned works.

2.2.3 Affinity tags used for protein immobilization

As stated earlier, following paragraphs will discuss the most interesting affinity tags, which were used for enzyme immobilization. In many cases these tags are often used also for affinity chromatography enabling very effective purification of desired proteins. General principle of these methods is illustrated in the figure 5.

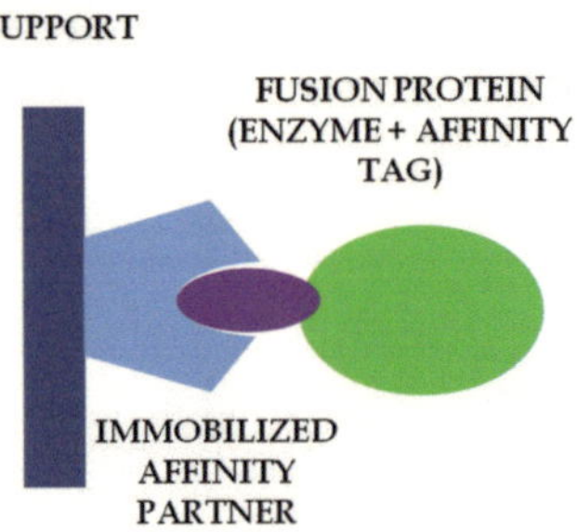

Fig. 5. Principle of immobilization methods enabled by affinity tags

2.2.3.1 Immobilization by polyhistidine tags

This method, based on the specific interaction of histidine-rich tags of different length attached to the protein of interest and of metal ions immobilized on the matrix (Ni^{2+}, Cu^{2+}, Zn^{2+} etc.), belongs to a group of immobilized metal affinity adsorption techniques. The attachment of polyhistidine tag to different protein partners was used in many cases for purification processes and also the interest of its possible employment in immobilizations is under intensive investigation. The reversibility of the method and thus the reuse of the supports can be ensured by using competitive ligand (e.g. imidazole or histidine) or by usage of metal ions chelator like EDTA. This method, although used in many immobilization experiments, can in some cases suffer from different complications. For example leaching of the metal ions from the support can cause product contamination. Also the interference of large number of metal-binding proteins or additional interactions between the immobilized protein and affinity support may represent a serious limitation. Several authors stated also relatively low binding affinity (similarly as for glutathione S-transferase tag or maltose-binding tag) as an important restriction of use of these affinity tags in immobilization processes. However, polyhistidine-tags exploiting immobilization techniques were successfully used e.g. for acetylcholine esterase, β-galactosidase, lactate dehydrogenase, β-glucuronidase or D-hydantoinase immobilization. During the development of methods exploiting the interaction with metal ions various tags of different composition, containing additional amino acid residues, were prepared too (Bucur et al., 2004; Daunert et al., 2007; Desai et al., 2002; Hernandez & Fernandez-Lafuente, 2011; Ho et al., 2004; Nilsson et al., 1997; Rao et al., 1998; Roy & Gupta, 2006; Saleemuddin, 1999; Takakura et al., 2010; Terpe, 2003).

2.2.3.2 Immobilization using avidin-biotin interaction

Another strategy for affinity immobilization exploits extraordinarily strong affinity interaction (probably one of the strongest noncovalent interactions between two biomolecules) between vitamin H - biotin (figure 6) and egg white glycoprotein avidin or its bacterial alternative streptavidin originating from *Streptomyces avidinii*. Biotinylated proteins may be attached to the support containing avidin as an interaction partner. The preparation of biotinylated conjugates is usually not accompanied by a significant loss of activity; however, chemical approaches result in heterogenous population of desired biotinylated protein. Methods of genetic engineering may represent a solution of above mentioned complication. β-Galactosidase may be given as an example of an enzyme immobilized by this strategy. In some cases (e.g. keratinkinase, β-galactosidase, trypsin) reversely fusion proteins of desired enzyme and streptavidin can be immobilized on a biotinylated support. During detailed research also other affinity tags with the ability to bind biotin were discovered. As an example tamavidin 2, a fungal avidin-like protein, used as affinity tag e.g. for sialyltransferase from *Photobacterium* sp. JT-ISH-224 immobilization on biotin carrying support, can be mentioned (Clare et al., 2001; Costa et al., 2005; Hernandez & Fernandez-Lafuente, 2011; Nilsson et al., 1997; Rao et al., 1998; Takakura et al., 2010; Turková, 1999).

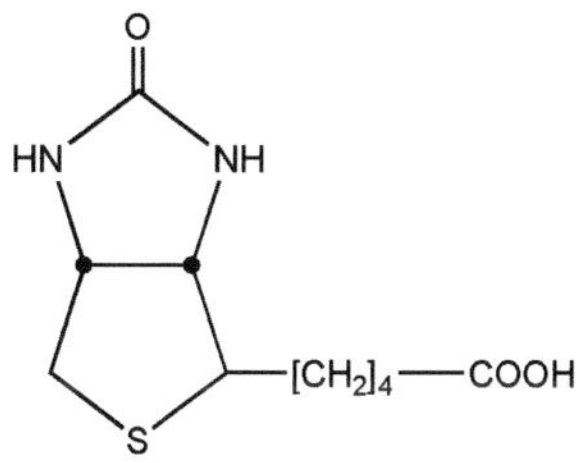

Fig. 6. Biotin

2.2.3.3 Immobilization by cellulose-binding domain

It is known that some bacterial or fungal proteins, e.g. exoglucanase and endoglucanase from *Cellulomonas fimi* or cellulase from *Trichoderma harzianum*, are able to specifically bind cellulose thanks to their cellulose-binding domain. These domains may be exploited as affinity tags for fusion proteins immobilization on cellulose supports, which are inert and exhibit only low non-specific affinity and are readily available. A genetically based fusion protein technique of this type was used e.g. for β-glucosidase from *Agrobacterium* sp., *Zymomonas mobilis* extracellular invertase or *Bacillus stearothermophilus* L1 lipase. Alternative method using chemical coupling by glutaraldehyde was utilized for preparation of a conjugate containing glucose oxidase and cellulose-binding domain. As an interesting inter-methodical example the immobilization technique used for alkaline phosphatase and β-glucosidase can be mentioned. In this case a chimeric protein containing cellulose-binding domain and streptavidin was prepared, attached on cellulose support and used for immobilization of biotinylated forms of above mentioned enzymes (Chern & Chao, 2005; Clare et al., 2001; Daunert et al., 2007; Hernandez & Fernandez-Lafuente, 2011; Hwang et al., 2004; Rao et al., 1998; Roy & Gupta, 2006; Saleemuddin, 1999; Terpe, 2003).

2.2.3.4 Immobilization by chitin-binding domain

Chitin represents another example of suitable affinity support, which is among other nontoxic, biodegradable and commonly occurring in the nature and thus available at relatively low cost. Chitin-binding domain, originating e.g. from chitinase A1 of *Bacillus circulans* WL-12, serves as appropriate affinity tag in these cases. D-Hydantoinase can be mentioned as an example of an enzyme successfully produced with a chitin-binding domain affinity tag and immobilized on chitin support. It can be supposed that also this type of affinity support will find a broader application field in the future (Chern & Chao, 2005; Daunert et al., 2007; Krajewska, 2004; Terpe, 2003).

2.2.3.5 Immobilization exploiting calmodulin as an affinity tag

Immobilization techniques utilizing the calcium-binding regulatory protein calmodulin as an affinity tag are based on the highly specific interaction of this molecule (attached to the protein of interest) with its phenothiazine ligand (figure 7). Oriented immobilized enzyme can be easily removed from the support thanks to Ca^{2+} dependence of the interaction process enabling the support regeneration. Covalent binding of the phenothiazine ligand to the support prevents the leaching of the ligand. This immobilization technique was used e.g. for organophosphorus hydrolase or β-lactamase (Daunert et al., 2007; Desai et al., 2002).

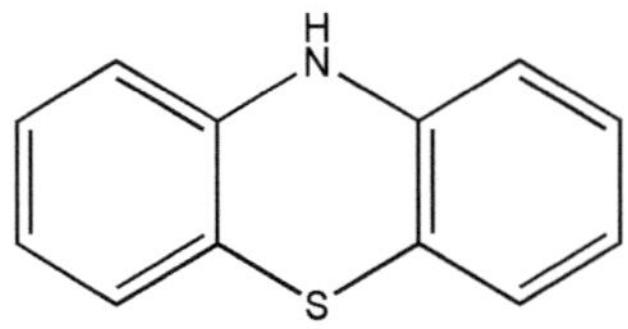

Fig. 7. Phenothiazine

We would like to end this part, devoted to the topic of utilization of affinity tags for protein immobilization, with directing the attention to the fact that also many others affinity tags were tested to be suitable as a tool for enzyme immobilization. However, mentioning all of them exceeds possibilities of this communication. As examples we can mention immobilization by polystyrene-binding peptides with their ability to specifically bind hydrophilic polystyrene supports, immobilization by DNA hybridization, exploiting the ability of polynucleotide chains of DNA to pair according to the rules of complementarity, or by using glutathione S-transferase tag, which is able to specifically bind to glutathione (Daunert et al., 2007; Kumada et al., 2010; Paternolli et al., 2002).

2.2.4 Other approaches for affinity immobilization

Although the methods of molecular biology and genetic engineering for fusion protein production are very useful and popular in immobilization strategies, in some cases native features of proteins of interest are possible to be exploit for the immobilization, too. A few examples of this type of techniques are mentioned in the following paragraphs. In some cases a very tight relations between these methods and earlier described principles based on fusion protein preparation are evident.

2.2.4.1 Glycoproteins immobilization

For the immobilization of glycoproteins (proteins posttranslationally modified by carbohydrate attachment to hydroxyl groups of serine and threonine or to amide group of asparagine) their specific interaction with carbohydrate-binding proteins i.e. lectins can be used. This type of posttranslational modification is very common in nature and was observed in many eukaryotes, archaebacteria or even in some prokaryotes. The specificity of above mentioned interactions was many times exploited in affinity based purification procedures. Concanavalin A originating from Jack bean (*Canavalia ensiformis*), displaying selectivity for defined mannosylated and glucosylated structures, is the mostly used lectin in immobilization processes. Considering the fact that carbohydrate chains are usually not involved in catalytical processes, this method of immobilization as a rule doesn´t evince significant influence on the catalytic activity of the enzyme. Promising results obtained for a glycoprotein acetylcholine esterase by Bucur and co-workers were published recently. Also other glycoenzymes, as for example glucose oxidase, invertase, β-galactosidase, carboxypeptidase Y or amyloglucosidase, were immobilized by this way. However, high cost for lectin supports can represent a limitation of application of these techniques and thus new ways for inexpensive carriers´ preparation are searched. Even on this field new methodologies were achieved. As an example the immobilization utilizing an affinity boronate gel also called as "general lectin", which is able to interact with many polar functional groups of suitable structure, can be mentioned. Although having a lower specificity for recognized carbohydrate structure, its advantages lie in its stability and lower cost. This technique was successfully used e.g. for horseradish peroxidase immobilization. Methods exploiting glycosyl specific antibodies supports, which are highly specific for a glycoenzyme of interest, were also reported as an alternative for glycoproteins immobilization (Ahmad et al., 2001; Bucur et al., 2004; Bucur et al., 2005; Liu & Scouten, 1996; Nelson & Cox, 2005; Rao et al., 1998; Roy & Gupta, 2006; Saleemuddin, 1999; Tischer & Wedekind, 1999; Turková, 1999).

2.2.4.2 Immobilization methods based on antibodies-antigens interactions

Several immobilization methods exploit the interaction specificity of antibodies (monoclonal or polyclonal) and their antigens. It is almost impossible to describe all evolved variations ranging from the simplest immobilization of protein on a support covered with appropriate antibody (used for e.g. carboxypeptidase A, trypsin, chymotrypsin), continuing with exploiting secondary antibodies (e.g. immobilization of horseradish peroxidase via monoclonal enzyme specific antibody and anti-Fc antibodies) to preparation of fusion protein and its interaction with anti-tag antibody (used for alkaline phosphatase). An important role in these techniques is played also by two antibodies-binding proteins, i.e. Protein A from *Staphylococcus aureus* with the ability to specifically bind the Fc region of immunoglobulins and streptococcal Protein G with broader binding abilities. Both mentioned proteins enable oriented immobilization of antibodies on the surface, thus ensuring a correct orientation of antibody binding sites for further interaction. Next to favourable possibility to control the orientation of immobilized enzyme also other advantages are connected with antibody-antigen interaction using immobilization techniques, e.g. the fact that antibodies represent a spacer separating the enzyme and the matrix ensuring thus enough space required for correct enzymatic activity. Some limitations may be represented by relatively high costs of monoclonal antibody preparation. As mentioned at the beginning of the paragraph, many different variations and improvements, encompassing e.g. using two simultaneously attached affinity tags, preparation of Fc-binding unit of Protein A or coimmobilization of two enzymes, were evolved. Also methods, where e.g. IgG or Protein A were used as affinity tags, were reported (Daunert et al., 2007; Nilsson et al., 1997; Rao et al., 1998; Roy & Gupta, 2006; Saleemuddin, 1999; Solomon et al., 1991; Turková 1999).

For methods based on the interaction of histidine-rich enzyme parts and metal chelate supports an alternative that doesn´t exploit strategies of genetic engineering exists, too. Some proteins, as for example alkaline phosphatase, lysozyme or ribonuclease A, are able to bind these supports also in their native form. Chemical procedure increasing the amount of surface histidine of glycoproteins was tested for *Penicillium chrysogenum* glucose oxidase and horseradish peroxidase. Similar interactions, observed between phosphorylated proteins and immobilized metal ions (Ga^{3+}, Fe^{3+} etc.), may be utilized e.g. for enrichment of phosphoproteins in studied samples or for phosphorylated proteins and peptides purification. For phosphoproteins immobilization their ability to bind alumina can be exploit as confirmed e.g. for pepsine, which contains a phosphoserine residue. Also for this type of interaction an alternative method converting enzymes of interest to phosphoproteins was evolved and successfully used for e.g. subtilisin (Chaga, 2001; Hernandez & Fernandez-Lafuente, 2011; Machida et al., 2007; Rao et al., 1998; Saleemuddin, 1999; Schmidt et al., 2007).

At the end of this chapter it is again necessary to emphasize that different variations and combinations of all above mentioned methods are possible and new improvements are published every year. As one nicely illustrative example a combined method used for alkaline phosphatase immobilization may serve. In this technique a fusion protein strategy and antibody-antigen interaction were connected, because a tagged protein was immobilized by specially prepared anti-tag antibodies (Hernandez & Fernandez-Lafuente, 2011).

3. Conclusion

The idea of enzyme immobilization has introduced a totally new look on using these excellent biocatalysts in technological processes. Immobilized enzymes have a big potential for biotechnological practice for their important economical, technological and ecological advantages. Some weak points in immobilization techniques, as e.g. possible lowering of enzyme activity, was a challenge for the modern research and has led to development of a broad range of immobilization methods.

Utilization of affinity interactions for protein immobilization is a new trend, which is nowadays intensively examined. Many interesting variations of these strategies are based on natural specific interactions of molecules; many others are exploiting new findings of molecular biology and genetic engineering. Though affinity immobilization techniques are nowadays studied by many research teams, there are still new possibilities in finding simple, effective and inexpensive methods that would enable the use of immobilized enzymes in common technological practice. Many successful results from recent years indicate the importance of affinity immobilization techniques in these innovative methodologies.

4. Acknowledgment

Publishing of this work was supported by the Ministry of Education, Youth and Sports (MSM6046137305).

5. References

Aehle, W. (2007). *Enzymes in industry* (third edition), Wiley-VCH, ISBN 978-3-527-31689-2, Weinheim

Ahmad, S.; Anwar, A. & Saleemuddin, M. (2001). Immobilization and stabilization of invertase on *Cajanus cajan* lectin support. *Bioresource Technology*, Vol.79, No.2, pp. 121-127, ISSN 0960-8524

Altenbuchner, J. & Mattes, R. (2005). *Escherichia coli*, In: *Production of recombinant proteins: novel microbial and eukaryotic expression systems*, G. Gellissen, (Ed.), 7-43, Wiley-VCH, ISBN 3-527-31036-3, Weinheim

Andreescu, S.; Bucur, B. & Marty, J.-L. (2006). Affinity immobilization of tagged enzymes, In: *Immobilization of enzymes and cells.* (second edition), J. M. Guisan, (Ed.), Humana Press Inc., ISBN 1-59745-053-7, New Jersey

Arechaga, I.; Miroux, B.; Runswick, M. J. & Walker, J. E. (2003). Over-expression of *Escherichia coli* F_1F_0-ATPase subunit a is inhibited by instability of the *uncB* gene transcript. *FEBS Letters,* Vol.547, No.1-3, pp. 97-100, ISSN 0014-5793

Arnau, J.; Lauritzen, C.; Petersen, G. E. & Pedersen, J. (2006). Current strategies for the use of affinity tags and tag removal for the purification of recombinant proteins. *Protein Expression and Purification,* Vol.48, No.1, pp. 1-13, ISSN 1046-5928

Baneyx, F. (1999). Recombinant protein expression in *Escherichia coli. Current Opinion in Biotechnology,* Vol.10, No.5, pp. 411-421, ISSN 0958-1669

Brena, B. M. & Batista-Viera, F. (2006). Immobilization of enzymes, In: *Immobilization of enzymes and cells.* (second edition), J. M. Guisan, (Ed.), Humana Press Inc., ISBN 1-59745-053-7, New Jersey

Bucur, B.; Danet, A. F. & Marty, J.-L. (2004). Versatile method of cholinesterase immobilisation via affinity bonds using Concanavalin A applied to the construction of a screen-printed biosensor. *Biosensors and Bioelectronics,* Vol.20, No.2, pp. 217-225, ISSN 0956-5663

Bucur, B.; Danet, A. F. & Marty, J.-L. (2005). Cholinesterase immobilisation on the surface of screen-printed electrodes based on concanavalin A affinity. *Analytica Chimica Acta,* Vol.530, No.1, pp. 1-6, ISSN 0003-2670

Carlsson, J.; Janson, J.-C. & Sparrman, M. (1998). Affinity chromatography, In: *Protein purification. Principles, high resolution methods, and applications.* (second edition). J.-C. Janson & L. Rydén, (Ed.), Viley-VCH, ISBN 0-471-18626-0, New York

Cao, L. (2005). *Carrier-bound immobilized enzymes. Principles, Application and Design* (first edition), Wiley-VCH, ISBN 978-1-61583-208-8, Weinheim

Chaga, G. S. (2001). Twenty five years of immobilized metal ion affinity chromatography: past, present and future. *Journal of Biochemical and Biophysical Methods,* Vol.49, No.1-3, pp. 313-334, ISSN 0165-022x

Chern, J.-T. & Chao, Y.-P. (2005). Chitin-binding domain based immobilization of D-hydantoinase. *Journal of Biotechnology,* Vol.117, No.3, pp. 267-275, ISSN 0168-1656

Clare, D. A.; Valentine, V. W.; Catignani, G. L. & Swaisgood, H. E. (2001). Molecular design, expression, and affinity immobilization of a trypsin-streptavidin fusion protein. *Enzyme and Microbial Technology,* Vol. 28, No. 6, pp. 483-491, ISSN 0141-0229

Costa, S. A.; Azevedo, H. S. & Reis, R. L. (2005). Enzyme immobilization in biodegradable polymers for biomedical applications, In: *Biodegradable systems in tissue engineering and regenerative medicine.* R. L. Reis & J. S. Román, (Ed.), CRC Press LLC, ISBN 978-0-203-49123-2, London

Cowan, D. A. & Fernandez-Lafuente, R. (2011). Enhancing the functional properties of thermophilic enzymes by chemical modification and immobilization. *Enzyme and Microbial Technology,* Vol.49, No.4, pp. 326-346, ISSN 0141-0229

Daunert, S.; Bachas, L. G.; Schauer-Vukasinovic, V.; Gregory, K.J.; Schrift, G. & Deo, S. (2007). Calmodulin-mediated reversible immobilization of enzymes. *Colloids and Surfaces B: Biointerfaces,* Vol.58, No.1, pp. 20-27, ISSN 0927-7765

Demain, A. L. & Vaishnav, P. (2009). Production of recombinant proteins by microbes and higher organism. *Biotechnology Advances,* Vol.27, No. 3, pp. 297-306, ISSN 0734-9750

Desai, U, A.; Sur, G.; Daunert, S.; Babbitt, R. & Li, Q. (2002). Expression and affinity purification of recombinant proteins from plants. *Protein Expression and Purification*, Vol.25, No.1, pp. 195-202, ISSN 1046-5928

García-Junceda, E.; García-García, J. F.; Bastida, A. & Fernández-Mayoralas, A. (2004). Enzymes in the synthesis of bioactive compounds: the prodigious decades. *Bioorganic & Medicinal Chemistry*, Vol.12, No.8, pp. 1817-1834, ISSN 0968-0896

Gellissen, G.; Strasser, A. W. M. & Suckow, M. (2005). Key and criteria to the selection of an expression platform, In: *Production of recombinant proteins: novel microbial and eukaryotic expression systems*, G. Gellissen, (Ed.), 1-5, Wiley-VCH, ISBN 3-527-31036-3, Weinheim

Guisan, J. M. (2006). Immobilization of enzymes as the 21st century begins, In: *Immobilization of enzymes and cells*. (second edition), J. M. Guisan, (Ed.), Humana Press Inc., ISBN 1-59745-053-7, New Jersey

Hernandez, K. & Fernandez-Lafuente, R. (2011). Control of protein immobilization: Coupling immobilization and side-directed mutagenesis to improve biocatalyst or biosensor performance. *Enzyme and Microbial Technology*, Vol.48, No.2, pp. 107-122, ISSN 0141-0229

Ho, L.-F.; Li, S.-Y.; Lin, S.-C. & Hsu, W.-H. (2004). Integrated enzyme purification and immobilization processes with immobilized metal affinity adsorbents. *Process Biochemistry*, Vol.39, No.11, pp. 1573-1581, ISSN 1359-5113

Hwang, S.; Ahn, J.; Lee, S.; Lee, T. G.; Haam, S.; Lee, K.; Ahn, I.-S. & Jung, J.-K. (2004). Evaluation of cellulose-binding domain fused to a lipase for the lipase immobilization. *Biotechnology Letters*, Vol.26, No.7, pp. 603-605, ISSN 1573-6776

Jacquet, A.; Daminet, V.; Haumont, M.; Garcia, L.; Chaudoir, S.; Bollen, A. & Biemans, R. (1999). Expression of a recombinant *Toxoplasma gondii* ROP2 fragment as a fusion protein in bacteria circumvents insolubiliy and proteolytic degradation. *Protein Expression and Purification*. Vol.17, No.3, pp. 392-400, ISSN 1046-5928

Krajewska B. (2004). Application of chitin- and chitosan-based materials for enzyme immobilization: a review. *Enzyme and Microbial Technology*, Vol.35, No.2-3, pp. 126-139, ISSN 0141-0229

Kumada, Y.; Kuroki, D.; Yasui, H.; Ohse, T. & Kishimoto, M. (2010). Characterization of polystryrene-binding peptides (PS-tags) for site-specific immobilization of proteins. *Journal of Bioscience and Bioengineering*, Vol.109, No.6, pp. 583-587, ISSN 1389-1723

Liu, X.-C. & Scouten, W. H. (1996). Studies on oriented and reversible immobilization of glycoprotein using novel boronate affinity gel. *Journal of Molecular Recognition*, Vol.9, No.5-6, pp. 462-467, ISSN 1099-1352

Machida, M.; Kosako, H.; Shirakabe, K.; Kobayashi, M.; Ushiyama, M.; Inagawa, J.; Hirano, J.; Nakano, T.; Bando, Y.; Nishida, E. & Hattori, S. (2007). Purification of phosphoproteins by immobilized metal affinity chromatography and its application to phosphoproteome analysis. *FEBS Journal*, Vol.274, No.6, pp. 1576-1587, ISSN 1742-4658

Mateo, C.; Palomo, J. M.; Fernandez-Lorente, G.; Guisan, J. M. & Fernandez-Lafuente, R. (2007). Improvement of enzyme activity, stability and selectivity via immobilization techniques. *Enzyme and Microbial Technology*, Vol.40, No.6, pp. 1451-1463, ISSN 0141-0229

Murray, R. K.; Granner, D. K.; Mayes, P. A. & Rodwell, V. W. (2002). *Harperova biochemie* (twenty-third edition), H+H, ISBN 80-7319-013-3, Jinočany

Nelson, D. L. & Cox, M. M. (2005). *Principles of Biochemistry* (fourth edition), W. H. Freeman & Co., ISBN 0-7167-4339-6, New York

Nilsson, J.; Ståhl, S.; Lundeberg, J.; Uhlén, M & Nygren, P. (1997). Affinity fusion strategies for detection, purification and immobilization of recombinant proteins. *Protein Expression and Purification*, Vol.11, No.1, pp. 1-16, ISSN 1096-0279

Panesar, P. S.; Panesar, R.; Singh, R. S.; Kennedy, J. F. & Kumar, H. (2006). Microbial production, immobilization and applications of β-D-galactosidase. *Journal of Chemical Technology and Biotechnology*, Vol.81, No.4, pp. 530-543, ISSN 1097-4660

Paternolli, C.; Ghisellini, P. & Nicolini, C. (2002). Development of immobilization techniques of cytochrome P450-GST fusion protein. *Colloids and Surfaces B: Biointerfaces*, Vol.23, No.4, pp. 305-311, ISSN 0927-7765

Primrose, S. B.; Twyman, R. M. & Old, R. W. (2001). *Principles of Gene Manipulation* (sixth edition), Blackwell Science Ltd, ISBN 0-632-05954-0, Oxford

Rao, S. V.; Anderson, K. W. & Bachas, L. G. (1998). Oriented immobilization of proteins. *Microchimica Acta*, Vol.128, No.3-4, pp. 127-143, ISSN 1436-5073

Rodrigues, R. C.; Berenguer-Murcia, Á & Fernandez-Lafuente, R. (2011). Coupling chemical modification and immobilization to improve the catalytic performance of enzymes. *Advanced Synthesis & Catalysis*, Vol.353, No.13, pp.2216-2238, ISSN 1615-4150

Roy, I. & Gupta, M. N. (2006). Bioaffinity immobilization, In: *Immobilization of enzymes and cells*. (second edition), J. M. Guisan, (Ed.), Humana Press Inc., ISBN 1-59745-053-7, New Jersey

Saleemuddin, M. (1999). Bioaffinity based immobilization of enzymes. *Advances in biochemical engineering/biotechnology*, Vol.64, pp. 203-226, ISSN 1616-8542

Schmidt, S. R.; Schweikart, F. & Andersson, M. E. (2007). Current methods for phosphoprotein isolation and enrichment. *Journal of Chromatography*, Vol.849, No.1-2, pp.154-162, ISSN 1570-0232

Sheldon, R. A. (2007). Enzyme immobilization: The quest for optimum performance. *Advanced Synthesis & Catalysis*. Vol.349, No.8-9, pp. 1289-1307, ISSN 1615-4169

Solomon, B.; Hadas, E.; Koppel, R.; Schwartz, F. & Fleminger, G. (1991). Highly active enzyme preparations immobilized via matrix conjugates anti-Fc antibodies. *Journal of Chromatography*, Vol.539, No.2, pp. 335-341, ISSN 0021-9673

Sørensen, H. P. & Mortensen, K. K. (2005). Advanced genetic strategies for recombinant protein expression in *Escherichia coli*. *Journal of Biotechnology*, Vol.115, No.2, pp. 113-128, ISSN 0168-1656

Takakura, Y.; Oka, N.; Kajiwara, H.; Tsunashima, M.; Usami, S.; Tsukamoto, H.; Ishida, Y. & Yamamoto, T. (2010). Tamavidin, a versatile affinity tag for protein purification and immobilization. *Journal of Biotechnology*, Vol.145, No.4, pp. 317-322, ISSN 0168-1656

Terpe, K. (2003). Overview of tag protein fusions: from molecular and biochemical fundamentals to commercial systems. *Applied Microbiology and Biotechnology*, Vol.60, No. 5, pp. 523-533, ISSN 1432-0614

Tischer, W. & Wedekind, F. (1999). Immobilized enzymes: Methods and applications, In: *Biocatalysis - From discovery to application. Topics in Current Chemistry*, Vol.200, pp. 95-126. W.-D. Fessner, (Ed.), Springer, ISSN 0340-1022, Berlin

Turková, J. (1999). Oriented immobilization of biologically active proteins as a tool for revealing protein interactions and function. *Journal of Chromatography B*, Vol.722, No.1-2, pp. 11-31, ISSN 1570-0232

Vishwanath, S.; Bhattacharyya, D.; Huang, W. & Bachas, L. G. (1995). Site-directed and random enzyme immobilization on functionalized membranes: kinetic studies and models. *Journal of Membrane Science*, Vol.108, No. 1-2, pp. 1-13, ISSN 0376-7388

Vodrážka, Z. (1999). *Biochemie* (second edition), Academia, ISBN 80-200-0600-1, Praha

Waldo, G. S.; Standish, B. M.; Berendzen, J. & Terwilliger, T. C. (1999). Rapid protein-folding assay using green fluorescent protein. *Nature Biotechnology*, Vol.17, No.7, pp. 691-695, ISSN 1087-0156

Part 2

Lectin Affinity Chromatography

3

Affinity Chromatography of Lectins

Jure Pohleven, Borut Štrukelj and Janko Kos
Jožef Stefan Institute
Slovenia

1. Introduction

Affinity chromatography is a technique used to purify compounds, such as proteins, that have the ability to non-covalently and reversibly bind specific molecules, known as ligands. This method differs from the classical chromatography techniques in that the protein is purified on the basis of a unique biochemical property. In affinity chromatography, the ligand is covalently attached to a matrix, which must be chemically inert, porous, and have a variety of functional groups suitable for coupling with diverse ligands. Various matrices and ligands are used in affinity chromatography, depending on the protein to be purified (Voet & Voet, 1995). A particular affinity chromatography technique, which uses carbohydrate adsorbents (ligands or matrices) for purification of glycan-binding proteins or lectins, is called carbohydrate affinity chromatography.

Lectins are a diverse group of proteins that bind specifically various carbohydrates. They are present in every organism, suggesting their role in basic biological functions, including regulatory, adhesive, defence against pathogens, and many others (Varki et al., 2009). In an organism, a number of different lectins are usually present in various isoforms, called isolectins (Van Damme et al., 1998). These differ only in slight variations in primary structure and carbohydrate-binding specificity, but can show differences in their biological activities (Leavitt et al., 1977). Therefore, when isolating a lectin it is important to purify the individual isolectins.

Different types of lectin are distinguished according to their overall structure - merolectins contain a single carbohydrate-binding domain, hololectins are composed of two or more such domains and chimerolectins contain additional, non-lectin domains, usually catalytic in function (Peumans & Van Damme, 1995). The majority of lectins contain multiple binding sites, and can thus, by binding and cross-linking of specific glycoreceptors on cell surfaces, agglutinate cells such as erythrocytes. Lectins also act as recognition molecules in cell–molecule and cell–cell interactions, and are involved in cellular processes, including cell adhesion, migration, differentiation, proliferation and apoptosis. Furthermore, many lectins elicit diverse physiological responses in various organisms and possess immunomodulatory properties and thus have potential roles in cancer and metastasis (Perillo et al., 1998; Sharon & Lis, 1989), which makes them potentially useful in biotechnological and biomedical applications.

The binding of lectins to carbohydrates is noncovalent and reversible, involving hydrogen bonds, hydrophobic, electrostatic and van der Waals interactions and dipole attraction. The binding of sugars (mono- and disaccharides) to a lectin is relatively weak, with dissociation constants in the millimolar or micromolar range. On the other hand, interactions of multivalent lectins with complex, branched carbohydrates containing multiple epitopes, result in high-avidity binding with nanomolar or even picomolar dissociation constants (Varki et al., 2009).

In general, lectins possess biochemical and binding properties which are very convenient for their purification by carbohydrate affinity chromatography. They do not react catalytically with carbohydrates, modifying the ligand, unless the lectin is a subdomain of a modular protein that contains another catalytic (glycosidase) domain. As noted earlier, lectins bind carbohydrates noncovalently and reversibly and at least the most widely used ligands, mono- and disaccharides, are usually bound relatively weakly, so that the lectin is readily released from an affinity column by competitive elution using specific free carbohydrates. Moreover, lectins and carbohydrates are both usually stable compounds, therefore elution techniques using extreme conditions of pH and/or ionic strength can also be applied to release a lectin from the carbohydrate affinity column.

In this chapter, current carbohydrate affinity chromatography methods for purification of lectins exhibiting various carbohydrate-binding specificities are presented. The affinity ligands, matrices, ligand coupling methods and elution techniques used are described in detail. Throughout the chapter, all sugars are of the D-configuration, unless otherwise stated.

2. Lectin isolation using carbohydrate affinity chromatography

Any source of interest can be selected for isolation of lectins, from humans, animals, plants, fungi to microorganisms, including viruses. First, a lectin must be released from the source into solution, usually by liberating it from the cells. Several methods can be used, depending on the mechanical characteristics of the source tissue, such as homogenization by mechanical disruption or by lysis (Voet & Voet, 1995). Lectins are water-soluble proteins, therefore are extracted with aqueous buffers and then centrifuged and filtered if necessary to remove the cell debris. Extraction should be carried out at 4 °C to prevent denaturation and degradation, by proteolytic enzymes, of the lectin of interest. Moreover, addition of protease inhibitors to the extraction buffer is recommended if proteolytic activity is observed in the extract (Kvennefors et al., 2008; Matsumoto et al., 2011; Watanabe et al., 2007), although some lectins have been found to be resistant to proteolytic degradation (Pohleven et al., 2009). In addition, glycosidase inhibitors, such as glucosidase inhibitor 1-deoxynojirimycin (0.2 mM) and the mannosidase inhibitor 1-deoxymannojirimycin (0.2 mM), have also been added to the extract to prevent cleavage of immobilized carbohydrate ligands in the affinity column (Watanabe et al., 2007). Before the extract is applied to the carbohydrate affinity column, some of the impurities can be removed using fractional precipitation with ammonium sulphate and/or other classical chromatography techniques. To follow the lectin fractions throughout the purification procedure, agglutinating activity can be monitored using erythrocytes specific for the lectin; a given lectin does not agglutinate all types of red blood cells, only those exhibiting the lectin-specific glycans.

Once the extract exhibiting agglutinating activity is prepared, the carbohydrate affinity chromatography method appropriate for purifying the lectin has to be selected. This involves the adsorbent that can bind the lectin and separate it from the extract, and an elution technique that can release the bound lectin from the column. It is therefore important to know the specificity and stability of the lectin and also to consider the stability of the adsorbent, since extremes of pH and temperature may be used for ligand coupling and elution. Carbohydrates and, in general, lectins are relatively stable molecules. The specificity of the desired lectin in the extract can be determined by examining its agglutinating activity, using various classes of erythrocytes that expose diverse glycans on their surfaces. For example human blood group A expresses non-reducing terminal α-*N*-acetylgalactosamine, group B, terminal α-galactose and group O, fucosyl-galactose (Schenkel-Brunner, 2007). In addition, hapten inhibition of agglutination, using various carbohydrates or glycoproteins, can be used to determine the specificity of the lectin, enabling the lectin-specific adsorbent to be selected. The stability of the lectin can be assessed by examining agglutinating activity of the extract, pretreated under various conditions of pH. However, observations suggest that the lectin in an extract is more stable than the isolated one, probably because it is stabilized by bound carbohydrates.

A wide variety of adsorbents have been described for lectin isolation and can be divided into the following groups: carbohydrate (mono-, di, and polysaccharide) ligands and glycoprotein ligands, both immobilized to the carrier matrix, and adsorbent (polysaccharide) matrices (Sections 3 and 4). In the latter case, no ligand is required, since the polysaccharide matrices themselves bind lectins, while in the former cases, ligand is coupled to a previously activated matrix. Depending on the ligand, one of several methods can be applied to activate the matrix with various reagents and subsequently immobilize carbohydrate or glycoprotein ligands. After the coupling, the amount of carbohydrate or glycoprotein ligand is determined (Section 5).

After the extract containing the lectin has been applied to the carbohydrate affinity adsorbent and appropriately incubated (for example for several hours at 4 °C or 1 hour at room temperature), the impurities are washed off the column and the bound lectin then eluted. This is usually performed using gravity-flow column chromatography or fast protein liquid chromatography (FPLC). Depending on the specificity and stability of the lectin and adsorbent, various elution techniques can be used, such as competitive elution with specific carbohydrates, desorption by changing the pH and/or ionic strength of the eluent, or elution with urea solutions. When extremes of pH are used for elution, the protein fractions and column have to be neutralized by buffer as soon as possible (Section 6). The eluted protein fractions can be detected by measuring their absorbance at 280 nm and the affinity column capacity readily estimated by following the agglutinating activity of the flow-through.

Carbohydrate affinity chromatography is a simple, one-step method for purifying lectins. However, in some cases additional separation techniques have to be used subsequently to purify the lectin to homogeneity, such as ion-exchange chromatography and/or gel filtration. For example, isolectins cannot be separated by carbohydrate affinity chromatography, since they show very similar carbohydrate-binding specificities; ion-exchange chromatography has been used successfully to separate individual isolectins (Guzmán-Partida et al., 2004; Horibe et al., 2010; Leavitt et al., 1977; Mishra et al., 2004; Ren et al., 2008; Sultan et al., 2009). Moreover, different lectins can sometimes be co-isolated from the extract, especially when the ligand is immobilized to polysaccharide matrices, which also bind lectins. In that case, serial carbohydrate affinity chromatography can be applied using two or more adsorbents in series (Chen et al., 1999; Kato et al., 2011; Lavanya Latha et al., 2006; Moreira et al., 1998; Ooi et al., 2002; Pohleven et al., 2011; Trindade et al., 2006).

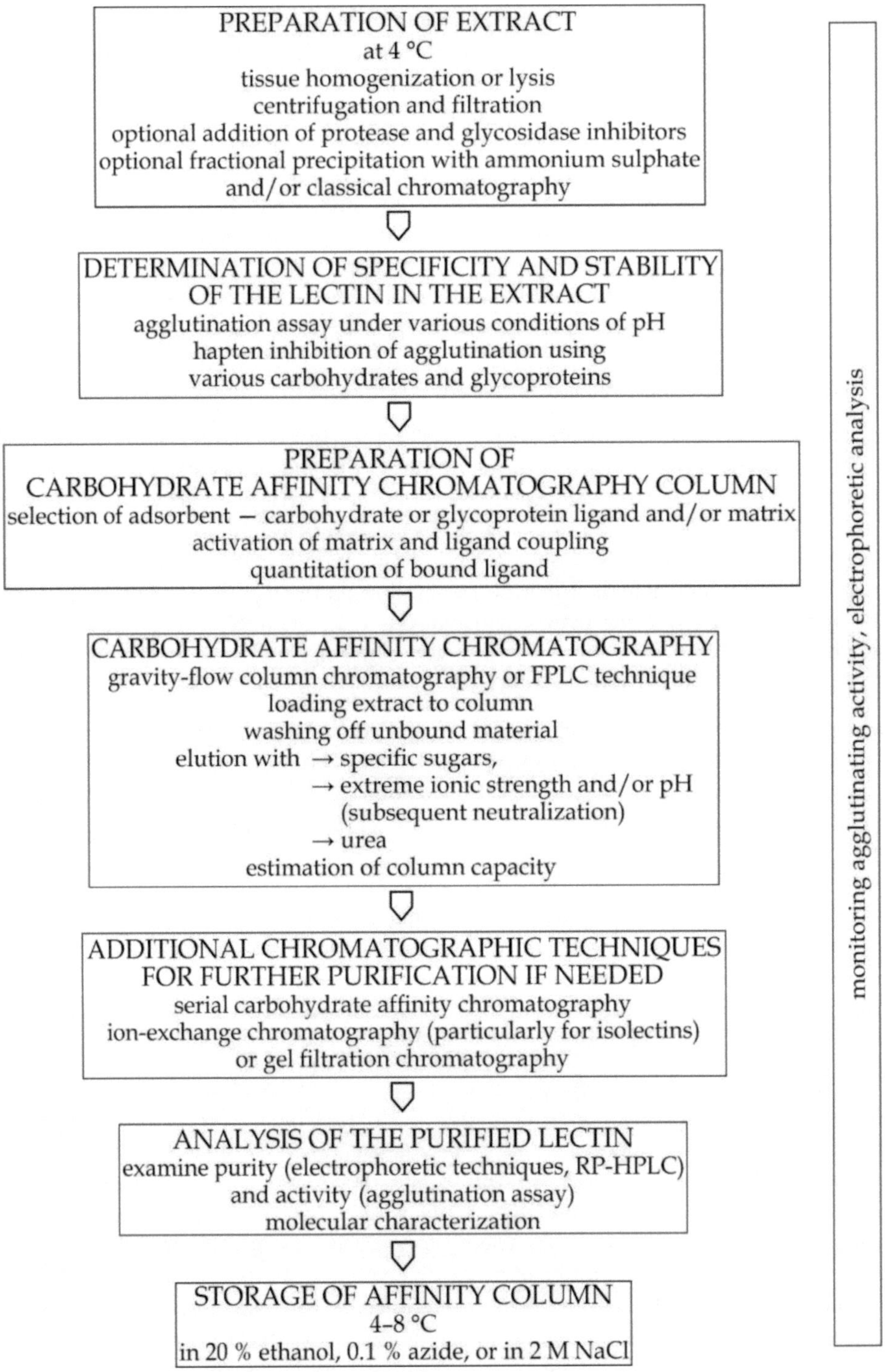

Fig. 1. Protocol for separation and purification of lectins using carbohydrate affinity chromatography

For example, the extract is applied to the affinity column that binds, besides the lectin of interest, non-desired lectins. Eluted lectins are then separated on a column that only binds non-desired lectins, and the lectin of interest is purified by simply washing the column.

Whatever the purification method used, the protein fractions are analyzed for purity using biochemical methods, such as electrophoretic techniques or reversed-phase high-performance liquid chromatography (RP-HPLC), and lectin activity is assessed by agglutination assay. After use, affinity chromatography columns should be stored at 4–8 °C, in the presence of a bacteriostat e.g. 20 % ethanol or 0.1 % azide (w/v) or in 2 M NaCl. The protocol for lectin purification using carbohydrate affinity chromatography is presented in Figure 1.

3. Ligands

In carbohydrate affinity chromatography, mono-, di-, and polysaccharides, as well as glycoproteins, can be used as matrix-immobilized ligands.

3.1 Mono- and disaccharides

Of all the ligands, mono- and disaccharides coupled to a Sepharose matrix are the most widely used adsorbents. Recently, isolation of lectins from various sources has been described using numerous sugars, such as lactose (Almanza et al., 2004; Braga et al., 2006; Fujii et al., 2011; Hamako et al., 2007; Kawsar et al., 2009; Mendonça-Franqueiro et al., 2011; Naeem et al., 2007b; Pohleven et al., 2009; Rocha et al., 2009; Silva et al., 2007), sucrose and glucose (Pohleven et al., 2011), galactose (Konami et al., 1991; Lavanya Latha et al., 2006; Mo et al., 2000; Moreira et al., 1998; Naeem et al., 2007b; Nagata, 2005), mannose (Andon et al., 2003; Ooi et al., 2010; Suseelan et al., 2002), fucose (Cammarata et al., 2007; Mansour & Abdul-Salam, 2009), L-rhamnose (Jimbo et al., 2007; Watanabe et al., 2008), and sugars with 2′-acetamido groups *N*-acetylgalactosamine (Gerlach et al., 2005; Perçin et al., 2009; Qureshi et al., 2006), *N*-acetylglucosamine (Kaur et al., 2005; Kim et al., 2006; Maheswari et al., 2002; Wang & Ng, 2003), and *N,N′*-diacetylchitobiose (Konami et al., 1991; Koyama et al., 2002).

Direct coupling of small ligand molecules, such as mono- and disaccharides, to the matrix can often result in binding of the lectin to sugar ligands being prevented, due to steric hindrance by the matrix. Spacer groups can minimize such interference and, furthermore, enable a variety of functional groups to be introduced, allowing the use of alternative techniques for ligand coupling to the matrix. A spacer group can be attached to the carrier (Section 4.2), or alternatively, sugar derivatives can be synthesized by attaching spacers, such as 6-amino-hexyl and ε-aminocaproyl groups (Lis & Sharon, 1981). Several carbohydrate derivatives have recently been synthesized and utilized in affinity chromatography for lectin isolations, such as the Galβ1–4Fuc derivative containing a hydrophilic spacer modified with a free amino group (Takeuchi et al., 2011), an iodoacetamidyl derivative of $Glc_1Man_9GlcNAc_2$ (Watanabe et al., 2007), and a derivative of L-fucose with a BSA spacer (Argayosa & Lee, 2009). In addition, a novel method for covalent immobilization of a range of carbohydrate derivatives onto polymeric resin beads has been described (Chen et al., 2007).

3.2 Polysaccharides

Recently, several polysaccharides have been immobilized to various matrices in order to isolate lectins. These include chitin, a polymer of a *N*-acetylglucosamine (Bovi et al., 2011; Narahari & Swamy, 2010; Santi-Gadelha et al., 2006; Trindade et al., 2006), mannan, a mannose polymer (Argayosa et al., 2011; Kvennefors et al., 2008; Naeem et al., 2007a; Ourth et al., 2005), alginate, a copolymer of β-mannuronate and α-L-guluronate (Roy et al., 2005), chitosan, a polysaccharide composed of glucosamine and *N*-acetylglucosamine (Chen & Xu, 2005), glucan laminarin (Chen et al., 1999) and exopolysaccharide extracted from *Azospirillum brasiliense* (Mora et al., 2008).

3.3 Glycoproteins

Carbohydrate moieties of glycoproteins, called glycans, can also be used as ligands for lectin isolation. Usually, glycans contain a variety of carbohydrates and therefore can be used for the isolation of lectins with various specificities (Lis & Sharon, 1981). Recently, numerous glycoproteins have been immobilized, usually to cyanogen bromide activated Sepharose. These include fetuin (Bhowal et al., 2005; Guzmán-Partida et al., 2004; Matsumoto et al., 2011; Naeem et al., 2007b; Ooi et al., 2002; Rittidach et al., 2007; Sun et al., 2007; Valadez-Vega et al., 2011; Yang et al., 2007), mucin from porcine (*Sus scrofa*) stomach (Chumkhunthod et al., 2006; Souza et al., 2010; Takahashi et al., 2008) and bovine (Bos taurus) submaxillary mucin (Imamichi & Yokoyama, 2010; Kawagishi et al., 1994; Naganuma et al., 2006), thyroglobulin (Ren et al., 2008; Wang et al., 2004) and ovalbumin (Mo et al., 1999). Besides, desialylated glycoproteins treated with mild acid hydrolysis or neuraminidases, which removes sialic acid, thus exposing other saccharides, have been used as ligands, such as asialofetuin (Bhat et al., 2010; Kaur et al., 2006; Naeem et al., 2007b; Nagre et al., 2010; Shangary et al., 1995) and asialomucin (Vega & Pérez, 2006).

Another method used for lectin isolation, based on immobilized glycoprotein ligands, is affinity chromatography on agarose-immobilized porcine (*Sus scrofa*) blood plasma proteins (Kajiya et al., 2003) or on stroma (Alpuche et al., 2010; Dresch et al., 2008; Vazquez et al., 1993; Veau et al., 1999) or membranes (Castillo-Villanueva et al., 2007) of various erythrocytes, fixed with 1 % glutaraldehyde and physically entrapped in various matrices.

4. Matrices

In carbohydrate affinity chromatography, polysaccharide matrices can be used in two ways - as adsorbents for binding lectins, or as carriers to which ligands can be attached.

4.1 Adsorbent matrices

A number of matrices are polysaccharide-based, and can thus be used as adsorbents for isolation of lectins with appropriate specificities, without prior immobilization of carbohydrate ligands. The most widely used is Sepharose, but Sephadex is also used. Both are commercially available (GE Healthcare). Sepharose is an agarose-based, bead-formed matrix, containing 2, 4, or 6 % agarose, hence designated Sepharoses 2B, 4B, or 6B. Furthermore, Sepharose CL gels are cross-linked derivatives of Sepharose, and thus chemically and physically more resistant, i.e. stable over a wider pH range, and can be applied when extreme conditions are used for coupling or elution. Agarose is a galactose-anhydrogalactose copolymer (Angal & Dean, 1977), so Sepharose has recently been used extensively as an adsorbent for isolation of lectins specific for galactosides and related carbohydrates (Cao et al., 2010; De-Simone et al., 2006; Kato et al., 2011; Moura et al., 2006; Pohleven et al., 2011). In contrast, Sephadex is a beaded gel prepared by cross-linking dextran, a polysaccharide composed of glucose. It has been used for purifying glucose/mannose-specific lectins from various sources (Biswas et al., 2009; Rangel et al., 2011; Roh & Park, 2005).

Many galactose-specific lectins recognize terminal non-reducing galactosyl residues, rather than internal ones, in galactan polysaccharides (Lis & Sharon, 1981). Therefore, partial, mild hydrolysis of Sepharose by acid treatment - for example incubation in 1 M HCl for three hours at 50 °C (Voss et al., 2006) - can be used to cleave the polysaccharide chains, thus exposing additional terminal galactosyl residues, without completely degrading the Sepharose (Ersson et al., 1973; Lis & Sharon, 1981). Recently, numerous lectins have been isolated on acid treated Sepharose (Jimbo et al., 2000; Kawagishi et al., 1994; Mishra et al., 2004; Stirpe et al., 2007; Voss et al., 2006).

In addition to Sepharose and Sephadex, commercially available Affi-Gel blue gel (Bio-Rad), composed of beaded, cross-linked agarose, has been used as an adsorbent matrix (Lin & Ng, 2008; Shao et al., 2011; Sharma et al., 2010; Wong et al., 2010). Further, naturally occurring cross-linked polysaccharides, which are inexpensive and simple to prepare with epichlorhydrin or divinyl sulphone, have also been extensively used as adsorbent matrices for isolating galactose-specific lectins (Lis & Sharon, 1981). These include guar gum or guaran (Sigma-Aldrich), a galactomannan polysaccharide that binds lectins with anomeric preference for α-galactose (Alencar et al., 2010; Santos et al., 2009; Souza et al., 2011; Sultan et al., 2009), a polysaccharide isolated from *Spondias purpurea* (Teixeira et al., 2007), a cross-linked seed gum matrix prepared from plant *Leucaena leucocephala* (Seshagirirao et al., 2005), a galactomannan from *Adenanthera pavonina* (Moreira et al., 1998; Teixeira-Sá et al., 2009; Trindade et al., 2006), and a cross-linked galactoxyloglucan from *Mucuna sloanei* (Teixeira-Sá et al., 2009). Moreover, adsorbent matrices from yeast glucan or curdlan, a polymer of glucose, were prepared to isolate glucose-specific lectins (Mikes & Man, 2003).

Since polysaccharide matrices like Sepharose and Sephadex bind lectins, when using such matrices as carriers for carbohydrate ligand immobilization a variety of lectins can be co-isolated; ones specific for the matrix and one for the ligand. To purify the latter lectin, additional methods, such as gel filtration or ion-exchange chromatography have to be used. Alternatively, serial carbohydrate affinity chromatography can be introduced, in which two or more affinity columns are used in tandem. For example, in the first step, the extract is applied to the affinity column (with matrix-immobilized ligand) which, besides the lectin of interest, also binds non-desired (matrix-specific) lectins. The eluted lectin mixture is then applied to a second column (with matrix alone) that only binds the non-desired lectins. The non-bound lectin of interest is washed off the column and thus purified. Several authors have purified lectins from a variety of sources using serial carbohydrate affinity chromatography (Chen et al., 1999; Kato et al., 2011; Lavanya Latha et al., 2006; Moreira et al., 1998; Ooi et al., 2002; Pohleven et al., 2011; Trindade et al., 2006) or Ultrogel-A4 agarose beads to remove non-specific agarose-binding proteins from the extract prior to subjecting it to carbohydrate affinity chromatography (Kvennefors et al., 2008). In contrast, non-polysaccharide matrices, such as silica carriers (Synsorb) (Kaur et al., 2006; Mo et al., 2000;

Naeem et al., 2007b; Shangary et al., 1995), poly(2-hydroxyethyl methacrylate) matrix of Spheron 300 (Nahálková et al., 2001), fractogel affinity matrix (Guzmán-Partida et al., 2004) or polyacrylamide beads (Dresch et al., 2008; Hořejší & Kocourek, 1978; Veau et al., 1999), can be used to avoid non-specific binding of lectins to the polysaccharide matrix.

4.2 Carrier matrices

Carrier matrices are used as supports to which various carbohydrate or glycoprotein ligands are coupled. These matrices have to be previously activated for the covalent coupling to take place. Various matrices are commercially available (Sepharose, Mini-leak agarose, Toyopearl, Synsorb, Seralose, and Spheron besides others), some of them being pre-activated and others already coupled with carbohydrate ligands.

Numerous matrices, to which diverse ligands were immobilized by various coupling methods, have been used recently to purify lectins. The most widely used have been differently activated Sepharoses – divinyl sulphone (Almanza et al., 2004; Lavanya Latha et al., 2006; Pohleven et al., 2011) and epoxy (epichlorohydrin) activated Sepharose have been used to immobilize carbohydrates (Chen et al., 1999; Kaur et al., 2005; Maheswari et al., 2002; Mora et al., 2008; Nagata, 2005), and cyanogen bromide activated Sepharose for glycoproteins (Bhat et al., 2010; Bhowal et al., 2005; Kajiya et al., 2003; Mo et al., 1999; Naganuma et al., 2006; Nagre et al., 2010; Vega & Pérez, 2006; Yang et al., 2007). *N*-hydroxysuccinimide Sepharose was used for coupling with sugar derivatives containing amino groups (Takeuchi et al., 2011), and thiopropyl Sepharose for immobilizing iodoacetamidyl glycan derivatives (Watanabe et al., 2007). Mini-Leak agarose (Kem-En-Tec), a divinyl sulphone activated matrix (Valadez-Vega et al., 2011), and Seralose (Konozy et al., 2002) have also been used. Moreover, several kinds of Toyopearl affinity resins (Tosoh) exist that use various coupling chemistries for the attachment of ligands to formyl, carboxy, amino, epoxy or tresyl groups. For lectin isolation, Toyopearl AF-Amino 650 was used to immobilize glycoproteins (Imamichi & Yokoyama, 2010; Kawagishi et al., 1994; Takahashi et al., 2008).

Commercially available agarose matrices pre-coupled with carbohydrates or glycoproteins have also been used, such as agarose beads with immobilized lactose (EY Laboratories) (Hamako et al., 2007), galactose (Pierce) (Trindade et al., 2006), mannose (Sigma-Aldrich) (Ooi et al., 2002; Suseelan et al., 2002), *N*-acetylgalactosamine (Gerlach et al., 2005; Qureshi et al., 2006), *N*-acetylglucosamine (Sigma-Aldrich) (Kim et al., 2006; Wang & Ng, 2003), *N,N'*-diacetylchitobiose (Sigma-Aldrich) (Koyama et al., 2002), mannan (Sigma-Aldrich) (Naeem et al., 2007a; Ourth et al., 2005), and fetuin (Sigma-Aldrich) (Ooi et al., 2002; Rittidach et al., 2007).

In contrast, non-polysaccharide-based matrices have also been used, such as silica carrier Synsorb with immobilized galactose (Mo et al., 2000; Naeem et al., 2007b), amino activated silica beads with immobilized asialofetuin (Clifmar) (Kaur et al., 2006; Shangary et al., 1995), 2-acetamido-2-deoxy-β-glucopyranoside-glycosylated Spheron 300 (Nahálková et al., 2001), azlactone-activated fractogel coupled with fetuin (Guzmán-Partida et al., 2004), polyacrylamide gel with immobilized carbohydrates (Hořejší & Kocourek, 1978), poly(hydroxypropyl methacrylate-glycidyl methacrylate) beads with *N*-acetylgalactosamine attached through epoxy groups of glycidyl methacrylate (Perçin et al., 2009) and monolithic columns with various immobilized sugars (Kato et al., 2011; Tetala et al., 2007) and the polysaccharide mannan (Bedair & El Rassi, 2004).

For immobilization of small carbohydrate ligands, such as mono- or disaccharides, to carrier matrices, it is necessary to use spacer arms to separate the ligand from the support thus making immobilized small ligand molecules more accessible to the lectin binding. In addition, spacers can be used to introduce a variety of functional groups, which allows the use of alternative coupling techniques (Lis & Sharon, 1981). A spacer group can be attached either to the sugar derivative ligand (Section 3.1) or to the carrier matrix and then carbohydrate ligand is coupled to the immobilized spacer to obtain affinity column for lectin purification (Lis & Sharon, 1981). In addition, activated matrices, such as divinyl sulphone activated, epoxy activated, *N*-hydroxysuccinimide activated Sepharose, CH Sepharose, and Thiopropyl Sepharose, already contain spacer arms and are prepared for ligand coupling.

5. Coupling methods

There are various methods for covalently attaching a carbohydrate or glycoprotein ligand to a carrier matrix. First, the matrix is activated, which can be done in several ways using various reagents and conditions. Commercially available pre-activated matrices are also available. During the activation, spacer arms of various lengths can be introduced to the matrix. Ligands are then covalently coupled via their functional groups to differently activated matrices, resulting in links with different stabilities (Pepper, 1992). After coupling, the amount of carbohydrate ligand coupled to the beads can be determined by the phenol-sulphuric acid method (Dubois et al., 1956) or the 3,5-dinitrosalicylic acid method (Bailey et al., 1992). On the other hand, glycoprotein ligands can be determined using protein quantitation methods, such as the Bradford (Bradford, 1976) and BCA protein assay (Pierce) methods.

5.1 The divinyl sulphone method

This method of activation is appropriate for a wide variety of polymer matrices, such as Sepharose, Sephadex, Sephacryl, Fractogel, Ultrogel, cellulose, and dextran. Commercially available divinyl sulphone activated agarose can also be obtained. Such matrices are suitable for coupling with carbohydrate and (glyco)protein ligands via their hydroxyl or amino groups. With this method, a highly reactive vinyl group is introduced to the matrix simultaneously with a spacer. The activation using divinyl sulphone is fairly rapid (70 minutes at room temperature; (Levi & Teichberg, 1981)) and the activated material is stable for up to 12 months in aqueous suspension at 4 °C. The resulting vinyl groups are highly reactive with hydroxylic compounds, therefore coupling takes place at lower pH and temperatures than with other methods. The coupling with carbohydrates takes 15 hours at pH 10-11, at room temperature in the dark (Levi & Teichberg, 1981) while for proteins, at even lower pH values in the range of 6.5-10. However, the link formed is unstable in alkaline solutions (Lis & Sharon, 1981; Pepper, 1992), so acidic elution of bound lectins is advised. The method has recently been used for coupling sugars to Sepharose in order to isolate lectins from various sources (Almanza et al., 2004; Lavanya Latha et al., 2006; Pohleven et al., 2011).

5.2 The epoxide method

This method is widely used for coupling hydroxyl groups of carbohydrates to epoxy activated matrices. In addition, amino and sulfhydryl groups of ligands can also be used for (glyco)protein immobilization (Murthy & Moudgal, 1986). The activation of the matrix provides an active epoxy group for coupling and a flexible, extended (12-atom) hydrophilic spacer arm. However, these reactions require high pH values (12-13) and temperatures above 40 °C (Lis & Sharon, 1981); the reaction is also slow. The aggressiveness of the method does not allow the use of silica or glass beads that are destroyed under such conditions (Pepper, 1992). On the other hand, increasing ionic strength facilitates the ligand immobilization, therefore very high salt concentrations allow coupling at pHs as low as 8.5 (Murthy & Moudgal, 1986). The resulting ether bond is stable, which ensures low leakage of ligands from the column (Lis & Sharon, 1981; Pepper, 1992). This method has been used for the isolations of lectins by employing epichlorohydrin for carbohydrate coupling to Sepharose (Chen et al., 1999; Kaur et al., 2005; Maheswari et al., 2002; Mora et al., 2008; Nagata, 2005) or for cross-linking polysaccharides to prepare an adsorbent matrix (Teixeira et al., 2007). Commercially available matrices, such as epoxy activated Sepharose (GE Healthcare) can be obtained.

5.3 The cyanogen bromide method

A cyanogen bromide activated matrix can be used for immobilizing ligands containing an amino group. The coupling reaction is rapid (2 hours at room temperature) and takes place at pH values between 8 and 10, most frequently at pH 8.3 (Pepper, 1992), however coupling at pH 7.4 was also reported (Kajiya et al., 2003). In carbohydrate affinity chromatography, this method is often used to immobilize glycoproteins, since carbohydrates rarely contain amino groups (galactosamine, glucosamine). The method is not suitable for coupling small ligands, since cyanogen bromide activation does not introduce a spacer. Alternatively, sugar derivatives containing spacers with amino groups can be prepared and immobilized using this method (Lis & Sharon, 1981). For lectin isolation, numerous glycoproteins have been immobilized to cyanogen bromide activated Sepharose (Bhat et al., 2010; Bhowal et al., 2005; Kajiya et al., 2003; Mo et al., 1999; Naganuma et al., 2006; Nagre et al., 2010; Vega & Pérez, 2006; Yang et al., 2007).

5.4 The *N*-hydroxysuccinimide method

Matrices with immobilized *N*-hydroxysuccinimide, such as *N*-hydroxysuccinimide activated Sepharose and CH Sepharose, form stable amide bonds with ligands containing an amino group. They also provide an eight carbon spacer arm. In carbohydrate affinity chromatography, the method is appropriate for coupling glycoproteins and carbohydrates or, more often, carbohydrate derivatives containing an amino group (Takeuchi et al., 2011).

5.5 The thiopropyl method

Thiopropyl matrices, such as Thiopropyl Sepharose, react reversibly under mild conditions with ligands containing thiol groups to form mixed disulphides and, in addition, provide a spacer. Alternatively, iodoacetamidyl carbohydrate derivatives can be conjugated to thiopropyl matrices by alkylation of the thiol groups, as described (Watanabe et al., 2007).

5.6 The azlactone method

Azlactone-activated matrices, such as Fractogel, react with amine groups of (glyco)proteins, forming a stable amide bond under physiological conditions. In carbohydrate affinity chromatography, it has been used for immobilizing the glycoproteins to a matrix (Guzmán-Partida et al., 2004).

6. Elution techniques

Two main elution techniques, depending on the ligand and the adsorbent, can be applied to release a bound lectin from the affinity column, namely competitive elution and elution at different pH and/or ionic strength. When choosing the most appropriate technique to desorb a bound lectin, its specificity, its stability, and the type of adsorbent, namely sugar or complex polysaccharide, must be considered. When a mono- or disaccharide is used as a ligand, the lectin is readily released from the affinity column by competitive elution, using specific carbohydrates, if possible ones with higher affinity for the lectin than the adsorbent. Recently, the great majority of authors report the use of competitive elution of lectins using specific mono- or disaccharides. For example, when a lectin was isolated on Sepharose, galactose was used for desorption, while glucose was used with Sephadex. The concentrations of sugars used are usually 0.2 M (ranging from 0.01 - 0.5 M), while sugar gradients (from 0.1 to 0.3 or 1 M) have also been used (Table 1). Subsequently, these lectin-bound sugars must be removed to free the binding sites of the lectin, usually by extensive dialysis. Alternatively, to avoid the latter step, lectins can be desorbed from the column by changing the conditions to extremes of ionic strength or pH. The latter technique depends on the chemical stability of the matrix, ligand and adsorbed substances and is not suitable for lectins and adsorbents that are destroyed in such conditions. However, carbohydrates and lectins are usually stable molecules; nevertheless, care must be taken not to damage them irreversibly. Therefore, fractions containing the proteins should be neutralized immediately, usually with 2 M or 1 M Tris-HCl buffer, pH 7.5 (Table 1). The column also has to be equilibrated with the binding buffer. When complex, branched oligo- or polysaccharides with high avidity for lectins are used as adsorbents, the lectins cannot be readily eluted using monovalent sugars with lower affinity for lectins. In this case, extreme pH and/or ionic strength conditions should also be used. Several authors report the elution of lectins using acidic (20-100 mM glycine-HCl or β-alanine buffer, pH ~2.5, or 1-3 % acetic acid) or alkaline (10 mM NaOH, 0.1 M triethanolamine buffer, pH 11, or Tris-OH buffer, pH 11.4) solutions as well as buffers containing 1 M NaCl or 3 M $MgCl_2$ (gradient was also used) (Table 1). In some cases, lectins have been eluted using 6 M (Matsumoto et al., 2011) or 8 M urea (Suseelan et al., 2002). The carbohydrate binding activity of some lectins, such as C-type lectins, depends on divalent metal ions (Ca^{2+}), therefore can be eluted using buffers containing chelating agents, such as 2–100 mM ethylenediaminetetraacetic acid (EDTA) (Table 1).

Carbohydrate-specific lectin	Ligand	Matrix	Elution technique	Reference
Galactoside-specific lectins showing specificity for galactose, lactose, *N*-acetyl-galactosamine, *N*-acetyl-lactosamine, or *N,N'*-diacetyl-lactosediamine	Galactose	Synsorb	0.2 M lactose	Mo et al., 2000
	Lactose	Divinyl sulphone activated Sepharose	0.2 M lactose, 0.01 M NaOH neutralized with 2M Tris, pH 7.5, or 0.1 M glycine buffer, pH 2.6	Almanza et al., 2004; Pohleven et al., 2009; Rocha et al., 2009
	N-acetyl-galactosamine	Agarose	0.2 M galactose or 0.01 M *N*-acetylgalactosamine	Gerlach et al., 2005; Qureshi et al., 2006
	Fetuin	Agarose	6 M urea	Matsumoto et al., 2011
	Bovine submaxillary gland mucin	Cyanogen bromide activated Sepharose 4B	3 M $MgCl_2$	Naganuma et al., 2006
	Porcine stomach mucin	Toyopearl AF-Amino-650M	3 M sodium thiocyanate (NaSCN)	Takahashi et al., 2008
	Porcine stomach mucin	Sepharose 4B	20 mM 1, 3-diaminopropane neutralized with 1 M Tris–HCl, pH 7.0	Chumkhunthod et al., 2006
	Asialofetuin	Sepharose CL	0.2 M lactose	Naeem et al., 2007b
	Asialofetuin	Amino activated silica beads	0.1 M glycine-HCl buffer, pH 2.5, neutralized with 2 M Tris-HCl, pH 8.3	Kaur et al., 2006
	Asialomucin	Cyanogen bromide activated Sepharose 4B	0.05 M Tris-OH, pH 11.4, neutralized with 2 N HCl	Vega & Pérez, 2006
	/	Sepharose	0.2 M galactose, 0.01 M EDTA, 0.1 M isopropyl-β-D-thiogalactoside, or 1 mM HCl, pH 3.0	Cao et al., 2010; De-Simone et al., 2006
	/	Acid treated Sepharose	0.1 M or 0.2 M galactose, or 0.2 M lactose	Jimbo et al., 2000; Mishra et al., 2004; Stirpe et al., 2007; Voss et al., 2006
	/	Affi-gel blue gel	1 M NaCl	Lin & Ng, 2008
	Alginate	Guar gum	0.1 M galactose	Roy et al., 2005
	/	Guar gum	1 M NaCl, 0.1 M or 0.2 M lactose, or 0.1 M glycine buffer, pH 2.6	Alencar et al., 2010; Santos et al., 2009; Souza et al., 2011; Sultan et al., 2009
	/	Cross-linked *Leucaena leucocephala* seed gum	0.2 M lactose	Seshagirirao et al., 2005

Carbohydrate-specific lectin	Ligand	Matrix	Elution technique	Reference
	/	Cross-linked *Adenanthera pavonina* galactomannan	0.2 M galactose or 0.1 M acetate buffer, pH 2.6	Moreira et al., 1998; Teixeira-Sá et al., 2009
	/	cross-linked *Spondias purpurea* polysaccharide	0.2 M galactose or 0.1 M β-alanine buffer, pH 2.6	Teixeira et al., 2007
Glucose/mannose-specific lectins showing specificity for *N*-acetyl-glucosamine, chitooligosaccharides, or chitin	Glucose	Divinyl sulphone activated Sepharose	0.2 M glucose or 0.01 M NaOH neutralized with 2M Tris, pH 7.5	Pohleven et al., 2011
	Mannose	Agarose	0.2 M mannose, 0.5 M mannopyranoside or 8 M urea	Andon et al., 2003; Ooi et al., 2010; Suseelan et al., 2002
	N-acetyl-glucosamine	Epoxy activated Sepharose 6B	0.1 or 0.2 M *N*-acetylglucosamine	Kaur et al., 2005; Maheswari et al., 2002
	Ovalbumin	Cyanogen bromide activated Sepharose 4B	0.2 M methyl α-D-mannoside	Mo et al., 1999
	/	Sephadex	0.2 M or 0.1 M glucose, 0.1 M glycine buffer, pH 2.6 or glucose	Biswas et al., 2009; Rangel et al., 2011; Roh & Park, 2005
	Mannan	Agarose	0.05 M mannose, 0.2 M methyl α-D-mannopyranoside or 2 mM EDTA	Argayosa et al., 2011; Naeem et al., 2007a; Ourth et al., 2005
	Chitin	/	0.1 M triethanol-amine, pH 11	Santi-Gadelha et al., 2006
Fucose-specific lectins	L-fucose	Agarose	0.1 – 1 M L-fucose gradient or 0.2 M L-fucose	Cammarata et al., 2007; Mansour & Abdul-Salam, 2009
L-rhamnose-specific lectin	L-rhamnose	Sepharose 4B	0.2 M L-rhamnose	Jimbo et al., 2007; Watanabe et al., 2008
Sialic acid-specific lectins showing specificity for neuraminyl oligosaccharides, or *N*-acetylneuraminic acid	Fetuin	Agarose	0.1 M *N*-acetyl-glucosamine	Rittidach et al., 2007; Sun et al., 2007
	Fetuin	Cyanogen bromide activated Sepharose 4B	0.05 M citrate buffer, pH 5.0, devoid of Ca^{2+} ions	Bhowal et al., 2005
	Rat erythrocyte stroma	Sephadex G-25	3 % acetic acid	Vazquez et al., 1993

Table 1. Carbohydrate affinity chromatography methods used to isolate lectins according to their carbohydrate-binding specificities.

7. Other methods for the purification of lectins

In addition to carbohydrate affinity chromatography, other methods have also been used to purify lectins. For example, some lectins have been purified using fractional precipitation with ammonium sulphate, followed by classical chromatographic methods, such as ion-exchange chromatography and gel filtration (Horibe et al., 2010; Pan et al., 2010). Lectins are usually glycoproteins, therefore some authors describe their isolation by affinity chromatography on the Sepharose-immobilized lectin Concanavalin A (Absar et al., 2005; Charungchitrak et al., 2011; Konkumnerd et al., 2010; Petnual et al., 2010; Yan et al., 2010). Moreover, ferromagnetic particles with immobilized polysaccharide levan have been prepared and used for isolation of lectins. The magnetic property of the particles favoured the washing of impurities using a magnetic field; the sugars were then used to release the lectins and recover the particles (Angeli et al., 2009). To purify ion-dependent lectins, which show affinity for metal ions, affinity precipitation using metal charged EGTA or affinity chromatography using iminodiacetic acid-Sepharose charged with metal ions was used. After adsorption, lectins were eluted using buffer containing EDTA (Naeem et al., 2006). A novel affinity chromatographic method for purifying lectins using glycosylated nanofibrous membrane has also been described. An affinity membrane with immobilized glucose ligands showed strong and reversible lectin binding capability (Che et al., 2011).

8. Conclusions

Carbohydrate affinity chromatography is by far the most widely used method for purifying lectins. It is a simple method that takes advantage of their specific properties. However, there is no generally applicable protocol. Selection of the most appropriate matrix, ligand and elution technique, apart from matrix activation and ligand coupling methods, can be difficult. These have to be selected according to the individual lectin. A variety of currently used methods are presented in Table 1, for isolating lectins according to their carbohydrate-binding specificity. To purify any lectin of interest, one of the described methods at least should be effective.

9. Acknowledgement

The authors would like to thank Dr. Jože Brzin and Prof. Roger Pain for critical reading of the manuscript. This work was supported by the Slovenian Research Agency Grant No. P4-0127.

10. References

Absar, N., Yeasmin, T., Raza, M. S., Sarkar, S. K. & Arisaka, F. (2005). Single Step Purification, Characterization and N-Terminal Sequences of a Mannose Specific Lectin from Mulberry Seeds. *The Protein Journal*, Vol. 24, No. 6, (August 2005), pp. 369-377, 1572-3887

Alencar, N. M. N., Oliveira, R. S. B., Figueiredo, J. G., Cavalcante, I. J. M., Matos, M. P. V., Cunha, F. Q., Nunes, J. V. S., Bomfim, L. R. & Ramos, M. V. (2010). An anti-inflammatory lectin from *Luetzelburgia auriculata* seeds inhibits adhesion and rolling of leukocytes and modulates histamine and PGE2 action in acute inflammation models. *Inflammation Research*, Vol. 59, No. 4, (April 2010), pp. 245-254, 1023-3830

Almanza, M., Vega, N. & Pérez, G. (2004). Isolating and characterising a lectin from *Galactia lindenii* seeds that recognises blood group H determinants. *Archives of Biochemistry and Biophysics*, Vol. 429, No. 2, (September 2004), pp. 180-190, 0003-9861

Alpuche, J., Pereyra, A., Mendoza-Hernández, G., Agundis, C., Rosas, C. & Zenteno, E. (2010). Purification and partial characterization of an agglutinin from *Octopus maya* serum. *Comparative Biochemistry and Physiology, Part B*, Vol. 156, No. 1, (May 2010), pp. 1-5, 1096-4959

Andon, N. L., Eckert, D., Yates, J. R. III & Haynes, P. A. (2003). High-throughput functional affinity purification of mannose binding proteins from *Oryza sativa*. *PROTEOMICS*, Vol. 3, No. 7, (July 2003), pp. 1270-1278, 1615-9853

Angal, S. & Dean, P. D. G. (1977). The effect of matrix on the binding of albumin to immobilized Cibacron Blue. *Biochemical Journal*, Vol. 167, No. 1, (October 1977), pp. 301-303, 0264-6021

Angeli, R., da Paz, N. V. N., Maciel, J. C., Araújo, F. F. B., Paiva, P. M. G., Calazans, G. M. T., Valente, A. P., Almeida, F. C. L., Coelho, L. C. B. B., Carvalho Jr., L. B., Silva, M. d. P. C. & Correia, M. T. d. S. (2009). Ferromagnetic Levan Composite: An Affinity Matrix to Purify Lectin. *Journal of Biomedicine and Biotechnology*, Vol. 2009, No. 179106, (June 2009), 1110-7243

Argayosa, A. M., Bernal, R. A. D., Luczon, A. U. & Arboleda, J. S. (2011). Characterization of mannose-binding protein isolated from the African catfish (*Clarias gariepinus* B.) serum. *Aquaculture*, Vol. 310, No. 3-4, (January 2011), pp. 274-280, 0044-8486

Argayosa, A. M. & Lee, Y. C. (2009). Identification of L-fucose-binding proteins from the Nile tilapia (*Oreochromis niloticus* L.) serum. *Fish and Shellfish Immunology*, Vol. 27, No. 3, (September 2009), pp. 478-485, 1050-4648

Bailey, M. J., Biely, P. & Poutanen, K. (1992). Interlaboratory testing of methods for assay of xylanase activity. *Journal of Biotechnology*, Vol. 23, No. 3, (May 1992), pp. 257-270, 0168-1656

Bedair, M. & El Rassi, Z. (2004). Affinity chromatography with monolithic capillary columns I. Polymethacrylate monoliths with immobilized mannan for the separation of mannose-binding proteins by capillary electrochromatography and nano-scale liquid chromatography. *Journal of Chromatography A*, Vol. 1044, No. 1-2, (July 2004), pp. 177-186, 0021-9673

Bhat, G. G., Shetty, K. N., Nagre, N. N., Neekhra, V. V., Lingaraju, S., Bhat, R. S., Inamdar, S. R., Suguna, K. & Swamy, B. M. (2010). Purification, characterization and molecular cloning of a monocot mannose-binding lectin from *Remusatia vivipara* with nematicidal activity. *Glycoconjugate Journal*, Vol. 27, No. 3, (April 2010), pp. 309-320, 0282-0080

Bhowal, J., Guha, A. K. & Chatterjee, B. P. (2005). Purification and molecular characterization of a sialic acid specific lectin from the phytopathogenic fungus *Macrophomina phaseolina*. *Carbohydrate Research*, Vol. 340, No. 12, (September 2005), pp. 1973-1982, 0008-6215

Biswas, S., Saroha, A. & Das, H. R. (2009). A lectin from *Sesbania aculeata* (*Dhaincha*) roots and its possible function. *Biochemistry (Moscow)*, Vol. 74, No. 3, (March 2009), pp. 329-335, 0006-2979

Bovi, M., Carrizo, M. E., Capaldi, S., Perduca, M., Chiarelli, L. R., Galliano, M. & Monaco, H. L. (2011). Structure of a lectin with antitumoral properties in king bolete (*Boletus edulis*) mushrooms. *Glycobiology*, Vol. 21, No. 8, (August 2011), pp. 1000-1009, 0959-6658

Bradford, M. M. (1976). A rapid and sensitive method for the quantitation of microgram quantities of protein utilizing the principle of protein-dye binding. *Analytical Biochemistry*, Vol. 72, No. 1-2, (May 1976), pp. 248-254, 0003-2697

Braga, M. D. M., Martins, A. M. C., Amora, D. N., de Menezes, D. B., Toyama, M. H., Toyama, D. O., Marangoni, S., Barbosa, P. S. F., de Sousa Alves, R., Fonteles, M. C. & Monteiro, H. S. A. (2006). Purification and biological effects of C-type lectin isolated from *Bothrops insularis* venom. *Toxicon*, Vol. 47, No. 8, (June 2006), pp. 859-867, 0041-0101

Cammarata, M., Benenati, G., Odom, E. W., Salerno, G., Vizzini, A., Vasta, G. R. & Parrinello, N. (2007). Isolation and characterization of a fish F-type lectin from gilt head bream (*Sparus aurata*) serum. *Biochimica et Biophysica Acta*, Vol. 1770, No. 1, (January 2007), pp. 150-155, 0304-4165

Cao, X., Huo, Z., Lu, M., Mao, D., Zhao, Q., Xu, C., Wang, C. & Zeng, B. (2010). Purification of lectin from larvae of the fly, *Musca domestica*, and in vitro anti-tumor activity in MCF-7 Cells. *Journal of Insect Science*, Vol. 10, No. 164, (September 2010), pp. 1-13, 1536-2442

Castillo-Villanueva, A., Caballero-Ortega, H., Abdullaev-Jafarova, F., Garfias, Y., del Carmen Jiménez-Martínez, M., Bouquelet, S., Martínez, G., Mendoza-Hernández, G. & Zenteno, E. (2007). Lectin from *Phaseolus acutifolius var. escumite*: Chemical Characterization, Sugar Specificity, and Effect on Human T-Lymphocytes. *Journal of Agricultural and Food Chemistry*, Vol. 55, No. 14, (July 2007), pp. 5781-5787, 0021-8561

Charungchitrak, S., Petsom, A., Sangvanich, P. & Karnchanatat, A. (2011). Antifungal and antibacterial activities of lectin from the seeds of *Archidendron jiringa* Nielsen. *Food Chemistry*, Vol. 126, No. 3, (June 2011), pp. 1025-1032, 0308-8146

Che, A.-F., Huang, X.-J. & Xu, Z.-K. (2011). Polyacrylonitrile-based nanofibrous membrane with glycosylated surface for lectin affinity adsorption. *Journal of Membrane Science*, Vol. 366, No. 1-2, (January 2011), pp. 272-277, 0376-7388

Chen, C., Rowley, A. F., Newton, R. P. & Ratcliffe, N. A. (1999). Identification, purification and properties of a β-1,3-glucan-specific lectin from the serum of the cockroach, *Blaberus discoidalis* which is implicated in immune defence reactions. *Comparative Biochemistry and Physiology, Part B*, Vol. 122, No. 3, (March 1999), pp. 309-319, 0305-0491

Chen, G., Tao, L., Mantovani, G., Geng, J., Nyström, D. & Haddleton, D. M. (2007). A Modular Click Approach to Glycosylated Polymeric Beads: Design, Synthesis and Preliminary Lectin Recognition Studies. *Macromolecules*, Vol. 40, No. 21, (October 2007), pp. 7513-7520, 0024-9297

Chen, H.-P. & Xu, L.-L. (2005). Isolation and Characterization of a Novel Chitosan-Binding Protein from Non-Heading Chinese Cabbage Leaves. *Journal of Integrative Plant Biology*, Vol. 47, No. 4, (April 2005), pp. 452-456, 1672-9072

Chumkhunthod, P., Rodtong, S., Lambert, S. J., Fordham-Skelton, A. P., Rizkallah, P. J., Wilkinson, M. C. & Reynolds, C. D. (2006). Purification and characterization of an *N*-acetyl-D-galactosamine-specific lectin from the edible mushroom *Schizophyllum commune*. *Biochimica et Biophysica Acta*, Vol. 1760, No. 3, (March 2006), pp. 326-332, 0304-4165

De-Simone, S. G., Netto, C. C. & Silva Jr., F. P. (2006). Simple affinity chromatographic procedure to purify β-galactoside binding lectins. *Journal of Chromatography B*, Vol. 838, No. 2, (July 2006), pp. 135-138, 1570-0232

Dresch, R. R., Zanetti, G. D., Lerner, C. B., Mothes, B., Trindade, V. M. T., Henriques, A. T. & Vozári-Hampe, M. M. (2008). ACL-I, a lectin from the marine sponge *Axinella corrugata*: Isolation, characterization and chemotactic activity. *Comparative Biochemistry and Physiology, Part C*, Vol. 148, No. 1, (July 2008), pp. 23-30, 1532-0456

Dubois, M., Gilles, K. A., Hamilton, J. K., Rebers, P. A. & Smith, F. (1956). Colorimetric Method for Determination of Sugars and Related Substances. *Analytical Chemistry*, Vol. 28, No. 3, (March 1956), pp. 350-356, 0003-2700

Ersson, B., Aspberg, K. & Porath, J. (1973). The phytohemagglutinin from sunn hemp seeds (*Crotalaria juncea*). Purification by biospecific affinity chromatography. *Biochimica et Biophysica Acta*, Vol. 310, No. 2, (June 1973), pp. 446-452, 0006-3002

Fujii, Y., Kawsar, S. M. A., Matsumoto, R., Yasumitsu, H., Ishizaki, N., Dogasaki, C., Hosono, M., Nitta, K., Hamako, J., Taei, M. & Ozeki, Y. (2011). A D-galactose-binding lectin purified from coronate moon turban, *Turbo (Lunella) coreensis*, with a unique amino acid sequence and the ability to recognize lacto-series glycosphingolipids. *Comparative Biochemistry and Physiology, Part B*, Vol. 158, No. 1, (January 2011), pp. 30-37, 1096-4959

Gerlach, D., Schlott, B., Zähringer, U. & Schmidt, K.-H. (2005). *N*-acetyl-D-galactosamine/*N*-acetyl-D-glucosamine - recognizing lectin from the snail *Cepaea hortensis*: purification, chemical characterization, cloning and expression in *E. coli*. *FEMS Immunology and Medical Microbiology*, Vol. 43, No. 2, (February 2005), pp. 223-232, 0928-8244

Guzmán-Partida, A. M., Robles-Burgueño, M. R., Ortega-Nieblas, M. & Vázquez-Moreno, I. (2004). Purification and characterization of complex carbohydrate specific isolectins from wild legume seeds: *Acacia constricta* is (vinorama) highly homologous to *Phaseolus vulgaris* lectins. *Biochimie*, Vol. 86, No. 4-5, (April-May 2004), pp. 335-342, 0300-9084

Hamako, J., Suzuki, Y., Hayashi, N., Kimura, M., Ozeki, Y., Hashimoto, K. & Matsui, T. (2007). Amino acid sequence and characterization of C-type lectin purified from the snake venom of *Crotalus ruber*. *Comparative Biochemistry and Physiology, Part B*, Vol. 146, No. 3, (March 2007), pp. 299-306, 1096-4959

Hořejší, V. & Kocourek, J. (1978). Studies on lectins XXXVI. Properties of some lectins prepared by affinity chromatography on O-glycosyl polyacrylamide gels. *Biochimica et Biophysica Acta*, Vol. 538, No. 2, (January 1978), pp. 299-315, 0006-3002

Horibe, M., Kobayashi, Y., Dohra, H., Morita, T., Murata, T., Usui, T., Nakamura-Tsuruta, S., Kamei, M., Hirabayashi, J., Matsuura, M., Yamada, M., Saikawa, Y., Hashimoto, K., Nakata, M. & Kawagishi, H. (2010). Toxic isolectins from the mushroom *Boletus venenatus*. *Phytochemistry*, Vol. 71, No. 5-6, (April 2010), pp. 648-657, 0031-9422

Imamichi, Y. & Yokoyama, Y. (2010). Purification, characterization and cDNA cloning of a novel lectin from the jellyfish *Nemopilema nomurai*. *Comparative Biochemistry and Physiology, Part B*, Vol. 156, No. 1, (May 2010), pp. 12-18, 1096-4959

Jimbo, M., Usui, R., Sakai, R., Muramoto, K. & Kamiya, H. (2007). Purification, cloning and characterization of egg lectins from the teleost *Tribolodon brandti*. *Comparative Biochemistry and Physiology, Part B*, Vol. 147, No. 2, (June 2007), pp. 164-171, 1096-4959

Jimbo, M., Yanohara, T., Koike, K., Koike, K., Sakai, R., Muramoto, K. & Kamiya, H. (2000). The D-galactose-binding lectin of the octocoral *Sinularia lochmodes*: characterization and possible relationship to the symbiotic dinoflagellates. *Comparative Biochemistry and Physiology, Part B*, Vol. 125, No. 2, (February 2000), pp. 227-236, 1096-4959

Kajiya, A., Koyama, Y., Mita, T., Mega, T., Hase, S., Kawakami, T., Honda, E., Munakata, H. & Isemura, M. (2003). Isolation of Lectins by Affinity Chromatography with Porcine Plasma Proteins Immobilized on Agarose. *Bioscience, Biotechnology, and Biochemistry*, Vol. 67, No. 9, (September 2003), pp. 2051-2054, 0916-8451

Kato, H., Uzawa, H., Nagatsuka, T., Kondo, S., Sato, K., Ohsawa, I., Kanamori-Kataoka, M., Takei, Y., Ota, S., Furuno, M., Dohi, H., Nishida, Y. & Seto, Y. (2011). Preparation and evaluation of lactose-modified monoliths for the adsorption and decontamination of plant toxins and lectins. *Carbohydrate Research*, Vol. 346, No. 13, (September 2011), pp. 1820-1826, 0008-6215

Kaur, A., Singh, J., Kamboj, S. S., Sexana, A. K., Pandita, R. M. & Shamnugavel, M. (2005). Isolation of an *N*-acetyl-D-glucosamine specific lectin from the rhizomes of *Arundo donax* with antiproliferative activity. *Phytochemistry*, Vol. 66, No. 16, (August 2005), pp. 1933-1940, 0031-9422

Kaur, M., Singh, K., Rup, P. J., Kamboj, S. S., Saxena, A. K., Sharma, M., Bhagat, M., Sood, S. K. & Singh, J. (2006). A Tuber Lectin from *Arisaema jacquemontii* Blume with Anti-insect and Anti-proliferative Properties. *Journal of Biochemistry and Molecular Biology*, Vol. 39, No. 4, (July 2006), pp. 432-440, 1225-8687

Kawagishi, H., Yamawaki, M., Isobe, S., Usui, T., Kimura, A. & Chiba, S. (1994). Two lectins from the marine sponge *Halichondria okadai*. An *N*-acetyl-sugar-specific lectin (HOL-I) and an *N*-acetyllactosamine-specific lectin (HOL-II). *Journal of Biological Chemistry*, Vol. 269, No. 2, (January 1994), pp. 1375-1379, 0021-9258

Kawsar, S. M. A., Matsumoto, R., Fujii, Y., Yasumitsu, H., Dogasaki, C., Hosono, M., Nitta, K., Hamako, J., Matsui, T., Kojima, N. & Ozeki, Y. (2009). Purification and biochemical characterization of a D-galactose binding lectin from Japanese sea hare (*Aplysia kurodai*) eggs. *Biochemistry (Moscow)*, Vol. 74, No. 7, (July 2009), pp. 709-716, 0006-2979

Kim, G. H., Klochkova, T. A., Yoon, K.-S., Song, Y.-S. & Lee, K. P. (2006). Purification and characterization of a lectin, bryohealin, involved in the protoplast formation of a marine green alga *Bryopsis plumosa* (Chlorophyta). *Journal of Phycology*, Vol. 42, No. 1, (February 2006), pp. 86-95, 1529-8817

Konami, Y., Yamamoto, K. & Osawa, T. (1991). Purification and Characterization of Two Types of Cytisus sessilifolius Anti-H(O) Lectins by Affinity Chromatography. *Biological Chemistry Hoppe-Seyler*, Vol. 372, No. 1, (February 1991), pp. 103-111, 0177-3593

Konkumnerd, W., Karnchanatat, A. & Sangvanich, P. (2010). A thermostable lectin from the rhizomes of *Kaempferia parviflora*. *Journal of the Science of Food and Agriculture*, Vol. 90, No. 11, (August 2010), pp. 1920-1925, 1097-0010

Konozy, E. H. E., Mulay, R., Faca, V., Ward, R. J., Greene, L. J., Roque-Barriera, M. C., Sabharwal, S. & Bhide, S. V. (2002). Purification, some properties of a D-galactose-binding leaf lectin from *Erythrina indica* and further characterization of seed lectin. *Biochimie*, Vol. 84, No. 10, (October 2002), pp. 1035-1043, 0300-9084

Koyama, Y., Katsuno, Y., Miyoshi, N., Hayakawa, S., Mita, T., Muto, H., Isemura, S., Aoyagi, Y. & Isemura, M. (2002). Apoptosis induction by lectin isolated from the mushroom *Boletopsis leucomelas* in U937 cells. *Bioscience, Biotechnology, and Biochemistry*, Vol. 66, No. 4, (April 2002), pp. 784-789, 0916-8451

Kvennefors, E. C. E., Leggat, W., Hoegh-Guldberg, O., Degnan, B. M. & Barnes, A. C. (2008). An ancient and variable mannose-binding lectin from the coral *Acropora millepora* binds both pathogens and symbionts. *Developmental and Comparative Immunology*, Vol. 32, No. 12, pp. 1582-1592, 0145-305X

Lavanya Latha, V., Nagender Rao, R. & Nadimpalli, S. K. (2006). Affinity purification, physicochemical and immunological characterization of a galactose-specific lectin from the seeds of *Dolichos lablab* (Indian lablab beans). *Protein Expression and Purification*, Vol. 45, No. 2, (February 2006), pp. 296-306, 1046-5928

Leavitt, R. D., Felsted, R. L. & Bachur, N. R. (1977). Biological and biochemical properties of *Phaseolus vulgaris* isolectins. *Journal of Biological Chemistry*, Vol. 252, No. 9, (May 1977), pp. 2961-2966, 0021-9258

Levi, G. & Teichberg, V. I. (1981). Isolation and physicochemical characterization of electrolectin, a β-D-galactoside binding lectin from the electric organ of *Electrophorus electricus*. *Journal of Biological Chemistry*, Vol. 256, No. 11, (June 1981), pp. 5735-5740, 0021-9258

Lin, P. & Ng, T. B. (2008). Preparation and Biological Properties of a Melibiose Binding Lectin from *Bauhinia variegata* Seeds. *Journal of Agricultural and Food Chemistry*, Vol. 56, No. 22, (November 2008), pp. 10481-10486, 0021-8561

Lis, H. & Sharon, N. (1981). Affinity chromatography for the purification of lectins (a review). *Journal of Chromatography A*, Vol. 215, No. 1-3, (October 1981), pp. 361-372, 0021-9673

Maheswari, R., Mullainadhan, P. & Arumugam, M. (2002). Isolation and characterization of an acetyl group-recognizing agglutinin from the serum of the Indian white shrimp *Fenneropenaeus indicus*. *Archives of Biochemistry and Biophysics*, Vol. 402, No. 1, (June 2002), pp. 65-76, 0003-9861

Mansour, M. H. & Abdul-Salam, F. (2009). Characterization of fucose-binding lectins in rock- and mud-dwelling snails inhabiting Kuwait Bay. *Immunobiology*, Vol. 214, No. 1, (January 2009), pp. 77-85, 0171-2985

Matsumoto, R., Shibata, T. F., Kohtsuka, H., Sekifuji, M., Sugii, N., Nakajima, H., Kojima, N., Fujii, Y., Kawsar, S. M. A., Yasumitsu, H., Hamako, J., Matsui, T. & Ozeki, Y. (2011).

Glycomics of a novel type-2 *N*-acetyllactosamine-specific lectin purified from the feather star, *Oxycomanthus japonicus* (Pelmatozoa: Crinoidea). *Comparative Biochemistry and Physiology, Part B*, Vol. 158, No. 4, (April 2011), pp. 266-273, 1096-4959

Mendonça-Franqueiro, E. d. P., Alves-Paiva, R. d. M., Sartim, M. A., Callejon, D. R., Paiva, H. H., Antonucci, G. A., Rosa, J. C., Cintra, A. C. O., Franco, J. J., Arantes, E. C., Dias-Baruffi, M. & Vilela Sampaio, S. (2011). Isolation, functional, and partial biochemical characterization of galatrox, an acidic lectin from *Bothrops atrox* snake venom. *Acta Biochimica et Biophysica Sinica*, Vol. 43, No. 3, (March, 2011), pp. 181-192, 1672-9145

Mikes, L. & Man, P. (2003). Purification and characterization of a saccharide-binding protein from penetration glands of *Diplostomum pseudospathaceum* – a bifunctional molecule with cysteine protease activity. *Parasitology*, Vol. 127, No. 1, (July 2003), pp. 69-77, 0031-1820

Mishra, V., Sharma, R. S., Yadav, S., Babu, C. R. & Singh, T. P. (2004). Purification and characterization of four isoforms of Himalayan mistletoe ribosome-inactivating protein from *Viscum album* having unique sugar affinity. *Archives of Biochemistry and Biophysics*, Vol. 423, No. 2, (March 2004), pp. 288-301, 0003-9861

Mo, H., Meah, Y., Moore, J. G. & Goldstein, I. J. (1999). Purification and characterization of *Dolichos lablab* lectin. *Glycobiology*, Vol. 9, No. 2, (February 1999), pp. 173-179, 0959-6658

Mo, H., Winter, H. C. & Goldstein, I. J. (2000). Purification and Characterization of a Neu5Acα2–6Galβ1–4Glc/GlcNAc-specific Lectin from the Fruiting Body of the Polypore Mushroom *Polyporus squamosus*. *Journal of Biological Chemistry*, Vol. 275, No. 14, (April 2000), pp. 10623-10629, 0021-9258

Mora, P., Rosconi, F., Franco Fraguas, L. & Castro-Sowinski, S. (2008). *Azospirillum brasilense* Sp7 produces an outer-membrane lectin that specifically binds to surface-exposed extracellular polysaccharide produced by the bacterium. *Archives of Microbiology*, Vol. 189, No. 5, (May 2008), pp. 519-524, 0302-8933

Moreira, R. A., Castelo-Branco, C. C., Monteiro, A. C. O., Tavares, R. O. & Beltramini, L. M. (1998). Isolation and partial characterization of a lectin from *Artocarpus incisa* L. seeds. *Phytochemistry*, Vol. 47, No. 7, (April 1998), pp. 1183-1188, 0031-9422

Moura, R. M., Queiroz, A. F. S., Fook, J. M. S. L. L., Dias, A. S. F., Monteiro, N. K. V., Ribeiro, J. K. C., Moura, G. E. D. D., Macedo, L. L. P., Santos, E. A. & Sales, M. P. (2006). CvL, a lectin from the marine sponge *Cliona varians*: Isolation, characterization and its effects on pathogenic bacteria and Leishmania promastigotes. *Comparative Biochemistry and Physiology, Part A*, Vol. 145, No. 4, (December 2006), pp. 517-523, 1095-6433

Murthy, G. S. & Moudgal, N. R. (1986). Use of epoxysepharose for protein immobilisation. *Journal of biosciences*, Vol. 10, No. 3, (September 1986), pp. 351-358, 0250-5991

Naeem, A., Ahmad, E., Ashraf, M. T. & Khan, R. H. (2007a). Purification and characterization of mannose/glucose-specific lectin from seeds of *Trigonella foenumgraecum*. *Biochemistry (Moscow)*, Vol. 72, No. 1, (January 2007), pp. 44-48, 0006-2979

Naeem, A., Haque, S. & Khan, R. H. (2007b). Purification and Characterization of a Novel β-D-Galactosides-Specific Lectin from *Clitoria ternatea*. *The Protein Journal*, Vol. 26, No. 6, (September 2007), pp. 403-413, 1572-3887

Naeem, A., Khan, R. H. & Saleemuddin, M. (2006). Single step immobilized metal ion affinity precipitation/chromatography based procedures for purification of concanavalin A and *Cajanus cajan* mannose-specific lectin. *Biochemistry (Moscow)*, Vol. 71, No. 1, (January 2006), pp. 56-59, 0006-2979

Naganuma, T., Ogawa, T., Hirabayashi, J., Kasai, K., Kamiya, H. & Muramoto, K. (2006). Isolation, characterization and molecular evolution of a novel pearl shell lectin from a marine bivalve, *Pteria penguin*. *Molecular Diversity*, Vol. 10, No. 4, (November 2006), pp. 607-618, 1381-1991

Nagata, S. (2005). Isolation, characterization, and extra-embryonic secretion of the *Xenopus laevis* embryonic epidermal lectin, XEEL. *Glycobiology*, Vol. 15, No. 3, (March 2005), pp. 281-290, 0959-6658

Nagre, N. N., Chachadi, V. B., Sundaram, P. M., Naik, R. S., Pujari, R., Shastry, P., Swamy, B. M. & Inamdar, S. R. (2010). A potent mitogenic lectin from the mycelia of a phytopathogenic fungus, *Rhizoctonia bataticola*, with complex sugar specificity and cytotoxic effect on human ovarian cancer cells. *Glycoconjugate Journal*, Vol. 27, No. 3, (April 2010), pp. 375-386, 0282-0080

Nahálková, J., Asiegbu, F. O., Daniel, G., Hřib, J., Vooková, B., Pribulová, B. & Gemeiner, P. (2001). Isolation and immunolocalization of a *Pinus nigra* lectin (PNL) during interaction with the necrotrophs—*Heterobasidion annosum* and *Fusarium avenaceum*. *Physiological and Molecular Plant Pathology*, Vol. 59, No. 3, (September 2001), pp. 153-163, 0885-5765

Narahari, A. & Swamy, M. J. (2010). Rapid affinity-purification and physicochemical characterization of pumpkin (*Cucurbita maxima*) phloem exudate lectin. *Bioscience Reports*, Vol. 30, No. 5, (October 2010), pp. 341-349, 0144-8463

Ooi, L. S. M., Ho, W.-S., Ngai, K. L. K., Tian, L., Chan, P. K. S., Sun, S. S. M. & Ooi, V. E. C. (2010). *Narcissus tazetta* lectin shows strong inhibitory effects against respiratory syncytial virus, influenza A (H1N1, H3N2, H5N1) and B viruses. *Journal of biosciences*, Vol. 35, No. 1, (March 2010), pp. 95-103, 0250-5991

Ooi, L. S. M., Yu, H., Chen, C.-M., Sun, S. S. M. & Ooi, V. E. C. (2002). Isolation and Characterization of a Bioactive Mannose-Binding Protein from the Chinese Chive *Allium tuberosum*. *Journal of Agricultural and Food Chemistry*, Vol. 50, No. 4, (February 2002), pp. 696-700, 0021-8561

Ourth, D. D., Narra, M. B. & Chung, K. T. (2005). Isolation of mannose-binding C-type lectin from *Heliothis virescens* pupae. *Biochemical and Biophysical Research Communications*, Vol. 335, No. 4, (October 2005), pp. 1085-1089, 0006-291X

Pan, S., Tang, J. & Gu, X. (2010). Isolation and characterization of a novel fucose-binding lectin from the gill of bighead carp (*Aristichthys nobilis*). *Veterinary Immunology and Immunopathology*, Vol. 133, No. 2-4, (February 2010), pp. 154-164, 0165-2427

Pepper, D. S. (1992). Some Alternative Coupling Chemistries for Affinity Chromatography, In: *Methods in Molecular Biology, Vol. 11: Practical Protein Chromatography*, Kenney, A. & Fowell, S., pp. (173-196), The Humana Press Inc., 978-089-603-213-2, Totowa, NJ

Perçin, I., Yavuz, H., Aksöz, E. & Denizli, A. (2009). *N*-acetyl-D-galactosamine-specific lectin isolation from soyflour with poly(HPMA-GMA) beads. *Journal of Applied Polymer Science*, Vol. 111, No. 1, (January 2009), pp. 148-154, 1097-4628

Perillo, N. L., Marcus, M. E. & Baum, L. G. (1998). Galectins: versatile modulators of cell adhesion, cell proliferation, and cell death. *Journal of Molecular Medicine*, Vol. 76, No. 6, (May, 1998), pp. 402-412, 0946-2716

Petnual, P., Sangvanich, P. & Karnchanatat, A. (2010). A lectin from the rhizomes of turmeric (*Curcuma longa* L.) and its antifungal, antibacterial, and α-glucosidase inhibitory activities. *Food Science and Biotechnology*, Vol. 19, No. 4, (August 2010), pp. 907-916, 1226-7708

Peumans, W. J. & Van Damme, E. J. M. (1995). Lectins as plant defense proteins. *Plant Physiology*, Vol. 109, No. 2, (October, 1995), pp. 347-352, 0032-0889

Pohleven, J., Brzin, J., Vrabec, L., Leonardi, A., Čokl, A., Štrukelj, B., Kos, J. & Sabotič, J. (2011). Basidiomycete *Clitocybe nebularis* is rich in lectins with insecticidal activities. *Applied Microbiology and Biotechnology*, Vol. 91, No. 4, (August 2011), pp. 1141-1148, 0175-7598

Pohleven, J., Obermajer, N., Sabotič, J., Anžlovar, S., Sepčić, K., Kos, J., Kralj, B., Štrukelj, B. & Brzin, J. (2009). Purification, characterization and cloning of a ricin B-like lectin from mushroom *Clitocybe nebularis* with antiproliferative activity against human leukemic T cells. *Biochimica et Biophysica Acta*, Vol. 1790, No. 3, (March 2009), pp. 173-181, 0304-4165

Qureshi, I. A., Dash, P., Srivastava, P. S. & Koundal, K. R. (2006). Purification and characterization of an *N*-acetyl-D-galactosamine-specific lectin from seeds of chickpea (*Cicer arietinum* L.). *Phytochemical Analysis*, Vol. 17, No. 5, (September 2006), pp. 350-356, 0958-0344

Rangel, T. B. A., Assreuy, A. M. S., Pires, A. d. F., de Carvalho, A. U., Benevides, R. G., Simões, R. d. C., da Silva, H. C., Bezerra, M. J. B., do Nascimento, A. S. F., do Nascimento, K. S., Nagano, C. S., Sampaio, A. H., Delatorre, P., da Rocha, B. A. M., Fernandes, P. M. B. & Cavada, B. S. (2011). Crystallization and Characterization of an Inflammatory Lectin Purified from the Seeds of *Dioclea wilsonii*. *Molecules*, Vol. 16, No. 6, (June 2011), pp. 5087-5103, 1420-3049

Ren, J., Shi, J., Kakuda, Y., Kim, D., Xue, S. J., Zhao, M. & Jiang, Y. (2008). Phytohemagglutinin isolectins extracted and purified from red kidney beans and its cytotoxicity on human H9 lymphoma cell line. *Separation and Purification Technology*, Vol. 63, No. 1, (October 2008), pp. 122-128, 1383-5866

Rittidach, W., Paijit, N. & Utarabhand, P. (2007). Purification and characterization of a lectin from the banana shrimp *Fenneropenaeus merguiensis* hemolymph. *Biochimica et Biophysica Acta*, Vol. 1770, No. 1, (January 2007), pp. 106-114, 0304-4165

Rocha, B. A. M., Moreno, F. B. M. B., Delatorre, P., Souza, E. P., Marinho, E. S., Benevides, R. G., Rustiguel, J. K. R., Souza, L. A. G., Nagano, C. S., Debray, H., Sampaio, A. H., de Azevedo Jr., W. F. & Cavada, B. S. (2009). Purification, Characterization, and Preliminary X-Ray Diffraction Analysis of a Lactose-Specific Lectin from *Cymbosema roseum* Seeds. *Applied Biochemistry and Biotechnology*, Vol. 152, No. 3, (March 2009), pp. 383-393, 0273-2289

Roh, K. S. & Park, N. Y. (2005). Characterization of the lectin purified from *Canavalia ensiformis* shoots. *Biotechnology and Bioprocess Engineering*, Vol. 10, No. 4, (July-August 2005), pp. 334-340, 1226-8372

Roy, I., Sardar, M. & Gupta, M. N. (2005). Cross-linked alginate-guar gum beads as fluidized bed affinity media for purification of jacalin. *Biochemical Engineering Journal*, Vol. 23, No. 3, (May 2005), pp. 193-198, 1369-703X

Santi-Gadelha, T., de Almeida Gadelha, C. A., Aragão, K. S., de Oliveira, C. C., Lima Mota, M. R., Gomes, R. C., de Freitas Pires, A., Toyama, M. H., de Oliveira Toyama, D., de Alencar, N. M. N., Criddle, D. N., Assreuy, A. M. S. & Cavada, B. S. (2006). Purification and biological effects of *Araucaria angustifolia* (Araucariaceae) seed lectin. *Biochemical and Biophysical Research Communications*, Vol. 350, No. 4, (December 2006), pp. 1050-1055, 0006-291X

Santos, A. F. S., Luz, L. A., Argolo, A. C. C., Teixeira, J. A., Paiva, P. M. G. & Coelho, L. C. B. B. (2009). Isolation of a seed coagulant *Moringa oleifera* lectin. *Process Biochemistry*, Vol. 44, No. 4, (April 2009), pp. 504-508, 1359-5113

Schenkel-Brunner, H. (2007). Blood group antigens, In: *Comprehensive Glycoscience*, Kamerling, J. P., Lee, Y. C., Boons, G.-J., Suzuki, A., Taniguchi, N. & Voragen, A. G. J., pp. (343-372), Elsevier, 978-044-452-746-2, Oxford

Seshagirirao, K., Leelavathi, C. & Sasidhar, V. (2005). Cross-linked leucaena seed gum matrix: an affinity chromatography tool for galactose-specific lectins. *Journal of Biochemistry and Molecular Biology*, Vol. 38, No. 3, (May 2005), pp. 370-372, 1225-8687

Shangary, S., Singh, J., Kamboj, S. S., Kamboj, K. K. & Sandhu, R. S. (1995). Purification and properties of four monocot lectins from the family araceae. *Phytochemistry*, Vol. 40, No. 2, (September 1995), pp. 449-455, 0031-9422

Shao, B., Wang, S., Zhou, J., Ke, L. & Rao, P. (2011). A novel lectin from fresh rhizome of *Alisma orientale (Sam.) Juzep*. *Process Biochemistry*, Vol. 46, No. 8, (August 2011), pp. 1554-1559, 1359-5113

Sharma, A., Wong, J. H., Lin, P., Chan, Y. S. & Ng, T. B. (2010). Purification and Characterization of a Lectin from the Indian Cultivar of French Bean Seeds. *Protein and Peptide Letters*, Vol. 17, No. 2, (February 2010), pp. 221-227, 0929-8665

Sharon, N. & Lis, H. (1989). Lectins as cell recognition molecules. *Science*, Vol. 246, No. 4927, (October, 1989), pp. 227-234, 0036-8075

Silva, J. A., Damico, D. C. S., Baldasso, P. A., Mattioli, M. A., Winck, F. V., Fraceto, L. F., Novello, J. C. & Marangoni, S. (2007). Isolation and Biochemical Characterization of a Galactoside Binding Lectin from *Bauhinia variegata* Candida (BvcL) Seeds. *The Protein Journal*, Vol. 26, No. 3, (April 2007), pp. 193-201, 1572-3887

Souza, B. W. S., Andrade, F. K., Teixeira, D. I. A., Mansilla, A. & Freitas, A. L. P. (2010). Haemagglutinin of the Antarctic seaweed *Georgiella confluens* (Reinsch) Kylin: isolation and partial characterization. *Polar Biology*, Vol. 33, No. 10, (October 2010), pp. 1311-1318, 0722-4060

Souza, J. D., Silva, M. B. R., Argolo, A. C. C., Napoleão, T. H., Sá, R. A., Correia, M. T. S., Paiva, P. M. G., Silva, M. D. C. & Coelho, L. C. B. B. (2011). A new *Bauhinia monandra* galactose-specific lectin purified in milligram quantities from secondary roots with antifungal and termiticidal activities. *International Biodeterioration and Biodegradation*, Vol. 65, No. 5, (August 2011), pp. 696-702, 0964-8305

Stirpe, F., Bolognesi, A., Bortolotti, M., Farini, V., Lubelli, C., Pelosi, E., Polito, L., Dozza, B., Strocchi, P., Chambery, A., Parente, A. & Barbieri, L. (2007). Characterization of highly toxic type 2 ribosome-inactivating proteins from *Adenia lanceolata* and *Adenia stenodactyla* (Passifloraceae). *Toxicon*, Vol. 50, No. 1, (July 2007), pp. 94-105, 0041-0101

Sultan, N. A. M., Kavitha, M. & Swamy, M. J. (2009). Purification and physicochemical characterization of two galactose-specific isolectins from the seeds of *Trichosanthes cordata*. *IUBMB Life*, Vol. 61, No. 4, (April 2009), pp. 457-469, 1521-6543

Sun, J., Wang, L., Wang, B., Guo, Z., Liu, M., Jiang, K. & Luo, Z. (2007). Purification and characterisation of a natural lectin from the serum of the shrimp *Litopenaeus vannamei*. *Fish and Shellfish Immunology*, Vol. 23, No. 2, (August 2007), pp. 292-299, 1050-4648

Suseelan, K. N., Mitra, R., Pandey, R., Sainis, K. B. & Krishna, T. G. (2002). Purification and characterization of a lectin from wild sunflower (*Helianthus tuberosus* L.) tubers. *Archives of Biochemistry and Biophysics*, Vol. 407, No. 2, (November 2002), pp. 241-247, 0003-9861

Takahashi, K. G., Kuroda, T. & Muroga, K. (2008). Purification and antibacterial characterization of a novel isoform of the Manila clam lectin (MCL-4) from the plasma of the Manila clam, *Ruditapes philippinarum*. *Comparative Biochemistry and Physiology, Part B*, Vol. 150, No. 1, (May 2008), pp. 45-52, 1096-4959

Takeuchi, T., Nishiyama, K., Yamada, A., Tamura, M., Takahashi, H., Natsugari, H., Aikawa, J.-i., Kojima-Aikawa, K., Arata, Y. & Kasai, K.-i. (2011). *Caenorhabditis elegans* proteins captured by immobilized Galβ1-4Fuc disaccharide units: assignment of 3 annexins. *Carbohydrate Research*, Vol. 346, No. 13, (September 2011), pp. 1837-1841, 0008-6215

Teixeira-Sá, D. M. A., Reicher, F., Braga, R. C., Beltramini, L. M. & de Azevedo Moreira, R. (2009). Isolation of a lectin and a galactoxyloglucan from *Mucuna sloanei* seeds. *Phytochemistry*, Vol. 70, No. 17-18, (December 2009), pp. 1965-1972, 0031-9422

Teixeira, D. M. A., Braga, R. C., Horta, A. C. G., Moreira, R. A., de Brito, A. C. F., Maciel, J. S., Feitosa, J. P. A. & de Paula, R. C. M. (2007). *Spondias purpurea* Exudate polysaccharide as affinity matrix for the isolation of a galactose-binding-lectin. *Carbohydrate Polymers*, Vol. 70, No. 4, (November 2007), pp. 369-377, 0144-8617

Tetala, K. K. R., Chen, B., Visser, G. M. & van Beek, T. A. (2007). Single step synthesis of carbohydrate monolithic capillary columns for affinity chromatography of lectins. *Journal of Separation Science*, Vol. 30, No. 17, (November 2007), pp. 2828-2835, 1615-9306

Trindade, M. B., Lopes, J. L. S., Soares-Costa, A., Monteiro-Moreira, A. C., Moreira, R. A., Oliva, M. L. V. & Beltramini, L. M. (2006). Structural characterization of novel chitin-binding lectins from the genus *Artocarpus* and their antifungal activity. *Biochimica et Biophysica Acta*, Vol. 1764, No. 1, (January 2006), pp. 146-152, 1570-9639

Valadez-Vega, C., Guzmán-Partida, A. M., Soto-Cordova, F. J., Álvarez-Manilla, G., Morales-González, J. A., Madrigal-Santillán, E., Villagómez-Ibarra, J. R., Zúñiga-Pérez, C., Gutiérrez-Salinas, J. & Becerril-Flores, M. A. (2011). Purification, Biochemical Characterization, and Bioactive Properties of a Lectin Purified from the Seeds of White Tepary Bean (*Phaseolus Acutifolius* Variety Latifolius). *Molecules*, Vol. 16, No. 3, (March 2011), pp. 2561-2582, 1420-3049

Van Damme, E. J. M., Peumans, W. J., Pusztai, A. & Bardocz, S. (February 25, 1998). *Handbook of plant lectins: Properties and biomedical applications*, John Wiley & Sons, 978-047-196-445-2, New York

Varki, A., Cummings, R. D., Esko, J. D., Freeze, H. H., Stanley, P., Bertozzi, C. R., Hart, G. W. & Etzler, M. E. (Eds.). (October 15, 2008). *Essentials of glycobiology*, Cold Spring Harbor Laboratory Press, 978-087-969-770-9, New York

Vazquez, L., Massó, F., Rosas, P., Montaño, L. F. & Zenteno, E. (1993). Purification and characterization of a lectin from *Macrobrachium rosenbergh* (Crustacea, Decapoda) hemolymph. *Comparative Biochemistry and Physiology, Part B*, Vol. 105, No. 3-4, (July-August 1993), pp. 617-623, 0305-0491

Veau, B., Guillot, J., Damez, M., Dusser, M., Konska, G. & Botton, B. (1999). Purification and characterization of an anti-(A+B) specific lectin from the mushroom *Hygrophorus hypothejus*. *Biochimica et Biophysica Acta*, Vol. 1428, No. 1, (June 1999), pp. 39-44, 0304-4165

Vega, N. & Pérez, G. (2006). Isolation and characterisation of a *Salvia bogotensis* seed lectin specific for the Tn antigen. *Phytochemistry*, Vol. 67, No. 4, (February 2006), pp. 347-355, 0031-9422

Voet, D. & Voet, J. G. (January 15, 1995). *Biochemistry* (Second Edition), John Wiley & Sons, 0-471-58651-X, New York

Voss, C., Eyol, E., Frank, M., von der Lieth, C.-W. & Berger, M. R. (2006). Identification and characterization of riproximin, a new type II ribosome-inactivating protein with antineoplastic activity from *Ximenia americana*. *The FASEB Journal*, Vol. 20, No. 8, (June 2006), pp. 1194-1196, 0892-6638

Wang, H. & Ng, T. B. (2003). Isolation of a novel *N*-acetylglucosamine-specific lectin from fresh sclerotia of the edible mushroom *Pleurotus tuber-regium*. *Protein Expression and Purification*, Vol. 29, No. 2, (June 2003), pp. 156-160, 1046-5928

Wang, S., Zhong, F.-D., Zhang, Y.-J., Wu, Z.-J., Lin, Q.-Y. & Xie, L.-H. (2004). Molecular Characterization of a New Lectin from the Marine Alga *Ulva pertusa*. *Acta Biochimica et Biophysica Sinica*, Vol. 36, No. 2, (February 2004), pp. 111-117, 0582-9879

Watanabe, T., Matsuo, I., Maruyama, J.-i., Kitamoto, K. & Ito, Y. (2007). Identification and Characterization of an Intracellular Lectin, Calnexin, from *Aspergillus oryzae* Using *N*-Glycan-Conjugated Beads. *Bioscience, Biotechnology, and Biochemistry*, Vol. 71, No. 11, (November 2007), pp. 2688-2696, 0916-8451

Watanabe, Y., Shiina, N., Shinozaki, F., Yokoyama, H., Kominami, J., Nakamura-Tsuruta, S., Hirabayashi, J., Sugahara, K., Kamiya, H., Matsubara, H., Ogawa, T. & Muramoto, K. (2008). Isolation and characterization of L-rhamnose-binding lectin, which binds to microsporidian *Glugea plecoglossi*, from ayu (*Plecoglossus altivelis*) eggs. *Developmental and Comparative Immunology*, Vol. 32, No. 5, pp. 487-499, 0145-305X

Wong, J. H., Ng, T. B., Zhang, K. Y. B., Sze, S. C. W. & Wang, H. X. (2010). Isolation of a Mitogenic Agglutinin with Relatively High Thermostability from Seeds of the Variegated Shell Ginger. *Protein and Peptide Letters*, Vol. 17, No. 1, (January 2010), pp. 38-43, 0929-8665

Yan, Q., Zhu, L., Kumar, N., Jiang, Z. & Huang, L. (2010). Characterisation of a novel monomeric lectin (AML) from *Astragalus membranaceus* with anti-proliferative activity. *Food Chemistry*, Vol. 122, No. 3, (October 2010), pp. 589-595, 0308-8146

Yang, H., Luo, T., Li, F., Li, S. & Xu, X. (2007). Purification and characterisation of a calcium-independent lectin (PjLec) from the haemolymph of the shrimp *Penaeus japonicus*. *Fish and Shellfish Immunology*, Vol. 22, No. 1-2, (January-February 2007), pp. 88-97, 1050-4648

The Difference of Lectin Recovery by Sugar-Bound Resin

Mitsuru Jimbo[1], Shin Satoh[1], Hirofumi Hasegawa[1], Hiroshi Miyake[1], Takao Yoshida[2], Tadashi Maruyama[2] and Hisao Kamiya[1]
[1]*School of Marine Biosciences, Kitasato University*
[2]*Japan Agency for Marine-Earth Science and Technology*
[1,2]*Japan*

1. Introduction

Lectins are sugar binding proteins and glycoproteins, and are widely distributed from bacteria to humans. Lectins are easily detected by agglutination of erythrocytes, known as hemagglutination. Stillmark first reported in his doctoral thesis that the extract from beans of the castor tree, *Ricinus communis*, contains lectin by the method of hemagglutination (Stillmark, 1888). Since each lectin binds to a specific kind of sugar, and sugar chains on the cell surface differ according to cell type and animals species, lectins can distinguish types of cells (Landsteiner & Raubitschek, 1907). Moreover, the lectin of red kidney beans has mitogenic activity, in which leucocytes are transformed into undifferentiated cells and initiate mitosis (Nowell, 1960). Thus, the lectins not only recognize specific cell types, but also affect cell physiology.

The lectins are defined as sugar binding proteins but are excluded from sugar binding antibodies and enzymes. They are classified into many groups based on amino acid sequences. For example, galectin, and C-type lectin are popular lectins (Kasai & Hirabayashi, 1996; Drickamer, 1999), and even today, various lectins with new structures are still being found in many animals, bacteria, and fungi (Jimbo et al., 2005; Tateno et al., 1998; Jimbo et al., 2000; Sato et al., 2011).

Lectins are involved in various biological phenomena such as self-defense (ficolins) (Matsushita, 2009), differentiation (Kawaguchi et al., 1991), and mineralization (Kamiya, et al., 2002). Self-defense is a well-known function of animal lectins. Lectins in invertebrates' hemolymph recognize and inactivate infecting bacteria, viruses, and so on. We found lectin activity in the hemolymph of the deep-sea bivalve *Calyptogena okutanii* (Jimbo et al., 2009).

Purification of a lectin is usually carried out using sugar affinity chromatography, since they bind to specific sugars. The scheme of lectin purification was shown in Figure 1. When a lectin-containing extract was applied to a resin to which the specific sugar was bound, the lectin binds to the resin and other proteins do not. After washing the resin, proteins except the lectin is flowed through the resin and the lectin only remains to be bound to the resin. When the specific sugar containing buffer is applied to the resin, the lectin detaches from the resin and the purified lectin are obtained from extract.

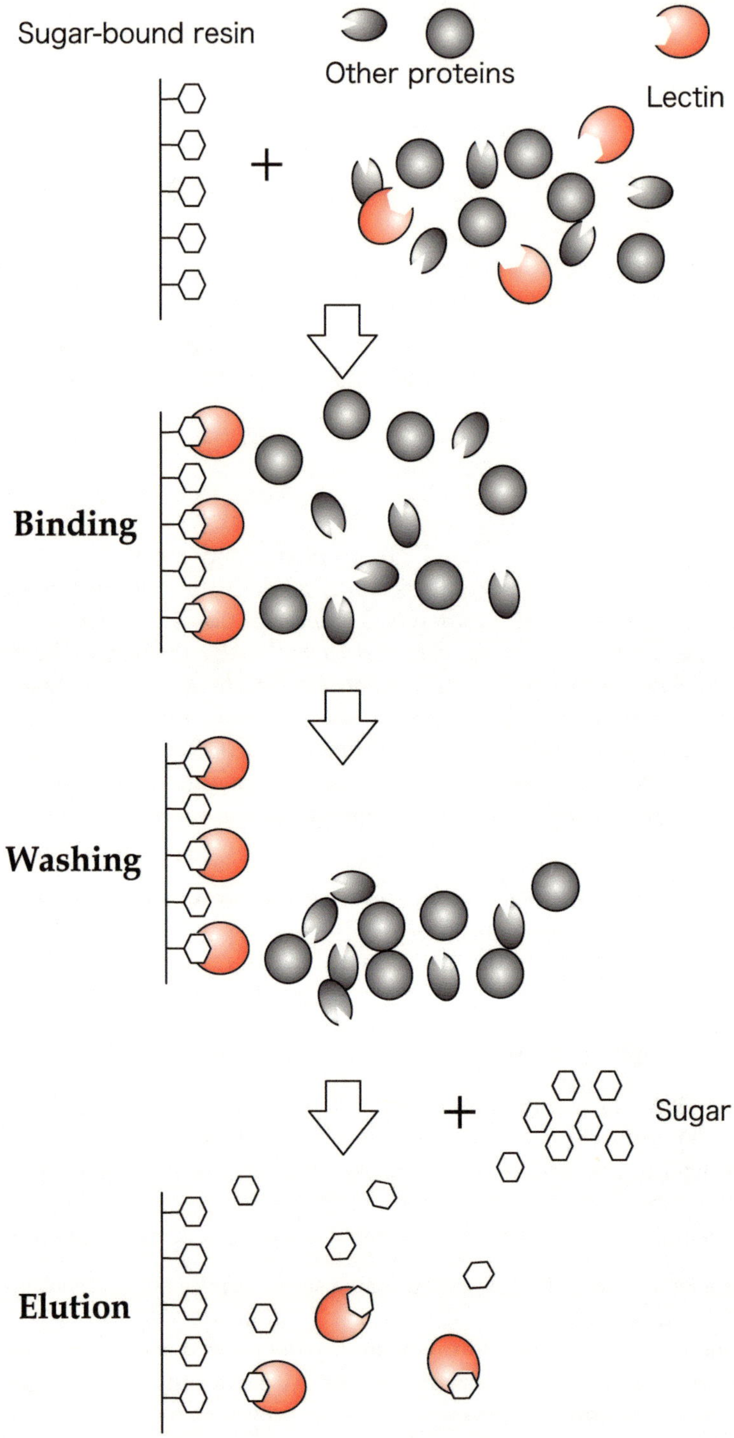

Fig. 1. A lectin purification scheme

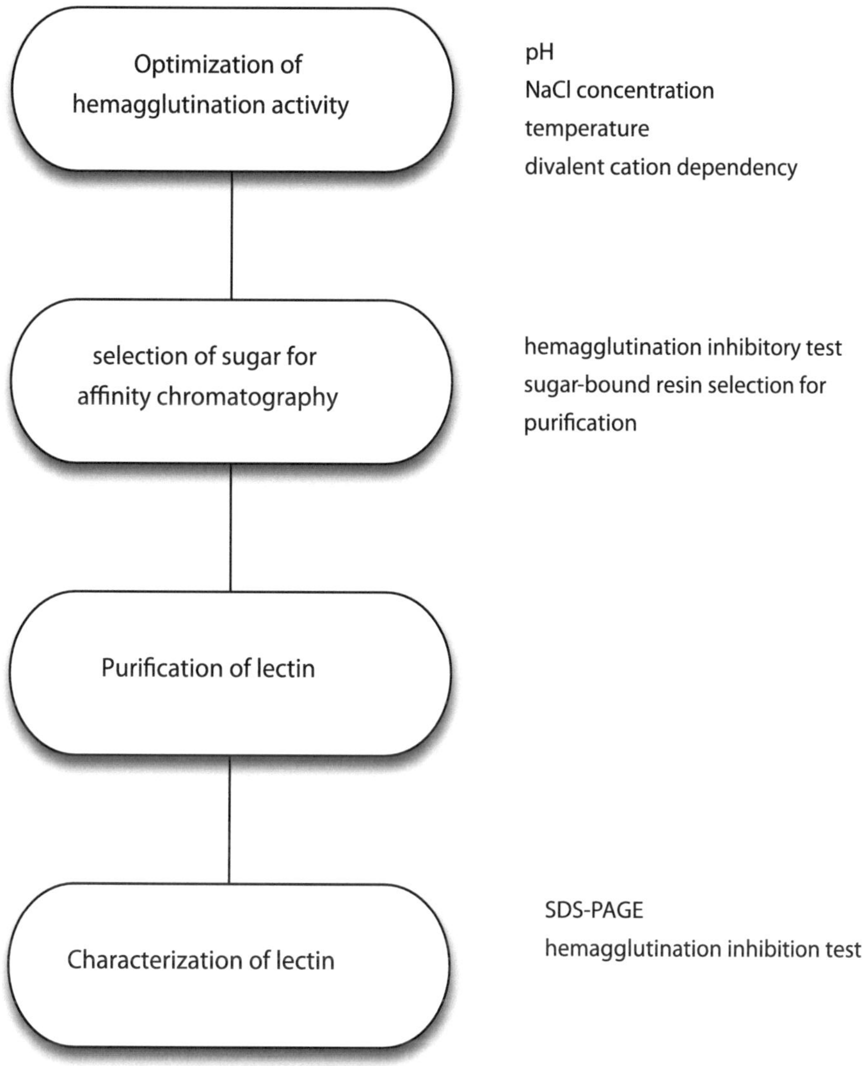

Fig. 2. A strategy of lectin purification

In the process of lectin purification, the usage of various types of resin results in drastically different lectin recovery with dissimilar properties of the purified lectin. In this chapter, we describe the optimal conditions for hemagglutination by lectin from *C. okutanii*, the employment of resin for lectin purification, and also the effects of resins on lectin recovery and the properties of the lectin (Fig. 2).

2. Collection of hemolymph of *Calyptogena okutanii*

Calyptogena okutanii belongs to the family Vesicomyidae and forms colonies at a depth of 750–2,100 m around Sagami Bay, Nankai Trough, and Okinawa Trough in Japan (Kojima & Ohta, 1997; Fujikura et al., 2000). The habitat around *C. okutanii* is inclement: low temperatures of about 2°C, lack of sunlight, and high pressure (more than 80 atm). Thus, many animals including *C. okutanii* living in this habitat are supported by chemosynthesis. *C. okutanii* harbors symbiotic sulfur-oxidizing bacteria in their gill epithelial cells. As a consequence of the symbiosis with these bacteria, their digestive tracts are degenerated. It is interesting how their proteins have adapted to a habitat of high pressure and cold temperatures. Since we found hemagglutination activity in the hemolymph of *C. okutanii*, we tried to purify it to compare the protein structure of the lectin of a deep-sea chemosynthesis-based bivalve with those in other animals.

C. okutanii was collected using the remotely operated vehicle Hyper-Dorphine of the Japan agency of Marine Science and Technology (JAMSTEC) at NT08-03 and NT08-24 operation, or the submarine Shinkai 6500 of JAMSTEC at the YK05-12 operation. The sampling point was a seep off Hatsushima, Sagami Bay, at depth of 850 m (35° 0.9441 N, 139° 13.3181 E). After collection, samples were kept at 4°C in filtrated sea water. Since *C. okutanii* and *C. soyoae* colonize together at the same place, the species of each bivalve was identified by sequence determination of the partial amplification of the cytochrome oxidase I gene (Folmeret al., 1994). The hemolymph was collected from the bivalve by cutting an adductor muscle using a scalpel, and was centrifuged at 2,000 rpm for 5 min at room temperature to remove blood cells. The supernatant was stored at -80°C before use.

3. Measurement of hemagglutination activity

Lectins are easily detected by measuring hemagglutination activity, because sugar chains of peptidoglycans, extracellular matrix, glycoproteins, or glycolipids cover the cell surface of blood cells. However, the structure of sugar chains of blood cells is different from organism to organism. In addition, the incubation temperature and buffer composition employed are known to affect the lectin activity. Thus, we examined the optimal conditions for hemagglutination activity for *C. okutanii* hemolymph using various animal erythrocytes.

3.1 Methods

4% animal erythrocyte suspensions were prepared as follows: Blood (chicken, guinea pig, goose, horse, bovine, sheep, rabbit, and human types A, B, AB, and O) was centrifuged at 2,000 rpm for 5 min at room temperature. The pellet was resuspended with 50 ml of 0.85% NaCl, and the solution was centrifuged. After washing three times, the pellet was resuspended to make a 4% erythrocyte suspension.

The hemagglutination activity was measured as follows: A sample (20 μl) was diluted by two-fold serial dilutions with dilution buffer (600 mM NaCl, 20 mM Tris-HCl, pH 8.5, 10 mM $CaCl_2$) in a 96-well titerplate. After 20 μl of 4% animal erythrocyte suspension was added to the dilution and mixed, the plate was incubated at 4°C, 25°C, or 37°C for 30 min. The hemagglutination activity (HU) was expressed as a titer defined as the reciprocal of the maximum dilution giving positive hemagglutination.

3.2 Condition determination

First, the hemagglutination activity was tested using several animal erythrocytes. As shown in Figure 3, only human erythrocyte suspensions gave high hemagglutination activities, while others gave less than one-tenth as much. The hemagglutination activity was not blood-type specific, and the differences in hemagglutination activity among blood types were small. So the hemagglutination activity of *C. okutanii* hemolymph was tested using human type A erythrocytes in following experiments.

Next, the effects of experimental conditions on hemagglutination activity were evaluated. We examined the effect of temperature on hemagglutination activity. The activity was the highest at 4°C and decreased at higher temperatures (Fig. 4A). For pH, the hemagglutination activity of *C. okutanii* was the highest from pH 7.0 to 9.0 and decreased at less than pH 7.0 or at more than pH 9.0 (Fig. 4B). The increase of NaCl concentrations resulted in higher hemagglutination activity. The activity increased up to 600 mM NaCl, did not changed at higher concentrations (Fig. 4C).

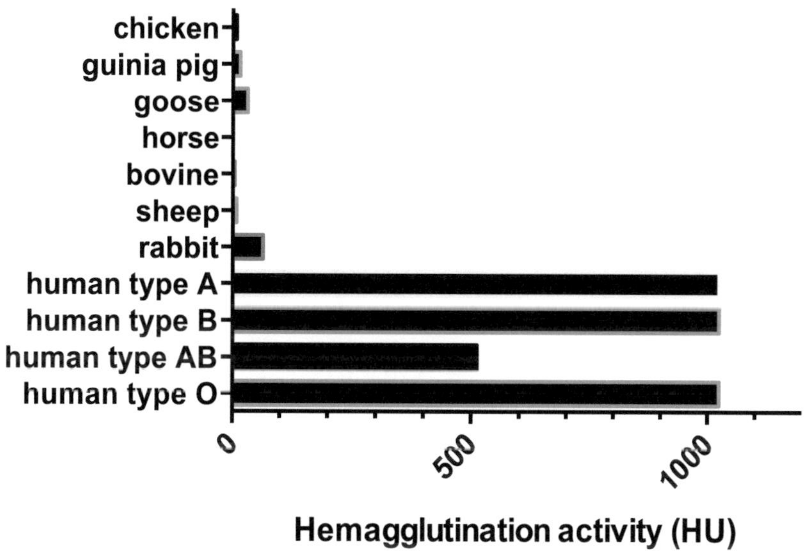

Fig. 3. Differences of hemagglutination activities of the *C. okutanii* hemolymph to various animal erythrocytes.

Divalent cations are important for the activity and structural stability of proteins, and some groups of lectins, like C-type lectins, need calcium ions for their activity (Drickamer, 1999). The hemagglutination activity of the hemolymph of *C. okutanii* also needs calcium ions. Interestingly, the hemagglutination activity showed a maximum titer at 10 mM $CaCl_2$ and rapidly decreased at higher concentrations (Fig. 4D).

The concentrations of NaCl and $CaCl_2$ in sea water are 600 mM and 10 mM, respectively. *C. okutanii* lives at a depth of 750-2,100 m, and the composition of sea water is almost constant in the range of that depth, and the temperature at that depth is below 4°C. This suggests that the lectin is adapted to the environment in which *C. okutanii* lives.

Given these results, the hemagglutination activity of *C. okutanii* hemolymph was measured at 4°C after dilution with 10 mM $CaCl_2$, 50 mM Tris-HCl, pH 8.5, 600 mM NaCl, using 4% human type A erythrocyte suspension.

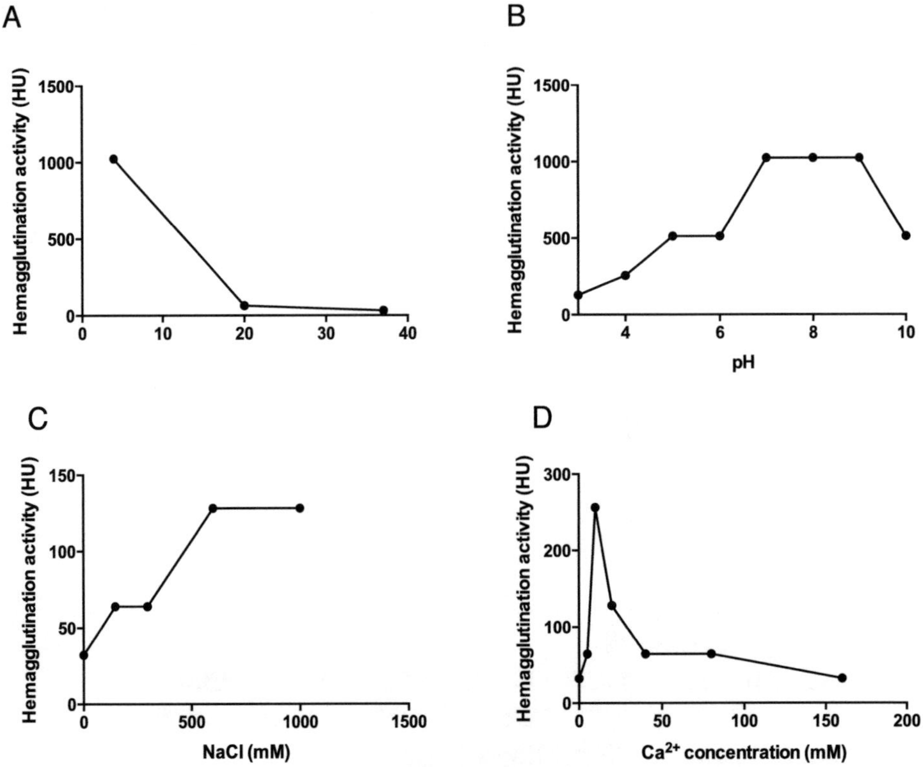

Fig. 4. Optimization of hemagglutination activity of the *C. okutanii* hemolymph.

The hemagglutination activity was measured by using different buffers for dilutions of hemolymph, or by using different incubation conditions. The hemagglutination test was conducted at 4, 25, and 37°C (A). For dilutions of hemolymph, solutions at different pHs containing a wide range buffers, glycine-acetate-MOPS pH 4, 5, 7, 8, 9, 10, or phosphate buffer at pH 6.0, were used (B). NaCl concentrations of dilution buffers for hemolymph were 0, 200, 400, 600, and 1000 mM (C). $CaCl_2$ concentrations were 0, 5, 10, 20, 40, 80, and 160 mM (D).

4. Selection of sugars for affinity chromatography

The determination of the sugar specificity of a lectin is needed to select the resin for using affinity purification. One test for determining sugar-binding specificity is the so-called hemagglutination inhibition test. A lectin is incubated with a sugar beforehand, and then the hemagglutination activity is examined. If the lectin has an affinity to the added sugar, hemagglutination is inhibited because the lectin already binds to the sugar. On the other hand, if the lectin does not have an affinity to the sugar, hemagglutination should occur. Using a variety of sugars, a sugar affinity profile of the lectin can be obtained.

4.1 Methods

10 μl of 0.2 M sugars (D-glucose, D-glucosamine, *N*-acetyl-D-glucosamine, D-galactosamine, *N*-acetyl-D-glucosamine, D-galacturonic acid, L-arabinose, D-ribose, D-fucose, D-xylose, D-

fructose, D-mannose, L-rhamnose, L-sorbose, *N*-acetyl-D-neuramic acid, maltose, lactose, or melibiose), or 0.5% of fetuin or mannnan were diluted with two-fold serial dilutions in a V-shaped 96-well titerplate. The hemagglutination titer of the hemolymph was adjusted to 16 HU by dilution with 600 mM NaCl, 10 mM $CaCl_2$, 50 mM Tris-HCl, pH 8.5. 10 μl of the diluted hemolymph was added to the V-shaped 96-well titerplate, and incubated at 4°C for 1 h. After incubation, 20 μl of 4% human type A erythrocyte suspension was mixed and incubated at 4°C for 30 min. The result was expressed as the minimum inhibitory concentration of the sugar that completely inhibited hemagglutination.

4.2 Hemagglutination inhibition of hemolymph

The hemagglutination activity of hemolymph was inhibited by *N*-acetyl neuramic acid (3.12 mM), *N*-acetyl glucosamine (6.25 mM), *N*-acetyl galactosamine (12.5 mM), and D-glucosamine (25 mM). Among the four inhibitory sugars, all except D-glucosamine share an *N*-acetyl group. It is expected that the lectin in the hemolymph can be purified using sugar-bound affinity chromatography using these sugars.

4.3 Preparation of sugar-bound resins

For lectin purification, many sugar-bound resins were used. To date, insoluble sugar polymers such as cellulose, dextran, or agarose (Ersson et al., 1973; Kamiya et al., 1988), glycoprotein-bound resin (Tunkijjanukij et al., 1997), and sugar-bound resin (Jimbo et al., 2007) have been used. To purify the hemolymph lectin, we chose to make a sugar bound resin using epoxy-activated Sepharose 6B (GE healthcare Bioscience). The resin was prepared as follows: 3.6 g of epoxy-activated Sepharose 6B was swelled with 500 ml of distilled water, and was washed with 700 ml of distilled water. After the addition of ligand solution (20 mg/ml *N*-acetyl neuramic acid, *N*-acetyl glucosamine, *N*-acetyl galactosamine, or L-fucose at pH 13), the Sepharose 6B was incubated at 45°C for 20 h with shaking. The suspension was filtrated using a glass filter, and the Sepharose 6B was collected. It was washed with 100 ml of distilled water, 100 ml of bicarbonate buffer (500 mM NaCl, 100 mM sodium bicarbonate, pH 8.0), and 100 ml of acetate buffer (500 mM NaCl, 100 mM sodium acetate, pH 4.0). The Sepharose 6B was suspended with 100 ml of 1 M monoethanolamine, and incubated at 40°C for 16 h, and then alternately washed with 100 ml of acetate buffer and borate buffer (0.5 M NaCl, 100 mM borate, pH 8.0) three times. After washing, the resin was suspended in 10 mM $CaCl_2$, 600 mM NaCl, 50 mM Tris-HCl, pH 8.5.

4.4 Lectin binding to sugar-bound Sepharose 6B

To confirm whether the hemolymph lectin was bound to the prepared sugar-bound Sepharose 6B, we examined whether the lectin was bound to the resin as follows. The sugar-bound resin (100 μl) was transferred to a 1.5 ml microtube and then washed with 1 ml of the binding buffer (10 mM $CaCl_2$, 600 mM NaCl, 50 mM Tris-HCl, pH 8.5) three times. 100 μl of the hemolymph was added to the washed sugar-bound resin, and incubated on ice for 1 h with occasional mixing. The tube was centrifuged at 5,000 rpm, at 4°C for 1 min, and the supernatant was transferred to a new tube. The hemagglutination activity of the supernatant was determined.

After the hemolymph of *C. okutanii* was incubated with *N*-acetyl neuramic acid-bound Sepharose 6B, *N*-acetyl glucosamine-bound sepharose 6B, or *N*-acetyl galactosamine-bound

Sepharose 6B, the supernatant of the mixtures had less than 10% of the hemagglutination activity of the hemolymph before incubation (Fig. 5). Bacause more than 90% of the lectin in the hemolymph was estimated to be bound to these sugar-bound resins, these resins can be used for the purification of hemolymph lectin. We selected *N*-acetyl neuramic acid- and *N*-acetyl glucosamine-bound Sepharose 6B as candidate resins for affinity purification because of their effective lectin binding.

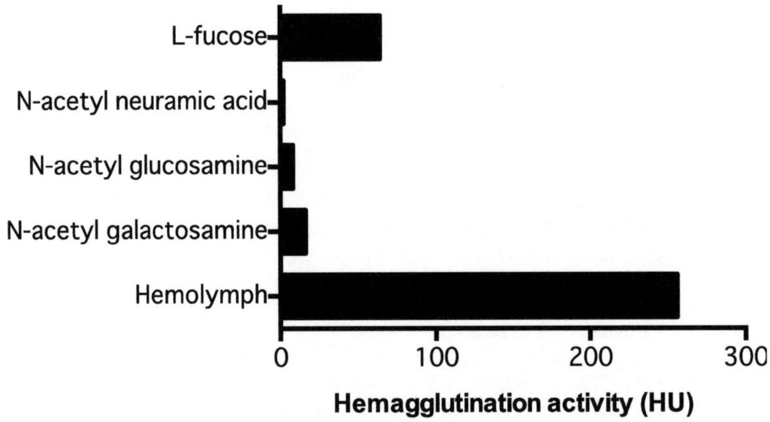

Fig. 5. Hemagglutination activity adsorption to sugar-bound resin.

The hemolymph was mixed with sugar-bound resin, and then the hemagglutination activity of the supernatant was measured. L-fucose, *N*-acetyl neuramic acid, *N*-acetyl glucosamine, and *N*-acetyl galactosamine indicate L-fucose-, *N*-acetyl neuramic acid-, *N*-acetyl glucosamine-, and *N*-acetyl galactosamine-bound Sepharose 6B, respectively. Hemolymph indicates that the hemagglutination activity was directly measured.

5. Affinity purification of the hemolymph lectin by using sugar-bound resin

The optimal condition for hemagglutination activity as described in Section 3 would be also suitable for affinity purification. The hemagglutination inhibition test and hemagglutination binding test to sugar-bound resin showed that the lectin in the hemolymph of *C. okutanii* can be purified by using *N*-acetyl neuramic acid- or *N*-acetyl-glucosamine-bound Sephrose 6B. In this section, we describe how the lectin was purified using these resins, and compare the activity recovery or property of each purified lectin.

5.1 Methods

5 ml of *N*-acetyl neuramic acid-bound resin was equilibrated with binding buffer (10 mM $CaCl_2$, 600 mM NaCl, 50 mM Tris-HCl, pH 8.5), and then 10 ml of the hemolymph was applied to the resin. After washing with binding buffer, the bound lectin was eluted by elution buffer (40 mM *N*-acetyl neuramic acid, 10 mM $CaCl_2$, 600 mM NaCl, 50 mM Tris-HCl, pH 8.5). In the case of using *N*-acetyl glucosamine-bound sepharose 6B, the bound lectin was eluted by elution buffer containing 0.2 M *N*-acetyl glucosamine instead of 40 mM *N*-acetyl neuramic acid. The eluate was dialyzed against the binding buffer overnight.

5.2 *N*-acetyl neuramic acid-bound Sepharose 6B

The chromatogram is shown in Figure 6. The hemagglutination activity was completely adsorbed to the *N*-acetyl neuramic acid-bound Sepharose 6B, and a single peak of the hemagglutination activity was eluted by the addition of 40 mM *N*-acetyl neuramic acid in the binding buffer. After dialysis against the binding buffer overnight, the recovery was only about 4% in terms of hemagglutination activity (Table 1). Prolonged dialysis decreased hemagglutination activity of the lectin. No other protein was eluted by 8 M urea, indicating that all the lectin bound to the resin was eluted by *N*-acetyl neuramic acid. Although EDTA, which is a calcium chelator, or ethyleneglycol was used for elution of the lectin to improve the recovery of hemagglutination activity, the eluates by them showed no activity.

The purified lectin was analyzed by SDS-PAGE (Figure 7). Only one component with an apparent molecular mass of 42.5 kDa was observed under reducing conditions, while three components with apparent molecular masses of 235, 208, and 114 kDa, respectively, were found under non-reducing conditions. This indicated that the lectin was composed of 6 or 3 subunits. The amino terminal amino acid sequence of the component was ENAXXIINIQCGYGAGCGAA. The lectin eluted with *N*-acetyl neuramic acid was named as COL-N.

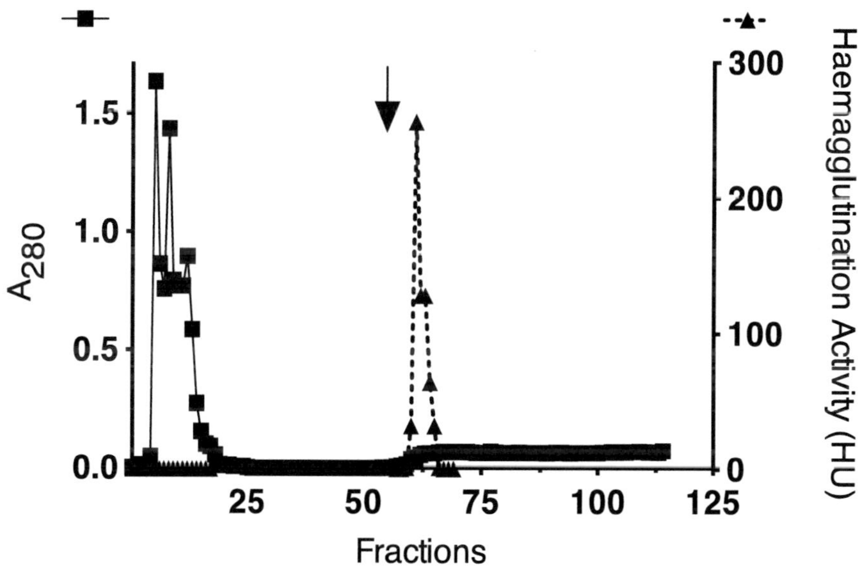

Fig. 6. Affinity purification using *N*-acetyl neuramic acid-bound resin.

The arrow indicated the addition of elution buffer. Closed squares and closed triangles indicate absorbance at 280 nm and hemagglutination activity, respectively.

	Protein (mg)	Total activity (total HU)	Recovery (%)	Specific activity (HU/mg protein)	Purification (Fold)
Hemolymph	222.3	307.2		1.38	1
COL-N	0.0624	12.2	3.96	196	142

Table 1. Purification of COL-N

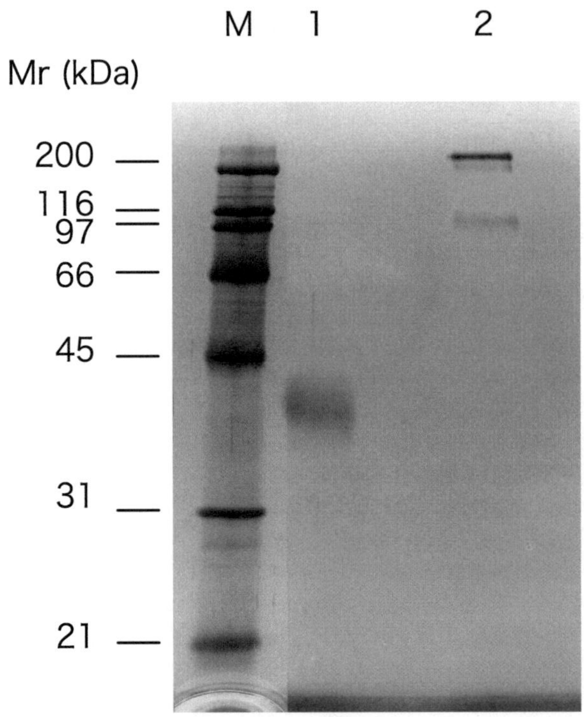

Fig. 7. SDS-PAGE of purified lectin

The purified lectin was analyzed by SDS-PAGE under reducing conditions (lane 1) or non-reducing conditions (lane 2). M indicates molecular standard proteins.

5.3 *N*-acetyl glucosamine-bound Sepharose 6B

When the lectin was eluted with *N*-acetyl glucosamine-bound resin, the hemagglutination activity was eluted by *N*-acetyl glucosamine. The chromatogram of this purification was similar to that using *N*-acetyl neuramic acid, but the recovery of hemagglutination activity using *N*-acetyl glucosamine-bound Sepharose 6B was much higher (Table 2). SDS-PAGE showed that the eluted peak was a single component with an apparent molecular mass of 42.5 kDa, similar to COL-N. Moreover, the amino terminal amino acid sequence of the lectin was identical to that of COL-N. So the lectin purified with *N*-acetyl glucosamine was named as COL-G. These results indicated that COL-N and COL-G were identical. However, the recovery of COL-G activity was drastically different. The recovery of hemagglutination activity of COL-G was as high as 85% and was about 24 times higher than that of COL-N. Moreover, the specific activity of COL-G was also higher than that of COL-N.

	Protein (mg)	Total activity (total HU)	Recovery (%)	Specific activity (HU/mg protein)	Purification (Fold)
Hemolymph	518	443		0.86	1
COL-G	1.24	384	85	310	360

Table 2. Purification table of COL-G

5.4 Sugar Inhibition profiles of COL-N and COL-G

We characterized the properties of COL-N and COL-G. Their optimal hemagglutination activities were the same as those of the hemolymph with regard to pH, Ca^{2+} concentration, and NaCl concentration. However, the hemagglutination inhibition profiles of these lectins were different (Table 3). The hemagglutination activity of COL-G was inhibited by D-glucosamine, *N*-acetyl glucosamine, *N*-acetyl galactosamine, and *N*-acetyl neuramic acid. The minimum inhibitory concentrations of these sugars to the hemagglutination activity of COL-G were identical to that of the hemolymph. On the other hand, the hemagglutination activity of COL-N was inhibited by D-glucosamine, *N*-acetyl glucosamine, and *N*-acetyl galactosamine like COL-G and the hemolymph of *C. okutanii*. However, the minimum inhibitory concentrations of these sugars to COL-N were two or four times higher than those to COL-G and the hemolymph. Moreover, the hemagglutination activity of COL-N was not inhibited by *N*-acetyl neuramic acid, even at a concentration of less than 100 mM, although the minimum inhibitory concentration of *N*-acetyl neuramic acid to the hemolymph was 3.12 mM.

Sugars	Minimum inhibitory concentration (mM)		
	hemolymph	COL-G	COL-N
D-glucose	-*	-	-
D-glucosamine	25	25	100
N-acetyl-D-glucosamine	12.5	12.5	25
D-galactose	-	-	-
D-galactosamine	-	-	-
N-acetyl-D-galactosamine	6.25	6.25	25
D-galacturonic acid	-	-	-
L-arabinose	-	-	-
D-ribose	-	-	-
D-fucose	-	-	-
D-xylose	-	-	-
D-fructose	-	-	-
D-mannose	-	-	-
L-rhamnose	-	-	-
L-sorbose	-	-	-
N-acetyl-D-neuramic acid	3.12	3.12	-
Maltose	-	-	-
Sucrose	-	-	-
Lactose	-	-	-
Melibiose	-	-	-
Fetuin	-	-	-
mannan	-	-	-

* Hemagglutination activity was not inhibited at less than 100 mM.

Table 3. Sugar Inhibition Test

Our hypothesis to explain these results is that the lectin COL-N tightly binds to *N*-acetyl neuramic acid. In affinity chromatography, lectins are adsorbed to a sugar-bound resin, and then eluted by a sugar that is identical to the sugar bound to the resin. Thus, the lectin forms a complex with the sugar. The sugar bound to the lectin is usually removed by dialysis, but this is difficult if the lectin is strongly bound to the sugar. COL-N was purified with *N*-acetyl neuramic acid–bound Sepharose 6B, which was the strongest inhibitor of the hemagglutination activity of the hemolymph among the sugars tested. Thus, it is possible that COL-N remains bound to *N*-acetyl neuramic acid and that overnight dialysis does not remove it. It is also possible that some of the lectins in the hemolymph that have affinity to *N*-acetyl neuramic acid remain bound to the resin after elution. But no protein was eluted by 8 M urea. It thus seems that no part of lectin bound to *N*-acetyl neuramic acid remained to be adsorbed to the resin.

If COL-N binds to *N*-acetyl neuramic acid, why did COL-N show hemagglutination activity in our experiments? One hypothesis is that there are two subunits that have different sugar binding affinities. As shown above, *C. okutanii* lectin is composed of 3 or 6 subunits. When the *C. okutanii* lectin was treated with glycopeptidase F, which removes sugar chains from N-glycosylated proteins, *C. okutanii* lectin became two components with apparent molecular masses of 32.7 and 34.6 kDa (data not shown), suggesting that *C. okutanii* lectin is composed of two different subunits. Since the amino terminal amino acid sequences of the two components were identical over 20 amino acid residues, each subunit must be closely related.

There are lectins that have a high similarity to each other, but their sugar binding profiles are different. Ficolins are serum lectins and are components of the innate immune system. Ficolin homologues are widely distributed in the animal kingdom (Lu & Le, 1998). Mice have two ficolins, ficolin A and ficolin B (Endo et al., 2005). The amino acid sequences of these lectins have high identity to each other (69%), but their sugar binding affinities are different. Ficolin A binds to *N*-acetyl glucosamine and *N*-acetyl galactosamine, while ficolin B binds to *N*-acetyl glucosamine, *N*-acetyl galactosamine, and *N*-acetyl neuramic acid. The sugar binding properties of the *C. okutanii* lectin subunits may be similar to those of ficolins. If we hypothesize that *C. okutanii* lectin is composed of two subunits like mouse ficolins, and that one has high affinity to *N*-acetyl neuramic acid while the other does not, the fact that COL-N has no affinity to *N*-acetyl neuramic acid can be explained as follows. In the case of COL-N, one subunit of the *C. okutanii* lectin is occupied by *N*-acetyl neuramic acid, while the other can still bind to *N*-acetyl glucosamine and *N*-acetyl galactosamine. The low recovery of the hemagglutination activity of COL-N is explainable assuming that COL-N is a complex with *N*-acetyl neuramic acid. If COL-N binds to *N*-acetyl neuramic acid, the total binding sites of COL-N should decrease, and the hemagglutination activity should also decrease.

6. Conclusion

For the purification of *C. okutanii* hemolymph lectin, *N*-acetyl neuramic acid-bound Sepharose 6B was poorly effective, and the properties of *C. okutanii* lectin were different from those of native lectin. When we used *N*-acetyl glucosamine-bound resin, the recovery of the lectin activity was 85%, and the sugar-binding property of the purified lectin was identical to that of the hemolymph. When a sugar that strongly binds to a lectin is used as an affinity resin, it is possible that the lectin is not well recovered. On the other hand, when a sugar that binds weakly to the lectin is used, the lectin bound to the resin is small and much of it is unbound, resulting in low recovery. To purify lectin effectively with its natural properties, careful selection of sugar-bound resin is important.

Moreover, this method is also applied to the others purified by affinity chromatography. To purify enzymes or cells with native activity, they can be purified by the affinity chromatography and it is possible that the activity of something purified is different from that of crude extract. Using a different resin for the purification, something purified with native activity can be obtained.

7. Acknowledgments

We thank the operating teams of the remotely operating vehicle Hyper-dorphine and the submarsible Shinkai 6500, and the crews of the research vessels Natsushima and Yokosuka that were used to collect *C. okutanii* specimens.

8. References

Drickamer, K. (1999). C-type lectin-like domains *Current opinion in structural biology*, Vol. 9, No.5, pp. 585–590

Endo, Y., Nakazawa, N., Liu, Y., Iwaki, D., Takahashi, M., Fujita, T., Nakata, M., et al. (2005). Carbohydrate-binding specificities of mouse ficolin A, a splicing variant of ficolin A and ficolin B and their complex formation with MASP-2 and sMAP *Immunogenetics*, Vol. 57, No. 11, pp. 837–844.

Ersson, B., Aspberg, K., & Porath, J. (1973). The phytohemagglutinin from sunn hemp seeds (*Crotalaria juncea*). Purification by biospecific affinity chromatography. *Biochimica et biophysica acta*, Vol. 310, No. 2, pp. 446–452.

Folmer, O., Black, M., Hoeh, W., Lutz, R., & Vrijenhoek, R. (1994). DNA primers for amplification of mitochondrial cytochrome c oxidase subunit I from diverse metazoan invertebrates. *Molecular marine biology and biotechnology*, Vol. 3, No. 5, pp. 294–299.

Fujikura, K., Kojima, S, & Fujiwara, Y. (2000). New Distribution Records of Vesicomyid Bivalves from Deep-sea Chemosynthesis-based Communities in Japanese Waters. *Japanese Journal of Malacology*, Vol. 59, No. 2, pp. 103–121.

Jimbo, M., Koike, K., Sakai, R., Muramoto, K, & Kamiya, H. (2005). Cloning and characterization of a lectin from the octocoral *Sinularia lochmodes. Biochemical and biophysical research communications*, Vol. 330, No. 1, 157–162.

Jimbo, M., Usui, R., Sakai, Ryuichi, Muramoto, Koji, & Kamiya, Hisao. (2007). Purification, cloning and characterization of egg lectins from the teleost *Tribolodon brandti. Comparative Biochemistry and Physiology Part B: Biochemistry and Molecular Biology*, Vol. 147, No. 2, pp. 164–171.

Jimbo, M, Yamaguchi, M., Muramoto, K, & Kamiya, H. (2000). Cloning of the *Microcystis aeruginosa* M228 lectin (MAL) gene. *Biochemical and biophysical research communications*, Vol. 273, No. 2, pp. 499–504.

Jimbo, M., Yoshida, T., Nogi Y. (2009) Characterization of Calyptogena okutanii and Bathymodiolus japonicas lectin, In: *Cruise Report NT08-24*. 10, Oct, 2011, Available from: http://docsrv.godac.jp/MSV2_DATA/23/NT08-24_all.pdf

Kamiya, H, Jimbo, M, Yako, H., Muramoto, K, & Nakamura, O. (2002). Participation of the C-type hemolymph lectin in mineralization of the acorn barnacle *Megabalanus rosa. Marine Biology*, Vol. 140, No. 6, pp. 1234–1240.

Kamiya, H, Muramoto, K, & Goto, R. (1988). Purification and properties of agglutinins from conger eel, *Conger Myriaster* (Brevoort), skin mucus. *Developmental and comparative immunology*, Vol. 12, No. 2, pp. 309-318.

Kasai, K., & Hirabayashi, J. (1996). Galectins: a family of animal lectins that decipher glycocodes. *Journal of biochemistry*, Vol. 119, No. 1, pp. 1-8.

Kawaguchi, N., Kawaguchi, N., Komano, H., Komano, H., Natori, S., & Natori, S. (1991). Involvement of Sarcophaga lectin in the development of imaginal discs of *Sarcophaga peregrina* in an autocrine manner. Developmental Biology, Vol. 144, No. 1, pp. 86–93.

Kojima, Shigeaki, & Ohta, S. (1997). *Calyptogena okutanii* n. sp., a Sibling Species of Calyptogena soyoae Okutani, 1957 (Bivalvia : Vesicomyidae). *The Japanese journal of malacology. Venus*, Vol. 56, No. 3, pp. 189–195.

Landsteiner, K., & Raubitschek, H. (1907). Beobachtungen uber Hämolyse und Hämagglutination. *Zbl. Bakt. I. Abt. Orig.*, pp. 600–607.

Lu, J., & Le, Y. (1998). Ficolins and the fibrinogen-like domain. *Immunobiology*, Vol. 199, No. 2, pp. 190–199.

Matsushita, M. (2009). Ficolins: complement-activating lectins involved in innate immunity. *Journal of innate immunity*, Vol. 2, No. 1, pp. 24–32.

Nowell, P. C. (1960). Phytohemagglutinin: an initiator of mitosis in cultures of normal human leukocytes. *Cancer research*, Vol. 20, pp. 462–466.

Sato, Y., Hirayama, M., Morimoto, K., Yamamoto, N., Okuyama, S., & Hori, K. (2011). High mannose-binding lectin with preference for the cluster of alpha1-2-mannose from the green alga *Boodlea coacta* is a potent entry inhibitor of HIV-1 and influenza viruses. *The Journal of biological chemistry*, Vol. 286, No. 22, pp. 19446-19458, ISSN 1083-351X

Stillmark, P. H. (1988). *Über Ricin, ein giftiges Ferment aus den Samen von Ricinus comm. L. und einigen anderen Euphorbiaceen*. (R. Kobert, Ed.). University of Dorpat, Estonia.

Tateno, H., Saneyoshi, A., Ogawa, T., Muramoto, K., Kamiya, H., & Saneyoshi, M. (1998). Isolation and characterization of rhamnose-binding lectins from eggs of steelhead trout (*Oncorhynchus mykiss*) homologous to low density lipoprotein receptor superfamily. *The Journal of biological chemistry*, Vol. 273, No 30, pp. 19190–19197.

Tunkijjanukij, S., Mikkelsen, H. V., & Olafsen, J. A. (1997). A heterogeneous sialic acid-binding lectin with affinity for bacterial LPS from horse mussel (*Modiolus modiolus*) hemolymph. *Comparative Biochemistry and Physiology Part B: Biochemistry and Molecular Biology*, Vol. 117, No. 2, pp. 273–286.

Part 3

Immunoaffinity Purification

5

Immunoaffinity Chromatography: A Review

Daad A. Abi-Ghanem and Luc R. Berghman
Texas A&M University
USA

1. Introduction

Affinity chromatography is a high resolution, high capacity, and one of the most powerful and diverse methods for separating proteins and other biological molecules of interest on the basis of a highly specific, reversible biological interaction between two molecules: an affinity ligand attached to a solid matrix to create a stationary phase, and a target molecule in a mobile phase. Specifically, immunoaffinity chromatography (IAC) relies on a solid stationary phase consisting of an antibody coupled to a chromatographic matrix or to magnetic beads, and harnesses the selective and strong binding of antibodies to their targets (Hage, 1998). Accordingly, any molecule that can be bound effectively by an antibody can be purified using IAC (Lesney, 2003). Purified antibodies are coupled to the inert solid phase and mixed with the antigen solution under conditions that favor adsorption. Following antigen capture, unwanted antigens are removed by washing, and the purified antigen is released by switching to conditions that favor desorption. Purification (often greater than 1000-fold) and simultaneous concentration of the target protein are thus achieved (Fitzgerald et al., 2011). One of the first uses of IAC was reported in 1951 by Campbell et al. who used immobilized bovine serum albumin on *p*-aminobenzyl cellulose to purify anti-albumin antibodies. Since then, there has been a great expansion in the applications of IAC for analytical, clinical, and diagnostic purposes.

2. Basic components

2.1 Antibodies

2.1.1 Antibody structure

The typical Y-shaped structure of an IgG molecule consists of two identical heavy (H) and two identical light (L) chains (50 and 25 kDa each, respectively), linked by disulfide bonds (Fig.1). All four chains consist of constant (C) and variable (V) domains. The lower part of the molecule, called the Fc region, is highly conserved between antibody classes, and mediates effector functions of antibodies. The upper arms of the antibody are referred to as the Fab regions. The V regions of both heavy and light chains combine to form two identical antigen binding sites. Within each V domain, amino acid sequence variation, and hence antigen recognition, is predominantly focused around three "hypervariable" regions. These residues are referred to as "complementarity determining regions" (CDRs). CDRs from the variable heavy and variable light chain domains are juxtaposed to create the antigen binding site that recognizes the antigenic epitope, a specific location on the antigen (Elgert, 1996).

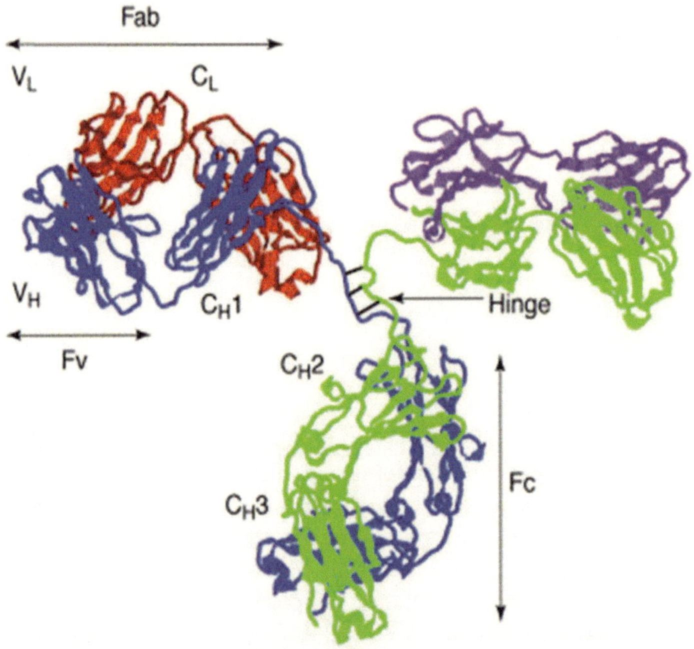

Fig. 1. Structure of an IgG molecule (Modified from Little et al., 2000).

Immunoaffinity chromatography relies on the exquisite binding between an antibody and an antigen, the result of four different types of non-covalent (and therefore reversible) interactions: ionic interactions, hydrogen bonds, van der Waals interactions, and hydrophobic interactions (Harlow and Lane, 1999b; Fitzgerald et al., 2011). Our ability to manipulate antibodies and antibody-antigen interactions offers great potential for the use of IAC in research as well as for therapeutic and diagnostic applications. Moreover, the advent of recombinant antibody production has paved the way for even more advances in manipulating antibodies to our advantage.

2.1.2 Choice of antibody

The primary isolation of specific antibodies is necessary for the subsequent purification of specific antigens. Antibodies used as ligands can be purified by precipitation with dextran or ammonium sulfate, or by isolation on a Protein A, Protein G, or Protein L column. The ideal antibody for use in immunoaffinity chromatography should possess two properties: (a) High intrinsic affinity, since an antibody attached to a solid phase has no room for cooperative binding. This is especially important when using a diluted antigen source, where quantitative antigen capture is hard to achieve. Quantitative binding of antigen to the immunoadsorbent along with a low background (non-specific interactions) are insured when using an antibody with an affinity $\geq 10^8$ and two hours of antigen-antibody contact. When the antibody affinity is $\leq 10^6$, some antigen will be left in solution, and exposure to the antibody column will have to be repeated; and (b) Ease of elution: This depends on the type and number of antigen-antibody bonds: the fewer types of interactions involved, the easier all of them can be destabilized (Harlow and Lane, 1999c).

Polyclonal antibodies (Pabs) are produced by multiple B-cell clones, and as a population can recognize and bind with varying affinities to a variety of independent epitopes on a single antigen (Fig.2) (Michnick and Sidhu, 2008). In a typical antiserum, only 5% of the Pabs are target-specific (Harlow and Lane, 1999a). The presence of several antibodies directed at different epitopes of the same antigen makes elution difficult and may damage the chromatography column and denature the antigen.

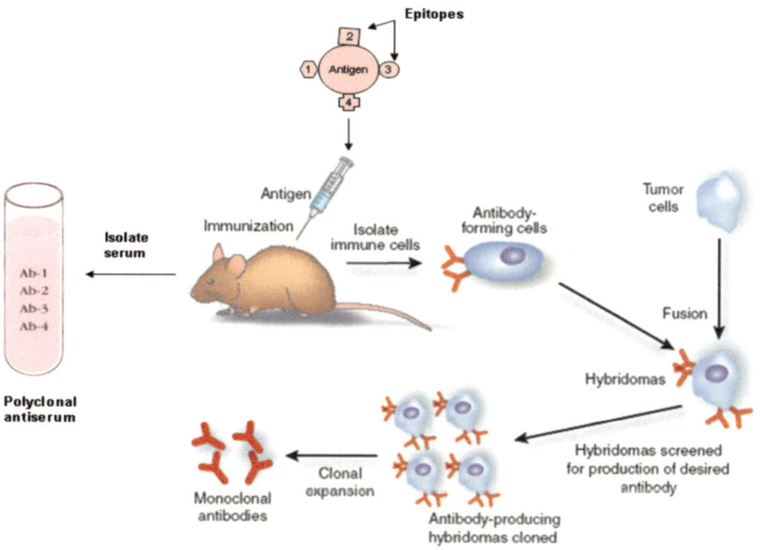

Fig. 2. Overview of poly- and monoclonal antibody production (Modified from Michnick and Sidhu, 2008, and Kuby, 1992)

Pabs are commonly obtained from sera of immunized animals, and are thus available in limited supply; even when a Pab proves suitable for use in affinity chromatography, it is often difficult to obtain multiple lots with consistent quality. To avoid these problems, Pabs can be purified by affinity chromatography over a column of antigen to obtain antigen-specific antibodies, as well as the elution profile of the antigen-antibody interaction, the same conditions of which can then be used to purify the antigen from a crude source. This is seldom practically achievable because it assumes that antigen is available for the purification of antibodies, which are then used to purify the antigen. This may seem like a "Catch-22", but this technology is often used when the host is immunized against a synthetic peptide (conjugated with a carrier protein) that mimics a B-cell epitope on a larger protein. In such a case, the peptide is commercially available in milligram quantities and will allow the isolation of an antigen-specific antibody population that can be used for IAC.

Monospecific Polyclonal Antibodies can alternatively be generated using recombinantly produced (typically) human protein fragments known as protein epitope signature tags (PrESTs). PrESTs are 100–150 amino acid fragments that are selected based on their relative low homology to other proteins in the human proteome (Agaton et al., 2004), thus minimizing cross-reactivity by the generated antibodies (Lindskog et al., 2005). The size of PrESTs is selected to be small enough for easy PCR handling and cloning, and large enough to provide conformational epitopes. PrEST selection also avoids certain restriction enzyme sites, transmembrane regions (which are poorly expressed in *Escherichia coli*), and signal peptides that are cleaved off during translocation in *E. coli* (Lindskog et al., 2005). PrESTs are expressed as fusion proteins to an albumin-binding protein and to a His_6 tag. The former functions as a "carrier protein" and confers an increased immune response (Libon et al., 1999), while the latter facilitates purification under denaturing conditions from *E. coli* inclusion bodies (Crowe et al., 1994). Following PrEST selection, expression, and purification, mass spectrometry and SDS-PAGE analysis are used to verify sequence accuracy and to provide protein purity analysis, respectively. The bicinchoninic acid assay is then used for determination of protein concentration of the purified PrEST antigens (Gunneras et al., 2008). PrEST-specific polyclonal antibodies are then obtained by immunizing animals with the purified PrEST proteins, and are in turn purified by using the PrEST proteins as affinity chromatography ligands. PrEST-specific polyclonal antibodies are extremely useful for expression and localization studies in both normal and diseased tissue using tissue microarrays (Larsson et al., 2006; Kampf et al., 2004).

A monoclonal antibody (Mab) is the product of a single immortal hybridoma cell line (a clone), and is thus available in unlimited supply. Mabs possess exquisite, well-defined specificity to a single epitope, and constitute a homogeneous binder population (Fig.2). High-affinity Mabs can bind to a large proportion of antigen. All antibodies bind to the same epitope, making elution conditions easy and gentle. The use of a pool of different Mabs is not recommended, as different epitopes on the antigen will be recognized, making desorption difficult and hence possibly denaturing the antigen and damaging the antibody column (Harlow and Lane, 1999c). Note that it is not necessary to have pure antigen to produce a monospecific Mab (Gustafsson, 1990). Therefore, a seemingly paradoxical approach becomes feasible: first making the specific Mab and then using it to isolate the corresponding antigen from the immunogen mix afterwards.

Recombinant antibodies and antibody fragments are produced *in vitro* by antibody phage display, bypassing the need to immortalize immune B-cells, as antibody genes are immortalized instead (Winter and Milstein, 1991). Antibodies can be produced as Fab, $F(ab')_2$ (two Fab units and the hinge region), single-chain antibody fragments (scFv), and diabodies (a dimeric scFv) (Rader and Barbas, 1997) (Fig.3). An scFv fragment is the smallest Ig fragment (one-sixth of a complete Ab) containing the whole antigen-binding site (Yokota et al., 1992). Following the cloning of the genes encoding the antibody heavy and light gene fragments, a large antibody repertoire can be constructed. Because heavy and light chains are combined randomly, each phage has the potential to display on its surface a unique antibody with a specific antigen-binding site (Rader and Barbas, 1997; Pini and Bracci, 2000). The genetic information encoding the displayed molecule is contained within the phage coat, thus providing a direct physical link between genotype and phenotype (Rader and Barbas, 1997). This linkage endows the protein with the two key characteristics of molecular evolution: replicability and mutability (Smith and Petrenko, 1997): It allows the selection, amplification, and manipulation of a specific clone from pools of millions. Moreover, the amino acid sequence of a selected phage can be deduced by deciphering the DNA sequence within (Barbas and Wagner, 1995). Because of the direct physical linkage between the DNA genotype and the antibody phenotype, recombinant antibodies are also easily optimized, and are amenable to fusions with proteins and peptides (drugs, toxins…) (Little et al., 2000;

Azzazy and Highsmith, 2002). Peptide tags can easily be introduced into recombinant antibodies, greatly facilitating purification and detection (Andris-Widhopf et al., 2000). Using phage display, antibodies can be expressed in *E. coli*, yeast, plants (plantibodies) against virtually any antigen, including conserved antigens, non-immunogenic molecules, and toxic molecules (Little et al., 2000; Hoogenboom and Chames, 2000). In addition, recombinant human or humanized antibodies circumvent the human response elicited by murine mAbs (Maynard and Georgiou, 2000). Regarding IAC, recombinant antibodies have the same advantages as monoclonal antibodies, i.e. monospecificity against a single epitope.

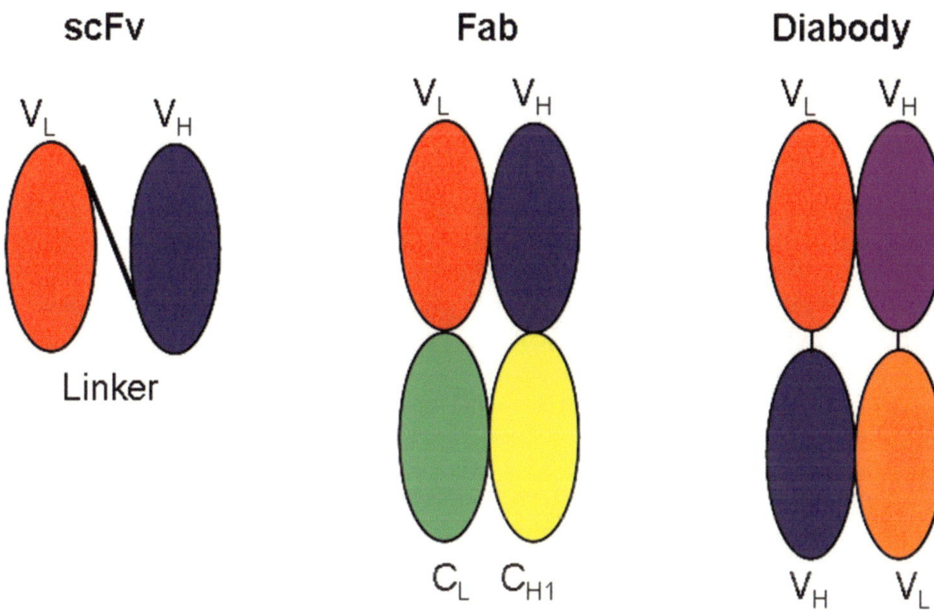

Fig. 3. Structure of scFv, Fab, and diabody antibody fragments.

The use of avian antibodies, IgY, has several major advantages. Chickens are phylogenetically very distant from mammals, and hence can be used to produce antibodies against highly conserved or weakly immunogenic mammalian epitopes (Jensenius et al., 1981). Avian antibodies are most commonly produced in eggs. Because chickens are inexpensive to maintain and a high yield of antibody can be obtained from the eggs, they represent a relatively inexpensive source of antibodies (Berghman et al., 2005). Eggs are more easily collected than blood samples, and a few eggs per week can provide the same amount of immunoglobulin as repeated bleeding of an immunized rabbit (Chui et al., 2004). However, for IAC, avian antibodies have the same drawbacks as mentioned above for Pabs, in that they represent a mix of specificities and affinities, unless they can be first purified against the antigen.

2.2 Solid matrix

The immunoaffinity matrix onto which the antibody ligand will be attached should be inexpensive, readily available, easy to use, and highly stable: the support material and the attached ligand should not react with the solvents used in the purification process, and should be resistant to degradation or damage by enzymes and microbes that might be present in the sample. It should also be able to withstand physical stress (*i.e.* pressure), especially when packed into a column, should remain intact throughout the purification process, and be easily regenerated under relatively harsh conditions (Urh et al., 2009). The IAC matrix should be easily modified for antibody attachment, and should be macroporous with uniform particle and pore size and good flow properties (Urh et al., 2009). A compromise should be achieved between pore size and surface area, as supports with small pore size have a large surface area, much of which may not be available for immobilization of antibody (Gustavsson and Larsson, 2006). In contrast, large pore support systems do not have accessibility problems, but may result in a low level of antibody attachment due to the small surface area. Supports with pore sizes of 300-500Å, which is approximately three to five times the diameter of an antibody, allow for maximum antibody coverage, as well as for suitable binding of immobilized antibodies to many small or medium sized targets (100-150 kDa) (Clarke et al., 2000).

Conventional matrices for use in IAC systems have been based on low-pressure resistance, allowing their operation under gravity flow with a slight vacuum or peristaltic flow applied (Schuste et al., 2000). These matrices include carbohydrate-based media (agarose, dextrose, or cellulose), synthetic organic supports such as acrylamide polymers, polymethacrylate derivatives, polyethersulfone matrices, or inorganic materials such as silica and zirconia (Fitzgerald et al., 2011). Cross-linked agarose is an extremely popular matrix because it can usually withstand a wide pH range (e.g. pH 3–12), most aqueous solvents (including denaturants), many organic solvents or modifiers, and enzymatic treatments. However, agarose beads and other soft gel matrices are more susceptible to pressure, relative to stronger supports, such as silica, polystyrene and other highly cross-linked materials (Urh et al., 2009). Immunomagnetic beads represent an example of a solid support with better chemical and hydrodynamic properties than conventional supports. Advantages of magnetic separation include quick retrieval of affinity beads at each step, thereby significantly shortening the purification process; bypassing sample pre-treatment such as filtration or centrifugation; and the ability to use viscous materials that would otherwise clog traditional columns (Urh et al., 2009). Magnetic particles are also available conjugated with common affinity ligands (Protein A, Protein G, streptavidin), or with specific mono- and polyclonal antibodies (Koneracka et al., 2006).

2.3 Antibody immobilization methods

The key factor in antibody immobilization onto the affinity matrix is to tightly bind the antibody to the support medium without interfering with the activity and accessibility of the antigen binding site (Kim and Hage, 2006). There are two fundamentally different ways of immobilizing antibodies to a support:

2.3.1 Random chemical attachment

This easy method usually targets the Lysine ε- amino groups on the antibodies. However, a decrease in activity is observed if the antibodies have some of those amine groups in their binding sites (Kortt et al., 1997). Steric hindrance and a decrease in binding efficiency may also occur, because the antibodies are immobilized in a random orientation, which bears the risk of having the binding site blocked by the attachment (Turková, 1999).

Covalent linkage of antibodies to the reactive groups of activated, commercially available beads is a cheap, fast, and robust method of antibody attachment to the solid phase. A typical example is cyanogen bromide (CNBr)-activated Sepharose (Nisnevitch and Firer, 2001), but various alternatives are available [tosyl and epoxy activated beads] (Hage, 1998; Larsson, 1984). This simple, straightforward methodology is recommended when the amount of antibody available is not expected to be a limiting factor, *e.g.* when the researcher has access to a hybridoma that can theoretically produce unlimited amounts of antibody. As pointed out above, the antigen-binding capacity of the antibody can be lost in the process of covalent bonding to the support; statistically this happens in two out of three attachments, and only one out of three antibodies is optimally immobilized, maintaining the potential binding of two antigen molecules. Nevertheless, the remaining capacity can still be impressive compared to the concentration of the target molecule in the extract to be purified. For instance, if 1 ml of CNBr-activated Sepharose slurry is derivatized with 5 mg of antibody (a typical ratio) and if one in three antibodies remains fully active (and can thus bind two antigen molecules), its binding capacity for a 25 kDa molecule is still approximately 500 μg per run. Another potential downside of random chemical attachment of antibodies is that the coupling capacity of the beads is so high (up to 30 mg of protein per ml of CNBr-agarose), that multipoint attachment can occur, which is another mechanism through which antigen-binding capacity can be lost. This can be prevented by limiting the time of the coupling reaction, or by adding some "inert" protein to the antibody to be coupled, so as to create competition for the available binding sites, and thus lower the probability of multipoint attachment.

Antibodies can also be immobilized onto a matrix by using a secondary ligand. In this case, biotinylated antibodies are adsorbed to a support containing immobilized avidin, streptavidin, or neutravidin (Moser and Hage, 2010). However, and unless antibodies are biotinylated at their Fc carbohydrate groups (O'Shannessy and Quarles, 1987), a decrease in binding capacity and efficiency may result from biotin molecules attaching at or near the antigen binding site, and/or from biotinylated antibodies randomly attaching to the streptavidin support (Moser and Hage, 2010).

2.3.2 Directional attachment

The most common way to achieve oriented attachment is to covalently stabilize the (reversible) bond between the antibody and Protein A, Protein G, or Protein L beads (Moser and Hage, 2010; Urh et al., 2009). Proteins A and G bind to the Fc portion of the antibody, while Protein L interacts with kappa light chains. None of these proteins blocks the Fab sites. Stabilization of the bond between antibody and binding protein is achieved by incubation with a bifunctional reagent (a cross-linker) of the ideal length. A typical example is dimethyl pimelimidate (Schneider et al., 1982), but alternatives with different lengths are also available. The quality of the immunoadsorbent can easily be checked by saving 50 μg of beads before and after the cross-linking reaction. The former should yield a 50 kDa band and a 25 kDa band under reducing SDS-PAGE, while the latter should no longer show the 50 kDa band, since the Fc fragment is supposed to be cross-linked with protein A. In spite of their attractiveness, immunoglobulin-binding beads are expensive, will bind extraneous antibody that might be present in the sample extract, and their specificities are not universal, but rather isotype-dependent. Proteins A and G possess different affinities for IgG types from different species, and even for IgG isotypes within the same species. Protein L can only bind certain subtypes of kappa light chains (Urh et al., 2009).

Oriented attachment is also possible via targeting the antibody's carbohydrate or sulfhydryl groups in the Fc fragment. For instance, antibodies can be coupled to the carbohydrate moieties found on the Fc, but not the Fab fragment (Ruhn et al., 1994; Nisnevitch and Firer, 2001). The Fc carbohydrates are first oxidized by incubation with sodium metaperiodate. The aldehyde groups that result from this oxidation are then reacted with a hydrazide-activated matrix (which is commercially available). The alternative is to reduce the disulfide bridge that links the two immunoglobulin heavy chains using a mild reducing agent such as 2-mercaptoethylamine. This leaves the disulfide bridges of the Fabs intact, but results in two "half-antibodies" with exposed sulfhydryl groups. The latter are then reacted with an iodoacetyl- or maleimide-activated matrix to form a chemically stable immunoadsorbent with full antigen-binding capacity (Spitznagel et al., 1993; Mallik et al., 2007).

3. Sample extraction

Recombinant protein production permits high-level expression of foreign proteins in hosts such as *Escherichia coli*, *Saccharomyces cerevisiae*, and *Pichia pastoris*. Host cells are then disrupted and the expressed target protein is released. The choice of the method for cell disruption and sample extraction is an empirical process that is highly dependent on the uniqueness of the target protein, the different cell types, and the scale of purification. In any case, such methods should be rapid and efficient while conserving the target protein's conformation and activity by minimizing proteolysis or oxidation, and reducing or preventing unnecessary lysis of the cell nuclei to prevent an increase in viscosity generated by the presence of genomic DNA strands (Grabski, 2009).

3.1 Preparation of cell lysate

3.1.1 Chemical and enzymatic cell disruption

These methods are particularly suited for small, laboratory-scale protein purification, and employ detergent-based reagents (such as B-PER®, Thermo Scientific and BugBuster®, EMD Chemicals) for effective cell disruption. These reagents are very fast and easy to use, do not require expensive equipment, and can readily be combined with lytic enzymes and nucleases for more efficient cell lysis and protein extraction from bacteria, yeast, plant cells, insect cells, and higher eukaryotes (Grabski, 2009). Another class of detergent-based reagents (such as B-PER ® Direct, Thermo Scientific) allows for high throughput automative processing of samples. Use of these reagents alongside high-activity lysozyme and nuclease permits for cell growth, extraction, and purification to take place in a single test tube or well, bypassing the need to separate the host cells from the culture media. The active ingredients of these reagents are detergents that weaken the host cells for rupture by disrupting cell membrane and cell wall structures. Concurrent treatment with lysozyme further breaks down cell walls, whereas the use of nucleases limits viscosity. Enzymatic treatments are further advantageous because they are gentle, simple, require no specialized instrumentation, and do not generate shear, heat, or oxidative damage. However, the optimal conditions for lytic enzymes (pH, temperature) may not be compatible with the target protein, and the presence of the enzyme in the extract may interfere with downstream purification and processing. The limited availability and cost of lytic enzymes further curbs their use at an industrial scale (Grabski, 2009).

3.1.2 Mechanical cell disruption

Mechanical methods of cell disruption are highly effective, rapid, cheap, and thus suited for large scale purification. The most popular mechanical disruption methods include sonication and high pressure homogenization (Harrison, 1991). These are not suitable for small culture volumes (<5 ml) because of inherent excess heat generation and oxidation damage (Grabski, 2009). Glass bead homogenization can be used at both laboratory and industry scale, and is effective for use with bacteria, plant and animal cells, yeast, spores, and fungi. Grinding with glass beads (on a vortex mixer or blender) creates abrasive and shear forces between the cells, the beads, and the reaction chamber itself. Care must be taken, however, to avoid excessive viscosity generated by the release of nucleic acids (Grabski, 2009; Harrison, 1991). Mechanical methods can be combined with an enzymatic method to increase the rate of cell disruption and extraction yield, reduce viscosity, and minimize product damage (Grabski, 2009).

3.2 Extraction buffer composition and volume

Besides being critical to effective cell disruption, the extraction buffer greatly affects subsequent purification steps and the protein's stability and recovery. An ideal extraction buffer promotes fast and efficient binding of the analyte to the immobilized antibodies, and will leave proteins in their native conformation (Moser and Hage, 2010; Grabski, 2009). The important criteria to consider are pH, ionic strength, buffer to cell pellet ratio, and the use of additives. To prevent precipitation, the buffer pH should be one unit above or below the isoelectric point of the target protein. An ionic strength of 20-50 mM and the presence of 50-100 mM sodium chloride will maintain buffering capacity, and will minimize ionic interactions in the cytoplasm that might lead to adsorption of target protein to charged particulates and its subsequent loss upon centrifugation or filtration (Grabski, 2009).

Proteolysis, dephosphorylation, and denaturation of protein occur as soon as the host cells are lysed. These processes can be slowed down dramatically by keeping the samples on ice and adding appropriate enzyme inhibitors to the lysis buffer just before use. For convenience, cocktails of different protease and phosphatase inhibitors (Aprotinin, PMSF, EDTA...) are commercially available. Other additives that can improve stability of the extracted protein include reducing agents to maintain reduced disulfide bonds, detergents to increase solubility of hydrophobic proteins, kosmotropes (such as glycerol, trehalose, and glycerol) to stabilize intermolecular interactions, and nucleases to reduce sample viscosity. However, the potential interference of these additives with downstream purification and detection must be considered and evaluated (Grabski, 2009).

To ensure effective disruption and adequate recovery, the volume of buffer used to resuspend cell pellets should be at least three times the volume of the original pellet. This will ensure at least 85% recovery of the liquid fraction obtained after removal of insoluble cell debris. A more soluble and less viscous protein extract can be obtained by using 5-10 volumes of extraction buffer, since highly concentrated extracts are susceptible to aggregation and will decrease diffusion rate and capture by the immunoaffinity matrix (Grabski, 2009).

3.3 Pre-clearing of lysate

To remove proteins that bind nonspecifically to the affinity matrix, the extract can be pre-incubated with the support matrix, or pre-cleared by incubation with an irrelevant antibody of the same species or with normal serum. This pre-clearing step will result in a lower background and an improved signal-to-noise ratio (Harlow and Lane, 1999c).

3.4 Removal of extraneous matter

Ahead of purification, particulate matter and contaminants must be removed from the extract by centrifugation and/or filtration, in order to avoid clogging of the chromatographic column. It may also be necessary to perform a desalting and buffer exchange step in order to transfer the sample to the correct buffer conditions (pH and salt concentration), and to remove unwanted small molecules. If the sample is reasonably clean after centrifugation, this last step can be omitted and replaced with a mere adjustment of the sample pH and ionic strength to that of the application buffer (Harlow and Lane, 1999c). Finally, if the sample extract represents a diluted protein solution, concentration of the sample before purification may be necessary to enhance the probability of quantitative recovery of the target molecule.

4. Sample adsorption

Because the antibody is bound to a solid phase, adsorption conditions should maximize antigen-antibody interaction (Harlow and Lane, 1999c). The efficiency of binding is related to the strength and the kinetics of this interaction, which in turn depend on the amount of immobilized antibody, the concentration of applied target, and the flow rate used for binding (Grabski, 2009). Binding can be performed in column or batch format (where the sample extract serves to keep the gel beads in suspension). The former allows for adjustment of flow rates, and therefore for extending the time of antigen-antibody interaction. Generally, a higher flow rate will reduce the binding efficiency, especially when the antibody-target interaction is weak, and/or the mass-transfer rate in the column is slow. In batch purification, the resin and sample are constantly mixed, thus promoting a maximum contact between the target and immobilized antibody. It often saves time, especially when dealing with large sample volumes, but requires optimisation of the amount of resin used. Because excess resin can result in an increase in nonspecific binding, as well as reduced target recovery due to readsorption during the elution step, it is preferable to saturate the resin with bound target (Harlow and Lane, 1999c; Grabski, 2009).

Optimal binding between antibodies and their targets typically occurs under physiological conditions, so the application buffer used in IAC is of neutral pH (7.0-7.4). This promotes fast and efficient binding of the desired analyte to the immobilized antibodies, with equilibrium constants for antibody binding ranging between 10^6 and 10^{12} M^{-1} (Moser and Hage, 2010).

Following binding, protein bound by nonspecific interactions is removed by washing. Increasing salt (0.1–0.5 M) or changing pH values will reduce ionic interactions, while decreasing salt, altering pH, or adding surfactants (such as Triton X-100) will remove proteins bound by nonspecific hydrophobic interactions. Contaminants with weak affinity to the ligand or to the support matrix itself can be removed by application of low amounts of competitive reagents. It is crucial to determine the appropriate flow rate and volume (e.g., 5–10 column bed volumes) of the wash buffer that will maximally remove contaminants while minimizing loss of target (Grabski, 2009; Harlow and Lane, 1999c).

5. Sample desorption

Elution of the antigen, often viewed as the most delicate step of an IAC protocol, should ideally be carried out in a way that keeps the antibody on the immunoadsorbent intact and maintains antigen activity (*e.g.* enzymatic or hormonal activity), if present, while still allowing later regeneration of the column (Firer, 2001). This is especially important if the column is to be used for a large number of samples. The sample can always be desorbed from the antibody because the four forces that stabilize the antigen-antibody complex (ionic, hydrogen bonding, van der Waals interactions, and hydrophobic bonds) are all reversible. Thus, the antigen-antibody complex can be destabilized by counteracting the forces at work in a particular antibody-epitope interaction. Desorption is thus essentially the reverse process of binding, where conditions are optimized to weaken the antibody-target interaction. Unfortunately, there is no way to figure out *a priori* what will be an effective eluent from a particular immunoadsorbent; this can only be determined empirically. The elution method of choice is often the use of low pH (2.0-2.5) which disrupts both ionic and hydrogen bonds between antigen and antibody (Narhi et al., 1997). If that procedure is not effective, the next best choice may be to resort to a commercially available eluent such as Gentle Elution Buffer (Thermo Scientific), the composition of which is proprietary and is reported to destabilize the antigen-antibody complex without damaging either partner of the complex. This solution contains very high concentrations of salts and other agents and requires thorough dialysis of the sample prior to downstream processing.

Denaturing agents (8 M urea or 6 M guanidinium hydrochloride) affect elution by promoting protein unfolding, while chaotropic salts (3 M sodium thiocyanate, magnesium chloride) disrupt the water molecules around the affinity interaction (Singh et al., 2003). Both methods, however, can disrupt protein structure and damage labile proteins, resulting in very low yields of active, purified protein. They may also decrease the lifetime of the antibody column (Burgess and Thompson, 2002).

The ultimate eluent is probably SDS denaturation of the antibody-antigen complex. While this may seem excessively harsh, it makes the antigen available for SDS-PAGE and ultimately for mass spectrometric analysis and *de novo* protein sequence determination. It may therefore be worth sacrificing some antibody (and antigen) in a single use procedure, if it allows one, for instance, to determine with certainty the molecule that a new monoclonal antibody recognizes.

Elution can also be performed in a specific way, by using a displacer agent which will compete with the target protein for binding to the immobilized antibody (Fitzgerald et al., 2011). For example, proteins containing a hemagglutinin (HA) tag can be purified on an anti-HA column, and eluted with an excess of HA. While this elution method is advantageous because of its specificity and mild conditions, the dissociation will ultimately depend on the affinity of antibody-target interaction. Also, large molar excesses of the competitor are required, elution is slow and results in broad elution peaks (Urh et al., 2009).

Following elution, the column should always be washed with the application buffer to allow for regeneration of the antibodies before another sample application (Urh et al., 2009).

6. Applications

Immunoaffinity chromatography is a versatile, powerful purification method based on well-characterized antibody-target interactions, making it amenable for use in many applications, including sample cleanup, and clinical and diagnostic assays for drugs, toxins, and biomarkers. The power of IAC can also be harnessed for use in immunoassays, including sandwich, competitive, and non-competitive binding assays. IAC has also been coupled with other methods such as HPLC, gas chromatography, mass spectrometry (MS), and capillary electrophoresis (CE). In the classic IAC on/off mode (where the analyte is bound to the column, impurities are washed away, and the analyte is then eluted off the column), IAC is commonly used for the selective purification of target compounds (proteins, glycoproteins, carbohydrates, lipids, bacteria, viral particles, drugs) from complex samples. (Weller, 2000; Gallant, 2004). Moreover, if a suitable detector (UV/visible absorbance, fluorescence) is placed after the column, direct detection of the analyte will be possible, provided the analyte is present at a relatively high concentration and is eluted in a sharp, well-defined peak that allows a good detection limit (Moser & Hage, 2010).

IAC methods are extensively used for sample clean-up prior to analysis of foods for mycotoxins, veterinary drug residues, pesticides, and environmental contaminants (Şenyuva & Gilbert, 2010). Undesirable components are removed from the sample, preceding the analysis by a second analytical method such as HPLC. This method has been used in the analysis of urine, food, water and soil extracts. A related method, immunodepletion, is used in proteomics for highly selective depletion of multiple high-abundance proteins, prior to analysis of minor sample components.

In recent years, IAC has been integrated with other analytical methods such as CE and MS. CE immunoassays (where antibodies are immobilized in CE capillaries) are utilized because they are easily automated, require small amounts of sample and reagents and still maintain a good detection limit, and offer relatively fast separation (Moser & Hage, 2008; Phillips and Wellner, 2007).

7. Conclusion

There are various methods of enriching or purifying a protein of interest from a complex mixture of other proteins and components. Immunoaffinity chromatography is the most powerful and versatile of these methods, an advantage bestowed by the specific binding properties between an immobilized antibody and its target. A single pass through an immunoaffinity column can achieve a 1,000- to 10,000-fold purification of a target from a crude mixture. Before planning an immunoaffinity purification procedure, and assuming an appropriate immobilized antibody is available, one needs to consider the following: (1) the sample source, which will dictate the necessary sample extraction conditions and pre-treatment; (2) the scale of purification required; (3) the purity required for the final application; and (4) the economic feasibility, including time and expense.

8. References

Agaton, C.; Falk, R.; Hoiden Guthenberg, I.; Gostring, L.; Uhlen, M. & Hober, S. (2004). Selective enrichment of monospecific polyclonal antibodies for antibody-based proteomics efforts. *J Chromatogr A* Vol.1043, No.1 (July 2004), pp. 33-40, ISSN 0021-9673.

Andris-Widhopf, J.; Rader, C.; Steinberger, P.; Fuller, R. & Barbas, C. F., 3rd (2000). Methods for the generation of chicken monoclonal antibody fragments by phage display. *J Immunol Methods* Vol. 242, No.1-2, (August 2000), pp. 159-181, ISSN 0022-1759.

Azzazy, H.M. & Highsmith, W.E., Jr. (2002). Phage display technology: clinical applications and recent innovations. *Clin Biochem* Vol..35, No.6, (September 2002), pp. 425-445, ISSN 0009-9120.

Barbas, C.F. 3rd & Wagner, J. (1995). Synthetic Human Antibodies: Selecting and Evolving Functional Proteins. *Methods: A Companion to Methods in Enzymology* Vol.8, No.2, (October 1995), pp. 94-103, ISSN 1046-2023.

Berghman, L.R.; Abi-Ghanem, D.; Waghela, S.D. & Ricke, S.C. (2005). Antibodies: an alternative for antibiotics? *Poult Sci* Vol. 84, No.4 (April 2004), pp. 660-666, ISSN 0032-5791.

Burgess, R.R. & Thompson, N.E. (2002). Advances in gentle immunoaffinity chromatography. *Curr Opin Biotechnol* Vol.13, No.4, (August 2002), pp. 304-30, ISSN 0958-1669.

Campbell, D.H.; Luescher, E. & Lerman, L. S. (1951). Immunologic Adsorbents: I. Isolation of Antibody by Means of a Cellulose-Protein Antigen. *Proc Natl Acad Sci USA* Vol.37, No.9 (September 1951), pp. 575-578, ISSN 0027-8424.

Chui, L.W.; King, R.; Chow, E.Y. & Sim, J. (2004). Immunological response to Mycobacterium avium subsp. paratuberculosis in chickens. *Can J Vet Res* Vol.38, No.4, pp. 302-308, ISSN 0830-9000.

Clarke, W.; Beckwith, J.D.; Jackson, A.; Reynolds, B.; Karle, E.M. & Hage, D.S. (2000). Antibody immobilization to high-performance liquid chromatography supports. Characterization of maximum loading capacity for intact immunoglobulin G and Fab fragments. *J Chromatogr A* Vol.888, No. 1-2, (August 2000), pp. 13-22, ISSN0021-9673.

Crowe, J.; Dobeli, H.; Gentz, R.; Hochuli, E.; Stuber, D. & Henco, K. (1994). 6xHis-Ni-NTA chromatography as a superior technique in recombinant protein expression/purification. *Methods Mol Biol* Vol.31, pp. 371-387, ISSN 1064-3745.

Elgert, K.D. (1996). Antibody Structure and Function. In: *Immunology: Understanding The Immune System*, K.D. Elgert, pp. 58-78, Wiley-Liss, ISBN 0-471-11680-7 New York, USA.

Firer, M.A. (2001). Efficient elution of functional proteins in affinity chromatography. *J Biochem Biophys Methods* Vol.49, No.1-3, (October 2001), pp. 433-442, ISSN 0165-022X.

Fitzgerald, J; Leonard P; Darcy, E & O'Kennedy R. (2011). Immunoaffinity chromatography. *Methods Mol Biol.* Vol.681, Part1, pp. 35-59, ISSN 1064-3745.

Gallant, S. R. (2004). Immunoaffinity chromatography of proteins. *Methods Mol Biol.* Vol.251, pp. 103-110, ISSN 1064-3745.

Grabski, A. C. (2009). Advances in preparation of biological extracts for protein purification. *Methods Enzymol.* Vol.463, pp. 285-303, ISSN 1557-7988.

Gunneras, S.; Agaton, C.; Djerbi, S. & Hansson, M. (2008). Prestige Antibodies--monospecific antibodies designed for immunohistochemical analysis. *Biotechniques* Vol.44, No.6 (May 2008), pp. 825-828, ISSN 0736-6205.

Gustafsson, B. (1990). Fusion Protocol for the Production of Mouse Hybridomas. In: *Methods in Molecular Biology, Vol. 5: Animal Cell Culture,* J.W. Pollard & J.M. Walker (Eds.), pp. 601-607, Humana Press, ISBN 0-8960-3150-0, New Jersey, USA.

Gustavsson, P-E. & Larsson, P-O. (2006). Support materials for affinity chromatography. In: *Handbook of Affinity Chromatography,* Hage, D.S. (Ed.). pp. 15-34, Taylor & Francis, ISBN 978-0-8247-4057-3, New York, USA.

Hage, D.S. (1998). Survey of recent advances in analytical applications of immunoaffinity chromatography. *J Chromatogr B Biomed Sci Appl* Vol.715, No.1 (September 1998), pp. 3-28, ISSN 1387-2273.

Harlow, E. & Lane, D. (1999a). Antibody Structure and Function. In: *Using Antibodies,* Harlow and Lane, pp. 1-20, Cold Spring Harbor laboratory Press, ISBN 978-087969544-6, New York, USA.

Harlow, E. & Lane, D. (1999b). Antibody-Antigen Interactions. In: *Using Antibodies,* Harlow and Lane, pp. 21-37, Cold Spring Harbor laboratory Press, ISBN 978-087969544-6, New York, USA.

Harlow, E. & Lane, D. (1999c). Immunoaffinity Purification. In: *Using Antibodies,* Harlow and Lane, pp. 311-343, Cold Spring Harbor laboratory Press, ISBN 978-087969544-6, New York, USA.

Harrison, S.T. (1991). Bacterial cell disruption: a key unit operation in the recovery of intracellular products. *Biotechnol Adv* Vol.9, No.2, pp. 217-240, ISSN 0734-9750.

Hoogenboom, H.R. & Chames, P. (2000). Natural and designer binding sites made by phage display technology. *Immunol Today* Vol. 21, No.8, (August 2000), pp. 371-378, ISSN 0167-5699.

Jensenius, J.C.; Andersen, I.; Hau, J.; Crone, M. & Koch, C. (1981). Eggs: conveniently packaged antibodies. Methods for purification of yolk IgG. *J Immunol Methods* Vol.46, No.1, pp. 63-68, ISSN 0022-1759.

Kampf, C.; Andersson, A.C.; Wester, K.; Björling, E.; Uhlen, M. & Ponten, F. (2004) Antibody-based tissue profiling as a tool for clinical proteomics. *Clinical Proteomics* Vol.1, No.3, (September 2004), pp. 285-299, ISSN 1542-6416.

Kim, H. & Hage, D.S. (2006). Immobilization methods for affinity chromatography. In: *Handbook of Affinity Chromatography,* Hage, D.S. (Ed.). pp. 36-77, Taylor & Francis, ISBN 978-0-8247-4057-3, New York, USA.

Koneracka, M.; Kopcansky, P.; Timbo, M.; Ramchand, C.N.; Saiyed, Z. M. & Trevan, M. (2006). Immobilization of enzymes on magnetic particles. In: *Immobilization of Enzymes and Cells,* J.M. Guisan (Ed.), pp. 217–228, Humana Press, ISBN 978-1-58829-290-2, New Jersey, USA.

Kortt, A.A.; Oddie, G.W.; Iliades, P.; Gruen, L.C. & Hudson, P.J. (1997). Nonspecific amine immobilization of ligand can be a potential source of error in BIAcore binding experiments and may reduce binding affinities. *Anal Biochem* Vol.253, No.1 (November 1997), pp. 103-111, ISSN 0003-2697.

Kuby, J. (1992). Hybridomas and Monoclonal Antibody. In: *Immunology,* J. Kuby, pp. 141-156, W.H. Freeman and Company, ISBN 0-7167-2257-7, New York, USA.

Larsson, K.; Wester K.; Nilsson, P.; Uhlen, M.; Hober, S. & Wernerus, H. (2006). Multiplexed PrEST immunization for high-throughput affinity proteomics. *J Immunol Methods* Vol.315, No.1-2 (August 2006), pp. 110-120, ISSN 0022-1759.

Larsson, P.O. (1984). High-performance liquid affinity chromatography. *Methods Enzymol* Vol.104, pp. 212-223, ISSN 0076-6879.

Lesney, M. (2003). Sticking with affinity chromatography. *Today's Chemist at Work,* (August 2003), pp. 38-40, ISSN 1062-094X.

Libon, C.; Corvaia, N.; Haeuw, J.F.; Nguyen, T.N.; Stahl, S; Bonnefoy, J.Y. & Andreoni, C. (1999). The serum albumin-binding region of streptococcal protein G (BB) potentiates the immunogenicity of the G130-230 RSV-A protein. *Vaccine* Vol.17, No.5, (February 1999), pp. 406-414, ISSN 0264-410X.

Lindskog, M.; Rockberg, J.; Uhlen, M. & Sterky, F. (2005). Selection of protein epitopes for antibody production. *Biotechniques* Vol.38, No.5, pp. 723-727, ISSN 0736-6205 .

Little, M.; Kipriyanov, S. M.; Le Gall, F. & Moldenhauer, G. (2000). Of mice and men: hybridoma and recombinant antibodies. *Immunol Today* Vol.21, No.8 (August 2000), pp. 364-370, ISSN 0167-5699.

Mallik, R.; Wa, C. & Hage, D.S. (2007). Development of sulfhydryl-reactive silica for protein immobilization in high-performance affinity chromatography. *Anal Chem* Vol.79, No.4, (February 2007), pp. 1411-1424, ISSN 0003-2700.

Maynard, J. & Georgiou, G. (2000). Antibody engineering. *Annu Rev Biomed Eng* Vol. 2, pp. 339-376, ISSN 1523-9829.

Michnick, S.W. & Sidhu, S.S. (2008). Submitting antibodies to binding arbitration. *Nat Chem Biol* Vol.4, No.6 (June 2008), pp. 326-329, ISSN 1552-4469.

Moser, A. C. & Hage, D.S. (2008). Capillary electrophoresis-based immunoassays: principles and quantitative applications. *Electrophoresis* Vol.29, No.16 (August 2008), pp. 3279-3295, ISSN 1557-7988.

Moser, A. C. & Hage, D.S. (2010). Immunoaffinity chromatography: an introduction to applications and recent developments. *Bioanalysis* Vol.2, No.4 (April 2010), pp. 769-790, ISSN 1757-6199.

Narhi, L.O.; Caughey, D.J.; Horan, T.; Kita, Y.; Chang, D. & Arakawa, T. (1997). Effect of three elution buffers on the recovery and structure of monoclonal antibodies. *Anal Biochem* Vol.253, No.2, (November 1997), pp. 236-245, ISSN 0003-2697.

Nisnevitch, M. & Firer, M.A. (2001). The solid phase in affinity chromatography: strategies for antibody attachment. *J Biochem Biophys Methods* Vol.49, No.1-3, pp. 467-480, ISSN 0165-022X.

O'Shannessy, D.J. & Quarles, R.H. (1987). Labeling of the oligosaccharide moieties of immunoglobulins. *J Immunol Methods* Vol.99, No.2 (May 1987), pp. 153-161, ISSN 0022-1759.

Phillips, T.M. & Wellner, E.F. (2007). Analysis of inflammatory biomarkers from tissue biopsies by chip-based immunoaffinity CE. *Electrophoresis* Vol.28, No.17 (August 2007), pp. 3041-3048, ISSN 0173-0835.

Pini, A. & Bracci, L. (2000). Phage display of antibody fragments. *Curr Protein Pept Sci* Vol. 1, No.2, (September 2000), pp. 155-169, ISSN 1389-2037.

Rader, C. & Barbas, C.F. 3rd (1997). Phage display of combinatorial antibody libraries. *Curr Opin Biotechnol* Vol.8, No.4 (August 1997), pp. 503-508, ISSN 0958-1669.

Ruhn, P.F.; Graver, S. & Hage, D.S. (1994). Development of dihydrazide-activated silica supports for high-performance affinity chromatography. *J Chromatogr A* Vol.669, No.1-2, (May 1994), pp. 9-19, ISSN 0021-9673.

Schneider, C.; Newman, R.A.; Sutherland, D.R.; Asser, U. & Greaves, M.F. (1982). A one-step purification of membrane proteins using a high efficiency immunomatrix. *J Biol Chem* Vol.257, No.18, (September 1982), pp. 10766-10769, ISSN 0021-9258.

Schuste, M.; Wasserbauer, E.; Neubauer, A. & Jungbauer, A. (2000). High speed immuno-affinity chromatography on supports with gigapores and porous glass. *Bioseparation* Vol.9, No.5, pp. 259-268, ISSN 0923-179X.

Şenyuva, H.Z. & Gilbert, J. (2010). Immunoaffinity column clean-up techniques in food analysis: A review. *J Chromatogr B Analyt Technol Biomed Life Sci* Vol.878, No.2 (January 2010), pp. 115-132, ISSN 1873-376X.

Singh, K.V.; Kaur, J.; Raje, M.; Varshney, G.C. & Suri, C.R. (2003). An ELISA-based approach to optimize elution conditions for obtaining hapten-specific antibodies. *Anal Bioanal Chem* Vol.377, No.1, (September 2003), pp. 220-224, ISSN 1618-2642.

Smith, G.P. & Petrenko, V.A. (1997). Phage Display. *Chem Rev* Vol. 97, No.2, (April 1997), pp. 391-410, ISSN 1520-6890.

Spitznagel, T.M.; Jacobs, J.W. & Clark, D.S. (1993). Random and site-specific immobilization of catalytic antibodies. *Enzyme Microb Technol* Vol.15, No.11, (November 1993), pp. 916-921, ISSN 0141-0229.

Turková, J. (1999). Oriented immobilization of biologically active proteins as a tool for revealing protein interactions and function. *J Chromatogr B Biomed Sci Appl* Vol.722, No.1-2, (February 1999), pp. 11-31. ISSN 1387-2273.

Urh, M.; Simpson, D. & Zhao, K. (2009). Affinity chromatography: general methods. *Methods Enzymol* Vol.463, pp. 417-438, ISSN 1557-7988.

Weller, M. G. (2000). Immunochromatographic techniques--a critical review. *Fresenius J Anal Chem* Vol.366, No.6-7 (March-April 2000), pp. 635-645, ISSN 0937-0633

Winter, G. & C. Milstein (1991). Man-made antibodies. *Nature* Vol.349, No.6307 (January 1991), pp. 293-299, ISSN 0028-0836.

Yokota, T.; Milenic, D.E.; Whitlow, M. & Schlom, J. (1992). Rapid tumor penetration of a single-chain Fv and comparison with other immunoglobulin forms. *Cancer Res* Vol. 52, No.12 (June 1992), pp. 3402-3408, ISSN 0008-5472.

6

Affinity Chromatography for Purification of IgG from Human Plasma

Lucia Hofbauer, Leopold Bruckschwaiger,
Harald Arno Butterweck and Wolfgang Teschner
Baxter Innovations GmbH, Vienna, Austria

1. Introduction

Edwin J. Cohn et al. developed cold ethanol fractionation for isolating different blood plasma fractions on an industrial scale during the first half of the last century. The fractionation process uses different solubility of plasma proteins by varying the pH, ethanol concentration, temperature, ionic strength and protein concentration (Cohn et al., 1946). Initially, the main reason for developing plasma fractionation on a large scale was for purification of albumin. Albumin is used for treating shock, hypoproteinemia (Janeway et al., 1944), acute or chronic nephritis (Thorn & Armstrong, 1945) and hepatic cirrhosis (Thorn et al., 1946) as well as other disorders.

Colonel Bruton was the first to use polyclonal immunoglobulin G (IgG) as a treatment (Bruton, 1952). During his work at the Walter Reed Army Hospital in 1952, he used IgG to successfully cure a young boy who had recurrent sepsis. Bruton discovered that the γ-globulin content in blood plasma can be enhanced by monthly subcutaneous (SC) injection of immune human serum globulin. Subsequently, immunoglobulin administration became the standard treatment for patients with hypogammaglobulinemia.

In the late 1970s, intravenous (IV) administration of IgG became the method of choice because the large volumes of the immunoglobulin product that are necessary for providing the physiologic levels of IgG for the effective treatment of various diseases such as primary immune deficiencies (PID), immune (idiopathic) thrombocytopenic purpura (ITP) or Kawasaki syndrome can be applied by this route (Weiler, 2004). New purification techniques were established to produce IVIG (intravenously administrable immunoglobulin G) preparations which would not give rise to the adverse effects such as fever, headache, arthralgia, serum sickness, aseptic meningitis, myocardial infarction and thromboembolic events typically seen after IV administration of immunoglobulin products intended for the intramuscular or subcutaneous route of administration.

Chromatography for protein purification on an industrial scale was developed supplementary to the use of different precipitating agents (e.g. polyethylene glycol (PEG) (Polson et al., 1964)) or batch-adsorption on DEAE Sephadex (Hoppe et al., 1967), starting from plasma itself or intermediates derived from Cohn's, Oncley's or Kistler & Nitschmann's processes (Falksveden & Lundblad, 1980, Hoppe et al., 1967, Kistler &

Nitschmann, 1962, Oncley et al., 1949, Suomela, 1980). Later, caprylic acid precipitation (Audran & Pejaudier, 1975, Steinbruch & Audran, 1969) was used for the purification of IgG. Further processes were developed which combined some of these methods. The resulting products were of high quality containing functionally intact IgG that showed a similar subclass distribution to that of plasma (Ballow, 2002). Suomela as well as Falksveden and Lundblad presented ion exchange chromatographic processes for the isolation of IgG in the book "The methods of plasma protein fractionation" edited by Curling. While Falksveden used PEG as a precipitation agent and in addition a combination of cation and anion exchange chromatography, Suomela added a Lysine Sepharose affinity chromatographic step after ion exchange chromatography to remove proteolytic activities, thus enhancing the quality of the final preparation (Falksveden & Lundblad, 1980, Suomela, 1980). Travis et al. described the advantages of removing albumin by mimetic dye affinity chromatography prior to fractionation in different fractionation schemes (Travis et al., 1976). Later Gianazza and Arnaud developed a method for purifying plasma proteins, which included the use of Cibacron Blue Sepharose affinity chromatography resin. They studied the behavior of 27 different plasma proteins and suggested affinity chromatography as a useful initial step in plasma fractionation (Gianazza & Arnaud, 1982).

Since the early nineteen seventies, affinity chromatography has been investigated for its use in the purification of many biomolecules. The hurdles which need to be overcome were already identified in the nineteen seventies: Ligand leaching, low capacity and harsh elution conditions (Travis et al., 1976). On the other hand, this method had and still has the advantages of a several thousand-fold enrichment of the target protein out of large volumes of crude starting materials, combined with high recoveries as well as the targeted separation of active and inactive material of denatured or functionally different forms. Roque et al. (Roque et al., 2007) and Low et al. (Low et al., 2007) have written excellent reviews of the development of affinity ligands used for antibody purification from biological to bioengineered or fully synthetic ligands.

Affinity chromatography has become the method of choice for the purification of monoclonal antibodies (Kelley, 2007, Kelley et al., 2009, Low et al., 2007), while the purification procedures for polyclonal plasma-derived antibodies traditionally incorporate precipitation steps combined with ion exchange chromatography.

In this chapter we describe the potential advantages and draw-backs of affinity chromatography for capturing IgG from clarified crude polyclonal IgG fractions of human plasma. IgSelect affinity media and Protein G Sepharose 4 Fast Flow (FF), both from GE Healthcare, were investigated for this purpose. Cohn fraction II+III paste, which mainly consists of α-, β- and γ-globulin (Cohn et al., 1944), was used as the starting material. Ideally the process involves the clarification of dissolved II+III paste by filtration and a one-step affinity chromatography process leading to an intravenously administrable IgG. A reduction in the complexity of the manufacturing should also lead to an improved IgG yield, which would increase the market supply needed for new indications like neurological disorders (e.g. Alzheimer's disease (Relkin et al., 2009)).

2. Affinity chromatography resins for polyclonal human IgG capture

Most commercially available affinity chromatography resins are protein A based. Hahn and coworkers (Hahn et al., 2003, Hahn et al., 2005, Hahn et al., 2006) did a comprehensive study of 15 currently available protein A affinity media, including investigating their mass transfer characteristics and selectivity. The 3 resins that came out with the top dynamic binding capacities (DBC) at low residence times were MabSelect Xtra ™, MabSelect SuRe™, both from GE Healthcare, and ProSep®-vA Ultra from Millipore, all with a DBC at 10% of about 40 mg/mL IgG at a residence time of 4 minutes. The residence time (e.g. in minutes) is the bed height divided by the linear flow rate (e.g. expressed in cm bed height per minute). It is preferable that the target protein binds at lower residence times, because manufacturing productivity is increased. Although the residence times and capacities of these protein A resins are better than those of first generation protein A resins like Protein A Sepharose 4 Fast Flow (GE Healthcare), the capacities of cation exchange resins used in large scale IgG capturing of about 150 mg/mL at a DBC of 10% (data not shown) are still superior. Protein A, a bacterial surface protein isolated from *Staphylococcus aureus,* interacts with the Fc (Fragment, crystallizable) part of the antibody. The Fc region is the tail of the antibody interacting with cell surface receptors and complement proteins to modulate the immune system. Even though the Fc part is regarded as constant across IgGs, not all IgG subclasses have identical Fc parts. This is the reason why protein A ligands have only low affinity to IgG_3. Protein-A-based media are not suitable for polyclonal IgG purification because the European Pharmacopoeia requires IgG preparations to have a subclass distribution similar to that found in human plasma (Morell et al., 1972, Schauer et al., 2003).

The remaining candidates suitable for large scale manufacturing were found to be the bacterial surface protein G and a novel camelid-antibody-based ligand from BAC company.

A 17-kDa recombinant protein G fragment manufactured in an *E. coli* where the albumin binding region of native protein G has been genetically deleted is used as a ligand for protein G affinity resins. Compared with protein A, which has only weak affinity to IgG_3, the protein G fragment has a high binding affinity to all human IgG subclasses, including IgG_3 (GE Healthcare, 2007b). Therefore the experiments described below were performed with Protein G Sepharose 4 Fast Flow, which has a capacity of 17 mg/mL at 7 min residence time according to the claim of the manufacturer. This resin is an example of a protein G resin belonging to the BioProcess Media which GE Healthcare (Björkgatan 30, 75109 Uppsala, Sweden) offers for industrial scale purification processes.

IgSelect is a new affinity resin, also manufactured by GE Healthcare, which uses technology from BAC (BAC B.V. Huizerstraatweg 28, 1411 GP Naarden, The Netherlands). According to BAC their CaptureSelect® ligands offer a unique affinity purification solution based on camelid-derived single domain antibody fragments. These small 14-kDa affinity ligands, which are produced in *Saccharomyces cerevisiae,* can be used as a platform solution for any biopharmaceutical purification challenge and, also according to BAC, have been proven in many applications to result in a high yield and purity of the biopharmaceutical as well as fewer purification steps than needed in conventional chromatography methods. All these factors affect the cost of the biopharmaceutical products. In addition, CaptureSelect ligands can be tailored to guarantee mild elution conditions, thereby maintaining the native state of the biopharmaceutical molecule of interest. The ligand is coupled to the matrix (cross-linked high flow agarose) by a long hydrophilic spacer (GE Healthcare, 2007a, GE Healthcare, 2007b).

According to BAC and GE Healthcare the resin has the following benefits:

- Binds to all subclasses of human IgG
- Rigid base matrix to allow high flow rates
- Animal-free production, generally recognized by authorities as safe (GRAS status)
- Mild elution conditions
- Capacity: 17 mg/mL at 2.4 min residence time

Low et al. (Low et al., 2007) emphasized in their review that CaptureSelect was the only ligand which showed a selectivity comparable to protein A and that it has the advantage of a higher elution pH than other resins that had been investigated such as protein-A-based resin (MabSelect®, GE Healthcare) and the synthetic ligands based on mimetic dye resin (Mabsorbent® A1P and A2P, Prometic Biosciences). Because of the advantages reported, we also selected this resin for capturing IgG from clarified dissolved Cohn fraction II+III paste.

3. Description of the IgG purification process

The Cohn separation methods result in five main precipitates. Fraction I, which mainly consists of fibrinogen, is obtained from either plasma or cryosupernatant after separation of cryoprecipitate or after additional adsorption of blood coagulation factors and inhibitors (as, for example, described for the Baxter product KIOVIG/Gammagard Liquid), adding 8% alcohol and adjusting the temperature to approximately –2°C. Fraction II, which mainly consists of IgG, is purified from fractionation II+III paste generated by separating raw immunoglobulin and raw albumin. Fraction III is a waste fraction containing, for example, lipid-bearing β-globulins and IgA. Fraction IV, which consists of α-globulins, can also be obtained in two steps: Fraction IV-1 enriched with α-1 antitrypsin, and fraction IV-4, which is used for further purification of transferrin or, most recently, butyrylcholinesterase (Weber et al., 2011). Fraction V is mainly composed of albumin.

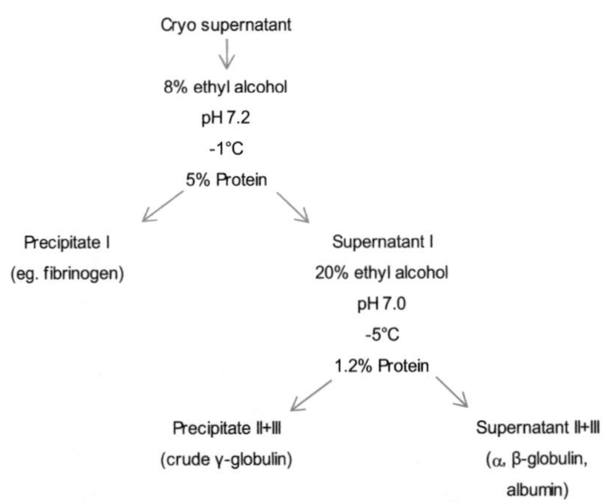

Fig. 1. Cold ethanol fractionation scheme from cryo-supernatant to fraction II+III

As the production of fraction III by the Cohn method leads to a considerable yield loss in the range of 20% (Buchacher & Iberer, 2006), most modern immunoglobulin purification methods that use plasma start with II+III paste and apply chromatographic purification methods after resuspending the paste and clarifying the suspension by filtration or centrifugation.

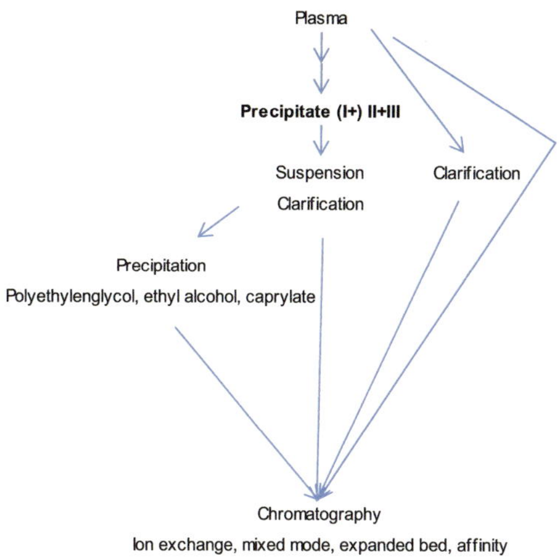

Fig. 2. Overview of modern IgG purification processes from plasma (see also Buchacher & Iberer, 2006)

We selected the II+III paste for IgG purification with affinity resins to compare yield and purity with final containers manufactured with conventional ion exchange chromatography, and also to avoid influencing the albumin and α-1 antitrypsin purification. This is of particular importance as regulatory authorities can require expensive and time-consuming clinical studies before they accept changes in manufacturing schemes.

Prevention of virus transmission has been a key aspect in the development of intravenous products as a large number of blood donations are required to produce a single immunoglobulin lot out of human plasma. In addition to donor selection and plasma testing, virus removal or inactivation steps during the manufacturing process are critical for assuring the safety of the product. Solvent-detergent (S/D) treatment, which works by irreversible destruction of the lipid envelope of viruses and is otherwise very gentle on sensitive proteins, is the most reliable and acceptable method for inactivating lipid-enveloped viruses. The concentrations of S/D reagents are low, typically 0.3% Tri-n-Butyl Phosphat (TnBP) and 1% detergent. Non-ionic detergents such as Tween 80 or Triton X-100 are preferred as they are easier to remove in subsequent chromatography steps. Because of these advantages of the S/D treatment, we subjected dissolved and filtered II+III paste to this virus inactivation step before affinity chromatography. The conditions used for the affinity chromatography are summarized below.

IgSelect affinity resin:

- Equilibration and washing: 20 mM NaH_2PO_4 + 150 mM NaCl, pH 7.4
- Loading: 17.4 mS/cm, ~ 10 g IgG/mL resin, pH 7.4
- Elution: 250 mM glycine, pH 4.0

Protein G Sepharose FF:

- Equilibration and washing: 20 mM NaH_2PO_4, pH 7.0
- Loading: 5.15 mS/cm, ~ 10 g IgG/mL resin, pH 7.0
- Elution: 100 mM glycine, pH 3.5

After affinity chromatography the eluate was concentrated and diafiltered against 250 mM glycine leading to a final bulk with 100 mg/mL protein at pH 4.4 to 4.9, similar to the commercially available IGIV product Gammagard Liquid/KIOVIG. An in-depth final container characterization was performed after sterile filtration and the results compared with Gammagard Liquid / KIOVIG specifications (see Table 1).

Measure	Method / Unit	IgSelect	Protein G	Specifications Gammagard Liquid
IgG recovery in eluate	% of Cohn pool	78	85	-
CAE	% γ-globulin	98.6	99	≥ 98
Molecular size distribution	Aggregates (>450 kDa) [%]	1.0	1.9	≤ 2
	Monomer/Dimers (160 - 320 kDa) [%]	99.0	98.0	≥ 95
IgG subclass	% IgG1/2/3/4	58.4/37.1/**0.6**/3.9	65.8/31.0/8.7/3.5	-
Amidolytic Activity	nmol/mL min	< 10	< 10	< 10
PKA	IU/mL	< 4	8.2	< 10
ACA	%	48.2	fixed all complement	< 50
IgA	mg/mL (at 10% protein)	0.7	1.0	≤ 0.14
Ligand	µg/mL (at 10% protein)	1.45	1.76	-

Table 1. Final container comparison of IgSelect affinity resin and protein G affinity resin versus the Gammagard Liquid product specifications (red = unfavorable results)

Both resins showed promising IgG recoveries of greater than 75% in the eluate. Purity (γ-globulin content), as measured by cellulose acetate electrophoresis (CAE) which has for decades commonly been used to determine the protein composition in plasma fractions (Kawai, 1973, Putnam, 1975), met Gammagard Liquid specifications. The IgG subclass distribution of the final container produced with IgSelect resin had an unusually low IgG_3 content, notwithstanding GE Healthcare's claim that IgSelect will bind all human IgG subclasses. Zandian & Jungbauer also used polyclonal IgG for their IgSelect evaluation but did not scrutinize the IgG subclass selectivity (Zandian & Jungbauer, 2009). In contrast to IgSelect, we found that final containers produced with Protein G Sepharose FF showed an IgG subclass distribution similar to the normal IgG subclass distribution in human blood plasma, as expected (Morell et al., 1972, Schauer et al., 2003).

Amidolytic activity was below the detection limit for both final containers. Amidolytic activity is a sum criterion, where the consumption of the chromogenic substrate PL-1 (D-norleucyl-L-lysin-p-nitroanilide-dihydochloride) is photometrically determined.

The Protein G Sepharose FF process resulted in a higher prekallikrein activator activity (PKA) than IgSelect (8.2 IU/mL compared with < 4 IU/mL), but it was still within the specifications for Gammagard Liquid ($\leq$ 10 IU/mL) and the European Pharmacopoeia's specifications for IGIV ($\leq$ 35 IU/mL). The prekallikrein activator activity in the sample tested forms kallikrein out of purified prekallikrein. The kallikrein activity converts the chromogenic substrate Pk-1 (D-α-aminobutanoic acid-L-cyclohexyl alanyl-L-arginin-p-nitroanilin-diacetate) into p-nitroanilin (pNA), which is measured photometrically.

The IgSelect final container complied with the European Pharmacopoeia specifications for anticomplementary activity (ACA), which is a measure of the non-specific consumption of complement by the immunoglobulin preparation ($\leq$ 50%). By contrast, the final container obtained with protein G consumed all complement. All 3 activities measured are criteria to assess the tolerability of these potential immunoglobulin products.

Neither of the chromatography affinity methods were able to reach the high purity standards of Gammagard Liquid with respect to IgA content (≤ 0.14 mg/mL IgA). Additional chromatographic purification steps would be required to reach Gammagard Liquid specifications.

One of the main disadvantages of both affinity resins was the ligand leaching. As much as 1.45 µg/mL and 1.76 µg/mL ligand were detected in the final container concentrated to a 10% protein solution for IgSelect and protein G, respectively. Considering a dose of the polyclonal plasma IgG of 1.0 g/kg (10 mL) body weight, a patient (75 kg) would receive 1.45 µg/mL * 75 * 10 mL = 1087.5 µg. Compared with monoclonal antibodies which are given at a much lower dose, the high ligand content of the polyclonal preparation might lead to unpredictable long-term side effects.

4. Reusability of IgSelect and protein G Sepharose FF

We performed a reusability study to test the stability of the resins. A fresh IgSelect resin was packed into the column with a final bed height of 15 cm. The flow-through was collected in fractions after equilibration and loading with a protein solution to reach a final IgG load of approximately 25 mg IgG/mL resin. The resin was then washed with 30 column volumes of washing buffer, eluted with three column volumes of elution buffer and cleaned with three column volumes of phosphoric acid, acetic acid and butyl alcohol (PAB) (Millipore, 2011) solution. Finally, after a hold time of 24 min, another three column volumes of PAB solution and five column volumes of 1 M NaCl were applied. After the first break-through curve (BTC), consecutive serial runs and cleaning cycles were performed without any fractions being collected. S/D-treated loading material was used throughout this study (for break-through and serial runs). Furthermore, a filter with a pore size of 15 µm was used as a column guard.

Fig. 3 compares the break-through curves (BTC 1 to 3) performed with fresh resin and with resins after 90 and 180 runs and cleaning cycles. The dynamic binding capacity (DBC) at 1% IgG loss decreased after 90 runs and cleaning steps from 16.5 mg to 12.5 mg IgG/mL resin. During these cycles a pressure rise was observed even though only S/D treated loading solutions were used. The pressure rise indicated fouling of the chromatographic media and showed that the recommended PAB cleaning was an insufficient cleaning step for the loading material used. A change of cleaning reagents or longer hold times may increase the cleaning efficiency.

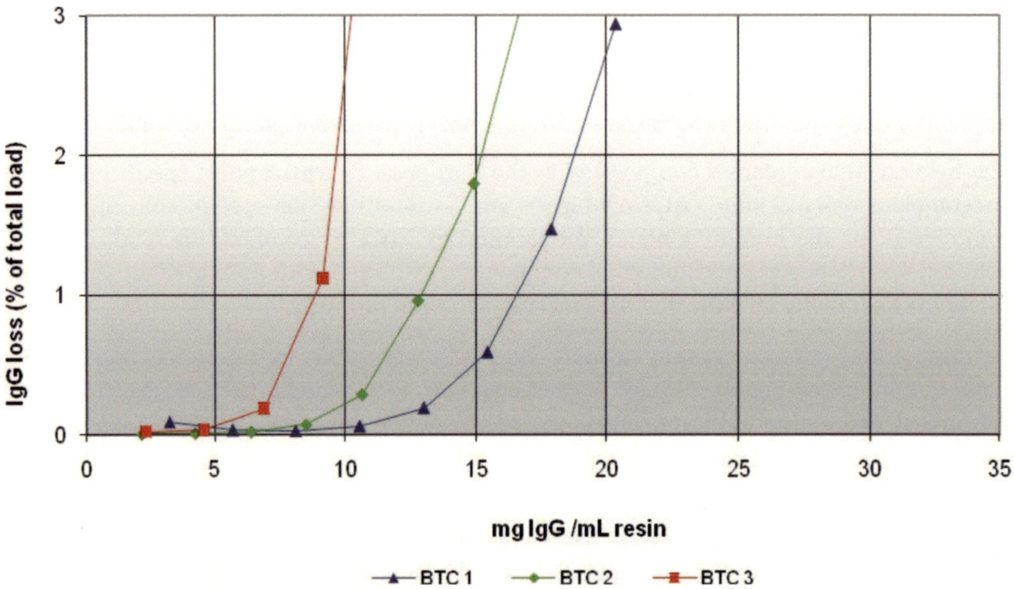

Fig. 3. IgSelect affinity media: break-through curves (BTC). BTC 1 = first run before PAB cleaning, BTC 2 = after 90 runs and PAB cleaning cycles, and BTC 3 = after 180 runs and PAB cleaning cycles

The protein yields in the eluates decreased concomitantly with the decline in resin capacity, which can be explained by the considerable ligand leaching or fouling of the chromatographic media, as already mentioned. 3 µg IgSelect ligand/mL 10% protein solution were found in the eluate after the first run, diminishing to 2.3 µg IgSelect ligand/mL 10% protein solution after 90 cycles. After 180 cycles the eluate contained 5 µg IgSelect ligand/mL 10% protein solution. The details are given in Table 2.

	IgG yield		*IgSelect Ligand*	
	[%]	*[g/L Plasma]*	*[ng/mL]*	*[µg/mL 10% protein solution]*
BTC 1 - Eluate	101.1	4.72	142	3.0
BTC 2 - Eluate	88.7	3.93	134	2.3
BTC 3 - Eluate	67.7	2.92	120	5.0

Table 2. IgG yield and IgSelect ligand concentration in eluate fraction

Our reusability study with Protein G Sepharose FF showed similar characteristics. The loading capacity decreased from a DBC at 1% of 28 mg per mL with fresh resin (see Fig. 4, BTC 1) to a DBC at 1% of approximately 20 mg per mL of resin after 57 runs and cleaning cycles (see Fig. 4, BTC 3). The pH of the loading material seemed to play only a minor, if any, role as shown by the results after 3 runs and cleaning cycles (see Fig. 4, BTC 2).

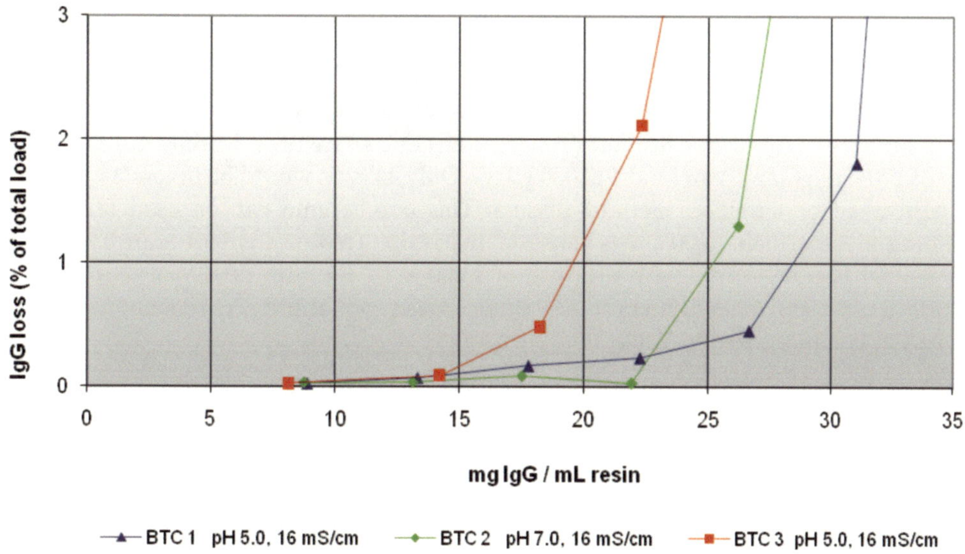

Fig. 4. Protein G Sepharose FF break-through curves (BTC): BTC 1 = without cleaning, BTC 2 = after 3 runs and cleaning cycles, and BTC 3 = after 57 runs and cleaning cycles (different loading pH)

5. Cost-effectiveness of affinity chromatography for the purification of polyclonal IgG from plasma

Reusability and loading capacity are the main criteria for calculating the cost-effectiveness of chromatography steps for IgG purification from human plasma. In the following example cost effectiveness is estimated based on the results gathered with IgSelect affinity resin and Protein G Sepharose FF. The example assumes a manufacturing through-put of one million liters of human plasma per year. This through-put is the same as initially planned for the new fractionation plant in Barcelona by the Spanish company Grifols, one of the world's leading producers of plasma products (reported by PRNewswire on February 18th, 2011).

As shown in Section 4, with a loading capacity of 10 g IgG/L resin the IgSelect affinity resin can be used for a maximum of 120 runs (with 120 PAB cleaning cycles) without a major loss of IgG during the loading and washing procedure. Assuming an IgG yield of 5.2 g/L plasma (US source plasma) in the IgSelect starting material, 5.2 million grams IgG have to be bound to the resin. The calculation further assumes a maximum column size for large-scale manufacturing of 1,200 L resin, which can be delivered as a radial flow column by Proxcys (Proxcys BV, Bedrijvenweg 4, NL-7833 JH Nieuw-Amsterdam, The Netherlands). 12,000 g IgG can be bound in one purification run on such a column. 434 runs would be needed to process the equivalent of 1 million liters of plasma (5.2 million g IgG/12,000 g IgG per run = 434 runs).

As the column can be used for a maximum of 120 times, repacking the column with fresh resin (4,800 L of resin) needs to be done four (3.62) times for 434 runs. The resin costs approximately 8,500 €/L. A total expenditure of approximately 36.9 million € would be required solely for the resin to process the equivalent of one million liters of plasma. Assuming a final container yield of 4.5 g/L plasma and a revenue of 50 €/g IgG, a total revenue of 225 million € would be expected. Therefore approximately 16.5% of the product revenue would be spent on the affinity media.

On the other hand one 1,200 L column of a conventional ion exchange column as described for Gammagard Liquid purification (Teschner et al., 2007) with a loading capacity of the resin of 100 mg IgG/mL can bind 120,000 g IgG. Only 43 runs would be necessary to process the equivalent of 1 million liters of plasma. This one column can be used for 23 years assuming a more than 1,000 times reuse of the resin. This means less than 0.05 column changes per year. Assuming a resin price of 2,000 €/L, the resin price per year would be 120,000 € or 300-fold less than the variant based on affinity chromatography. This calculation is summarized in Table 3.

	IgSelect	**Protein G Sepharose FF**	**Cation exchange media**	**Unit**
Starting material		1,000,000		L plasma
IgG concentration		5.2		g IgG / L plasma
IgG amount		5,200,000		g IgG / 1 million L plasma
Column size		1200		L resin
Load	10	20	100	mg IgG / mL resin
IgG load/run	12,000	24,000	120,000	g IgG / column
Reusability of one column	120	60	>1000	runs / column and resin
Total runs	434	217	43	runs / 1 million L plasma
New resin	3.62	3.62	<0.05	refilled columns / year
Resin costs	8500	8500	2000	€ / L resin
Resin costs per year	36.9	36.9	0.12	price (million €) / column
Water needed for SD removal (30CV)	15.6	7.8	1.55	(million L) /year
Water costs for SD removal (0.37€/L)	5.78	2.9	0.57	million € / year

Table 3. Calculation of cost-effectiveness: Data for IgSelect affinity resin and Protein G Sepharose FF compared with cation exchange resin

Additionally, the greater column resin and run requirements of the affinity column variant implies more buffer consumption, more labor and a higher environmental burden. This is illustrated in the following calculation. Usually 20 to 30 column volumes are needed to remove solvent and detergent reagents from the IgG bound to the column. In the above-mentioned example for IgSelect affinity resin, 434 runs with a 1,200 L column require 15.6 million L of water, alone for the washing process. The same removal process for S/D reagents will only consume 43 runs*1,200 L*30 = 1,548 million L of water in the ion exchange variant—10 times less. Assuming a water price of 0.37 € per L, just for S/D removal, additional water cost of 5 million € have to be added for the IgSelect process. Considering water is also used for cleaning, regeneration and equilibration, the additional water cost for the IgSelect process compared with ion exchange chromatography would be much higher than those estimated for S/D removal.

6. State of the art IgG preparations and their use

Traditionally human plasma immunoglobulin is mainly used for primary immunodeficiency and severe combined immunodeficiency (Haeney, 1994, Toubi & Etzioni, 2005, Weiler, 2004), but also patients with autoimmune diseases are treated with IgG (Kaveri et al., 1991, Schwartz, 1990). More recently, neurological disorders come into focus (Blaes et al., 1999, Elovaara et al., 2008, Wiles et al., 2002). Currently more than 20 million liters of plasma are fractionated worldwide per year and more than 50 million grams of IgG are manufactured thereof (Buchacher & Iberer, 2006, Burnouf, 2011). These numbers are only one reason to continuously improve productivity and keeping an eye on cost effectiveness.

The safety of the final product is another key aspect which is monitored with increasing attention. During the last decades the control of the donors, the test of the plasma donations and the incorporation of dedicated virus reduction steps into the manufacturing schemes was regularly improved to a very high standard (Furuya et al., 2006, Kreil et al., 2004, Poelsler et al., 2008, Stucki et al., 2008, Trejo et al., 2003) in order to reduce the risk of virus transmission with blood products as often reported in the early days (Bresee et al., 1996, Farrugia & Poulis, 2001).

Very recently, a growing number of reported adverse events for one IGIV product on the market led to the withdrawal of the product and prompted the manufacturer, the authorities and the competitors to investigate the root cause (Roemisch et al., 2011). The investigation resulted in a further requirement for normal human immunoglobulin to show that the manufacturing process has steps incorporated removing the thromboembolic potential from the IgG (as announced by EMEA).

Another actual trend in the immunoglobulin market is the transition from the intravenous (IV) to the subcutaneous (SC) route of administration which allows more flexibility for the patient. This trend is reflected by new IGSC products on the market or in the clinical stage (Jolles et al., 2011, Teschner et al., 2009).

Most of the immunoglobulin products from human plasma are manufactured conventionally combining precipitation and ion exchange chromatography (Ballow et al., 2003, Stein et al., 2009, Teschner et al., 2007), but there are also attempts to introduce new techniques like expanded bed chromatography (Anspach et al., 1999, Barnfield Frej et al., 1997, Hubbuch et al., 2001) or affinity chromatography in the manufacturing of plasma derived therapeutic proteins (Suomela, 1980). In view of the limitations of affinity chromatography the successful implementation is reported for low aboundant plasma proteins (Weber et al., 2011) and hyperimmune IgG preparations (Bryant et al., 2005) which are produced in lower amounts and given at lower doses intramuscularly. Some plasma derived hyperimmune products soon may even be replaced by recombinant products (Frandsen et al., 2011).

Another more general trend is the replacement of IGIV by monoclonal antibodies tailored for specific autoimmune diseases or cancer (Waldmann, 2006) and more recent developments of chimeric molecules and biosimilars (Goldsmith et al., 2007, Kaneko & Niwa, 2011, Wozniak-Knopp et al., 2010). As the number of monoclonal antibodies on the market and their demand is continuously growing, the manufacturers already took the first hurdle and dramatically improved the cell culture yield. The next bottleneck is the downstream processing of the cell culture supernatant and interestingly with higher volumes and IgG concentrations similar restrictions are encountered as for the plasma fractionation processes shown above (Birch & Racher, 2006).

These developments are supported by efforts undertaken by the producers of chromatography resins to offer new resins with ligands depicting affinity properties combined with a higher stability (Baines et al., 2009, ProMetic Bioscience, 2005, Shi et al., 2009). In a certain respect alkali stable mixed mode resins might be the answer to the challenge as there is a trend to more complex ligand structures resulting in the combination of ionic, hydrophobic hydrogen bounding and thiophilic interaction properties resembling at the end affinity like interactions.

7. Conclusion

Affinity chromatography is known to be a highly efficient purification step for trace proteins and monoclonal IgG. We did a detailed evaluation of the feasibility of this technique for purifying polyclonal human IgG using IgSelect affinity media and Protein G Sepharose FF.

Starting with II+III paste and including solvent-detergent treatment, the main advantage of the two affinity chromatography media investigated was found to be the potential of a single step purification process to reach a high γ-globulin purity (almost 100%) as well as an excellent process efficiency of ≥ 75%. However, we encountered drawbacks with this purification technique during the optimization experiments.

Reusability experiments showed considerable ligand leaching, which resulted in a deterioration of the binding capacity of the resin from run to run. This disadvantage was even more severe as the initial binding capacity was already low at 10 mg IgG/mL affinity resin for the use of IgSelect affinity resin and 20 mg IgG/mL affinity resin for the use of Protein G Sepharose FF. This is five to ten times lower than what is usually seen for ion exchange resins. Even with a 2-fold improvement in the binding capacity or reusability, the affinity chromatography option is more expensive and labor intensive than the classic ion exchange purification technique.

Affinity resins cannot be cleaned with sodium hydroxide without a significant reduction in binding capacities. Cleaning steps with either 0.1%Triton X-100 or PAB were not sufficient to exclude fouling of the resin, as indicated by a pressure rise in our experiments. This emphasizes the importance of developing a potent cleaning step or a more resistant resin. Such a cleaning step is mandatory for resins used in the blood plasma fractionation industry to prevent cross contamination from batch to batch.

Both affinity resins investigated were 300-fold more expensive in terms of resin cost, reusability, and binding capacity and at least 5- to 10-fold more expensive in terms of water consumption than common ion exchange resins, e.g. CM Sepharose FF.

From a large scale perspective, intravenously administrable polyclonal IgG produced from human plasma by one-step affinity chromatography using IgSelect or Protein G Sepharose FF as a single purification step was not found to be feasible in terms of IVIG specifications, ligand leaching and because of the significantly higher cost. However, it is an option for the purification of considerable lower amounts of hyperimmunes from human plasma, suitable for subcutaneous or intramuscular administration.

The key step for the introduction of affinity chromatography into large scale plasma fractionation remains the development of a more stable affinity or pseudo-affinity resin without the loss of specificity.

8. Acknowledgment

The work reported in this article was funded by Baxter BioScience and all authors are employees of Baxter BioScience.

The contributions of the entire Plasma Product Development/Product Support Department, notably Reinhard Grausenburger and Theresa Bauer are gratefully acknowledged. The authors especially thank Professor Hans-Peter Schwarz for his continuous support of the work, Elise Langdon-Neuner for editing and Patrick Eagleman for reviewing the article.

9. References

Anspach, F. B., Curbelo, D., Harmann, R., Garke, G. & Deckwer, W. D. (1999). Expanded-bed chromatography in primary protein purification. *Journal of Chromatography A 865*, 129-144.

Audran, R. & Pejaudier, L. (1975). Obtention d'une Préparation d'immunoglobulines G.A.M (IgGAM) à Usage Thérapeutique. *Rev. Fr. Transfus. Immunohematol. 18*, 119-135.

Baines, D., Williams, S. & Burton, S. (2009). Alternatives to protein A capture. *Bioprocessing 8*, 35-37.

Ballow, M. (2002). Intravenous immunoglobulins: clinical experience and viral safety. *J. Am. Pharm. Assoc. 42*, 449-458.

Ballow, M., Berger, M., Bonilla, F. A., Buckley, R. H., Cunningham-Rundles, C. H., Fireman, P., Kaliner, M., Ochs, H. D., Skoda-Smith, S., Sweetser, M. T., Taki, H. & Lathia, C. (2003). Pharmacokinetics and tolerability of a new intravenous immunoglobulin preparation, IGIV-C, 10% (Gamunex, 10%). *Vox Sang. 84*, 202-210.

Barnfield Frej, A. K., Johansson, H. J., Johansson, S. & Leijon, P. (1997). Expanded bed adsorption at production scale: Scale-uP verification, process example and sanitization of column and adsorbent. *Bioprocess Engineering 16*, 57-63.

Birch, J. R. & Racher, A. J. (2006). Antibody purification. *Advanced Drug Delivery Reviews 58*, 671-685.

Blaes, F., Strittmatter, M., Merkelbach, S., Jost, V., Klotz, M., Schimrigk, K. & Hamann, G. F. (1999). Intravenous immunoglobulins in the therapy of paraneoplastic neurological disorders. *J. Neurol. 246*, 299-303.

Bresee, J. S., Mast, E. E., Coleman, P. J., Baron, M. J., Schonberger, L. B., Alter, M. J., Jonas, M. M., Yu, M. Y., Renzi, P. M. & Schneider, L. C. (1996). Hepatitis C virus infection associated with administration of intravenous immune globulin: A cohort study. *JAMA. 276*, 1563-1567.

Bruton, O. C. (1952). Agammaglobulinemia. *Pediatrics 9*, 722-728.

Bryant, C., Baines, D., Carbonell, R., Chen, T., Curling, J., Hayes, T., Burton, S., & Hammond, D. (2005). A new, high yielding, affinity cascade for sequential isolation of plasma proteins of therapeutic value http://www.prometicbiosciences.com/assets/files/presentations/03%20-%20PPB%2005%20Bryant.pdf (15.09.2011)

Buchacher, A. & Iberer, G. (2006). Purification of intravenous immunoglobulin G from human plasma--aspects of yield and virus safety. *Biotechnol. J. 1*, 148-163.

Burnouf, T. (2011). Recombinant plasma proteins. *Vox Sang. 100*, 68-83.

Cohn, E. J., Oncley, J. L., Strong, L. E., Hughes, W. L. & Armstrong, S. H. (1944). Chemical, clinical and immunological studies on the products of human plasma fractionation. I. The characterization of the protein fractions of human plasma. *J. Clin. Invest 23*, 417-432.

Cohn, E. J., Strong, L. E., Hughes, W. L., Mulford, D. J., Ashworth, J. N., Melin, M. & Taylor, H. L. (1946). Preparation and Properties of Serum and Plasma Proteins. IV. A System for the Separation into Fractions of the Protein and Lipoprotein Components of Biological Tissues and Fluids. *J. Am. Chem. Soc. 68*, 459-475.

Elovaara, I., Apostolski, S., van Doorn, P., Gilhus, N. E., Hietaharju, A., Honkaniemi, J., van Schaik, I., Scolding, N., Soelberg Sørensen, P. & Udd, B. (2008). EFNS guidelines for the use of intravenous immunoglobulin in treatment of neurological diseases: EFNS task force on the use of intravenous immunoglobulin in treatment of neurological diseases. *Eur. J. Neurol. 15*, 893-908.

Falksveden, L. G. & Lundblad, G. (1980). Ion exchange and polyethylene glycol precipitation of immunoglobulin G. *Methods of Plasma Protein Fractionation,* edited by J. M. Curling, pp. 93-105. London: Academic Press.

Farrugia, A. & Poulis, P. (2001). Intravenous immunoglobulin: regulatory perspectives on use and supply. *Transfus. Med. 11*, 63-74.

Frandsen, T. P., Naested, H., Rasmussen, S. K., Hauptig, P., Wiberg, F. C., Rasmussen, L. K., Jensen, A. M. V., Persson, P., Wikén, M., Engström, A., Jiang, Y., Thorpe, S. J., Förberg, C. & Tolstrup, A. B. (2011). Consistent manufacturing and quality control of a highly complex recombinant polyclonal antibody product for human therapeutic use. *Biotechnology and Bioengineering 108*, 2171-2181.

Furuya, K., Murai, K., Yokoyama, T., Maeno, H., Takeda, Y., Murozuka, T., Wakisaka, A., Tanifuji, M. & Tomono, T. (2006). Implementation of a 20-nm pore-size filter in the plasma-derived factor VIII manufacturing process. *Vox Sang. 91*, 119-125.

GE Healthcare (2007a). IgSelect affinity media pp. 1-4. Data file 28-9257-92 AA

GE Healthcare (2007b). Protein G Sepharose 4 Fast Flow pp. 1-6. Data File 18-1012-91 AC

Gianazza, E. & Arnaud, P. (1982). A general method for fractionation of plasma proteins. Dye-ligand affinity chromatography on immobilized Cibacron blue F3-GA. *Biochem. J. 201*, 129-136.

Goldsmith, D., Kuhlmann, M. & Covic, A. (2007). Through the looking glass: the protein science of biosimilars. *Clin. Exp. Nephrol. 11*, 191-195.

Haeney, M. (1994). Intravenous immune globulin in primary immunodeficiency. *Clin. Exp. Immunol. 97 Suppl 1*, 11-15.

Hahn, R., Bauerhansl, P., Shimahara, K., Wizniewski, C., Tscheliessnig, A. & Jungbauer, A. (2005). Comparison of protein A affinity sorbents II. Mass transfer properties. *J. Chromatogr. A 1093*, 98-110.

Hahn, R., Schlegel, R. & Jungbauer, A. (2003). Comparison of protein A affinity sorbents. *J. Chromatogr. B Analyt. Technol. Biomed. Life Sci. 790,* 35-51.

Hahn, R., Shimahara, K., Steindl, F. & Jungbauer, A. (2006). Comparison of protein A affinity sorbents III. Life time study. *J. Chromatogr. A 1102,* 224-231.

Hoppe, H. H., Krebs, H. J., Mester, T. & Hennig, W. (1967). Production of anti-Rh gamma globulin for preventive immunization. *Munch. Med. Wochenschr. 109,* 1749-1752.

Hubbuch, J. J., Heeboll-Nielsen, A., Hobley, T. J. & Thomas, O. R. T. (2001). A new fluid distribution system for scale-flexible expanded bed adsorption. *Biotechnology and Bioengineering 78,* 35-43.

Janeway, C. A., Gibson, S. T., Woodruff, L. M., Heyl, J. T., Bailey, O. T. & Newhouser, L. R. (1944). Chemical, clinical and immunological studies on the products of human plasma fractionation. VII. Concentrated human serum albumin. *J Clin Invest 23,* 465-490.

Jolles, S., Bernatowska, E., de, G. J., Borte, M., Cristea, V., Peter, H. H., Belohradsky, B. H., Wahn, V., Neufang-Huber, J., Zenker, O. & Grimbacher, B. (2011). Efficacy and safety of Hizentra((R)) in patients with primary immunodeficiency after a dose-equivalent switch from intravenous or subcutaneous replacement therapy. *Clin. Immunol. 141,* 90-102.

Kaneko, E. & Niwa, R. (2011). Optimizing therapeutic antibody function: progress with Fc domain engineering. *BioDrugs. 25,* 1-11.

Kaveri, S. V., Dietrich, G., Hurez, V. & Kazatchkine, M. D. (1991). Intravenous immunoglobulins (IVIg) in the treatment of autoimmune diseases. *Clin. Exp. Immunol. 86,* 192-198.

Kawai, T. (1973). *Clinical Aspects of the Plasma Proteins,* 1 ed. Tokyo: Igaku Shoin LTD.

Kelley, B. (2007). Very Large Scale Monoclonal Antibody Purification: The Case of Conventional Unit Operations. *Biotechnol. Prog. 23,* 995-1008.

Kelley, B., Blank, G., & Lee, A. (2009). Downstream Processing of Monoclonal Antibodies: Current Practices and Future Opportunities. *Process Scale Purification of Antibodies,* edited by U. Gottschalk, pp. 1-23. USA: Wiley.

Kistler, P. & Nitschmann, H. (1962). Large Scale Production of Human Plasma Fractions. *Vox Sang. 7,* 414-424.

Kreil, T. R., Poelsler, G., Teschner, W., & Schwarz, H. (2004). Development of a new 10% liquid, triple virus reduced intravenous immune globulin product, new generation IGIV *J. Allergy Clin. Immunol. Volume 113, Number 2 (2004)* S128.

Low, D., O'Leary, R. & Pujyr, N. S. (2007). Future of antibody purification. *J. Chromatogr. B Analyt. Technol. Biomed. Life Sci. 848,* 48-63.

Millipore (2011). PAB - An Enhanced Sanitisation Option for ProSep Protein A Affinity Chromatography Media pp. 1-4. Technical Brief

Morell, A., Skvari, F. & Steinberg, A. (1972). Correlations between the Concentrations of the Four Subclasses of IgG and Gm Allotypes in Normal Human Sera. *J. Immunol. 108,* 195-206.

Oncley, J. L., Merlin, M., Richert, D. A., Cameron, J. W. & Gross, P. M. (1949). The Separation of the Antibodies, Isoagglutinins, Prothrombin, Plasminogen and beta 1-Lipoprotein into Subfractions of Human Plasma. *J. Am. Chem. Soc. 71,* 541-550.

Poelsler, G., Berting, A., Kindermann, J., Spruth, M., Hämmerle, T., Teschner, W., Schwarz, H. P. & Kreil, T. R. (2008). A new liquid intravenous immunoglobulin with three dedicated virus reduction steps: virus and prion reduction capacity. *Vox Sang. 94*, 184-192.

Polson, A., Potgieter, G. M., Largier, J. F., Mears, G. E. & Joubert, F. J. (1964). The Fractionation of Protein Mixtures by Linear Polymers of High Molecular Weight. *Biochim. Biophys. Acta 82*, 463-475.

ProMetic Bioscience (2005). Mimetic ligands pp. 1-5. http://wolfson.huji.ac.il/purification/PDF/affinity/PROMETIC_MimeticLigands.pdf (15.09.2011)

Putnam, F. W. (1975). *The plasma proteins*, 2 ed. New York: Academic Press.

Relkin, N. R., Szabo, P., Adamiak, B., Burgut, T., Monthe, C., Lent, R. W., Younkin, S., Younkin, L., Schiff, R. & Weksler, M. E. (2009). 18-Month study of intravenous immunoglobulin for treatment of mild Alzheimer disease. *Neurobiol. Aging 30*, 1728-1736.

Roemisch, J. R., Kaar, W., Zoechling, A., Kannicht, C., Putz, M., Kohla, G., Schulz, P., Pock, K., Huber, S., Fuchs, B., Buchacher, A., Krause, D., einberger, J. & empters, G. (2011). Identification of activated FXI as the major biochemical root cause in IVIG batches associated with thromboembolic events. Analytical and experimental approaches resulting in corrective and preventive measures implemented into the Octagam(R) manufacturing process. *WebmedCentral* 1-16.

Roque, A. C., Silva, C. S. & Taipa, M. A. (2007). Affinity-based methodologies and ligands for antibody purification: advances and perspectives. *J. Chromatogr. A 1160*, 44-55.

Schauer, U., Stemberg, F., Rieger, C. H., Borte, M., Schubert, S., Riedel, F., Herz, U., Renz, H., Wick, M., Carr-Smith, H. D., Bradwell, A. R. & Herzog, W. (2003). IgG subclass concentrations in certified reference material 470 and reference values for children and adults determined with the binding site reagents. *Clin. Chem. 49*, 1924-1929.

Schwartz, S. A. (1990). Intravenous immunoglobulin (IVIG) for the therapy of autoimmune disorders. *J. Clin. Immunol. 10*, 81-89.

Shi, Q. H., Cheng, Z. & Sun, Y. (2009). 4-(1H-imidazol-1-yl) aniline: a new ligand of mixed-mode chromatography for antibody purification. *Journal of Chromatography A 1219*, 6081-6087.

Stein, M. R., Nelson, R. P., Church, J. A., Wasserman, R. L., Borte, M., Vermylen, C. & Bichler, J. (2009). Safety and Efficacy of Privigen®, a Novel 10% Liquid Immunoglobulin Preparation for Intravenous Use, in Patients with Primary Immunodeficiencies. *J. Clin. Immunol. 29*, 137-144.

Steinbruch, M. & Audran, R. (1969). Isolation of IgG Immunoglobulin from Human Plasma Using Caprylic Acid. *Rev. Fr. Etud. Clin. Biol. 14*, 1054-1058.

Stucki, M., Boschetti, N., Schafer, W., Hostettler, T., Kasermann, F., Nowak, T., Gröner, A. & Kempf, C. (2008). Investigations of prion and virus safety of a new liquid IVIG product. *Biologicals. 36*, 239-247.

Suomela, H. (1980). An Ion Exchange Method for Immunoglobulin Production. *Methods of Plasma Protein Fractionation,* edited by J. M. Curling, pp. 107-113. London: Academic Press.

Teschner, W., Butterweck, H. A., Auer, W., Muchitsch, E. M., Weber, A., Liu, S. L., Wah, P. S. & Schwarz, H. P. (2007). A new liquid, intravenous immunoglobulin product (IGIV 10%) highly purified by a state-of-the-art process. *Vox Sang. 92,* 42-55.

Teschner, W., Butterweck, H. A., Pölsler, G., Kreil, T. R., Weber, A., Muchitsch, E. M., Ehrlich, H. J. & Schwarz, H. P. (2009). Preclinical characterization and viral safety of a new human plasma-derived liquid 20 percent immunoglobulin concentrate for subcutaneous administration. *2nd European Congress of Immunology.*

Thorn, G. W. & Armstrong, S. H. (1945). Chemical, clinical, and immunological studies on the products of human plasma fractionation; the use of salt-poor concentrated human serum albumin solution in the treatment of chronic Bright's disease. *J. Clin. Invest 24,* 802-828.

Thorn, G. W., Armstrong, S. H. & Davenport, V. D. (1946). Chemical, clinical, and immunological studies on the products of human plasma fractionation; the use of salt-poor concentrated human serum albumin solution in the treatment of hepatid cirrhosis. *J. Clin. Invest 25,* 304-323.

Toubi, E. & Etzioni, A. (2005). Intravenous immunoglobulin in immunodeficiency states: state of the art. *Clin. Rev. Allergy Immunol. 29,* 167-172.

Travis, J., Bowen, J., Tewksbury, D., Johnson, D. & Pannell, R. (1976). Isolation of albumin from whole human plasma and fractionation of albumin-depleted plasma. *Biochem. J. 157,* 301-306.

Trejo, S. R., Hotta, J. A., Lebing, W., Stenland, C., Storms, R. E., Lee, D. C., Li, H., Petteway, S. & Remington, K. M. (2003). Evaluation of virus and prion reduction in a new intravenous immunoglobulin manufacturing process. *Vox Sang. 84,* 176-187.

Waldmann, T. L. (2006). Effective cancer therapy through immunomodulation. *Annu. Rev. Med. 57,* 65-81.

Weber, A., Butterweck, H., Mais-Paul, U., Teschner, W., Lei, L., Muchitsch, E. M., Kolarich, D., Altmann, F., Ehrlich, H. J. & Schwarz, H. P. (2011). Biochemical, molecular and preclinical characterization of a double-virus-reduced human butyrylcholinesterase preparation designed for clinical use. *Vox Sang. 100,* 285-297.

Weiler, C. R. (2004). Immunoglobulin therapy: history, indications, and routes of administration. *Int. J. Dermatol. 43,* 163-166.

Wiles, C. M., Brown, P., Chapel, H., Guerrini, R., Hughes, R. A., Martin, T. D., McCrone, P., Newsom-Davis, J., Palace, J., Rees, J. H., Rose, M. R., Scolding, N. & Webster, A. D. (2002). Intravenous immunoglobulin in neurological disease: A specialist review. *J. Neurol. Neurosurg. Psychiatr. 72,* 440-448.

Wozniak-Knopp, G., Bartl, S., Bauer, A., Mostageer, M., Woisetschlager, M., Antes, B., Ettl, K., Kainer, M., Weberhofer, G., Wiederkum, S., Himmler, G., Mudde, G. C. & Ruker, F. (2010). Introducing antigen-binding sites in structural loops of immunoglobulin constant domains: Fc fragments with engineered HER2/neu-binding sites and antibody properties. *Protein Eng Des Sel 23,* 289-297.

Zandian, M. & Jungbauer, A. (2009). Engineering properties of a camelid antibody affinity sorbent for Immunoglobulin G purification. *J. Chromatogr. A. 1216*, 5548-5556.

Part 4

GST – Tagged Affinity Chromatography

7

A Study of the Glutathione Transferase Proteome of *Drosophila melanogaster*: Use of S- Substituted Glutathiones as Affinity Ligands

Ramavati Pal, Milana Blakemore, Michelle Ding and Alan G. Clark*
Victoria University of Wellington, Wellington
New Zealand

1. Introduction

Glutathione transferases (GSTs) are a widely distributed super-family of enzymes involved in detoxification, catalyzing the conjugation of a great range of electrophilic compounds with the tripeptide glutathione (Boyland and Chasseaud, 1969). Much of the interest in insect GSTs has, for many years, been focused on their role in the development of resistance to many insecticides (Enayati et al., 2005; Motoyama and Dauterman, 1980). The enzymes are involved in resistance to most of the major classes of insecticide. These include organophosphates (e.g. Lewis & Sawicki, 1971; Oppenoorth et al., 1977); organochlorines, especially DDT (Clark & Shamaan, 1984, Tang & Tu, 1994), chitin synthesis inhibitors (Sonoda & Tsumuki, 2005), and pyrethroids by both direct (Yamamoto et al., 2009b) and indirect (Vontas et al., 2001) mechanisms. There are many instances in which detoxication of insecticides has been shown to be catalyzed by GSTs of the Delta or Epsilon classes (Lumjuan et al., 2005; Tang & Tu, 1994; Wei et al., 2001) but GSTs from Omega and Zeta classes (Yamamoto et al., 2009a; Yamamoto et al., 2009b) have also been reported as being involved in insecticide resistance.

In addition to their well-established toxicological roles, it is becoming increasingly apparent that insect GSTs may be involved in a number of other important physiological processes. These include olfaction (Rogers et al., 1999), regulation of apoptosis (Adler et al., 1999; Udomsinprasert et al., 2004), eye pigment synthesis (Kim et al., 2006), haeme binding (Lumjuan et al., 2007) and wound healing (Li et al., 2002). In order to disentangle these multiple roles, it is desirable to develop methods to characterize as fully as possible the expression under differing conditions of the many insect glutathione transferases.

In a previous study (Alias & Clark, 2007) the glutathione conjugate of bromosulfophthalein (BSP), a strong inhibitor of several and a substrate of some GSTs (Prapanthadara et al., 2000) was employed as a ligand. The use of this matrix resulted in the purification of GSTs from Sigma, Delta and Epsilon families but many members from these families were not detected and no members of the Zeta and Omega families were detected at all. In the present work

* Corresponding Author

the use of different ligands is examined, to determine to what extent ligand choice influences the part of the GST proteome thus isolated.

Although the catalytic repertoires of GSTs tend to overlap, different families of GST are likely to have catalytic activity with substrate(s) characteristic principally of that family (Jakoby, 1978; Yu, 2002). Many, but not all, GSTs use 2,4-dinitrochlorobenzene (CDNB) as a substrate and the use of the product conjugate (S-2,4-(dinitrophenyl) GSH (DNP-SG)) as a ligand might be expected to isolate a wide range of isoforms. On the other hand, in *Drosophila*, the substrate 3,4-dichloronitrobenzene (DCNB) is used by a more restricted range of isoforms (Alias & Clark, 2007), so that a matrix employing its glutathione conjugate, S-(2-chloro-4-nitrophenyl) glutathione (CNP), might be expected to bind a narrower range of GSTs. Similarly, in using S-(4-nitrobenzyl) glutathione (NB-SG) as ligand, by analogy with mammalian results (Jakoby, 1978), it was anticipated that a different range of GSTs again would be selected. In the present work, we explore the potential of this approach to define the GST proteome.

2. Material and methods

2.1 Materials

Sepharose 6B, epichlorhydrin, L-glutathione (reduced), CDNB, DCNB, lactate dehydrogenase, "modified" sequencing-grade trypsin, Coomassie Brilliant Blue G-250, and Protease Inhibitor Cocktail (general use) were purchased from Sigma-Aldrich. Immobiline™ Drystrips, Destreak™ reagent, Vivaspin™ centrifugal concentrators and HiTrap™ desalting columns (5 ml) were obtained from GE Healthcare (NZ). Bradford Protein Assay reagent and SDS 2D PAGE protein standards were from BioRad Laboratories NZ. Aluminium-backed Silica Gel 60 plates were from Merck Ltd. (NZ). Benchmark™ protein ladder was obtained from Invitrogen (NZ) Ltd.

All other materials and chemicals used were of the highest purity available commercially. Chromatography was carried out using an Amersham Bioscience AKTA FPLC™.

Drosophila melanogaster, wild-type adult, 5-day post-emergence, flies were supplied by the School of Biological Science, Victoria University of Wellington. They were collected and stored at -20 °C until required.

2.2 Synthesis of glutathione conjugates

These conjugates were prepared by incubating CDNB and DCNB or p-nitrobenzyl chloride in ethanol with GSH in deionised water at pH 9.6, adapting the method of Vince *et al* (1971). The mixtures were kept at room temperature for 5-48 hr depending on the reactivity of the compound. After incubation, ethanol was removed from the mixture by rotary evaporation. The conjugates DNP-SG, CNP-SG or NB-SG were precipitated by decreasing the pH 9.6 from to 3.0. The precipitate was filtered, redissolved and reprecipitated twice and dried *in vacuo*. The final product was chromatographically homogeneous when examined on 0.25 mm layers of silica gel 60 in butanol: acetic acid: water (4:1:5 – upper phase) using ninhydin: collidine in ethanol (0.3: 5 : 95 (w:v:v)) as a location reagent (Lato et al., 1974). The conjugates were immobilised on epichlorhydrin-activated Sepharose 6B as described by for the BSP-SG conjugate (Clark et al., 1990).

The extent of substitution was estimated by dissolving aliquots of the gels by heating to 100°C in 6M HCl for 30 sec. The hydrolysates were neutralized with 1M-NaOH. The extent of substitution was calculated using extinction coefficients for DNP-SG and CNP-SG of 9600, 8400 $l.mol^{-1}cm^{-1}$ at 340 and 344 nm respectively (Habig et al., 1974b) and for NB-SG of 4210 $l.mol^{-1}cm^{-1}$at 280nm. The extent of substitution was 9.55 μmol/ ml of gel for DNP-SG, 9.65 μmol/ ml of gel for CNP-SG and 11.5 μmol/ ml of gel for NB-SG.

2.3 Enzyme assays

Enzyme activity was assayed by measuring the conjugation of CDNB, DCNB and ethacrynic acid (EA) with glutathione (Habig et al., 1974b). Assays with *p*-nitrophenol acetate (*p*-NPA) (Keen et al., 1976), *trans*-2-nonenal (TNE) (Brophy et al., 1989) and dehydroascorbic acid (DHA) (Kim et al., 2006) were also carried out to study the specificity of the partially purified GST preparations.

2.4 Protein determination

During the affinity chromatographic procedures, protein was monitored by measuring the extinction of the collected fractions at 280 nm. For specific activity determination, samples of known enzymatic activity were pooled and concentrated using VivaSpin centrifugal concentrators (10kD) and protein was assayed by using the Bio-Rad protein assay kit, with bovine serum albumin (BSA) as the standard protein (Bradford, 1976).

2.5 Enzyme preparation

Adult fruit flies (1 – 3 g) were homogenised in five volumes of cold 0.05 M phosphate buffer, pH 7.4 containing 0.1 mM phenylmethylsulfonyl fluoride (PMSF), 1 mM dithiothreitol (DTT), 1 mM EDTA and 0.1 mM phenylthiourea (PTU) by using a Polytron ™ homogeniser. Protease inhibitor cocktail was added to the concentration specified by the manufacturer and cysteine (2mg/ml) was added to prevent the oxidative darkening of the homogenate. The homogenate was centrifuged at 100,000 x g for 1h at 4°C in a Beckman XL-80 ultracentrifuge. The supernatant was filtered through glass wool and passed over a 5 ml HiTrap Desalting column which had been equilibrated with 0.05 M sodium phosphate buffer, pH 7.4, at a flow rate of 30 ml/h. This column bound potentially inhibitory pigments with sufficient affinity to separate them from the excluded proteins. In some experiments the supernatant was chromatographed on a column of Sephadex G-25 (5x 25 cm) to remove pigments. Fractions containing activity towards CDNB were pooled. Losses in activity were negligible during this procedure, but up to 20% of the activity was lost if the enzyme was allowed to stand for 24h at 4°C. For this reason, affinity chromatography followed immediately after this procedure.

2.6 Affinity chromatography

The initial stage of the analysis involved chromatography of the pooled fractions from the depigmentation stage on a GSH-Sepharose column (1.0 x 5 cm) at a flow rate of 1 ml/min. The enzyme sample was followed by 0.05 M sodium phosphate buffer (Buffer A) until the absorbance at 280 nm had fallen to a value close to zero. Five volumes of 1.0 M NaCl buffered as above at pH 7.4 were then applied and followed again by five volumes of buffer A. The bound GSTs were eluted with 20 mM glutathione in 0.05 M sodium phosphate buffer, pH 9.6. Fractions of 3 ml were collected. See Fig. 1. This column employed as its ligand glutathione, attached via its sulfhydryl group to epichlorohydrin-activated Sepharose (Clark et al., 1990). This matrix removes specifically, and almost quantitatively, glutathione transferases S1, D1 and D3 which enables a subsequent clearer separation of GSTs from other classes (Alias and Clark, 2007). The extent of substitution of this column routinely used was 10 µmol/ml gel but substitution levels of up to 30 µmol/ml gel were tested.

The aryl-substituted GSH matrices were packed under gravity in disposable plastic syringes to form columns of bed volume 5 ml. Void fractions from the GSH affinity matrix having significant activity with CDNB were pooled and applied to the columns. The bound GSTs were eluted with 20 mM GSH at pH 9.6 as described above. Aliquots from each stage of the purification were reserved for testing for activity with the test substrates. Experiments were performed in triplicate. The ligands for these matrices, having aryl substituents on the sulfydryl group of the glutathione and therefore acting as product inhibitors competing with glutathione, are linked to the epichlorohydrin-activated Sepharose via the glutamyl amino group, as has been described for the glutathione conjugate of bromosulfophthalein (Clark et al., 1990).

2.7 Gel electrophoresis

Proteins in the material eluted from the affinity column were examined by SDS PAGE using standard conditions (Laemmli, 1970) in a 12% polyacrylamide gel and by 2D electrophoresis using the standard methods described below. In the first dimension isoelectric focusing was carried out in a Multiphor™ II Electrophoresis Unit (Pharmacia Biotech) apparatus. The precipitated proteins were mixed with rehydration buffer (8 M Urea, 2 % (w/v) CHAPS, 15 mM DTT, 30 mM thiourea) containing 2% pH 3-10 Ampholyte solution and Destreak™ reagent. Immobiline™ Drystrips (pI 4-7 and 3-10 for all samples) were rehydrated overnight and were focused in a three phase voltage programme. The first phase was at 200V for 1 min; second, 200-3500V ascending for 1.5h and the third phase was 3500V for 1.5h. Proteins in the gel were reduced with DTT and alkylated with iodoacetamide using standard methods. In the second dimension, the focused strips were run on SDS PAGE to separate proteins based on molecular mass. Proteins were located by staining with colloidal Coomassie Blue G-250 (Neuhoff et al., 1988). 2D gels were calibrated with respect to molecular mass using the BenchMark™ Protein Ladder and with respect to isoelectric point with BioRad SDS 2D PAGE protein standards.

2.8 Protein identification by MALDI-TOF

Protein spots were excised from gels and destained with 50 mM ammonium bicarbonate: acetonitrile (MeCN) (1:1). When destained, gel pieces were dehydrated and completely dried by vacuum centrifugation. The gel pieces were then rehydrated with 2 µl of 50 mM ammonium bicarbonate solution containing 0.125 µg of sequencing grade trypsin at room temperature. After 2-3 h, 30 µl of 50 mM ammonium bicarbonate solution was added to assist diffusion of peptides at room temperature overnight. Peptides were later extracted by removing the ammonium bicarbonate solution, followed by two washes of 30 µl for 1 h each with a solution containing 0.2% TFA and then with 0.2% TFA:MeCN (1:1). The extracts were then dried in a Speed-Vac for 2 h. The samples were stored at 4^0C until required for analysis.

The MALDI-TOF analysis was performed using a PerSeptive Biosystem Voyager DE™ PRO Biospectrometry™ Workstation equipped with a delayed extraction unit. The mass spectra were obtained by using a Voyager Instrument Control Panel V 5.0 programme. The laser intensity was in the range of 1800-2100, and the accelerating voltage was 20 kV. Peptide spectra were obtained in positive reflectron mode in the range of 750 to 3500 *m/z*. The peptide solution was loaded onto the MALDI target plate by mixing with 2 µl of a matrix solution, prepared by dissolving 15 mg/ml alpha-cyano-4-hydroxycinnamic acid solution in 50% MeCN-0.1% trifluoroacetic acid (v/v), and allowed to dry. External calibration used Calibration Mixture 2 from Applied Biosystems. Each mass spectrum was generated by accumulating data from 150 to 200 laser shots. Database searches were carried out with the peptide masses against the non redundant NCBI database using the search programmes ProFound (http://prowl.rockefeller.edu/) (Zhang & Chait, 2000) and results were confirmed using MASCOT (www.matrixscience.com).

2.9 Assignment of GST family

GSTs identified by MALDI-TOF, which had not been explicitly assigned to particular families were matched against the *D. melanogaster* database in SwissProt Knowledgebase using the Basic Local Alignment Search Tool, BLAST (http://web.expasy.org/blast/). Family assignment was made on the basis of those known GSTs with which the test sequences had the greatest identity.

3. Results

Preliminary experiments showed that all three matrices bound GSTs from the Sigma, Delta, Epsilon and Omega classes. In order to reduce the complexity of the 2D gels obtained, an initial affinity purification of *D. melanogaster* extracts on GSH-agarose affinity was employed to remove the greater part of the Sigma and Delta sub-groups of GSTs. The GSTs removed had high specific activity for CDNB and 2D electrophoresis (not shown) confirmed earlier results that they consisted almost exclusively of GSTs S1, D1 and D3 (Alias and Clark, 2007). The use of matrix substituted with glutathione at up to 30 µmol/ml gel did not change the qualitative nature of the GSTs bound. As shown in Figure 1A and Table 1, a significant proportion of the material with CDNB activity (~40%) was not retained by the GSH affinity column.

The ability of the S-substituted matrices to adsorb the unbound activity was then tested. All of them did adsorb such activity but adsorption was not complete: between 25 and 50% of the applied activity (i.e. 10-25% of the original starting activity) was recovered in the void fractions. Material adsorbed and then eluted from these columns contained significant GST activity with CDNB. A typical elution profile is shown in Fig 1B in which the DNP-SG matrix was used. Material eluted from DNP-SG and CNP-SG columns contained significant activity with DCNB (Table 2) but the eluate from the NB-SG column did not show significant activity with this substrate.

Several model substrates were tested on the preparations and the results are shown in Table 2 below. All preparations showed activity with most of the substrates. The specific activities of the preparations from the DNP-SG and CNP-SG matrices with the different substrates were almost identical. That from the NB-SG matrix differed in that it showed no detectable activity with DCNB.

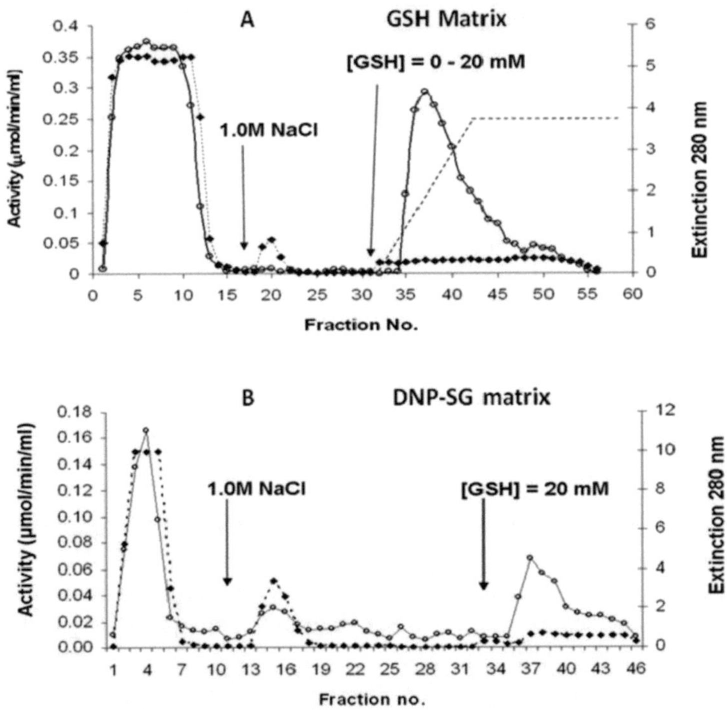

Fig. 1. Typical chromatography elution profiles from substituted glutathione affinity matrices. (A) shows chromatography on immobilised glutathione; (B) shows chromatography on the DNP-SG matrix. Open symbols, enzyme activity with CDNB. Closed symbols, extinction at 280 nm. Experiments were carried out in triplicate, as described in Section 2.6.

Procedure	Total activity (µmol/min) CDNB	Specific activity (µmol/min /mg) CDNB	Yield (%)	Purification (fold)
GSH matrix				
Applied	12.7 ± 1.4	0.132 ± 0.023	100	0.95
Unbound	6.00 ± 3.7	0.093 ± 0.04	43	0.67
Bound	7.00 ± 1.2	10.76 ± 4.67	51	77.9
DNP-GSH matrix				
Applied	3.00 ± 1.9	0.093 ± 0.04	100	1
Unbound	1.40 ± 0.5	0.051 ± 0.02	46	0.54
Bound	1.20 ± 0.6	3.529 ± 0.09	40	38
CNP-GSH matrix				
Applied	3.00 ± 1.9	0.093 ± 0.04	100	1
Unbound	1.70 ± 0.7	0.054 ± 0.02	56	0.58
Bound	1.30 ± 0.3	2.954 ± 0.77	43	32
NB-SG matrix				
Applied	0.86 ± 0.07	0.38 ± 0.03	100	1
Unbound	0.22 ± 0.03	0.24 ± 0.03	25	0.63
Bound	0.23 ± 0.01	7.9 ± 0.30	26	21

Table 1. Affinity chromatography of glutathione transferase activity from *D. melanogaster*. Chromatographic separations were carried out as described in Section 2.5 and 2.6. The results above are mean +/- S.D. from triplicate experiments.

Otherwise, the spectrum of activities was similar to that of the other two types of preparation. Of note were the high dehydroascorbate reductase activities that all preparations showed. The material eluted from the GSH column had the highest specific activity with CDNB.

The affinity-purified fractions obtained by using the DNP-SG and CNP-SG gels were concentrated, precipitated and subjected to 2D electrophoresis. The results from typical gels (pI = 4-7) after chromatography on these media are shown in Fig 2. Isoelectrofocusing over a 3-10 range of pI resulted in poorly resolved protein zones in the range 5-6 and these gels are not shown.

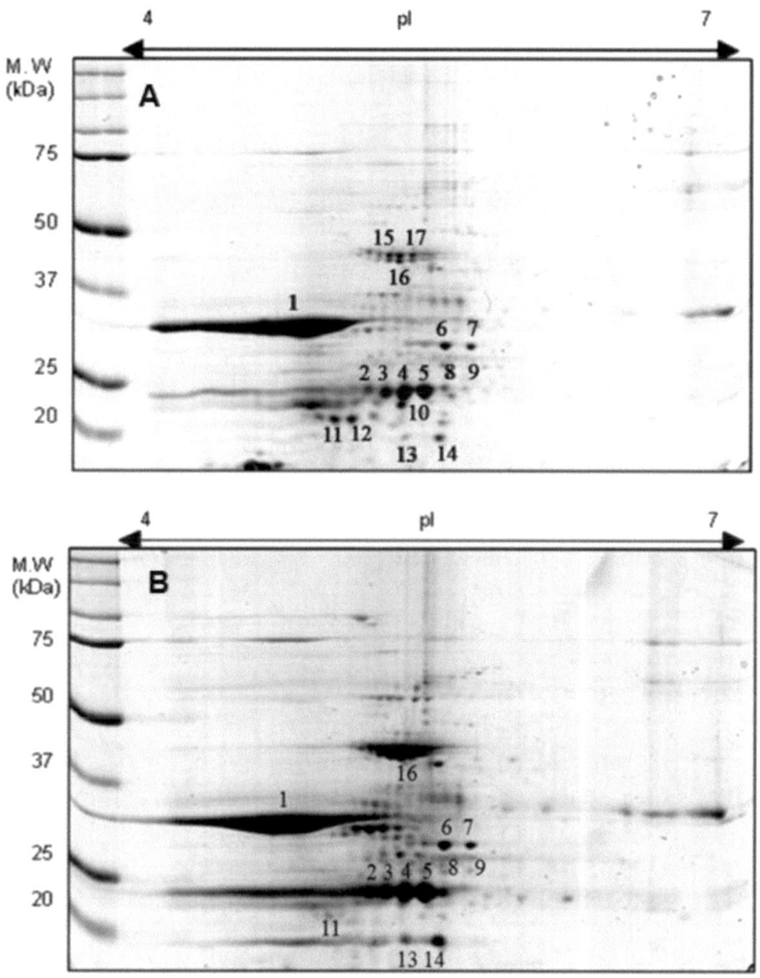

Fig. 2. Two dimensional gel electrophoresis on DNP-SG- (A) or CNP-SG- (B) affinity-purified *D. melanogaster* GSTs. Affinity-purified *D. melanogaster* GSTs were separated in the first dimension on a 7 cm pH 4-7 linear IPG strip and then in the second dimension on a 12% SDS-PAGE gel. The gel was stained with Coomassie Blue G250. The identification of the numbered spots can be found in Table 3 below.

Substrate	Specific activity (µmol/min/mg protein)			
	Eluted from GSH	Eluted from DNP-SG	Eluted from CNP-SG	Eluted from NB-SG
CDNB	13.60± 3.4	4.30 ± 0.90	3.10 ± 0.06	7.34 ± 1.13
DCNB	n.s.	0.18 ± 0.01	0.09 ± 0.06	n.s.
NPA	1.00 ± 0.40	1.30 ± 0.46	1.50 ± 0.52	1.99± 0.75
TNE	0.84 ± 0.48	0.35 ± 0.16	0.16 ± 0.04	0.41 ± 0.12
DHA	3.30 ± 0.70	11.1 ± 6.4	12.65 ± 0.43	27.4 ± 6.3
EA	7.10 ± 1.20	14.2 ± 0.05	17.22 ± 0.01	21.77± 2.88

Table 2. Substrate specificity of glutathione transferases preparations from *D. melanogaster*. Activities were determined from triplicate experiments as described in Section 2.3

Proteins were identified by MALDI-TOF and are listed below in Table 3. Most marked are zones corresponding to DmGSTS1, not all of which had been trapped by the GSH column, and GST CG16936 (spots 2-5). The latter appears at four loci of the same apparent molecular mass and of differing pI. At higher isoelectric points (pI 6.7 and 8.0, not visible in gels covering the pH range 4-7) two Epsilon class GSTs, E3 and E9, were identified in DNP-SG preparations.

Spot No.	**Flybase Annotation**	**Protein**	**Mr/pI (Theor.)**	**Mr./pI (Exp.)**	**P/C**	**Z score**
1	CG8938	GST S1	27.6/4.6	33/4.6	8/47	2.43
2	CG16936	Epsilon GST	25.4/5.9	24/5.0	6/33	2.43
3	CG16936	Epsilon GST	25.4/5.9	24/5.1	9/49	2.35
4	CG16936	Epsilon GST	25.4/5.9	24/5.2	8/47	2.35
5	CG16936	Epsilon GST	25.4/5.9	24/5.3	7/38	2.35
6	CG6673	Omega GST	28.7/6.5	28/5.4	13/48	2.43
7	CG6673	Omega GST	28.7/6.5	28/5.5	15/46	2.43
8	CG6776	Omega GST	27.7/6.5	26/5.4	5/24	1.90
9	CG6776	Omega GST	27.7/6.5	26/5.5	6/22	2.09
10	CG10045	GST D1	23.9/6.9	23/5.2	4/26	2.43
11	CG3269	Drab2	23.7/5.8	20/4.8	7/37	2.43
12	CG32671	Rab9Fa	23.1/6.0	20/4.9	4/21	2.43
13	CG1707	Glyoxallase	20.14/6.1	19/5.2	11/60	2.37
14	C CG1707	Glyoxallase	20.14/6.1	19/5.3	13/72	2.39
15	CG8725	CSN4	46.7/5.9	46/5.1	11/33	1.40
16	CG8725	CSN4	46.7/5.9	46/5.2	9/29	2.43
17	CG9042	GPDH	44.7/6.4	46/5.2	13/38	2.00

Table 3. Identification of *Drosophila* GSTs after chromatography on DNP-SG or CNP-SG matrices. Protein zones are numbered as in Fig. 2. The proteins were identified using MALDI TOF MS peptide mass fingerprinting as described in section 2.7. The P/C column shows the number of peptides identified (P) and the percent coverage of the identified protein (C). Also shown are the theoretical and experimental values for Mr and pI of the identified proteins. A Z-score of greater than 2.34 indicates a probability of less than 1% that the result could have been attained by chance.

These had previously been detected on the gels analysing the product of BSP-SG chromatography (Alias and Clark, 2007). Spots 6 and 7 were identified as Omega GST CG6673 (isoform B). Fainter protein zones (8 and 9) were identified, with lower confidence, as the Omega GST CG6776.

Non-GST proteins eluted from the DNP-SG matrix were also identified. Spots 13 and 14 were identified as CG1707, lactoyl glutathione lysase (Glyoxlase I). Also eluted from the matrix were Rab small GTP-binding proteins (spots 11, 12; CG3269 and CG32671) and higher molecular weight contaminants including glycerol 3-phosphate dehydrogenase (spot 17; CG9042) and the COP9 complex homolog subunit 4 (CSN4, spots 15,16).

The same experiments carried out using CNP-GSH as ligand produced very similar results: 2D gels were almost indistinguishable and specific activities of the enzyme preparations obtained from the two different procedures were almost identical (See Table 2).

Identities of corresponding protein zones in gels resulting from the two types of affinity chromatography were determined independently. These are shown in Table 3 below.

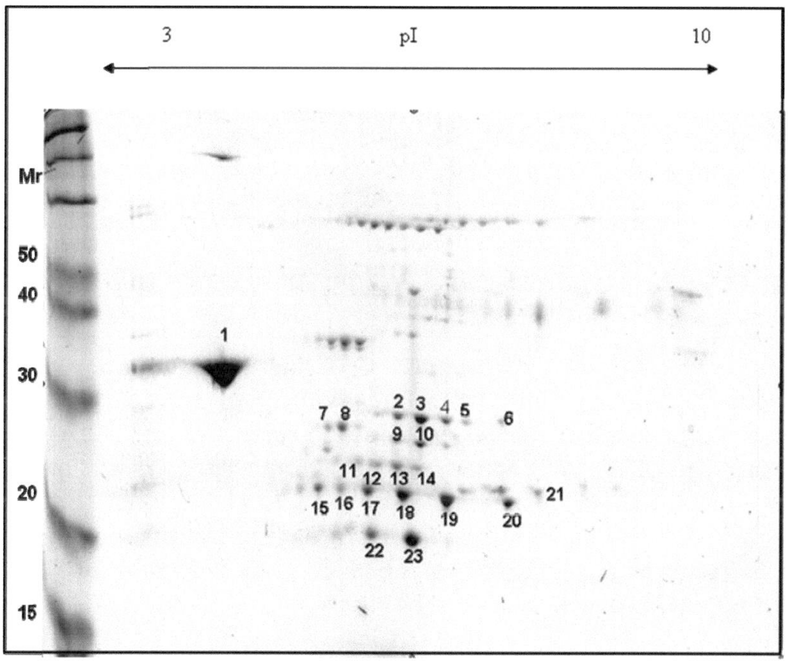

Fig. 3. Two dimensional gel electrophoresis of NB-SG affinity-purified *D. melanogaster* GSTs. Affinity-purified *D. melanogaster* GSTs were analysed as described in Section 2.7. The identification of the numbered spots can be found in Table 4 below.

When NB-SG was the affinity ligand, a rather different picture was seen (Fig. 3). As in the previous experiments Sigma GST had not been completely removed and there was detectable presence of GSTs D1 and D3. In addition to the Omega GSTs revealed by the other two matrices, the use of NB-SG showed the presence of CG6781, the Omega GST involved in synthesis of eye pigments (Kim et al. 2006).

Spot	Flybase Annotation	Protein	pI/Mr (Theor.)	pI/Mr (Exp.)	P/C	Z score
1	CG8938	GST S1	4.6/27.65	4.5/32.4	27/67	2.43
2	CG6673-PB	Omega GST	6.5/28.76	6.75/29.5	19/65	2.43
3	CG6673-PB	Omega GST	6.5/28.76	7.1/29.5	17/55	2.43
4	CG6673-PB	Omega GST	6.5/28.76	7.1/29.5	13/51	2.43
6	CG6673-PA	Omega GST	7.1/29.17	8.1/29.5	11/62	2.27
7	CG6781	Omega GST	5.5/24.87	5.9/286	5/22	2.43
8	CG6781	Omega GST	5.5/24.88	6.1/28.6	15/21	2.43
9	CG6776	Omega GST	6.5/27.76	6.75/27.4	7/18	1.35
10	CG6776	Omega GST	6.5/27.77	7.1/27.4	10/47	2.43
11	CG5224	Epsilon	5.6/25.36	6.25/26.0	9/30	1.71
12	CG11784	Epsilon	6.1/25.98	6.5/26.0	6/35	2.43
13	CG11784	Epsilon	6.1/25.99	6.75/26.0	9/48	2.43
14	CG11784	Epsilon	6.1/25.10	7.0/26.0	6/38	2.22
15	CG4381	GST D3	5.3/22.89	5.8/24.2	9/38	2.43
16	CG10045	GST D1	6.9/23.89	6.1/24.2	3/24	1.03
17	CG10045	GST D1	6.8/24.02	6.45/24.2	8/33	2.43
18	CG10045	GST D1	6.8/24.03	6.85/23.7	11/45	2.43
19	CG10045	GST D1	6.8/24.04	7.5/23.4	12/45	2.43
20	CG10045	GST D1	6.9/23.89	8.25/23.4	10/40	2.43
21	CG17524	GST E3	7.0/24.93	8.6/24.2	9/55	2.3
22	CG1707	Glyoxallase	6.1/20.14	6.5/21.7	8/44	2.43
23	CG1707	Glyoxallase	6.1/20.15	7.0/21.5	9/59	2.43

Table 4. Identification of *Drosophila melanogaster* GSTs after chromatography on the NB-SG matrix. Proteins identified from the NP-SG affinity-purified fractions of *D. melanogaster* separated over a pH range 3-10. Protein zones are numbered as in Fig. 3. The proteins were identified using MALDI TOF MS peptide mass fingerprinting as described in Methods. The P/C column shows the number of peptides identified (P) and the percent coverage of the identified protein (C). Also shown are the theoretical and experimental values for Mr and pI of the identified proteins. A Z-score of greater than 2.34 indicates a probability of less than 1% that the result could have been attained by chance.

Particularly noticeable was the absence of the putative GST CG16936 a protein which was prominent in gels of preparations isolated with the other two matrices. Multiple forms of the closely related protein, CG11784 were seen in these gels. These two proteins as well as CG5224 are closely related to the Epsilon class. Sequence analysis using the programme BLAST shows CG16936 to be closely aligned with GSTs E4 and E9, having 45 % identity with both: CG5224 was most similar to E9 and E10 (scores of 42% with both) and GCG11784 was most similar to E7 and E5 (scores of 37 and 36% respectively). As with the other matrices, the NB-SG matrix also bound higher molecular weight contaminants including glycerol 3-phosphate dehydrogenase (GPDH).

4. Discussion

The multiplicity and overlapping substrate specificities of the GSTs (Habig et al., 1974a; Jakoby 1978) renders them an extremely effective defence against many toxic compounds but this, along with overlapping cellular roles, may render it difficult to attribute particular functions to any individual or group of GSTs and in many contexts, it may in fact be more appropriate to regard them as acting as a complex, rather than as individual enzymes. The current work investigates the potential of the use of *S*-substituted glutathiones as affinity ligands to isolate and characterise the expression of the members of this super-family, selectively.

Several affinity chromatography systems have been used to isolate GSTs from many sources. Ligands tested have included glutathione immobilised through its sulphur group (Clark et al., 1990; Fournier et al., 1992) and glutathione (Yu, 1989) and *S*-substituted glutathione immobilised through the γ-glutamyl amino group. Ligands in the latter category have included the glutathione conjugate of the dye bromosulfophthalein (BSP-SG) (Alias & Clark, 2007; Clark et al., 1977), *S*-hexyl glutathione (Grant & Matsumura, 1989; Guthenberg et al., 1979). *S*-(2,4-dinitrophenyl) glutathione appears to have been employed as an affinity ligand only in the isolation of the ATP-dependent glutathione transporter (Awasthi et al., 1998).

In few studies on the affinity chromatography of insect GSTs has a detailed examination of the GST proteome thus isolated been made: the focus has usually been on the isolation of particular types of GST. Studies on *Musca domestica* (Clark et al., 1990; Fournier et al., 1992) and *D. melanogaster* (Alias & Clark, 2007) GSTs showed that the use of GSH immobilised through its thiol group led to the isolation and identification of a limited number of GSTs from only the Sigma and Delta families. These were, in the case of *D. melanogaster* S1, D1 and D3. The use of BSP-SG as ligand isolated more GSTs from the Delta and Epsilon families (Alias & Clark, 2007). These included D2 (elevated after treatment of the insects with phenobarbital), D1, D3, E3, E6, E7, E9 and an additional, putative Epsilon class GST, CG16936. A small fraction of the Sigma GST was also bound by this matrix. Several non-GST proteins were also bound. Of these, CG1707 (Glyoxallase I) was unsurprising since its substrate is lactoyl glutathione and it binds, presumably, via its affinity for the glutathione moiety. However, this matrix also bound a number of dehydrogenases including CG12262 (acyl CoA dehydrogenase), CG32954 (alcohol dehydrogenase), CG4665 (dihydropteridine reductase). The reason for binding of these proteins is not clear: they resist elution by high concentrations of salt and whether they bind through specific interactions with the matrix or by non-specific hydrophobic interaction or whether it may be through specific protein-protein interaction with bound GST proteins is not currently known.

Results from the present experiments show similarities to those obtained with the BSP matrix (Alias & Clark, 2007) but there are differences. There is, following use of the DNP-SG or CNP-SG materials a substantial zone of Sigma GST, larger than seen with use of the BSP matrix. This suggests that these materials have a stronger affinity for the Sigma GST than did the BSP matrix. The matrices currently under consideration appeared to have a differing selectivity for Epsilon class GSTs compared with the BSP-GSH matrix. Of this family, only E3, E9 and CG16936 are seen in common in preparations made with these matrices, the last-named protein being the most abundant.

A major difference from studies with the BSP-SG matrix is the appearance of four zones containing proteins from the Omega GST family, two identified as CG6673, isoform B (spots

6 and 7) and two (spots 8, 9) as CG6776. The Omega isoform CG6673B has a high dehydroascorbate reductase activity (Kim et al., 2006) and accounts for the high reductase activity in the material eluted from these columns. These isoforms also have minimal catalytic activity with CDNB (Kim et al., 2006) suggesting that GSTs may bind to GSH conjugates of a substrate with which they have little or no catalytic activity. This is supported by the observation that the matrices carrying conjugates of CDNB and DCNB, substrates that discriminate strongly between differing GSTs, appear to bind almost identical groups of proteins, both GSTs and non-GSTs. From this it may be concluded that differences in catalytic performance may be generated by relatively subtle changes in the active site that have little impact on the binding of product analogues. With regard to the Omega GSTs, this view is supported by the work of Whitbread et al. (2005) who demonstrated that the change of Cysteine 32 to Alanine resulted in an increase of activity with CDNB of over an order of magnitude.

A striking feature of the studies of the affinity chromatographic analysis of *D. melanogaster* GSTs is in the small number so far identified. In the works cited here, the number of GSTs positively identified is fourteen, only one third of the number possible (Tu & Agkul, 2005). This may reflect the modest expression of some isoforms. Heavy loading of gels indicates the presence of as yet unidentified proteins in the appropriate Mr range on 2D gels. It may also be that some isoforms have low affinity for the media so far studied. That might explain the absence of any members of Zeta or Theta families from our list of identified proteins.

5 Conclusions

5.1 The present work

The primary aim of this work was to determine whether affinity chromatography systems could be designed to characterise the proteome of the superfamily of glutathione S-transferases, with their multiple toxicological and metabolic roles, in a target organism. This aim has been partially successful. These experiments, in combination with the results from Alias & Clark, 2007 show that the use of different *S*-substituted glutathiones as affinity ligands results in the isolation of differing, but overlapping groups of GSTs from the Delta, Epsilon, Sigma and Omega families. It has not been possible to predict which will bind to a given matrix. The use of these media demonstrates the expression *in vivo* of a number of GSTs, particularly those related to the Epsilon class, the production of which has not previously been shown. Their use also establishes the presence of multiple forms of several GSTs. In the case of the Sigma and D1 isoforms it seems probable that these result from post-translational modifications (Alias & Clark, 2007; Pal & Clark, unpublished).

The proteins identified account for only about 50% of the Sigma, Delta, Epsilon and Omega families. This may be due to low levels of expression or to low affinity for the matrices of some isoforms. A further disappointing aspect of the work has been the failure to identify members of the Zeta or Theta families.

5.2 Future work

The overlap in specificities of the media discussed here suggests that their use in combination should lead to a greater coverage of the GST proteome than the use of any single medium. Alternatively, their use in sequence should lead to the preferential isolation of particular types of GST (CG16936 would be a possible target for such a study) and sequential use might also be employed to enrich preparations in isoforms expressed at too low a level to be identified in the current experiments.

It is possible that Theta and Zeta families have not been detected because of a low affinity for the S-aryl glutathiones. Different types of ligand need to be investigated for the isolation of members of these families.

This work has involved the study of *D. melanogaster* as a model insect but the techniques described have been developed with a view to applying them to insect pests that are of economic or health significance. This proteomic approach generates information that does not depend on knowledge of the genome of the organism in question. Arthropod genomic databases are sufficiently extensive so that cross-database matching of proteomic data may be sufficient to obtain a provisional characterisation of GSTs isolated in this fashion from a pest species. The approach may thus serve as a tool for preliminary studies in organisms for which there is no genomic database. It enables preliminary characterisation of GSTs that are produced to a significant extent and are thus, perhaps, most immediately worthy of individual study.

6. Acknowledgements

This work has been supported in part by funding from the School of Biological Sciences, Victoria University of Wellington and the Wellington Medical Research Foundation. The authors are grateful to Chris Thorn for the supply of insects and to Dr Paul Teesdale-Spittle for a critical reading of the manuscript. One of us, R.P. is grateful to the Victoria University of Wellington for a PhD scholarship.

7. References

Adler, V., Yin, Z.M., Fuchs, S.Y., Benezra, M., Rosario, L., Tew, K.D., Pincus, M.R., Sardana, M., Henderson, C.J., Wolf, C.R., Davis, R.J. & Ronai, Z., (1999). Regulation of JNK signalling by GSTp. *Embo Journal* 18, 1321-1334.

Alias, Z., & Clark, A.G., (2007). Studies on the glutathione S-transferase proteome of adult *D. melanogaster*. Responsiveness to chemical challenge. Proteomics 7, 3618-3628.

Awasthi ,S., Singhal, S.S., Srivastava,S.K., Torman, R.T. Zimniak, P., Bandorowicz-Pikula, J., Singh, S.V., Piper, J.T., Awasthi, Y.C. & Pikula, S. (1998), ATP-Dependent Human Erythrocyte Glutathione-Conjugate Transporter. I. Purification, Photoaffinity Labeling, and Kinetic Characteristics of ATPase Activity. *Biochemistry* 37, 5231-5238.

Boyland, E., & Chasseaud, L., (1969). Role of glutathione and glutathione S-transferases in mercapturic acid biosynthesis. *Advances in Enzymology and Related Areas of Molecular Biology* 32, 173-219.

Bradford, M.M., (1976). Rapid and sensitive method for quantitation of microgram quantities of protein utilizing principle of protein-dye binding *Analytical Biochemistry* 72, 248-254.

Brophy, P.M., Papadopoulos, A., Touraki, M., Coles, B., Korting, W. & Barrett, J., (1989). Purification of cytosolic glutathione transferases from *Schistocephalus solidus* (plerocercoid)-interaction with anthelmintics and products of lipid-peroxidation. *Molecular and Biochemical Parasitology* 36, 187-196.

Chelvanayagam, G., Parker, M.W., & Board, P.G., (2001). Fly fishing for GSTs: a unified nomenclature for mammalian and insect glutathione transferases. *Chemico-Biological Interactions* 133, 256-260.

Clark, A.G., Letoa, M., & Wong, S.T. (1977). Purification by affinity chromatography of a glutathione S-transferase from larvae of *Galleria mellonella. Life Sciences* 20, 141-147.

Clark, A.G., Marshall, S.N., & Qureshi, A.R., (1990). Synthesis and use of an isoform-specific affinity matrix in the purification of glutathione S-transferases from the housefly, *Musca domestica* (L.). *Protein Expression and Purification* 1, 121-126.

Clark, A.G. & Shamaan, N.A. (1984): Evidence that DDT-dehydrochlorinase from the house fly is a glutathione S-transferase. *Pesticide Biochemistry and Physiology* 22: 249-261.

Enayati, A.A., Ranson, H., & Hemingway, J., (2005). Insect glutathione transferases and insecticide resistance. *Insect Molecular Biology* 14, 3-8.

Fournier, D., Bride, J.M., Poirie, M., Berge, J.B., & Plapp, F.W., (1992). Insect glutathione S-transferases: Biochemical characteristics of the major forms from houseflies susceptible and resistant to insecticides. *Journal of Biological Chemistry* 267, 1840-1845.

Grant, D.F., & Matsumura, F., (1989). Glutathione S- transferase 1 and 2 in susceptible and insecticide resistant *Aedes aegypti. Pesticide Biochemistry and Physiology* 33, 132-143.

Guthenberg, C., Akerfeldt, K., & Mannervik, B., (1979). Purification of glutathione S-transferase from human placenta. *Acta Chemica Scandinavica Series B-Organic Chemistry and Biochemistry* 33, 595-596.

Habig, W.H., Pabst, M.J., Fleischner, G., Gatmaitan, Z., Arias, I.M., & Jakoby, W.B., (1974a). Identity of glutathione S-transferase B with ligandin,a major binding protein of liver. *Proceedings of the National Academy of Sciences of the United States of America* 71, 3879-3882.

Habig, W.H., Pabst, M.J., & Jakoby, W.B., (1974b). Glutathione S-transferases- first enzymatic step in mercapturic acid formation. *Journal of Biological Chemistry* 249, 7130-7139.

Jakoby, W.B. (1978). The glutathione S-transferases: a group of multifunctional detoxification proteins. *Advances in Enzymology* 46, 383-414.

Keen, J.H., Habig, W.H., & Jakoby, W.B., (1976). Mechanism for several activities of glutathione S-transferases. *Journal of Biological Chemistry* 251, 6183-6188.

Kim, J., Suh, H., Kim, S., Kim, K., Ahn, C., & Yim, J., (2006). Identification and characteristics of the structural gene for the *Drosophila* eye colour mutant sepia, encoding PDA synthase, a member of the Omega class glutathione S-transferases. *Biochemical Journal* 398, 451-460.

Laemmli, U.K., (1970). Cleavage of structural proteins during assembly of head of bacteriophage -T4 *Nature* 227, 680-685.

Lato, M., Rufini, S., Ghebregz.M, Ciuffini, G., & Mezzetti, T., (1974). Sensitive chromatographic technique for screening of amino acid metabolic defects in newborn. *Clinica Chimica Acta* 53, 273-280.

Lewis, J.B. & Sawicki, R.M. (1971). Characterization of the resistance mechanisms to diazinon, parathion and diazoxon in the organophosphorus resistant SKA strain of houseflies *Musca domestica. Pesticide Biochemistry and Physiology*, 1, 275-285.

Li, D., Scherfer, C., Korayem, A.M., Zhao, Z., Schmidt, O., & Theopold, U., (2002). Insect hemolymph clotting: evidence for interaction between the coagulation system and the prophenoloxidase activating cascade. *Insect Biochemistry and Molecular Biology* 32, 919-928.

Lipke, H., & Kearns, C. W. (1959). DDT dehydrochlorinase I. Isolation, chemical properties, and spectrophotometric assay *The Journal of Biological Chemistry* 234, 2123-2128.

Lumjuan, N., McCarroll, L., Prapanthadara, L.A., Hemingway, J., & Ranson, H., (2005). Elevated activity of an Epsilon class glutathione transferase confers DDT resistance in the dengue vector, *Aedes aegypti*. *Insect Biochemistry and Molecular Biology* 35, 861-871.

Lumjuan, N., Stevenson, B.J., Prapanthadara, L.A., Somboon, P., Brophy, P.M., Loftus, B.J., Severson, D.W., & Ranson, H., (2007). The *Aedes aegypti* glutathione transferase family. *Insect Biochemistry and Molecular Biology* 37, 1026-1035.

Motoyama, N., & Dauterman, W.C., (1980). Glutathione S-transferases: their role in the metabolism of organophosphorus insecticides. *Reviews in Biochemical Toxicology* 2, 49-69.

Neuhoff, V., Arold, N., Taube, D., & Ehrhardt, W., (1988). Improved staining of proteins in polyacrylamide gels including isoelectric-focusing gels with clear background at nanogram sensitivity using Coomassie Brilliant Blue G-250 and R-250. *Electrophoresis* 9, 255-262.

Oppenoorth, F. J., Smissaert, H. R., Welling, W., van der Pas, L. J. T., & Hitman, K. T. (1977). Insensitive acetylcholinesterase, high glutathione-*S*-transferase, and hydrolytic activity as resistance factors in a tetrachlorvinphos-resistant strain of house fly *Pesticide Biochemistry and Physiology* 7, 34-47 .

Prapanthadara, L., Promtet, N., Koottathep, S., Somboon, P., & Ketterman, A.J., (2000). Isoenzymes of glutathione S-transferase from the mosquito *Anopheles dirus* species B: the purification, partial characterization and interaction with various insecticides. *Insect Biochemistry and Molecular Biology* 30, 395-403.

Rogers, M.E., Jani, M.K., & Vogt, R.G., (1999). An olfactory-specific glutathione-S-transferase in the sphinx moth *Manduca sexta*. *Journal of Experimental Biology* 202, 1625-1637.

Tang, A.H., & Tu, C-P.D., (1994). Biochemical characterization of *Drosophila* glutathione S-transferases D1 and D21. *Journal of Biological Chemistry* 269, 27876-27884.

Sonoda, S., & Tsumuki, H (2005). Studies on glutathione *S*-transferase gene involved in chlorfluazuron resistance of the diamondback moth, *Plutella xylostella* L. (Lepidoptera: Yponomeutidae) *Pesticide Biochemistry and Physiology* 82 94–101

Tu, C-P.D., & Agkul, B. (2005) *Drosophila* Glutathione S-Transferases. *Methods in Enzymology*. 401, 204- 226.

Udomsinprasert, R., Bogoyevitch, M.A., & Ketterman, A.J., (2004). Reciprocal regulation of glutathione S-transferase spliceforms and the *Drosophila* c-Jun N-terminal kinase pathway components. *Biochemical Journal* 383, 483-490.

Vince, R., Daluge, S., & Wadd, W.D. (1971). Studies on the inhibition of glyoxalase I by *S*-substituted glutathiones. *Journal of Medicinal Chemistry* 14, 402-404.

Vontas, J.G., Small, G.J., & Hemingway, J. (2001). Glutathione S-transferases as antioxidant defence agents confer pyrethroid resistance in *Nilaparvata lugens*. *Biochemical Journal 357*, 65-72.

Wei, S.H., Clark, A.G., & Syvanen, M., (2001). Identification and cloning of a key insecticide-metabolizing glutathione S-transferase (MdGST-6A) from a hyper insecticide-resistant strain of the housefly *Musca domestica*. *Insect Biochemistry and Molecular Biology* 31, 1145-1153.

Whitbread, A.K., Masoumi, A., Tetlow, N., Schmuck, E., Coggan, M., & Board, P.G. (2005). Characterization of the Omega class of glutathione transferases. *Methods in Enzymology* 401, 78-99.

Yamamoto, K., Nagaoka, S., Banno, Y., & Aso, Y., (2009a). Biochemical properties of an Omega-class glutathione S-transferase of the silkmoth, *Bombyx mori*. *Comparative Biochemistry and Physiology C-Toxicology & Pharmacology* 149, 461-467.

Yamamoto, K., Shigeoka, Y., Aso, Y., Banno, Y., Kimura, M., & Nakashima, T., (2009b). Molecular and biochemical characterization of a Zeta-class glutathione S-transferase of the silkmoth. *Pesticide Biochemistry and Physiology* 94, 30-35.

Yu, S.J., (1989). Purification and characterization of glutathione transferases from 5 phytophagous Lepidoptera. *Pesticide Biochemistry and Physiology* 35, 97-105.

Yu, S.J., (2002) Substrate specificity of glutathione S-transferases from the fall armyworm. Pesticide Biochemistry and Physiology 74, 41-51.

Zhang, W. & Chait, B.T., (2000) ProFound - An expert system for protein identification using mass spectrometry peptide mapping information" *Analytical Chemistry* 72 2482-2489.

Part 5

Nucleic Acids Affinity Purification

8

RNA Affinity Chromatography

Nehal Thakor[1] and Martin Holcik[1,2]
[1]*Apoptosis Research Centre, Children's Hospital of Eastern Ontario*
[2]*Department of Pediatrics, University of Ottawa*
[1,2]*Canada*

1. Introduction

Translation of eukaryotic mRNAs to proteins is the final step of gene expression which involves three steps; initiation, elongation and termination. Translation initiation is a rate limiting and highly regulated process, likely because it is more effective to control the very first step of protein translation instead of dealing with the consequences of aberrant protein translation. Most eukaryotic mRNAs harbor 5′ m^7G cap structure and 3′ poly (A) tail. Typically, translation of eukaryotic mRNA starts with the association of eukaryotic intiation factor (eIF) 4F complex (eIF4E, eIF4G, eIF4A) with the 5′ m^7G cap structure via eIF4E. 43S initiation complex, comprising 40S ribosomal subunit, ternary complex (eIF2-GTP-$tRNAi^{Met}$) and the multi-subunit initiation factor eIF3, is then recruited to mRNA via interaction of eIF3 with the scaffolding protein eIF4G. Subsequently, this pre-initiation complex is believed to scan the mRNA in the 5′ to 3′ direction until an initiation start codon is recognized. eIF2 delivers initiator tRNA into the peptidyl (P) site of the ribosome where initiation codon is situated. Following recruitment of initiator tRNA, eIF5 binds to the resulting 48S initiation complex and induces GTPase activity of eIF2α. Initiation protein factors are released from the 48S initiation complex upon GTP hydrolysis by eIF2α, and the 60S ribosome subunit joins the 40S ribosome subunit to form 80S initiation complex in a process that is aided by eIF5B. Elongation of polypeptide chain synthesis commences following 80S initiation complex formation at AUG (for detail reviews see Gebauer & Hentze, 2004; Holcik & Sonenberg, 2005; Graber & Holcik, 2007; King et al., 2010) (Figure 1).

During physiological or pathophysiological stress conditions the cap-dependent translation initiation is compromised either due to the proteolytic cleavage of initiation factors or changes in the phosphorylation status of the initiation factors and their binding partners (reviewed in King et al., 2010). A class of mRNAs harbouring Internal Ribosome Entry Site (IRES) can bypass the global attenuation of protein translation. IRESes are believed to directly recruit ribosome to the vicinity of start codon thus bypassing the need for cap-binding and ribosome scanning. We and others have shown that the IRES mechanism is utilized preferentially during conditions when normal, cap-dependent translation initiation is attenuated and in fact represents a critical survival switch during oncogenesis (e.g. Holcik et al., 2000; Silvera & Schneider, 2009). Translational control by internal initiation thus represents a novel and unique regulatory mechanism that is critical for cell survival.

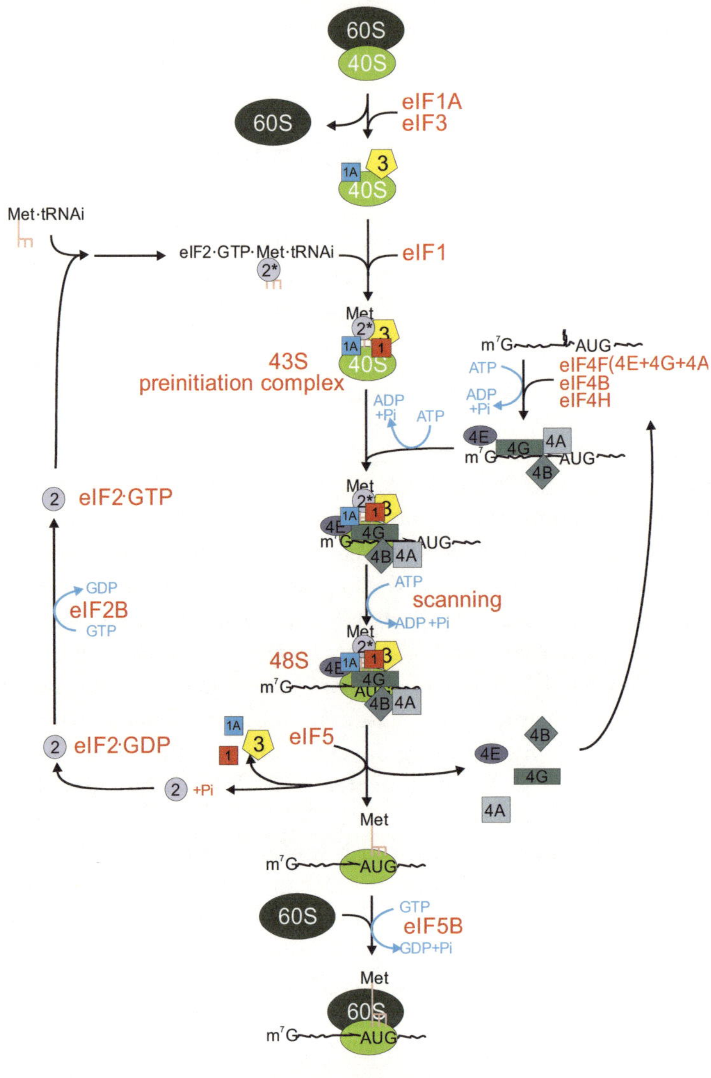

Only key initiation factors (eIFs) involved in the stepwise assembly of the 80S initiation complex are shown. Binding of the eIF4 factors on the mRNA is thought to melt the secondary structure in the 5′ untranslated region and the unfolded mRNA is then capable of interacting with the 43S preinitiation complex. The 40S ribosomal subunit, with associated initiation factors, is then thought to scan in the 3′ direction until an AUG initiation codon in a favorable context is found and the 48S initiation complex is formed. The recruitment of the 60S ribosomal subunit to form the 80S initiation complex completes the initiation step and protein synthesis continues with polypeptide elongation and termination steps (Adapted from Holcik & Sonenberg, 2005).

Fig. 1. Schematic diagram of cap dependent translation initiation

A sizeable proportion of cellular mRNAs, perhaps as much as 10%, has been shown to be translated by a cap-independent mechanism (Johannes & Sarnow, 1998; Blais et al., 2004). It is likely that most of these mRNAs contain an IRES, as IRES-mediated translation is the only validated cap-independent translational mechanism described to date. IRES elements were initially discovered in picornaviruses, where they initiate translation of naturally uncapped viral RNAs (Pelletier & Sonenberg, 1988). Cellular IRESes have been described in a small, but growing, number of mRNAs and often encode proteins that play key roles in cell growth and proliferation, differentiation and the regulation of apoptosis. Thus, we and others have proposed that the selective regulation of IRES-mediated translation is important for the regulation of cell death and survival. Indeed, the experimental data from many laboratories have now validated this hypothesis in a number of models (reviewed in Holcik & Sonenberg, 2005; Braunstein et al., 2007; Silvera et al., 2010; Spriggs et al., 2010; Blagden & Willis, 2011; Komar & Hatzoglou, 2011). In contrast to cap-dependent translation, the translation of IRES-containing cellular mRNA is poorly understood and requires the activity of auxiliary RNA-binding proteins that function as IRES trans-acting factors (ITAFs). Exactly how ITAFs modulate IRES activity is not clear; ITAFs were suggested to act as adapter proteins which act as a bridge between the ribosome and mRNA (Mitchell et al., 2005), or as RNA chaperones which remodel mRNA into a conformation that permits ribosome recruitment (Yaman et al., 2003). We have been studying IRES-mediated translation of cellular mRNAs during stress using two anti-apoptotic proteins, X chromosome-linked inhibitor of apoptosis (XIAP) and cellular inhibitor of apoptosis protein 1 (cIAP1), as model systems (Holcik et al., 1999; Graber et al., 2010; Riley et al., 2010; Thakor & Holcik, 2011). To gain full understanding of the regulation and function of XIAP and cIAP1 IRESes, it was necessary to isolate and identify ITAFs which modulate IRES activity. To this end we have used several complementary approaches that are described in detail below.

2. RNA-affinity chromatography

RNA-affinity chromatography is relatively simple technique which facilitates isolation and characterization of various RNA binding proteins. Numerous strategies have been employed for RNA affinity chromatography using variety of affinity matrices which immobilize RNA to the solid support either covalently or non-covalently, and are briefly described below. While all strategies can be used to successfully isolate RNA-binding proteins, they all present distinct challenges that need to be considered when designing the experiments. Cyanogen bromide activated sepharose and adipic acid dihyrazide agarose are widely used to covalently immobilize RNA (Kaminski et al., 1995; Caputi et al., 1999; Copeland & Driscoll, 1999; Sela-Brown et al., 2000; Hovhannisyan & Carstens, 2009). In the first approach, random sites of RNA attach to the matrix and therefore not all RNA molecules may maintain a homogeneous conformation which limits the number of accessible protein binding sites. Furthermore, cyanogen bromide activated sepharose can non-specifically capture RNA from the cell lysates used for the affinity chromatography which results in isolation of unspecific proteins. The second approach requires oxidation of RNA at 3′ end using sodium periodate which makes the process inconvenient.

The poly-A tailed RNA bound proteins can be isolated using poly-U sepharose which non-covalently immobilizes poly-A tailed RNA (Neupert et al., 1990; Siebel et al., 1994). This approach requires extensive processing of the cytoplasmic lysate, which is used for affinity chromatography, in order to remove endogenous poly-A tailed mRNAs which would otherwise compete for binding on poly-U sepharose.

The RNA which contains recognition sequences for MS2 (bacteriophage coat protein) can be non-covalently attached to amylose beads via a recombinant chimeric protein MS2-MBP (maltose binding protein) (Zhou et al., 2002). In this approach, RNA attachment on the solid matrix depends on the affinity of MS2-MBP protein for the RNA and for the amylose beads. This approach is relatively expensive and inconvenient because it requires purification of recombinant MS2-MBP protein.

Aptamers are functional oligonucleotide sequences which bind non-covalently with high affinity and specificity to proteins, peptides and other small molecules (Tombelli et al., 2005; Ravelet et al., 2006; Hutanu & Remcho, 2007; Peyrin, 2009). Non-covalent RNA attachment can be achieved by inserting aptamer sequences which specifically bind with streptomycin, so called StreptoTag, (Bachler et al., 1999) or streptavidin (Srisawat & Engelke, 2001) immobilized on the solid matrix. If this approach is used, the aptamer specific for streptomycin or streptavidin needs to be inserted at the site where it is exposed to the solvent. If the aptamer insertion site falls along the length of RNA, then it limits protein binding sites.

The high affinity binding of biotin by avidin ($K_d \sim 10^{-15}$ M) has made the biotin-avidin association an extremely powerful tool for affinity chromatography and a method of choice for many researchers (e.g. Rouault et al., 1989; Bayer & Wilchek, 1990; Ruby et al., 1990; Gerbasi & Link, 2007; Sharma, 2008). RNA can be biotinylated either at 5′ end by enzymatic reaction or internally by using biotinylated nucleotides during *in vitro* transcription. The 5′ end biotinylated RNA is preferred over RNA carrying biotin substitutions along its length because internally tagged RNA can bind to the matrix using random sites which limits the protein binding sites. Therefore, we have used biotin-avidin RNA-affinity chromatography approach followed by mass-peptide fingerprinting to isolate and indentify ITAFs which modulate XIAP and cIAP1 IRES activities (Baird et al., 2007; Lewis et al., 2007; Graber et al., 2010; Durie et al., 2011). A general scheme for avidin-biotin RNA affinity chromatography is illustrated in Figure 2. First, *in vitro* transcribed IRES RNA is biotinylated at 5′ end and conjugated to the avidin-agarose beads which are then incubated with pre-cleared cytoplasmic lysate. After stringent washing, captured proteins are separated on SDS-PAGE and stained with SYPRO Ruby or Silver stains. Specific bands can then be excised and analyzed by MALDI-TOF mass spectrometry and peptide mass fingerprinting. Identity of the proteins can be further confirmed by Western blot analysis.

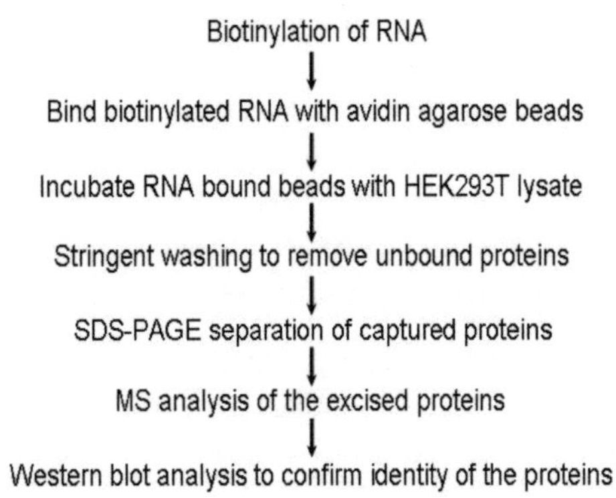

Fig. 2. Schematic diagram of avidin-biotin RNA affinity chromatography procedure

3. Avidin-biotin RNA affinity chromatography procedure

3.1 RNA biotinylation

In vitro transcribe the RNA of interest using Megashortscript kit (Ambion, Austin, TX, USA) and biotinylate at 5′ end by enzymatic reaction using EndTag™ Nucleic Acid Labeling System (Vector Laboratories, Burlingame, CA) as illustrated in Figure 3. Carry out dephosphorylation of RNA by combining 1 µl universal reaction buffer, 0.6 nmol RNA and 1 µl alkaline phosphatase. Bring the final reaction volume to 10 µl and incubate at 37°C for 30 min. Subsequently, add 2 µl universal reaction buffer, 1 µl ATPγS and 2 µl T4 polynucleotide kinase into the reaction. After bringing up the reaction volume to 20 µl, incubate the reaction further at 37°C for 30 min. Mix biotin maleimide (10 µl) with the reaction and incubate at 65°C for 45 min. Mix water (70 µl) and buffered phenol (100 µl) (Sigma-Aldrich, Oakville, ON, Canada) with the reaction and vortex briefly. Centrifuge the tube at 13,000 RPM for 1 min at 4°C and collect the upper aqueous layer. To this aqueous fraction add 5 µl precipitant and 263 µl of 95% ethanol and precipitate RNA overnight at -20°C. Next day centrifuge the tube at 13,000 RPM for 30 minutes at 4°C to obtain RNA pellet. Wash the pellet with 70% ethanol and dry it under vacuum. Resuspend the pellet in 20 µl RNAse free water and store at -80°C if necessary.

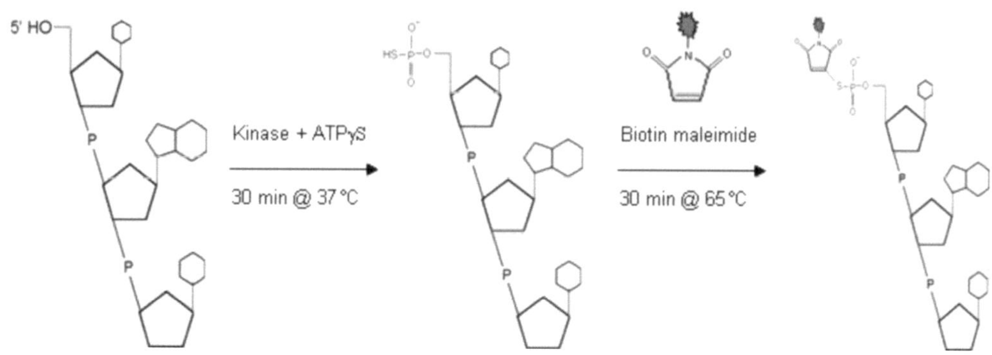

(Adapted from product manual, Vector Laboratories)

Fig. 3. Schematic outline of RNA biotinylation at 5′ end

3.2 Preparation of the cytoplasmic lysate

The following procedure was used to isolate RNA binding proteins from HEK293 cells, but it can be applied to any other cell line. Grow cells to ~80% confluency in five 10 cm plates and collect them in 15 ml Falcon tube in phosphate buffered saline (PBS). Chill the cells on ice and centrifuge at 800 g for 5 min and collect the cell pellet. Wash the cell pellet two times with cold PBS and centrifuge at 800 g at 4°C for 5 min. Resuspend the cells in 750 µl of homogenization buffer (10 mM Tris-HCl (pH 7.4), 1.5 mM $MgCl_2$, 10 mM KCl) containing 0.1 mM phenylmethylsulfonyl fluoride (PMSF), 0.5 mM dithiothretol (DTT) and 10 µg/ml leupeptin and lyse the cells by dounce-homogenizer (30 strokes, pestle 'B'). Centrifuge the lysate at 2,000g at 4°C for 10 min to pellet nuclei. Centrifuge the lysate further at 10,000 g at 4°C for 15 min and collect the supernatant. Determine protein concentration of the lysate by Bradford's or other suitable protein assay. Bring the final concentrations of glycerol and KCl in the lysate to 5 % (vol/vol) and 150 mM respectively, if necessary store the lysate at -80°C.

3.3 Pre-clearing of the cytoplasmic lysate

Mix the cytoplasmic lysate (2 mg) with 20 µl RNAsin (Promega, Madison, WI, USA) and 12 µg yeast tRNA (Sigma-Aldrich). Bring the final volume of the lysate to 825 µl by binding buffer (10 mM Tris-HCl (pH 7.4), 1.5 mM $MgCl_2$, 150 mM KCl, 0.1 mM PMSF, 0.5 mM dithiothretol (DTT), 10 µg/ml leupeptin and 0.05% NP-40). Mix the buffer-washed avidin agarose beads (225 µl) (Sigma-Aldrich) with the cytoplasmic lysate and incubate at 4°C for 2 h on a rotator. Centrifuge the tube at 6,500 rpm for 1 min to pellet the beads and collect the supernatant.

3.4 Capturing RNA bound proteins on avidin agarose matrix

Wash avidin agarose beads (225 µl) with the binding buffer and combine them with 120 µg (0.24 nmol) of biotinylated RNA and incubate at 4°C for 2 h on a rotator. Centrifuge the beads at 6,500 rpm for 1 min and wash two times with 500 µl binding buffer. Mix the RNA-bound beads with 2 mg of pre-cleared cytoplasmic lysate and incubate at room temperature (RT) on a rotator for 30 min. Incubate the tube at 4°C for further 2 h on a rotator. Wash beads 5 times with binding buffer and finally boil with 1X SDS-PAGE loading buffer. Separate captured proteins on SDS-PAGE and stain with SYPRO Ruby stain (Genomic Solutions, Ann Arbor, MI). Excise specific bands from the gel and subject them to mass peptide fingerprinting. Alternatively, confirm the protein identity by Western blot analysis with suitable antibodies.

3.5 Avidin-biotin RNA-affinity chromatography isolation of XIAP and cIAP1 IRES binding proteins

To illustrate the utility of the above-described approach we have previously used the minimal functional XIAP IRES (-162 to +1 nt of 5′ UTR of XIAP) (Lewis et al., 2007) and cIAP1 IRES (-150 to +1 nt of 5′ UTR of cIAP1) (Graber et al., 2010) elements to isolate and identify XIAP and cIAP1 ITAFs, respectively. When XIAP IRES RNA conjugated beads were used in an affinity pulldown, at least six distinct proteins were isolated (Figure 4A Lane # 2) (Lewis et al., 2007). A control reaction using avidin-agarose beads alone did not yield any proteins (Figure 4A Lane # 1) indicating that the proteins isolated with XIAP IRES RNA were specific. The mass peptide fingerprinting of p37 protein species captured on XIAP IRES RNA revealed that in fact p37 protein species is a mixture of hnRNPA1 and HuR (Figure 4B) (Lewis et al., 2007; Durie et al., 2011). The specificity of HuR and hnRNPA1 interaction with the XIAP IRES RNA was confirmed by Western blot analysis (Figure 4C) (Lewis et al., 2007; Durie et al., 2011). The protein species p52 and p44 were also subjected to mass peptide fingerprinting and were identified as La autoantigen (p52) and hnRNP C1/C2 (p44); both proteins were previously shown to interact with XIAP IRES RNA (Holcik & Korneluk, 2000; Holcik et al., 2003). In addition, Western blot analysis of the RNA chromatography eluate using antibodies against candidate proteins Baird et. al. confirmed the presence of La autoantigen, hnRNP C1/C2 and PTB in the XIAP IRES RNA-protein complex (Baird et al., 2007) (Figure 4C).

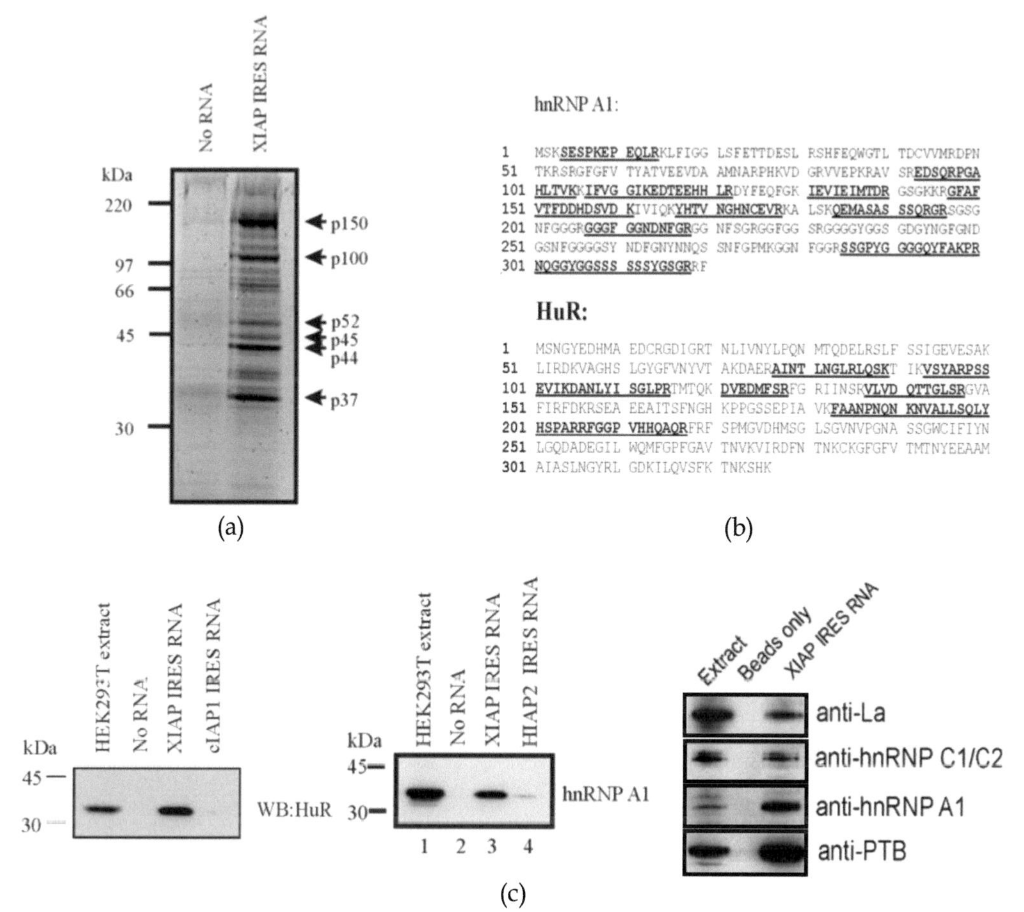

(A) Precleared HEK293T cytoplasmic lysate was incubated with avidin agarose coated with XIAP IRES RNA or avidin agarose only. Captured proteins were separated by SDS-PAGE and stained with SYPRO Ruby stain and indicated protein species were excised from the gel and subjected to mass spectrometry. (B) p37 was identified as a mixture of hnRNPA1 and HuR; underlined peptides were identified by mass spectroscopy. (C) Westernblot analysis of proteins captured on XIAP IRES RNA by avidin-biotin RNA affinity chromatography confirmed the specificity of the protein-XIAP RNA interaction. (Adapted from Baird et al., 2007; Lewis et al., 2007; Durie et al., 2011).

Fig. 4. Avidin-biotin RNA-affinity chromatography isolation of XIAP IRES binding proteins

In a similar approach, Graber et. al. used IRES containing portion of cIAP1 5′ UTR (Probe I), and non-IRES portion of cIAP1 5′ UTR as a control (Probe II) (Figure 5A), to isolate and identify proteins which specifically interact with the cIAP1 IRES. Four distinct proteins were isolated (Figure 5B)(Graber et al., 2010). The MALDI-TOF mass spectrometry analysis of cIAP1 IRES captured proteins identified them as RNA Helicase A (RHA), insulin-like growth factor 2 mRNA binding protein 1 (IGF2BP1), NF90 and NF45 (Figure 5C). Furthermore, Western blot analysis using antibodies against RHA, IGF2BP1, NF90 and NF45 confirmed that these proteins specifically interact with the cIAP1 IRES but not with the non-IRES portion of the cIAP1 5′ UTR (Figure 5D) (Graber et al., 2010).

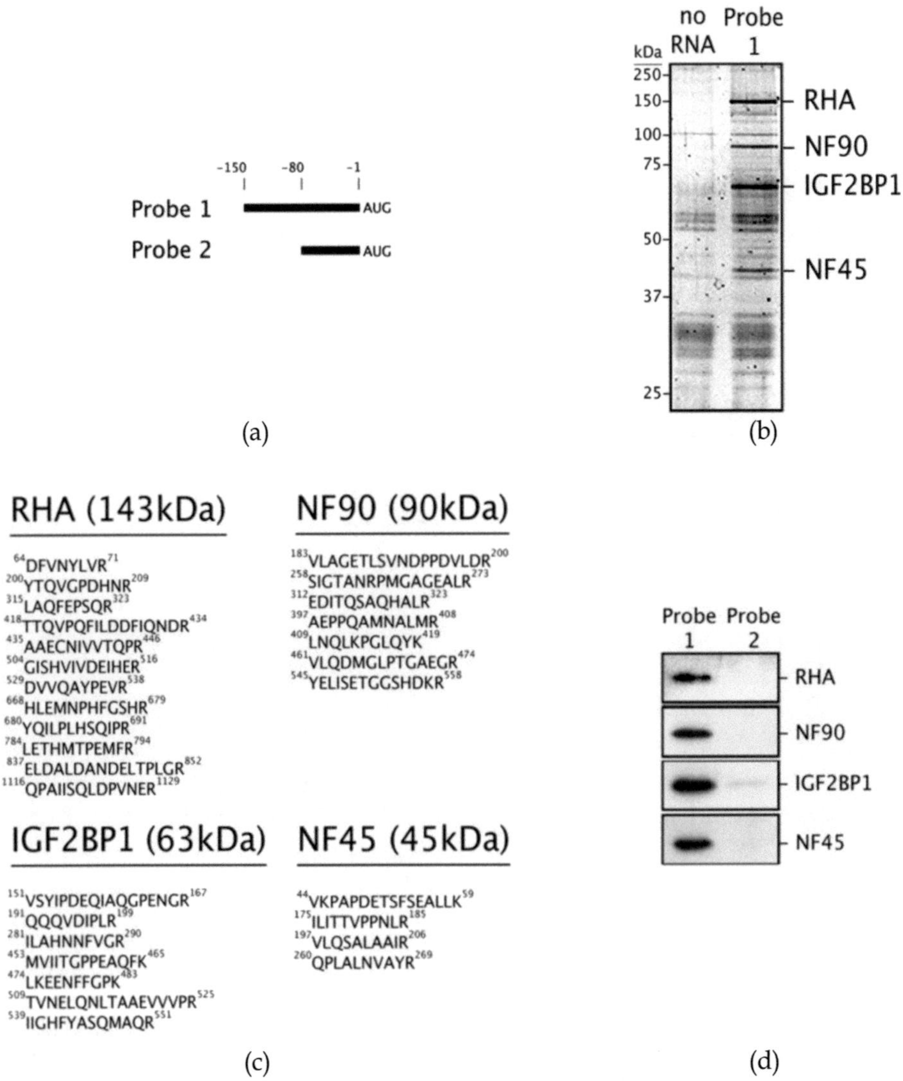

(A) Probe I: IRES containing portion of cIAP1 5′ UTR; Probe II: non-IRES portion of cIAP1 5′ UTR. (B) Pre-cleared HEK293T cytoplasmic lysate was incubated with avidin agarose beads coated with cIAP1 IRES RNA or avidin agarose beads only. Captured proteins were separated by SDS-PAGE and stained by SYPRO Ruby stain (C) Indicated protein species were excised from the gel and subjected to mass spectrometry. (D) Western blot analysis of proteins captured on cIAP1 IRES RNA by avidin-biotin RNA affinity chromatography confirmed the specificity of the interaction with the IRES portion of the cIAP1 5′ UTR. (Adapted from Graber et al., 2010)

Fig. 5. Avidin-biotin RNA-affinity chromatography isolation of cIAP1 IRES binding proteins

4. RNA-protein complex immunoprecipitation

While the RNA affinity chromatography will provide information about the *in vitro* interaction between the protein and given RNA, the biological relevance of this interaction needs to be further probed in a cellular setting. The first step in this process is to validate if the RNA-protein interaction occurs in cells, and this can be done by RNA immunoprecipitation (RIP) technique. We have used the *in vivo* crosslinking combined with co-precipitation of RNA-protein complexes as described (Niranjanakumari et al., 2002) to demonstrate *in vivo* interaction between the endogenous HuR and hnRNP A1 and the endogenous XIAP mRNA. Briefly, HEK293T cells were treated with formaldehyde to cross link RNA-protein complexes. Crosslinking reaction was then stopped by adding glycine and cells were lysed by sonication. Crosslinked RNA-protein complexes were immunoprecipitated using anti-hnRNPA1 (Santa Cruz Biotechnology, Santa Cruz, CA, USA), anti-La (Chan & Tan, 1987), anti-GAPDH (Advanced ImmunoChemicals, Long Beach, CA), anti-HuR (Santa Cruz Biotechnology) or anti-TIA-1/TIAR (clone 3E6; a gift from Dr. P. Anderson) antibodies. Following immunoprecipitation and crosslinking reversal the immunoprecipitated RNA was isolated using Trizol reagent (Invitrogen, Carlsbad, CA, USA). The isolated RNA was then reverse transcribed using an oligo dT_{18} primer and Superscript II (Invitrogen) to obtain cDNA. The resulting cDNA was analyzed by PCR amplification using gene specific primers and the PCR products were visualized on an ethidium bromide stained 1.5 % (wt/vol) agarose gel (Lewis et al., 2007; Durie et al., 2011).

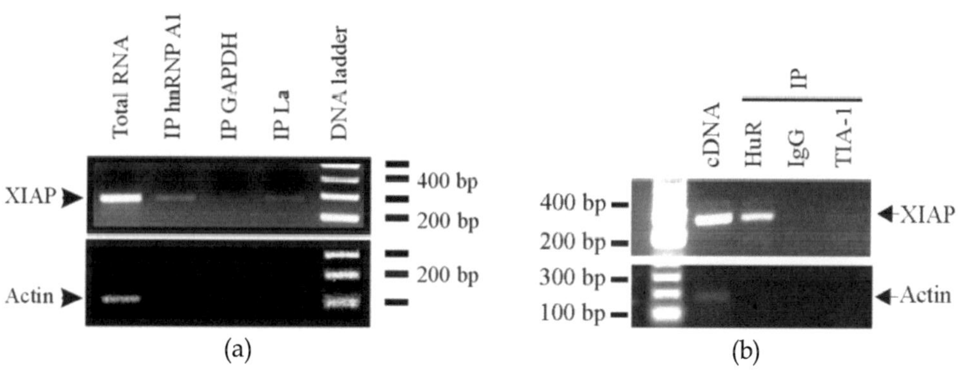

RNA-protein complexes were crosslinked using formaldehyde, isolated from cells and immunoprecipitated using antibodies against hnRNAP A1, GAPDH, La (A), HuR and TIA-1 (B). After immunoprecipation and crosslinking reversal the RNA was isolated and reverse transcribed. Gene-specific primers were used for PCR amplification of the resultant cDNA, and PCR products were visualized by gel electrophoresis. (Adapted from Lewis et al., 2007; Durie et al., 2011)

Fig. 6. hnRNPA1 and HuR interact with XIAP mRNA *in vivo*

XIAP mRNA was isolated by immunoprecipitation with La antibody which was previously shown to interact with XIAP mRNA (Holcik & Korneluk, 2000). Immunoprecipitation with hnRNAP A1 and HuR isolated XIAP mRNA confirmed their association with XIAP mRNA *in vivo* (Figure 6A; lane 2 and Figure 6B; lane 2). However, immunoprecipitation with GAPDH, TIA-1 (a known RNA binding protein) or IgG did not yield XIAP mRNA (Figure 6A; lane 3, Figure 6B; lanes 3 and 4) confirming that isolation of XIAP mRNA by immunoprecipitation with hnRNAPA1, HuR and La is specific. Importantly, β-actin mRNA (a non-specific and abundant mRNA) was not isolated by immunoprecipitation with La, hnRNP A1, GAPDH, HuR, IgG or TIA-1 antibodies.

5. UV crosslinking of RNA-protein complexes

Once candidate ITAFs are isolated by the RNA affinity chromatography, their association with the IRES of interest can be further studied using UV crosslinking of RNA-protein complexes. Unlike RNA chromatography, the UV crosslinking tests the direct interaction between the purified protein and RNA. To determine if hnRNP A1 and HuR bind directly to the XIAP IRES, UV crosslinking experiment was performed using ^{32}P labelled XIAP IRES RNA probe and purified recombinant GST-hnRNP A1 and GST-HuR. GST or increasing amounts of GST-hnRNP A1 and GST-HuR were mixed with the radiolabeled RNA in a 96 well plate and UV-irradiated on ice with a 254-nm UV light source at 400,000 μJ/cm^2. UV-irradiated RNA-protein complexes were then treated with RNAase T1, resolved on a SDS-PAGE gel and visualised by autoradiography.

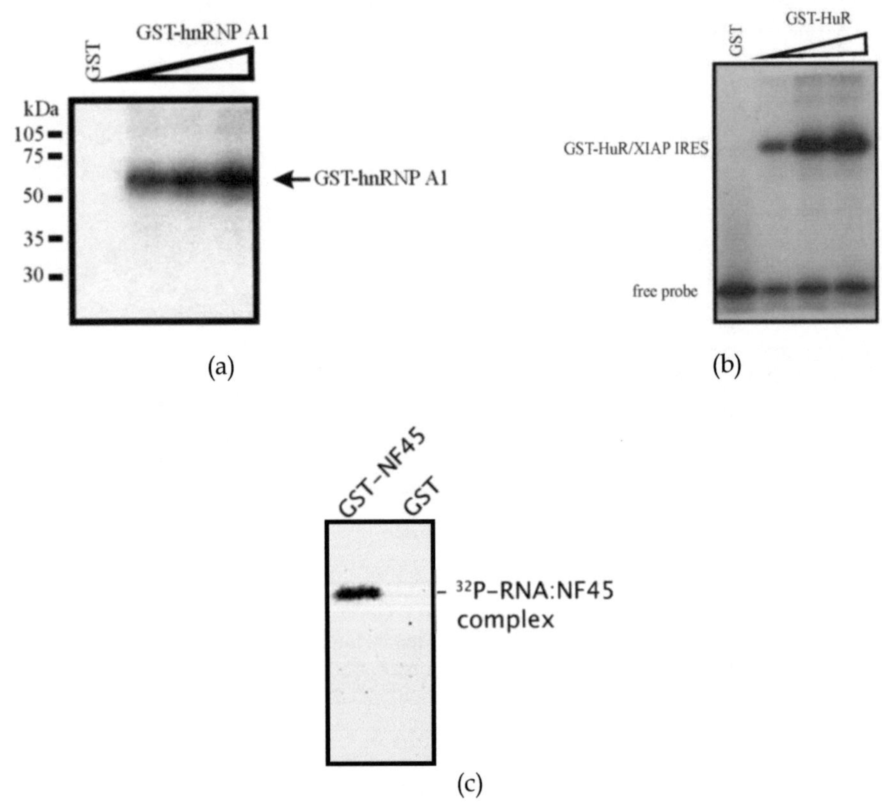

GST or increasing amounts of GST-hnRNP A1 (A) or GST-HuR (B) was mixed with radiolabeled XIAP IRES RNA probe and UV-crosslinked. It was then resolved on a SDS-PAGE gel and visualised by autoradiography. (C) GST-NF45 was mixed with radiolabeled cIAP1 IRES RNA probe and UV-crosslinked. It was then resolved on a SDS-PAGE gel and visualised by autoradiography (Adapted from Lewis et al., 2007; Graber et al., 2010; Durie et al., 2011).

Fig. 7. UV crosslinking of RNA-protein complexes

We found that GST-hnRNP A1 (Figure 7A) and GST-HuR (Figure 7B) was crosslinked to XIAP IRES RNA in a dose dependent manner which confirmed the irdirect interaction with

XIAP IRES RNA. Similarly, GST-NF45 directly interacts with IRES portion of cIAP1 5′ UTR (Figure 7C). GST did not crosslink with XIAP IRES RNA or cIAP1 IRES RNA indicating specific interactions of purified ITAFs with their respective IRES RNA.

Once the direct interaction of protein and RNA is established, it is important to define binding site(s) of the protein on the given RNA in order to understand how proteins, isolated by RNA affinity chromatography, modulate function of the given RNA. The protein binding site on the RNA can be delineated by oligonucleotide competition assay. In this approach, unlabeled oligonucleotides (competitors) are incubated with the protein of interest for 15 min before the addition of ^{32}P labelled target RNA. Subsequently, UV crosslinking is performed as described above and RNA-protein complexes are resolved on SDS-PAGE and visualized by autoradiography. We used oligonucleotide competition assay to map the hnRNP A1 and HuR binding sites on the XIAP IRES, and NF45 binding sites on the cIAP1 IRES. The direct interaction of XIAP IRES RNA with GST-hnRNP A1 was outcompeted by RNA oligonucleotides spanning -50 to -25 nt and -25 to 0 nt of XIAP IRES RNA (Figure 8A, top panel). This suggested that hnRNP A1 bound between -50 to 0 nt on XIAP IRES RNA, highlighted with grey box (Figure 8B). Similarly, GST-HuR binding sites were found to be between -68 to -88 nt and -115 to -135 nt of XIAP IRES RNA sequence which are located near the central core of domain I (Figure 8A, lower panel). Based on the previously defined consensus sequence (5′-NNUUNNUUU-3′) (Meisner et al., 2004) we proposed that GST-HuR binds domain Ib (-126 to -137 nt) of the XIAP IRES (Figure 8C) (Durie et al., 2011).

The protein binding sites on the target RNA can also be determined using slightly different approach; first the unlabelled DNA oligonucleotide competitors are hybridized to heat denatured ^{32}P labelled RNA and then it is UV crosslinked with the candidate protein, separated on SDS-PAGE and visualized by autoradiography. Using this approach we determined that AU-rich base of stem loop I (SL I) is essential for NF45 binding on cIAP1 IRES RNA (Graber et al., 2010). The NF45 binding site was further confirmed by mutating SL I of cIAP1 IRES and performing UV crosslinking experiment as described above (Graber et al., 2010).

Once the *in vitro* and *in vivo* interactions of ITAFs with the specific IRES RNA are established and the binding sites of ITAFs on specific IRES are delineated, the ability of ITAFs to modulate specific IRES mediated protein production can be studied using various cell culture techniques. These include overexpression or siRNA-mediated downregulation of candidate ITAFs followed by the determination of IRES activity using reporter constructs, and evaluation of the effect on the translation of endogenous protein by Western blot and polysome analyses. Using these techniques we have shown that cytoplasmic overexpression of FLAG-hnRNP A1 reduces endogenous XIAP expression and XIAP IRES activity (Lewis et al., 2007). Conversely, when expression of hnRNP A1 is reduced by targeting hnRNP A1 mRNA using specific siRNA, endogenous XIAP protein levels and IRES activity were significantly increased (Lewis et al., 2007). In contrast, GFP-HuR overexpression resulted in an increase of endogenous XIAP expression, whereas reduction in HuR expression by siRNA resulted in a decrease of endogenous XIAP expression and IRES activity (Durie et al., 2011). These findings suggest that hnRNP A1 negatively modulates XIAP IRES while HuR positively modulates XIAP IRES activity (Lewis et al., 2007; Durie et al., 2011).

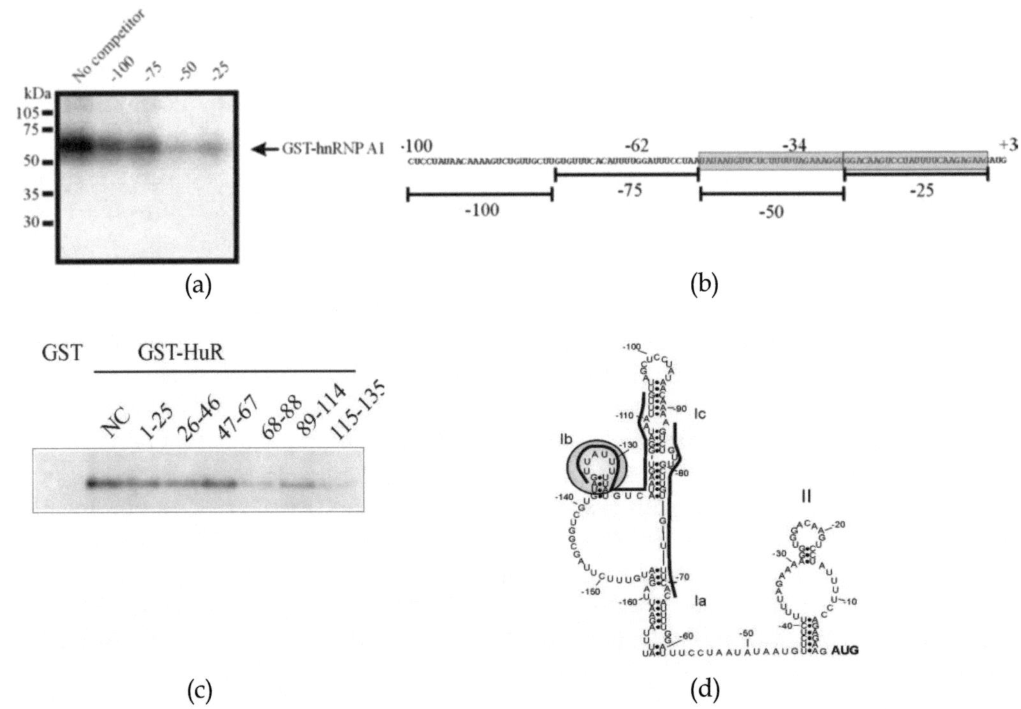

(A) ^{32}P labelled XIAP IRES RNA was UV crosslinked with hnRNAP A1 (top) or HuR (bottom) in the absence or the presence of 100-fold excess competitor RNA oligonucleotides, separated by SDS-PAGE and visualized by autoradiography. (B) Design of RNA oligonucleotide competitors and GST-hnRNP A1 binding site, indicated by grey box. (C) XIAP IRES secondary structure (Baird et al., 2007); black lines indicate competitor oligonucleotide which outcompete GST-HuR binding. Grey circle denotes the proposed HuR binding site in domain Ib. (Adapted from Lewis et al., 2007; Durie et al., 2011)

Fig. 8. Determination of GST-hnRNP A1 and GST-HuR binding sites on XIAP IRES RNA

7. Affinity purification of the XIAP IRES initiation complex

Avidin-biotin RNA affinity chromatography is a powerful tool to isolate proteins associated with RNA due to tight interaction between avidin and biotin. One limitation of this approach is that 5′ end of the RNA is attached to the avidin matrix prior to biological complex formation. In a specific example of translation initiation, this will not allow for cap dependent ribosome recruitment. Because, the 5′ end of the biotinylated RNA remains in a close proximity of the solid matrix, ribosome recruitment on short IRES RNA could also be hindered. MS2-MBP based RNA affinity chromatography approach was used for the isolation of 48S initiation complex formed on hepatitis C virus (HCV) IRES RNA (Ji et al., 2004). In this approach MS2 recognition hairpins were inserted near 5′ end of RNA which would limit cap-dependent ribosome recruitment and initiation complex formation on RNAs which require ribosome scanning. Locker et al has described StreptoTag-based RNA affinity chromatography to isolate 48S initiation complex formed on HCV IRES RNA (Locker et al., 2006). In this approach 3′ end-inserted StreptoTag sequence binds to sepharose coupled streptomycin with high affinity (Kd, 1 μM) which overcomes the above mentioned limitations.

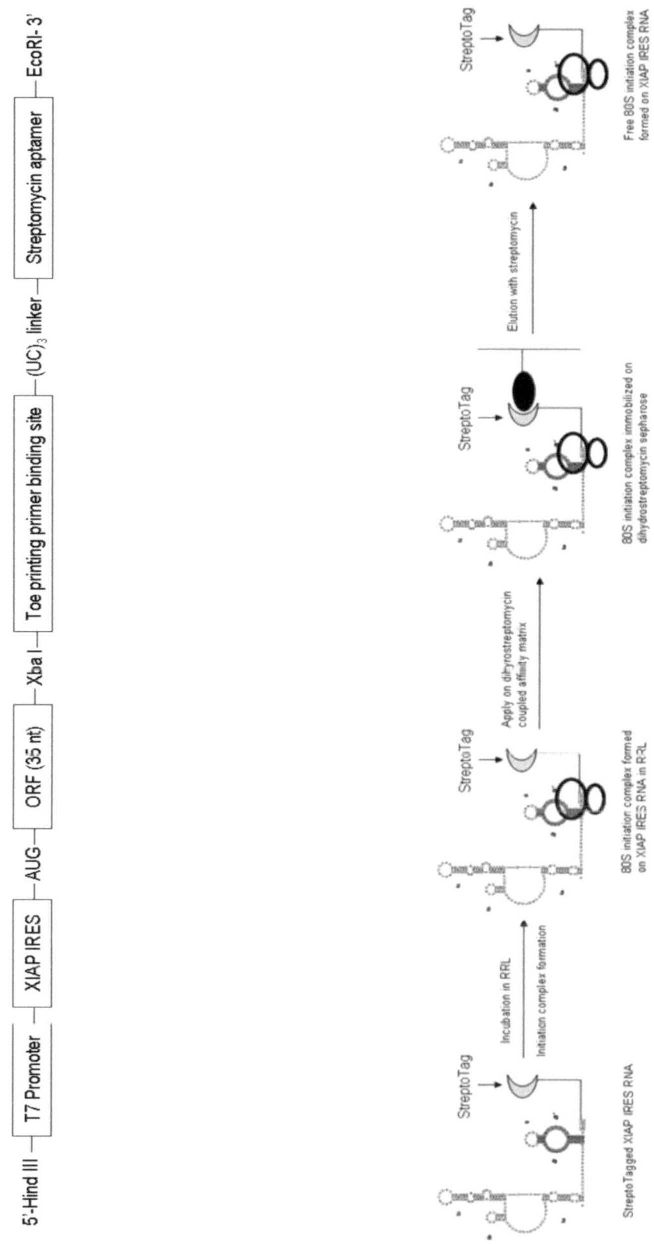

(a) (b)

(A) Schematic diagram of DNA construct from which StreptoTagged XIAP IRES RNA was transcribed. The RNA was used for toeprinting analysis and affinity chromatography. (Adapted from Thakor & Holcik, 2011) (B) A general scheme of isolation of initiation complex formed on StreptoTagged XIAP IRES RNA by streptomycin affinity chromatography.

Fig. 9. Streptomycin affinity chromatography

The interaction of StreptoTagged RNA and dihydrostreptomycin is reversible and StreptoTagged RNA can be dissociated from the solid support by competition with free streptomycin. This makes streptomycin-RNA affinity chromatography a powerful and a versatile tool to isolate various functional biological complexes including initiation complexes formed on cellular mRNAs. Although we have used this approach to isolate and characterize initiation complex formed on XIAP IRES RNA, this approach can also be used to isolate other RNA associated proteins or other complexes. We transcribed StreptoTagged XIAP IRES RNA from the construct showed in Figure 9A as described (Thakor & Holcik, 2011). Initiation complex was formed on StreptoTagged XIAP IRES RNA in RRL and applied on the dihydrostreptomycin sepharose column. After stringent washing the initiation complex formed on StreptoTagged XIAP IRES RNA was eluted using streptomycin solution (Figure 9B) and further analysed by agarose gel and toeprinting assay.

7.1 Coupling of dihydrostreptomycin on epoxy-activated sepharose 6B

Rehydrate epoxy-activated sepharose 6B (3g) (GE Healthcare, Glattbrugg, Switzerland) in 50 ml water for 30 min at room temperature (RT) and wash with 600 ml water using a sintered glass funnel. Suspend epoxy-activated sepharose 6B beads in 40 ml coupling solution (3 mM dihydrostreptomycin in 10 mM NaOH) and incubate on a rotator for 2 h at RT. Rotate the tube at 37°C overnight on a rotator. Decant the beads in a sintered glass funnel and wash with 200 ml of 10 mM NaOH solution. Subsequently, suspend the beads in 6% (wt/vol) ethanolamine solution prepared in water. Rotate the tube for 10 min at RT on a rotator. Rotate the tube further at 42°C overnight in the dark. Decant the beads in the sintered glass funnel and apply three rounds of washes with 50 ml alternating acidic buffer (0.1 M NaOAc and 0.5 M NaCl, pH 4) and basic buffer (0.1 M Tris-HCl and 0.5 M NaCl, pH 7.4). Finally, suspend the beads in 40 ml storage buffer (10 mMTris-HCl and 10 mM NaN_3, pH 7.4) and store at 4°C in the dark up to 3 weeks. Using a low pressure peristaltic pump pack the dihydrostreptomycin coupled beads in Poly-Prep® chromatography columns (Bio-Rad, Hercules, CA, USA).

7.2 Dihydrostreptomycin-RNA affinity chromatography

Incubate 4 ml of untreated RRL with 150 ng/ml poly I:C, 1 mM ATP, 10 µl ribonuclease inhibitor (Promega) and 0.1 mg/ml CHX (freezes 80S on the RNA) at 37°C for 20 min. Mix 12 ml binding buffer [20 mMTris (pH 7.6), 10 mM $MgCl_2$, 120 mM KCl, 8% sucrose, 2 mM dithiothreitol] containing EDTA-free protease inhibitor cocktail (Roche, Mississauga, ON, Canada) and 1 mM puromycin with RRL and incubate the reaction at 37°C for 10 min. Add an *in vitro* transcribed, uncapped strepto-tagged XIAP IRES RNA and 1 mM GTP into the reaction mixture. In order to achieve higher yield of XIAP IRES initiation complex, supplement the reaction with purified 40S and 60S from HeLa cells. Incubate the reaction further for 10 min at 37°C to form the 80S initiation complex. Perform the following steps at 4°C. While monitoring on UV recorder load 80S assembly reaction onto the dihydrosteptomycin coupled beads column (1 ml/min). Wash the column with the binding buffer (1 ml/min) until the base line is reached. Finally, elute XIAP IRES RNA initiation complex using 10 µM streptomycin solution prepared in the binding buffer (Figure 10A, top panel).

7.3 Toeprinting analysis of XIAP IRES initiation complex isolated by dihydrostreptomycin-RNA affinity chromatography

Dilute 5 µl of the eluate in 500 µl toeprinting buffer [20 mMTris-HCl (pH 7.6), 100 mM KOAc, 2.5 mM $MG(OAc)_2$, 5% (wt/vol) sucrose, 2 mM DTT and 0.5 mM spermidine] and concentrate back to 40 µl volume using microcon (Millipore, Billerica, MA, USA) and incubate at 30°C for 3 min. Subsequently, add 5 pmol of toeprinting primer (5-CTCGATATGTGCATCTGTA-3) (5′ end labeled with IRDye™800) and incubate reaction on ice for 10 min. Add 1.82 mM ATP, 1 mM dNTPs, 5 mM $Mg(OAc)_2$ and 1 µl of avian myeloblastosis virus reverse transcriptase (Promega) to the reaction and bring the final volume to 50 µl by toeprinting buffer. Allow primer extension to occur for 45 min at 30°C. Purify the cDNA products by phenol:chloroform extraction and analyze on a standard 6% sequencing gel using a model 4200 IR2 sequence analyzer (LI-COR, Lincoln, Nebraska, USA).

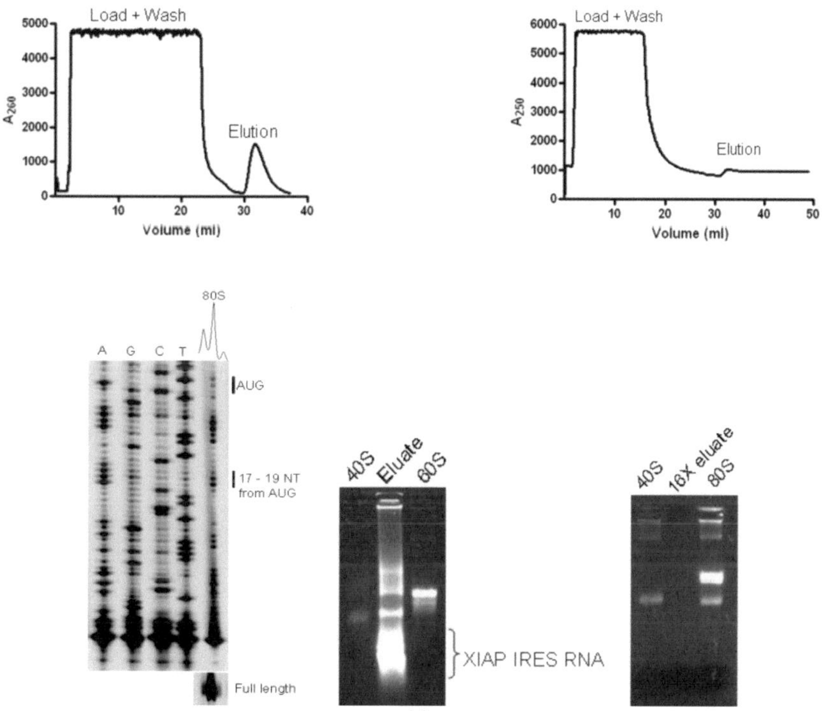

(c)

(A) RRL was supplemented with 40S and 60S ribosomal subunit and subsequently treated with poly I:C to induce eIF2α phosphorylation. XIAP IRES RNA initiation complex was purified by streptomycin-RNA affinity chromatography (top panel). Agarose gel analysis of the eluate was performed (lower panel). (B) Ribosome does not bind nonspecifically on dihydrostreptomycin coupled beads. 1 ml RRL was mixed with 3 ml binding buffer and loaded onto the dihydrostreptomycin column and washed with binding buffer, finally elution was carried out using streptomycin solution (top panel). Agarose gel analysis of the eluate was carried out (lower panel). (C) Eluate was further analysed by toeprinting assay. (Adapted from Thakor & Holcik, 2011).

Fig. 10. XIAP IRES RNA forms 80S initiation complex

7.4 Formation and isolation of XIAP IRES initiation complex by RNA-Streptomycin affinity chromatography

Physiological and pathophysiological stresses induce eIF2α phosphorylation which attenuates global protein translation (Holcik & Sonenberg, 2005). Several reports suggested that IRES mediated translation of XIAP is sustained during eIF2α phosphorylation conditions (Holcik et al., 2000; Warnakulasuriyarachchi et al., 2004; Gu et al., 2009; Muaddi et al., 2010; Riley et al., 2010). We induced eIF2α phosphorylation in RRL using poly I:C (a mimic of dsRNA virus). Using primer extension inhibition assay, the so-called toeprinting assay, we have showed that XIAP IRES RNA is able to form initiation complex in eIF2α phosphorylation condition (Thakor & Holcik, 2011). We wished to confirm the nature of XIAP IRES initiation complex formed in eIF2α phosphorylation condition. To this end, we performed RNA-Streptomycin affinity chromatography in eIF2α phosphorylation condition using the Locker method (Locker et al., 2006) modified as described above.

We performed agarose gel analysis of the eluate using standard TBE gel (Figure 10A, lower panel). 40S and 60S ribosomal subunits and XIAP IRES RNA were detected by ethidium bromide staining. A mock streptomycin-RNA affinitychromatography experiment, in which XIAP IRES RNA was omitted, did not yield 80S initiation complex (Figure 10B). These findings indicate that indeed 80S initiation complex was isolated using streptomycin-RNA affinity chromatography in eIF2α phosphorylation condition. The eluate was further subjected to toeprinting analysis as described (Thakor & Holcik, 2011). We observed toeprints +17 to +19 nt downstream of AUG (Figure 10C) which is a hallmark of ribosome recruitment to the AUG and stable ribosome-RNA complex formation. Depending on the distribution of fluorescence intensity (+17< +18 > +19) (Shirokikh et al., 2010) we further confirmed that indeed 80S initiation complex was isolated which was formed on XIAP IRES RNA in eIF2α phosphorylation condition (Figure 10C). The initiation complex can be further purified by sucrose density gradient separation. Purified initiation complex can be subjected to Western blot analysis and peptide mass finger printing to confirm the presence of canonical eukaryotic initiation factors and ITAFs.

8. Conclusion

We have presented several alternative strategies for the isolation, purification and characterization of RNA binding proteins and complexes using RNA affinity chromatography. The avidin-biotin RNA affinity chromatography is a versatile and commonly used approach which can be used to isolate RNP complexes formed on variety of RNAs. For example, this approach can be used to isolate spliceozome complexes, RNA stabilizing or destabilizing proteins or IRES Trans Acting Factors (ITAFs). In order to illustrate its usability, we have isolated and identified XIAP and cIAP1 IRES associated proteins. Additional biochemical, molecular biology and cell culture methods were then used to demonstrate that the isolated ITAFs regulate XIAP and cIAP1 expression through IRES. Alternately, dihydrostreptomycin-RNA affinity chromatography can be employed to isolate various RNP complexes and other biologically functional complexes formed on RNA. In this approach, unlike avidin-biotin RNA affinity chromatography, RNA anchors the matrix through the 3′ end. Furthermore, biological complex is formed on StreptoTagged RNA prior to its attachment to the dihydrostreptomycin matrix. Therefore, StreptoTag-ed RNA containing either 5′ m^7G structure or an IRES element is used in the first step to form initiation complex. In the subsequent step, initiation complex is isolated by dihydrostreptomycin-RNA affinity chromatography. For example, we have isolated functional XIAP IRES initiation complex formed inpoly I:C treated RRL which has been further characterized.

9. References

Bachler M, Schroeder R, von Ahsen U. 1999. StreptoTag: a novel method for the isolation of RNA-binding proteins. *Rna5*:1509-1516.

Baird SD, Lewis SM, Turcotte M, Holcik M. 2007. A search for structurally similar cellular internal ribosome entry sites. *Nucleic Acids Res 35*:4664-4677.

Bayer EA, Wilchek M. 1990. Application of avidin-biotin technology to affinity-based separations. *J Chromatogr 510*:3-11.

Blagden SP, Willis AE. 2011. The biological and therapeutic relevance of mRNA translation in cancer. *Nat Rev Clin Oncol 8*:280-291.

Blais JD, Filipenko V, Bi M, Harding HP, Ron D, Koumenis C, Wouters BG, Bell JC. 2004. Activating transcription factor 4 is translationally regulated by hypoxic stress. *Mol Cell Biol 24*:7469-7482.

Braunstein S, Karpisheva K, Pola C, Goldberg J, Hochman T, Yee H, Cangiarella J, Arju R, Formenti SC, Schneider RJ. 2007. A hypoxia-controlled cap-dependent to cap-independent translation switch in breast cancer. *Mol Cell 28*:501-512.

Caputi M, Mayeda A, Krainer AR, Zahler AM. 1999. hnRNP A/B proteins are required for inhibition of HIV-1 pre-mRNA splicing. *Embo J 18*:4060-4067.

Chan EK, Tan EM. 1987. Human autoantibody-reactive epitopes of SS-B/La are highly conserved in comparison with epitopes recognized by murine monoclonal antibodies. *J Exp Med 166*:1627-1640.

Copeland PR, Driscoll DM. 1999. Purification, redox sensitivity, and RNA binding properties of SECIS-binding protein 2, a protein involved in selenoprotein biosynthesis. *J Biol Chem 274*:25447-25454.

Durie D, Lewis SM, Liwak U, Kisilewicz M, Gorospe M, Holcik M. 2011. RNA-binding protein HuR mediates cytoprotection through stimulation of XIAP translation. *Oncogene 30*:1460-1469.

Gebauer F, Hentze MW. 2004. Molecular mechanisms of translational control. *Nat Rev Mol Cell Biol 5*:827-835.

Gerbasi VR, Link AJ. 2007. The myotonic dystrophy type 2 protein ZNF9 is part of an ITAF complex that promotes cap-independent translation. *Mol Cell Proteomics 6*:1049-1058.

Graber TE, Baird SD, Kao PN, Mathews MB, Holcik M. 2010. NF45 functions as an IRES trans-acting factor that is required for translation of cIAP1 during the unfolded protein response. *Cell Death Differ 17*:719-729.

Graber TE, Holcik M. 2007. Cap-independent regulation of gene expression in apoptosis. *Mol Biosyst 3*:825-834.

Gu L, Zhu N, Zhang H, Durden DL, Feng Y, Zhou M. 2009. Regulation of XIAP translation and induction by MDM2 following irradiation. *Cancer Cell 15*:363-375.

Holcik M, Gordon BW, Korneluk RG. 2003. The internal ribosome entry site-mediated translation of antiapoptotic protein XIAP is modulated by the heterogeneous nuclear ribonucleoproteins C1 and C2. *Mol Cell Biol 23*:280-288.

Holcik M, Korneluk RG. 2000. Functional characterization of the X-linked inhibitor of apoptosis (XIAP) internal ribosome entry site element: role of La autoantigen in XIAP translation. *Mol Cell Biol 20*:4648-4657.

Holcik M, Lefebvre C, Yeh C, Chow T, Korneluk RG. 1999. A new internal-ribosome-entry-site motif potentiates XIAP-mediated cytoprotection. *Nat Cell Biol 1*:190-192.

Holcik M, Sonenberg N. 2005. Translational control in stress and apoptosis. *Nat Rev Mol Cell Biol 6*:318-327.

Holcik M, Yeh C, Korneluk RG, Chow T. 2000. Translational upregulation of X-linked inhibitor of apoptosis (XIAP) increases resistance to radiation induced cell death. *Oncogene 19*:4174-4177.

Hovhannisyan R, Carstens R. 2009. Affinity chromatography using 2' fluoro-substituted RNAs for detection of RNA-protein interactions in RNase-rich or RNase-treated extracts. *Biotechniques 46*:95-98.

Hutanu D, Remcho VT. 2007. Aptamers as molecular recognition elements in chromatographic separations. *Adv Chromatogr 45*:173-196.

Ji H, Fraser CS, Yu Y, Leary J, Doudna JA. 2004. Coordinated assembly of human translation initiation complexes by the hepatitis C virus internal ribosome entry site RNA. *Proc Natl Acad Sci USA 101*:16990-16995.

Johannes G, Sarnow P. 1998. Cap-independent polysomal association of natural mRNAs encoding c-myc, BiP, and eIF4G conferred by internal ribosome entry sites. *Rna 4*:1500-1513.

Kaminski A, Hunt SL, Patton JG, Jackson RJ. 1995. Direct evidence that polypyrimidine tract binding protein (PTB) is essential for internal initiation of translation of encephalomyocarditis virus RNA. *Rna 1*:924-938.

King HA, Cobbold LC, Willis AE. 2010. The role of IRES trans-acting factors in regulating translation initiation. *Biochem Soc Trans 38*:1581-1586.

Komar AA, Hatzoglou M. 2011. Cellular IRES-mediated translation: the war of ITAFs in pathophysiological states. *Cell Cycle 10*:229-240.

Lewis SM, Veyrier A, Hosszu Ungureanu N, Bonnal S, Vagner S, Holcik M. 2007. Subcellular relocalization of a trans-acting factor regulates XIAP IRES-dependent translation. *Mol Biol Cell 18*:1302-1311.

Locker N, Easton LE, Lukavsky PJ. 2006. Affinity purification of eukaryotic 48S initiation complexes. *Rna 12*:683-690.

Meisner NC, Hackermuller J, Uhl V, Aszodi A, Jaritz M, Auer M. 2004. mRNA openers and closers: modulating AU-rich element-controlled mRNA stability by a molecular switch in mRNA secondary structure. *Chembiochem 5*:1432-1447.

Mitchell SA, Spriggs KA, Bushell M, Evans JR, Stoneley M, Le Quesne JP, Spriggs RV, Willis AE. 2005. Identification of a motif that mediates polypyrimidine tract-binding protein-dependent internal ribosome entry. *Genes Dev 19*:1556-1571.

Muaddi H, Majumder M, Peidis P, Papadakis AI, Holcik M, Scheuner D, Kaufman RJ, Hatzoglou M, Koromilas AE. 2010. Phosphorylation of eIF2alpha at serine 51 is an important determinant of cell survival and adaptation to glucose deficiency. *Mol Biol Cell 21*:3220-3231.

Neupert B, Thompson NA, Meyer C, Kuhn LC. 1990. A high yield affinity purification method for specific RNA-binding proteins: isolation of the iron regulatory factor from human placenta. *Nucleic Acids Res 18*:51-55.

Niranjanakumari S, Lasda E, Brazas R, Garcia-Blanco MA. 2002. Reversible cross-linking combined with immunoprecipitation to study RNA-protein interactions in vivo. *Methods 26*:182-190.

Pelletier J, Sonenberg N. 1988. Internal initiation of translation of eukaryotic mRNA directed by a sequence derived from poliovirus RNA. *Nature 334*:320-325

Peyrin E. 2009. Nucleic acid aptamer molecular recognition principles and application in liquid chromatography and capillary electrophoresis. *J Sep Sci 32*:1531-1536.

Ravelet C, Grosset C, Peyrin E. 2006. Liquid chromatography, electrochromatography and capillary electrophoresis applications of DNA and RNA aptamers. *J Chromatogr A1117*:1-10.

Riley A, Jordan LE, Holcik M. 2010. Distinct 5' UTRs regulate XIAP expression under normal growth conditions and during cellular stress. *Nucleic Acids Res 38*:4665-4674.

Rouault TA, Hentze MW, Haile DJ, Harford JB, Klausner RD. 1989. The iron-responsive element binding protein: a method for the affinity purification of a regulatory RNA-binding protein. *Proc Natl Acad Sci USA 86*:5768-5772.

Ruby SW, Goelz SE, Hostomsky Z, Abelson JN. 1990. Affinity chromatography with biotinylated RNAs. *Methods Enzymol 181*:97-121.

Sela-Brown A, Silver J, Brewer G, Naveh-Many T. 2000. Identification of AUF1 as a parathyroid hormone mRNA 3'-untranslated region-binding protein that determines parathyroid hormone mRNA stability. *J Biol Chem 275*:7424-7429.

Sharma S. 2008. Isolation of a sequence-specific RNA binding protein, polypyrimidine tract binding protein, using RNA affinity chromatography. *Methods Mol Biol 488*:1-8.

Shirokikh NE, Alkalaeva EZ, Vassilenko KS, Afonina ZA, Alekhina OM, Kisselev LL, Spirin AS. 2010. Quantitative analysis of ribosome-mRNA complexes at different translation stages. *Nucleic Acids Res 38*:e15.

Siebel CW, Kanaar R, Rio DC. 1994. Regulation of tissue-specific P-element pre-mRNA splicing requires the RNA-binding protein PSI. *Genes Dev 8*:1713-1725.

Silvera D, Formenti SC, Schneider RJ. 2010. Translational control in cancer. *Nat Rev Cancer 10*:254-266.

Silvera D, Schneider RJ. 2009. Inflammatory breast cancer cells are constitutively adapted to hypoxia. *Cell Cycle 8*:3091-3096.

Spriggs KA, Bushell M, Willis AE. 2010. Translational regulation of gene expression during conditions of cell stress. *Mol Cell 40*:228-237.

Srisawat C, Engelke DR. 2001. Streptavidin aptamers: affinity tags for the study of RNAs and ribonucleoproteins. *Rna 7*:632-641.

Thakor N, Holcik M. 2011. IRES-mediated translation of cellular messenger RNA operates in eIF2{alpha}- independent manner during stress. *Nucleic Acids Res.* doi 10.1093/nar/gkr701

Tombelli S, Minunni M, Mascini M. 2005. Analytical applications of aptamers. *Biosens Bioelectron 20*:2424-2434.

Warnakulasuriyarachchi D, Cerquozzi S, Cheung HH, Holcik M. 2004. Translational induction of the inhibitor of apoptosis protein HIAP2 during endoplasmic reticulum stress attenuates cell death and is mediated via an inducible internal ribosome entry site element. *J Biol Chem 279*:17148-17157.

Yaman I, Fernandez J, Liu H, Caprara M, Komar AA, Koromilas AE, Zhou L, Snider MD, Scheuner D, Kaufman RJ, Hatzoglou M. 2003. The zipper model of translational control: a small upstream ORF is the switch that controls structural remodeling of an mRNA leader. *Cell 113*:519-531.

Zhou Z, Sim J, Griffith J, Reed R. 2002. Purification and electron microscopic visualization of functional human spliceosomes. *Proc Natl Acad Sci USA 99*:12203-12207.

Part 6

Immobilized Metallic Ion Affinity Chromatography (IMAC)

9

Ion Metallic Affinity Chromatography and Purification of Bacterial Toxin

Luciano Moura Martins and Tomomasa Yano
State University of Campinas - UNICAMP, Institute of Biology, Department of Genetics, Molecular Biology and Bioagents
Brazil

1. Introduction

The history of the scientific development of the Chromatography as a chemical separation tool cannot be considered ancient, as the proper establishment of Chemistry as a science is considered modern when compared to the other sciences developed before it. The first literary description of the Chromatography separation technique development is attributed to the works or Mikhail Tswet in 1890 by some authors, where the chemical separation of plant extracts was conducted inside laboratory and first established, permitting the chemical separation of plant chemical substances according to their chemical characteristics of polarity and chemical affinity to the stationary phase. At that time, due to the color attributed to the plant extracts separated experimentally, the technique was named as "chromatography", where chromato means color and graphy means registering or writing. Although the many chemical substances and bio molecules that can be separated by this technique have no color, the name was kept and the technique recognized as one of the most efficient and applicable for laboratory purification and analytical processes.

The same way Chemistry is a modern and new science, the chromatography techniques also are, as over the years chromatography has proven to be an important tool for chemical substances separation. During this processes of development of the many techniques of chemical separations promoted by chromatography, one of them was structured as an affinity chromatography technique of immobilized ions, also known as Immobilized Metallic Ion Affinity Chromatography (IMAC), which is the main focus of this book and chapter.

The first description of this technique apply was related by the methods discovered and developed Cuatrecasas et al. (1968), by which Pedro Cuatrecasas and Mier Wilchek were awarded the wolf prize in medicine in 1987, who demonstrated that affinity chromatography is a really recent technique, where the immobilization of metallic ions was first established, permitting the chemical separation of chemical substances presenting affinity to this stationary phase.

In fact, this affinity technique works on separating proteins presenting amino acids like cysteine and histidine, which have chemical affinity to metallic ions. In general manner, there are many areas where IMAC can be applied and the practice has shown it to be a very useful tool when applied to proteins presenting such amino acids expressed on their surfaces. One good example of this is the purification of the Human Immunodeficiency Virus (HIV) reverse transcriptase by the use of IMAC.

The development on bacteria toxins purification and their characterization is one of the focus and goals of our developed works and IMAC has proven to be an excellent tool in separating toxins presenting this characteristic of ion metallic affinity, mainly by nickel (Ni^{++}) and cooper (Cu^{++}). Basically, as will be described afterwards, IMAC separates proteins by specific amino acids affinity, being selective to proteins presenting such amino acids on the surfaces of their tertiary or quaternary structures.

2. Applicability of the IMAC technique to molecular biology

Nowadays, there are many examples of the applicability of the IMAC associated to Molecular Biology on trying to purify bacteria proteins.

The Molecular Biology plays its important role in selecting the genes of the bacteria coding portions of interest proteins and toxins, where these portions are frequently selected and extracted by the action of restriction enzymes like Eco R1, among others.

After obtaining the desired restriction fragments, these ones are incorporated by complementary binding into commercial disposable bacteria plasmids, treated with the same restriction enzymes, in order to promote the correct complementary junctions.

More than the correct complementarities, these Molecular Biology tools give the assurance of a protein being synthesized with a poly histidine tag, generally introduced into the amino terminal of the protein structure to be transcribed and translated, which can be determined by the plasmids structure and the specific site of action of the restriction enzyme, promoting the target DNA (fragments obtained form bacteria) incorporation by nucleotide complementation of restricted DNA fragments.

It seems to be an easy process to be explained, but in practice, this has proven to be one of the most difficult parts of the work to be defined and conducted inside the laboratory, side by side to the problems of protein expression and purification, considering the many problems and fails that may occur in these steps. After a well succeed incorporation, it is necessary to insert these transformed plasmids into competent bacteria and verify the presence of the desired proteins expression, without forgetting that many fails during the steps of the plasmids insertion may also occur in a kind of work that may take a generous amount of time to be established in the case of unknown or not described proteins.

As a general manner, these plasmids carry a coding of an antibiotic resistance product. The target transformed bacteria can be checked by the exposition and growth in the presence of the specific antibiotic, showing the insertion step was well succeed when compared to the target bacteria not submitted to the transformation, which must be necessarily and naturally not resistant to the applied antibiotic, what gives the certainty, in the end of this step, that the bacteria submitted to the transformation process received or not the genetically inserted plasmids. Actually, nowadays these target plasmids, the restriction enzymes and the competent bacteria can be purchased in specific kits, which have the purpose of conducting and making possible this kind of work, in a manner and trying of making the process a little easier.

The fact of having the target protein or toxin being synthesized with a poly histidine tag improves the IMAC processes by the affinity of these matrixes to this amino acid.

There are also many examples of well succeed separations by this combination of Molecular Biology tools and IMAC (Catani et al., 2004), what gives the many possibilities of studying these proteins structures due to the many quantities of target protein produced by the bacteria, as the plasmids are structured and designed with a synthesis promoter portion, in a open reading frame (ORF) generally associated to the presence of a determined carbohydrate like lactose, galactose or mannose for instance.

After the protein purification steps, the products are submitted to a final specific enzymes action, which promote the poly histidine tag cleavage. Overall a final purification step is associated to this process, such as a gel filtration (molecular exclusion) chromatography, for example, to separate the target protein and the poly histidine tag fragments, or any other technique like a second dimensional electrophoresis (2D Gel Electrophoresis).

These techniques have permitted, in practice, many researches to reveal the third dimensional structures of many structural proteins, toxins and enzymes, associating the IMAC tool to other kind of analyses like Mass Spectrometry, Crystallography, Nuclear Magnetic Resonance, Nuclear Magnetic Resonance Spectroscopy, among the examples of possibilities of analytical processes offered by the recent Chemistry analysis developed technologies.

3. Why does IMAC Chromatography works on toxin purification of bacteria not genetically modified?

The concept of IMAC chromatography has been attributed nowadays as a chromatography technique applied to the areas where Molecular Biology have been also applied, where proteins are designed to be synthesised with poly histidine tags, making ion metallic affinity possible by its interaction. In fact this first explanation of applicability to this kind chromatography is the one which is the mostly used and scattered in the scientific world, but what has to be reminded is that these molecular tools are applied and adjusted to the proteins that do not present these characteristics of natural ion metallic affinity. Though, reminding also the history of the development of the technique, it must not be forgotten that there are many proteins, from the most different possible origins, presenting natural ion metallic affinity, attributed to the presence of histidine and cysteine in their structures, which can be purified by the applying of this technique as well.

By having this general overview, its easier to understand why sometimes it is naturally possible to have contaminants and interferences when conducting IMAC purification processes of products of genetically transformed microorganisms. Our practice in purifying bacteria toxins has proven that there are naturally microorganisms products that present also affinity to metallic ions, which may not only interfere in the purification processes, but must be separated from the target protein in order not to prejudice the following analyses steps, when the focus of studying a determined protein has or presents sample standards for structure determinations, for three dimensional shape determination, Crystallography and others.

The purification process is so important in these cases that its success must be followed in each step made, in order to verify the purity of the target protein before it is submitted to the following characterization sequential steps.

These verification stages generally are followed by gel electrophoresis specific for protein analyses like first and bi dimensional (2D) PAGE or SDS-PAGE, including also other verification assays when the target protein has an specific biological activity, for instance, if the target protein is an enzyme, its biological activity can be detected by assays offering its substrate and measuring the formation of the enzymatic product. When the focus is a toxin, it must be applied to a target of its toxicity, for example, haemolysins can be applied to haemolytic assays, enterotoxins can be applied to enterotoxicity assays, cytotoxins can be applied to cytotoxic *in vitro* assays, or proteases can be applied to target protein extracts. This step of verification not only defines if the target protein is pure enough and prepared to the next steps of chemical characterizations, but also defines and shows how drastic the purification processes have been to its integrity and maintenance of the biological activity, as chromatographic processes are always associated to minor or greater losses. This way, the verification processes are very useful for the losses data calculation and comparison of the best choice of chromatographic technique to be applied.

In general, IMAC is an excellent technique for protein purification due to its characteristics of not abolishing drastically the protein biological activities or promoting denature. That is why this technique was associated to the modern tools of Molecular Biology, where, with the help and knowledge of Microbiology in manipulating and transforming microorganisms, the applicability of this chromatography is such a respected success.

So, considering the many possibilities in the applicability of IMAC and the fact it is a kind of chromatography which is able to promote a good performance to the separation of raw and crude extracts, it is perfectly applicable to the purification of untransformed bacteria product proteins or peptides that present histidine and cysteine able to interact with the IMAC stationary phase.

4. Needs of association to other chromatography techniques

Nowadays it is known that the progress of proteomics depends on the development of protein separation techniques advances and technologies, being the IMAC a really robust method to be used for proteins and peptides or bacterium origin or not.

The immobilized metal affinity chromatography can be widely applied as a pre fractionation method as so as to the analytical separations, where it increases the resolution in protein separation.

Although being a robust technique, sometimes practice shows that it is necessary to combine IMAC to other protein analytical technologies. In fact this occurs also to genetically modified and expressed proteins the same way to the native proteins presenting natural ion affinity because of the presence of specific amino acids like histidine on their surfaces. So the natural occurrence of other proteins considered undesired or contaminants may easily occur and it is necessary to keep in mind the possible needs of applying a next step to the bacteria toxins purification processes to promote a high quality chemical separation. These chromatographic and chemical separation technologies can eliminate the undesired contaminants from the target protein or peptide and our observations have proven that the use of IMAC as a fractionation and analytical technique can be followed by any other separation step. Actually it depends most on the chemical characteristic of the target protein being purified and attention must be given to its chemical nature.

In our routine, many techniques have been applied after IMAC step, most of them proving to be really well succeed, like ion exchange, gel filtration, reverse phase, hydrophobic interaction, among other simple methods, like salting out or molecular weight ultra filtration separation techniques, which can be easily used and applied. The only thing to be considered while choosing the next step is which are the important conditions for keeping the protein or peptide structures because, depending on the separation technique chosen, denature may occur and depending also on the objective of the separation and purification work, it is not interesting to have a denatured product in the final of the processes, even if this final product is extremely pure in the end, because of the necessity of keeping the biological activity. That is why the needs of knowing the denaturizing agents and processes aligned to the knowledge of the chemical structure is important while making choose of the next purification steps and mainly choosing the right next step to be applied after IMAC fractionation.

Its also important to remember that IMAC is a common place technique in the modern protein purification and many other technologies are being developed worldwide in order to associate this technique to other chromatographic principles, and one good example of this is the effective function of IMAC supported on cationic exchangers, where the chromatographic separation permits two modes of purification in a single column. This proves that many efforts are being done in this area and that new technologies can be expected in the nearly future, making these needs of associating IMAC to a second chromatography technique easier to native or genetically cloned and expressed proteins.

5. Examples of IMAC applied in our work

In our laboratory (Bacterium Virulence Factors Laboratory of the Biology Institute of the State University of Campinas, São Paulo, Brazil), we work focused on the studies of human enteropathogenicity mechanisms of bacterium origin. Having this focus, with no doubt, purifying these enterotoxins is substantial to characterize biologically and chemically these toxins. Actually, during our trajectory, some bacterial toxins were purified experimentally in our laboratory, like some virulence factors of human pathogens like *Escherichia coli, Plesiomonas shigelloides, Aeromonas hydrophila, Aeromonas sobria, Aeromonas veronii biovar sobria* and others.

Among these purified toxins there are some examples of purification successes obtained by the use of IMAC when purifying native or genetically cloned bacterium toxins. Some of these examples include the works on the cloning, expression, and purification of the virulence-associated protein D from *Xylella fastidiosa* (Catani et al., 2004), a very important citric fruit pathogen associated to orange plantation production and financial harm in Brazil, where the cloning technology was applied while characterizing the virulence factors associated to the orange trees and fruit diseases and also the purification of the Vacuolating Cytotoxic Factor (VCF) of *Aeromonas veronii* biovar *sobria* (Martins et al., 2007), an enterotoxic cytotoxin associated to cases of human enteric diseases. IMAC showed to be extremely important to the development of these scientific works as in the first case we described proteins structured to present poly histidine tags and in the second case we presented a toxin having natural ion metallic affinity to cooper. In the Figure 1 there is an example of the typical chromatogram obtained by the IMAC process of purification of the VCF of *Aeromonas veronii* biovar *sobria*. It is important to notice that the chromatographic process to this specific bacterium culture crude extracts reveals that IMAC is able to remove most of the undesired contaminant materials from the target sample, being only necessary to make a second chromatographic step in a molecular weight exclusion column to finally obtain a toxin product of highly pure degree, what was confirmed by bi dimensional (2D) SDS-PAGE analysis which revealed the presence of a single spot protein of 50KDa.

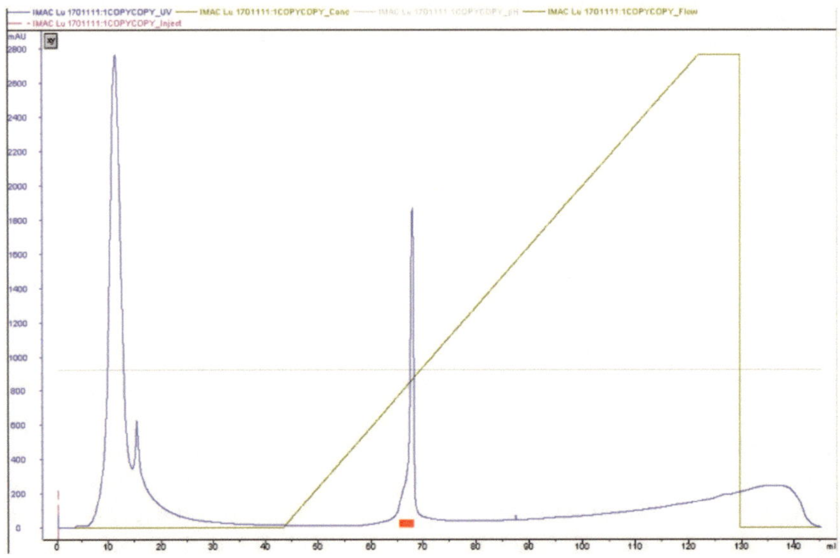

Fig. 1. Chromatographic profile observed while conducting the purification process of *Aeromonas veronii* biovar *sobria* VCF from crude extracts by IMAC charged with cooper (Cu^{++}). Assays were conducted by applying of linear Imidazole gradient elution (0 - 50mM) in Tris-HCl pH 8,0. The represented red line (▬) under the chromatogram is attributed to the portion which was adsorbed and eluted from the IMAC stationary phase and presented positive vacuolating cytotoxic activity over VERO (green kidney monkey) cells and positive enterotoxic activity over mice and rabbits.

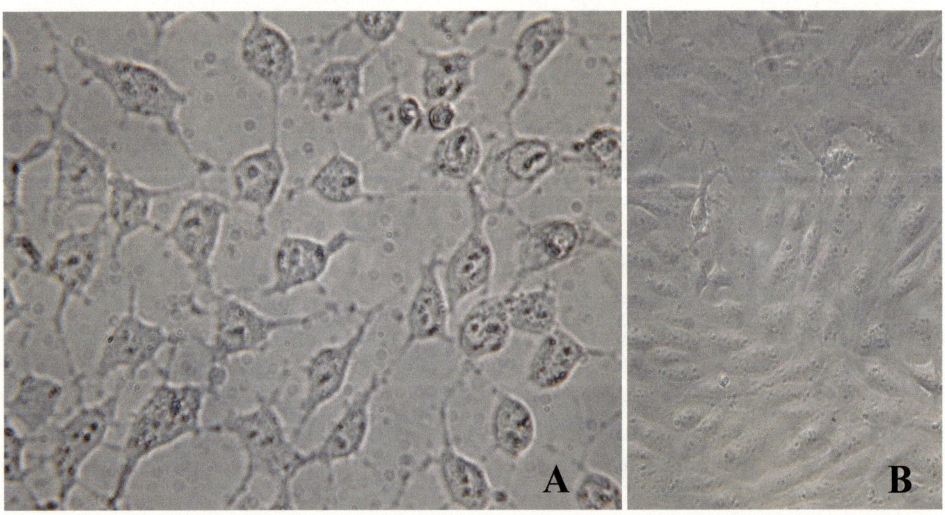

Fig. 2. Aspects of the cytotoxic effects of *A. caviae* thermo stable enterotoxic toxin (Ent-Ac) purified by IMAC charged with cooper ions over cultured Vero (green monkey kidney) cells. Fig A- Cells treated with the chromatographic eluted portion, presenting rounding, cell to cell leakage showing the loss of cellular membrane junctions and nuclear condensation. Fig. B- Cellular control. Magnifications 430X.

It is interesting to notice that among the studies that are being conducted in the moment in our laboratory, there are two low molecular weight bacterium toxins produced respectively by *Escherichia coli* and by *Aeromonas caviae*, both with enterotoxic and cytotoxic activities presenting strong cooper affinity. These toxins are under publication process and the aspects of ion metallic desorption is so strong that cooper is extracted from the stationary phase and carried out with the toxins, that, during the linkage to cooper, loose their biological activities.

Our findings reveal that only an effective treatment of these toxins with EDTA (Ethylenediamine tetraacetic acid) solutions after the chromatographic process is able to recover their biological activities by the chelating activity of this agent, unmaking this ions chemical linking, probably permitting the toxins to recover their active third dimensional structures configuration. As the stationary phase of IMAC support structure looses the immobilized ions, it is necessary to recover its structures with cooper solutions, preparing the stationary phases to subsequent chromatographic processes.

Our observations show that the immobilized metal affinity chromatography can be widely applied to experimental analyses of bacterium toxins, both for the pre fractionation or analytical separations, where it has shown a great resolution in protein separation.

6. Disadvantages or technical inconveniences

The selectivity or the adsorption capability of IMAC stationary phases depend not only on the chelating immobilized in the chromatographic matrixes, but also on the composition of the mobile phase. As mentioned, the retention of the adsorbed proteins into the IMAC stationary phase occurs due to the contribution of many physic-chemical interactions that can be intensified or minimized depending on the constitution of the mobile prepared phase. That is why it is of extreme importance to choose the right composition of mobile phases before starting IMAC separations to the best recovers of the target proteins and consequent obtaining of good results.

Basically, most of the traditional buffers applied to liquid chromatography of bio molecules may be applied as mobile phases to IMAC separations, where the ones possessing highly affinity to metallic ions, such as the ones presenting tricine and citrate in their composition, must be avoided. When applied, these kinds of buffers generally remove the metallic ions from the support of the stationary phase, prejudicing the chromatographic processes. The most applied mobile phases to IMAC processes are sodium phosphates, sodium acetates and the zwiterionics (considered as good buffers) as the examples of MOPS or 3-(N morfoline) propane sulfonic acid, MES 2-(N-moroline etanossulfonic acid and HEPES N-(2 hidrozietil) piperazine-N (2-etane sulfonic acid) (Winzerling et al., 1992).

The elution conditions to adsorbed bio molecules into the chromatographic matrixes of IMAC can be conducted by the use of different conditions, among them the use of pH gradual changes, the use of imidazole gradients and the most drastic of them, the apply EDTA (Ethylenediamine tetraacetic acid) solutions, when it is not possible to recover the target protein of the adsorption effect form the stationary phase. This last described method carries out all of the immobilized ions form the stationary phase support, what makes necessary the charging of the matrix with new ion metallic solutions, to the reconstitution of the stationary phases and extra work with the target protein. This process can be also used when a kind of metallic ion from the stationary phase is supposed to be substituted by another inside the laboratory ambient, for instance, when a stationary phase containing nickel (Ni^{++}) needs to be substituted by cooper (Cu^{++}), accordingly to the methodologies needs.

7. Conclusion

The ion metallic affinity chromatography is a trustable, reproducible and robust technique for fractionation and analysis, not only to genetically modified cloned and expressed proteins but also to natural proteins presenting ion metallic affinity.

Hopes on the development of this chromatographic tool technology in the future are integrated with the proteomic methods where it will greatly continue on contributing to the revolution of expression, cell mapping and structural proteomics.

8. Acknowledgments

We are grateful to Ana Stella Menegon Degrossoli for technical assistance over the works developed in the Bacterium Virulence Factors Laboratory, Department of Genetics, Molecular Biology and Bioagents of the Biology Institute of the State University of Campinas - UNICAMP, São Paulo, Brazil.

All the scientific works developed in our laboratory were and are supported by Fundação de Amparo à Pesquisa do Estado de São Paulo (FAPESP) and Conselho Nacional de Desenvolvimento e Tecnológico (CNPq), Brazil.

9. References

Belew, M.; Yip, T.T.; Andersson, L.; Ehrnström, R. (1987). High-performance analytical applications of immobilized metal ion affinity chromatography. *Anal Biochem*. Vol. 164, No. 2, pp. 457-465.

Bresolin, I.T.L.; Miranda, E.A. and Bueno, S.M.A. (2009). Cromatografia de afinidade por íons metálicos imobilizados (IMAC) de biomoléculas: aspectos fundamentais e aplicações tecnológicas. *Quim. Nova*. Vol. 32, pp. 1288-1296.

Catani, C.F.; Azzoni, A.R.; Paula, D.P.; Tada, S.F.S.; Rosseli, L.K.; Souza, A.P.; Yano, T. (2004). Cloning, expression, and purification of the virulence-associated protein D from *Xylella fastidiosa*. *Prot Expres Purification*. Vol. 37, pp. :320-326.

Charlton, A.; Zachariou, M. (2008). Immobilized metal ion affinity chromatography of native proteins. *Methods Mol Biol*. Vol. 421, pp. 25-35.

Cuatrecasas, P.; Wilcheck, M.; Anfinsen, C.B. (1968). Selective enzyme purification by affinity chromatography. Proc. Natl. Acad. Sci. USA. Vol. 61, pp. 636-643.

Gaberc-Porekar, V. and Menart, V. (2001). Perspectives of immobilized-metal affinity chromatography. *J.Biochem. Biphys. Methods*. Vol. 49, pp. 335-360.

Martins, L.M.; Catani, C.F.; Falcón, R.M.; Azzoni, A.A.; Carbonell, G.V.; and Yano, T. (2007). Induction of apoptosis in Vero cells by *Aeromonas* veronii biovar sobria vacuolating cytotoxic factor (VCF). *FEMS Immunol Med. Microbiol*. Vol. 49 pp. 197-204.

Nakagawa, Y.; Yip, T.T.; Belew, M.; Porath, J. (1988). High-performance immobilized metal ion affinity chromatography of peptides: analytical separation of biologically active synthetic peptides. *Anal Biochem*. Vol. 168, No. 1, pp. 75-81.

Sun, X.; Chiu. J.F.; He, Q.Y. (2005). Application of immobilized metal affinity chromatography in proteomics. *Expert Rev Proteomics*. Vol. 2, No. 5, pp. 649-657.

Part 7

Usage of Affinity Chromatography in Purification of Proteases, Protease Inhibitors, and Other Enzymes

10

Polyhistidine Affinity Chromatography for Purification and Biochemical Analysis of Fungal Cell Wall-Degrading Enzymes

Takumi Takeda
Iwate Biotechnology Research Center, Japan

1. Introduction

More than 10 years ago, proteins had to be purified to homogeneity by a combination of multiple chromatography steps, and their amino-terminus sequences were determined before their genes could be cloned. Nowadays, purification of native proteins is still an important step, but the utilization of recombinant proteins, which are produced by microbes, fungi, plants and animals by means of introducing an expression vector harboring a coding DNA sequence, provides a time-saving and useful way for analyses of protein characteristics and crystal structure, and for industrial applications. For detailed biochemical characterization, purified proteins (single molecular preparations) are preferentially used in order to avoid the effects of contaminant proteins.

A significant technique for quickly obtaining purified proteins is affinity chromatography using epitope-tags and antibodies. Epitope-tags, such as polyhistidine, hemagglutinin (HA) and FLAG among others, are easily fused to recombinant proteins by preparing a DNA structure containing the nucleotide sequence of the epitope-tag at the 5' and/or 3' end of the cDNA of the protein of interest. Following expression, the epitope-tagged recombinant protein can be specifically separated by affinity chromatography using epitope-tag binding and/or antibody-linked resins. Antibodies against epitope-tags are also used for immunoblot analysis and immune-precipitation. Although polyclonal or monoclonal antibodies against recombinant proteins are also precious tools for immune-precipitation and cellular localization, they may not be suitable for use with many kinds of recombinant proteins due to the requirement for antibodies corresponding to the individual protein.

The recent development of molecular biological techniques has contributed to elucidation of the genome DNA sequence of many species. The resultant sequence databases are being used in various fields, including gene expression analyses and protein engineering. To identify the function of proteins encoded by genome DNA sequences, the production and purification of recombinant proteins has become an essential strategy, in which the addition of an epitope-tag sequence enhances both the preparation of purified recombinant proteins and the following characterization.

An example for this epitope-tagged technique can be applied on the preparation of purified recombinant cell wall-degrading enzymes and the functional analyses. Microbial cell wall-degrading enzymes play a significant role in the degradation of both their own cell walls and plant cell walls during infection (Reese, et al., 1950; Henrissat, et al., 1985; Wood, 1992). In addition, plant cell wall-degrading enzymes are also involved in wall loosening during cell expansion, cell wall biosynthesis, and countermeasures against infectious pathogens (Walton, 1994; Nicol, et al., 1998; Takeda, et al., 2002). These enzymes, especially those from microbes, are one of the most important industrial products with applications in, for example, beer and wine, animal feed, paper, textile, laundry detergent, and food ingredient industries (Bhat, 2000). Hence, reducing cellulase manufacturing costs by increasing the productivity of cellulases with high specific activities through biotechnological modification is a desired research goal. To identify cell wall-degrading enzymes that are needed for various industrial applications, it is necessary to carry out a series of steps encompassing the production, purification, characterization and molecular modification of cell wall-degrading enzymes. Here, the production, purification and characterization of recombinant epitope-tagged proteins are described, with a particular focus on fungal cell wall-degrading enzymes.

2. Production of epitope-tagged recombinant proteins

2.1 Addition of an epitope-tag to a recombinant protein

To identify recombinant proteins produced by host cells, visualization by Coomassie brilliant blue 250-R (CBB) staining and immunoblotting after sodium dodecyl sulfate polyacrylamide gel electrophoresis (SDS-PAGE) are routinely carried out. Furthermore, purifying recombinant proteins is an important procedure for determining biochemical properties, such as substrate specificity, optimum reaction temperature and pH, and stability. Epitope-tags, which consist of the amino acids shown in Table 1, are useful and convenient tool both for purifying proteins using epitope-tag affinity chromatography and for immunoblot analysis using antibody.

Epitope-tags	Amino acid sequence
Heptahistidine	HHHHHHH
HA	YPYDVPDYA
c-Myc	EQKLISEEDL
FLAG	DYKDDDDK
VSV-G	YTDIEMNRLGK
HSV	QPELAPEDPED
V5	GKPIPNPLLGLDST

Table 1. Amino acid sequences of epitope-taggs

2.2 Expression of polyhistidine-tagged protein

For the detailed analysis of cell wall degradation, *Trichoderma reesei* endo-1,4-β-glucanase (TrCel12A) was produced in *Brevibacillus choshinensis* (Takara-Bio) and purified using a polyhistidine-binding resin (Clontech). The *TrCel12A* gene was cloned by PCR using GXL DNA polymerase (Takara-Bio) and specific DNA primers, which were designed on the basis of the *T. reesei* genome DNA sequence. A DNA sequence encoding seven contiguous histidines was added to the 3' end of the *TrCel12A* gene by PCR as shown in Fig. 1.

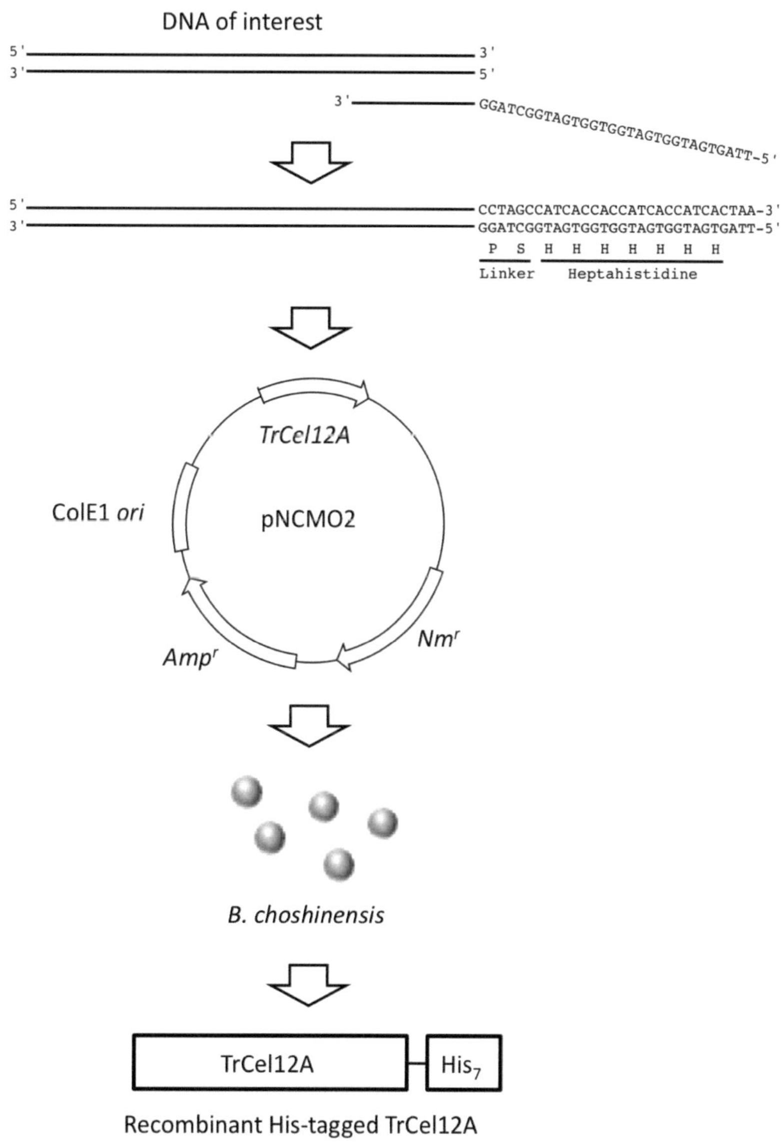

Fig. 1. Scheme of production of recombinant His_7-tagged TrCel12A in *B. choshinensis*. *Nm^r*; neomycin resistance gene, *Amp^r*; ampicillin resistance gene, ColE1 *ori*; *E. coli* replication origin.

The synthesized DNA was cloned into an expression vector (pNCMO2) for protein expression in *B. choshinensis*. The construct was transformed into *B. choshinensis* cells by electroporation. The transformants were screened on MT [1% (w/v) glucose, 1% (w/v) Tryptone, 2% (w/v) Beef extract, 2% Yeast Extract, $FeSO_4$ (65 μmol/L), $MnSO_4$ (41 μmol/L), $ZnSO_4$ (6 μmol/L) and $MgCl_2$ (43 μmol/L), pH7.0] agar plates supplemented with neomycin (10 μg/mL), and the transformants obtained were cultured for producing recombinant heptahistidine-tagged (His_7-tagged) TrCel12A as previously described (Takeda, et al., 2010). *B. choshinensis*, which is a Gram-positive bacteria, secretes synthesized proteins into the culture medium when a secretion signal peptide is fused to the recombinant protein. In contrast, *Escherichia coli*, which is widely used as a host cell, accumulates synthesized recombinant proteins in the cytosol. Because the amount of proteins secreted into the culture medium is very small as compared with that of cytosolic proteins, secretion of the recombinant protein will make any subsequent purification procedure easier.

2.3 Purification of His_7-tagged TrCel12A

The culture filtrate of *B. choshinensis* expressing His_7-tagged TrCel12A was subjected to ammonium sulfate (60%, w/v) precipitation, and the precipitate obtained by centrifugation was dissolved in Equilibration buffer (10 mM sodium phosphate, pH 7.0, 50 mM NaCl) for polyhistidine affinity chromatography. The solution was applied to a polyhistidine-binding resin charged with cobalt ion: on this column, His_7-tagged TrCel12A is captured by cobalt ion and released by buffer exchange with imidazol. After washing the column with Equilibration buffer, the proteins that bound to the resin were sequentially eluted with 0.1 X Elution buffer (10 mM sodium phosphate, pH 7.0, 50 mM NaCl, 20 mM imidazol), Elution buffer (50 mM sodium phosphate, pH 7.0, 50 mM NaCl, 200 mM imidazol) and Elution buffer containing 40 mM EDTA. The fractions were subjected to SDS-PAGE followed by CBB staining to verify the purity of the eluates (Fig. 2A).

Expressed His_7-tagged TrCel12A was detected at around 24 kDa (Fig. 2A). The eluate in 0.1 X Elution buffer (Fig. 2A, lane 1) contained His_7-tagged TrCel12A and several other proteins. The eluate in Elution buffer (Fig. 2A, lane 2) consisted of mostly His_7-tagged TrCel12A and a small quantity of other proteins. In contrast, the eluate in Elution buffer containing 40 mM EDTA (Fig. 2A, lane 3) contained His_7-tagged TrCel12A without other visible proteins. Thus, His_7-tagged TrCel12A and non-specific binding proteins are gradually eluted from the resin with increasing imidazol concentration. These results imply that most His_7-tagged TrCel12A remained bound to the resin after application of the Elution buffer recommended by the manufacturer, and that His_7-tagged TrCel12A was best eluted with Elution buffer containing EDTA. EDTA chelates cobalt ion from the resin, resulting in the release of TrCel12A that bound tightly to the resin. Subsequently, EDTA and cobalt ion were removed from the protein preparation by ultrafiltration.

2.4 Immunoblot analysis using antibody against polyhistidine-tag

Recombinant proteins with an epitope-tag are recognized by an antibody against the corresponding epitope-tag. For example, His_7-tagged TrCel12A expressed in *B. brevibacillus* was subjected to SDS-PAGE, followed by immunoblotting using an antibody against polyhistidine (Qiagen) (Fig. 2B). Immunoblot analysis showed that the molecular weight of the proteins that reacted with antibody was identical to that of the major proteins observed in the eluate from polyhistidine-binding resin by CBB staining (Fig. 2A). Purified protein in the eluate with Elution buffer containing EDTA was recognized by the antibody.

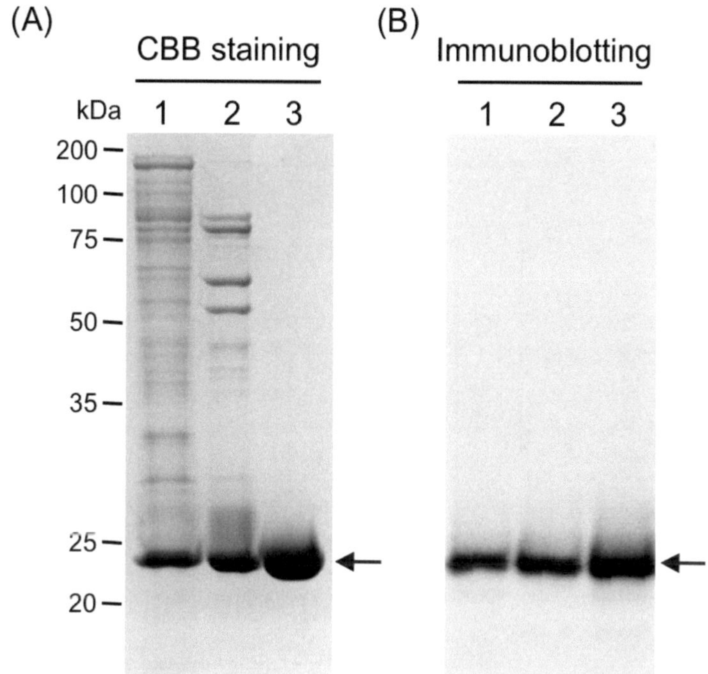

Fig. 2. Visualization of His_7-tagged TrCel12A with CBB (A) and immunoblotting using antibody against polyhistidine-tag (B) after separation by SDS-PAGE. Lane 1, eluate from polyhistidine-binding resin eluted with 0.1 X Elution buffer; lane 2, eluate with Elution buffer; lane 3, eluate with Elution buffer containing 40 mM EDTA. Arrows indicate His_7-tagged TrCel12A.

3. Biochemical analysis of His_7-tagged protein

3.1 Effect of EDTA on purified His_7-tagged TrCel12A

Native TrCel12A catalyzes the hydrolysis of 1,4-β-glucans such as carboxymethyl cellulose (CMC), crystalline cellulose, phosphoric acid-swollen cellulose (PSC), xyloglucan and 1,3-1,4-β-glucan (Sprey and Uelker, 1992). For determination of its biochemical properties, recombinant His_7-tagged TrCel12A must retain hydrolytic activity after elution with Elution buffer containing EDTA. To test the effect of EDTA treatment on hydrolytic activity, His_7-tagged TrCel12A was incubated with 1,3-1,4-β-glucan and sodium acetate (100 mM, pH 5.5) in the presence of different concentration of EDTA. Hydrolysis of 1,3-1,4-β-glucan by TrCel12A results in the increase in the number of the molecule with a reducing terminus. The reaction mixture was mixed with *p*-hydroxybenzoic acid hydrazide and incubated in a boiling water. Hydrolytic activity was determined by measuring the absorbance at 410 nm in a spectorphotometer as described previously (Miller, 1972) (Fig 3). The hydrolytic activity of His_7-tagged TrCel12A was not negatively influenced by EDTA. Similarly, the hydrolytic activities of recombinant proteins that we produced previously a xyloglucan-specific endoglucanase (XEG) in *B. choshinensis*, β-glucosidases in *Aspergillus oryzae* and cellobiohydrolase in *Magnaporthe oryzae* were not negatively affected by EDTA. However, not all His_7-tagged recombinant proteins are applicable to EDTA elution because some enzymes, such as peroxidases, require metal ions for their catalytic reaction, and these metal ions are removed by the chelater EDTA (Cohen, et al., 2002; Lundell, et al., 2010).

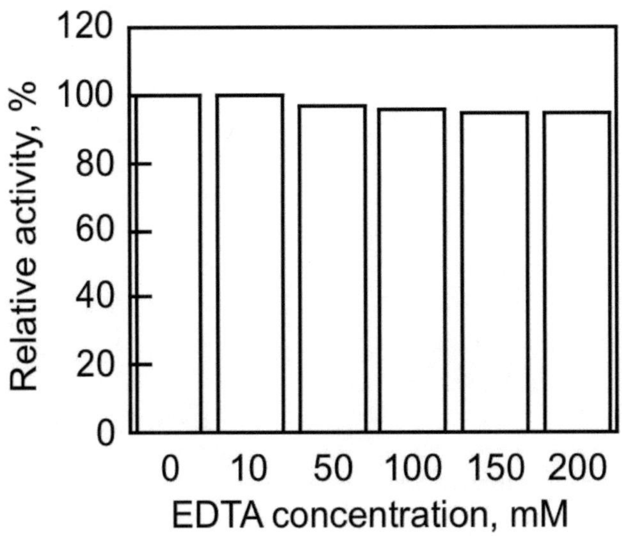

Fig. 3. Effect of EDTA on the hydrolytic activity of purified His_7-tagged TrCel12A. The hydrolytic activity of His_7-tagged TrCel12A toward 1,3-1,4-β-glucan was assayed in the presence of EDTA.

3.2 Substrate specificity and product analysis

To characterize recombinant His_7-tagged TrCel12A expressed by *B. choshinensis*, its hydrolytic activity was assayed as described above. His_7-tagged TrCel12A preferentially hydrolyzed water-soluble β-1,4-glucans such as xyloglucan and 1,3-1,4-β-glucan, and slightly cleaved crystalline cellulose, PSC and CMC, of which the greatest hydrolysis was observed toward 1,3-1,4-β-glucan (Fig. 4). HPLC (ICS-3000, Dionex) analysis of the hydrolyzed products showed that His_7-tagged TrCel12A produced mainly cellobiose, cellotriose and cellotetraose from PSC after a long incubation (Fig. 5). Thus, His_7-tagged TrCel12A purified by polyhistidine affinity chromatography was accurately characterized without effects from other contaminant proteins. Native TrCel12A in *T. reesei* culture filtrate has previously been purified to homogeneity by a combination of chromatography steps; in contrast, recombinant His_7-tagged TrCel12A with hydrolytic activity was purified by one-step polyhistidine affinity chromatography.

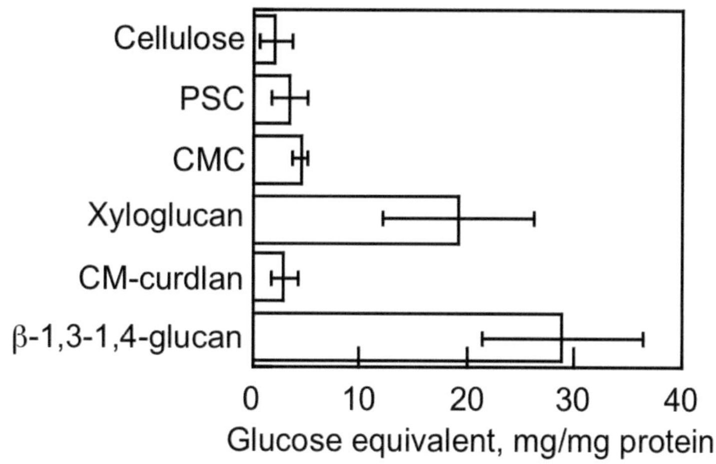

Fig. 4. Substrate specificity of His_7-tagged TrCel12A. Hydrolytic activity of His_7-tagged TrCel12A was determined using diverse polymers as indicated on the *y* axis. Data are the means ± SE of three determinations.

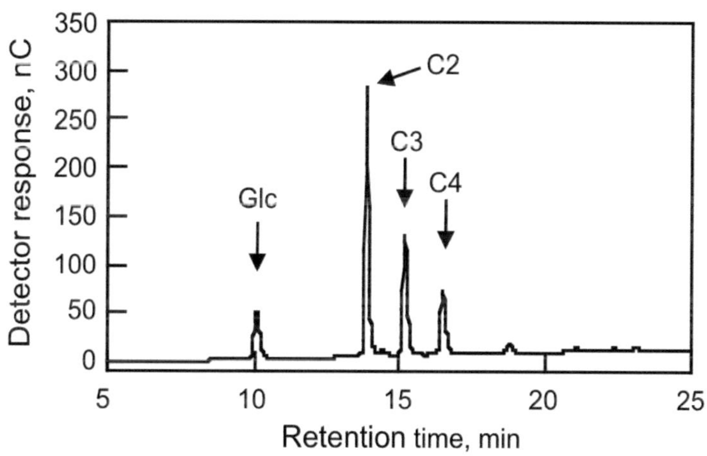

Fig. 5. Product analyses of hydrolysates from PSC. After incubation of PSC with His_7-tagged TrCel12A in sodium phosphate buffer (100 mM, pH 5.5), a portion of the reaction mixture was subjected to HPLC. Arrows indicate the position of glucose (Glc), cellobiose (C2), cellotriose (C3) and cellotetraose (C4) eluted during the separation.

4. Other cell wall-degrading enzymes with His_7-tag

Magnaporthe oryzae is the pathogen that causes rice blast, the most devastating fungal disease of rice, and it secretes a variety of cell wall-degrading enzymes during its invasion of rice. The complete genome DNA sequences of *Magnaporthe* genera have been elucidated (Dean et al., 2004; Yoshida et al., 2007). Therefore, genetic analysis and protein engineering can be facilitated by utilizing these DNA databases. Indeed, recombinant proteins with hydrolytic activities toward cell wall polymers and oligosaccharides have been produced after cloning the corresponding DNA by PCR from complementary DNA pools. For example, cellobiohydrolase (MoCel6A), which preferentially hydrolyzes PSC, cellulose and 1,3-1,4-β-glucan, has been produced from *M. oryzae* and *Aspergillus oryzae* (Takahashi, et al., 2010). β-Glucosidases (MoCel3A and MoCel3B), which produce glucose from β-1,3- and β-1,4-glucans, have been produced from *M. oryza*e (Takahashi, et al., 2011). A specific 1,3-1,4-β-glucanase has been produced from both *B. choshinensis* and *M. oryzae* (Takeda, et al., 2010). The purified proteins were obtained by using polyhistidine affinity chromatography coupled with an EDTA-containing elution buffer. In contrast, for the preparation of β-glucosidase (UeBgl3A) from *Ustilago esculenta*, whose genome DNA sequence has not been elucidated, the partially purified β-glucosidase was first obtained in low quantity after two steps of ion affinity chromatography and then subjected to trypsin-digestion followed by MS/LC/LC analysis to identify the partial amino acid sequence. On the basis of the deduced peptide sequence, the cDNA was then cloned by using degenerate DNA primers. Lastly, His_7-tagged UeBgl3A was expressed by *A. oryzae* and purified by polyhistidine affinity chromatography (Nakajima, et al., 2011).

5. Conclusion

Purifying native and recombinant proteins is an important procedure for the detailed analysis of properties such as mode of action and substrate specificity, and for determining optimal reaction conditions. From this viewpoint, the addition of epitope-tags to a protein facilitates easy purification by epitope-tag affinity chromatography; however, this method is applicable only to recombinant proteins. Especially, the polyhistidine-tag has the ability to bind metals attached to a resin and is liberated by exchange with imidazol, as demonstrated in this chapter. Furthermore, tightly bound proteins can be eluted by using EDTA, which chelates metals from the resin. In addition, antibodies against epitope-tags are a useful tool to detect epitope-tagged proteins. Because antibodies bind tightly to the corresponding antigen, however, it is necessary to use strong acid or SDS to release the antigens from the antibodies in which active enzymes would become inactivated. Thus, we have to proceed with caution when using affinity resins and antibodies for the purification and detection of epitope-tagged proteins.

So far, the large-scale production of recombinant proteins that catalyze the hydrolysis of cell walls has been carried out in *B. choshinensis*, *M. oryzae* and *A. oryzae* as a host cell in our lab because the heptahistidine-tag works well for purification, and non-specific binding of intact proteins are removed by increasing concentration of imidazol. On the other hand, His_7-tagged and HA-tagged TrCel12A have been produced in *Nicotiana benthamiana* by transient expression as described previously (Takken et al., 2000), and verified by immunoblotting using antibody against polyhistidine and HA, respectively (data not shown). However, purification of His_7-tagged TrCel12A expressed by *N. benthamiana* did not work well because it was difficult to separate His_7-tagged TrCel12A from many proteins that bound non-specifically to the polyhistidine-binding resin. Selecting the best epitope-tag to use in terms of not only purification but also immunoblotting should be done before the protein production procedure is established.

6. References

Bhat, M. K. (2000). Cellulases and related enzymes in biotechnology. *Biotechnol Adv,* Vol. 18, *No.* 5, pp. 355-383.

Cohen, R., Persky, L. & Hadar, Y. (2002). Biotechnological applications and potential of wood-degrading mushrooms of the genus *Pleurotus. Appl Microbiol Biotechnol,* Vol. 58, *No.* 5, pp. 582-594.

Dean, R. A. et al. (2005) The genome sequence of the rice blast fungus *Magnaporthe grisea. Nature,* Vol. 434, *No.* 7036, pp. 980-986.

Henrissat, B., Driguez, H., Viet, C. & Schulein, M. (1985). Synergism of cellulases from *Trichoderma reesei* in the degradation of cellulose. *Bio/Technology,* Vol. 3, pp. 722-726.

Lundell, T. K., Mäkelä, M. R. & Hildén, K. (2010). Lignin-modifying enzymes in filamentous basidiomycetes—ecological, functional and phylogenetic review. *J Basic Microbiol,* Vol. 50, *No.* 1, pp. 5-20.

Miller, M. (1972). A new reaction for colorimetric determination of carbohydrates. *Anal Biochem,* Vol. 47, *No.* 1, pp. 273-279.

Nakajima, M., Yamashita, T., Takahashi, M., Nakano, Y. & Takeda, T. (2011). Identification, cloning, and characterization of β-glucosidase from *Ustilago esculenta. Appl Microbiol Biotech,* (in print).

Nicol, F., His, I., Jauneau, A., Vernhettes, S., Canut, H. & Höfte, H. (1998). A plasma membrane-bound putative endo-1,4-β-glucanase is required for normal wall assembly and cell elongation in Arabidopsis. *EMBO J,* Vol. 17, *No.* 19, pp. 5563-5576.

Reese, E. T., Siu, R. G. H. & Levinson, H. S. (1950). The biological degradation of soluble cellulose derivatives and its relationship to the mechanism of cellulose hydrolysis. *J Bacteriol,* Vol. 59, *No.* 4, pp. 485-497.

Sprey, B. & Uelker, A. (1992). Isolation and properties of a low molecular mass endoglucanase from *Trichoderma reesei. FEMS Microbiol Lett,* Vol. 71, *No.* 3, pp. 253-258.

Takahashi, M., Takahashi, H., Nakano, Y., Konishi, T., Terauchi, R. & Takeda, T. (2010). Characterization of a cellobiohydrolase (MoCel6A) produced by *Magnaporthe oryzae. Appl Environ Microbiol,* Vol. 76, *No.* 19, pp. 6583-6590.

Takahashi, M., Konishi, T. & Takeda, T. (2011). Biochemical characterization of *Magnaporthe oryzae* β-glucosidases for efficient β-glucan hydrolysis. *Appl Microbiol Biotech,* Vol. 91, *No.* 4, pp. 1073-1082.

Takken F. L., Luderer, R., Gabriëls, S. H., Westerink, N., Lu, R., de Wit, P. J. & Joosten, M. H. (2000). A functional cloning strategy, based on a binary PVX-expression vector, to isolate HR-inducing cDNAs of plant pathogens. *Plant J,* Vol. 24, *No.* 2, pp. 275-283.

Takeda, T., Furuta, Y., Awano, T., Mizuno, K., Mitsuishi, Y. & Hayashi, T. (2002). Suppression and acceleration of cell elongation by integration of xyloglucans in pea stem segments. *Proc Natl Acad Sci USA,* Vol. 99, *No.* 13, pp. 9055-9060.

Takeda, T., Takahashi, M., Nakanishi-Masuno, T., Nakano, Y., Saitoh, H., Hirabuchi, A., Fujisawa, S. & Terauchi, R. (2010). Characterization of endo-1,3-1,4-β-glucanases in GH family 12 from *Magnaporthe oryzae. Appl Microbiol Biotech,* Vol. 88, *No.* 5, pp. 1113-1123.

Walton, J. D. (1994). Deconstructing the cell wall. *Plant Physiol,* Vol. 104, pp. 1113-1118.

Wood, T. M. (1992). Fungal cellulases. *Biochem Soc Trans,* Vol. 20, *No.* 1, pp. 46-53.

Yoshida, K., Saitoh, H., Fujisawa, S., Kanazaki, H., Matsumura, H., Yoshida, K., Tosa, Y., Chuma, I., Takano, Y., Win, J., Kamoun, S. & Terauchi, R. (2008). Association genetics reveals three novel avirulence genes from the rice blast fungal pathogen *Magnaporthe oryzae. Plant Cell,* Vol. 21, *No.* 5, pp. 1573-1591.

11

Novel Detection Methods Used in Conjunction with Affinity Chromatography for the Identification and Purification of Hydrolytic Enzymes or Enzyme Inhibitors from Insects and Plants

Alexander V. Konarev[1] and Alison Lovegrove[2]
[1]*All-Russian Institute for Plant Protection*
[2]*Rothamsted Research*
[1]*Russia*
[2]*UK*

1. Introduction

Plant proteinaceous α–amylase and proteinase inhibitors play many important roles *in planta* as endogenous enzyme regulators and as protective factors against harmful organisms. They are also of interest as antinutrients in human and animal feed, and as anticancer and antiviral agents in medicine. They may also be used as genetic markers in the study of plant diversity and evolution (Dunaevsky et al., 2005; Franco et al., 2002; Gatehouse, 2011; Habib & Fazili, 2007; Konarev, 1996; Konarev et al., 2002b; Mosolov and Valueva, 2005; Shewry & Lucas, 1997). Affinity chromatography is an effective method for the fast purification of various hydrolytic enzymes and their inhibitors. Its use in conjunction with detection methods that can quickly and easily identify components of proteins mixtures with the sought-after activity can appreciably simplify the search and purification of novel enzymes or enzyme inhibitors, especially those with low or non-typical substrate specificities. Advances have been made in fractionation and detection methods by many laboratories but we describe below approaches used by us including the use of affinity chromatography in combination with novel detection methods to identify and purify novel forms of insect and fungal proteinases and also proteinase inhibitors from plants (Conners et al., 2007; Konarev, 1985, 1986, 1990a, 1996; Konarev & Fomicheva, 1991a; Konarev et al., 1999a, 1999b, 2000, 2002a, 2004, 2008; 2011; Luckett et al., 1999). In some cases these have been used for the analysis of the diversity of α–amylase and proteinase inhibitors in various plant taxa (Konarev 1982a, 1986b, 1987b, 1996, Konarev et al., 1999c, 2000, 2002a, 2002b, 2004). The aim of the present chapter is to bring to a larger audience and summarize published works that are hard to access to non-Russian readers or described in scattered publications and to provide some examples using the approaches described. These data have been obtained in the All-Russian Institute for Plant Protection (VIZR) from 1981 and, after 1996, in collaborative work of VIZR with Rothamsted Research, University of Bristol (UK), National Institute of Agrobiological Sciences (Japan) and Hacettepe University (Turkey).

2. Methods for protein fractionation and the identification of enzymes and their inhibitors

2.1 General description of approaches

Affinity chromatography is widely used in the purification of α–amylases, proteinases and the inhibitors of these enzymes in plants, animals and insects (Buonocore et al., 1975; Nagaraj & Pattabiraman, 1985; Saxena et al., 2010). Various methods for the detection of α–amylase inhibitors (Giri & Kachole, 1996; Fontanini et al., 2007) and proteinase inhibitors (Pichare & Kachole, 1994; Mulimani et al., 2002) in protein mixtures are known. We have developed a set of universal and sensitive methods for the detection of inhibitors of various insect, plant, fungal and mammalian α–amylases and proteinases (Konarev 1981, 1982b, 1985, 1996; Konarev and Fomicheva, 1991a; Konarev et al., 1999b, 2000). These are most often used with polyacrylamide gel electrophoresis (PAGE), isoelectric focusing (IEF) and thin layer gel-filtration (TLGF) and in such variants may also be used to search for novel forms of enzyme inhibitors or to monitor the affinity chromatography and other protein fractionation techniques. The distinctive feature of the methods for detection of α–amylase inhibitors we describe is the use of polyacrylamide gels (PAG), containing both amylase and starch. At PAGE protein sample is separated and inhibitors are detected in such gel but at IEF or TLGF this gel serves as 'replica' for detection. In case of proteinase inhibitors sensitive 'gelatin replicas' are used allowing inhibitors for up to 4 different proteinases (or inhibitors of α-amylases and proteinases) to be detected among proteins separated within the same gel.

2.2 Protein samples

Plant seeds were obtained from the collection of the N.I. Vavilov Institute of plant industry, St. Petersburg, (VIR, accessions designated by number "k-.."), the Botanical institute, St. Petersburg, various seed companies (UK) or collected on expeditions by the author. Proteins were isolated from ground seeds or vegetative organs homogenized with different extractants depending upon which proteins were of interest (water, 2 M urea, 7.4 mM Tris - 57 mM glycine buffer pH 8.3 etc.) in the various ratios (w/v) ranging from 1:2 to 1:10. The extractions were carried out at 20° C for 0.5-1 h. For enzymes showing greater sensitivity to inhibitors, the extract was diluted by 50 to 1000 fold. Extracts were heated for 15 min at 80° C for inactivation of endogenous α– and β–amylases and proteinases and denatured proteins were removed by centrifugation. The sources of α–amylases and proteinases for analyses were fresh or freeze dried salivary glands, guts and homogenates of insects, human saliva and wheat grains germinated for 2 days. Commercial preparations (α–amylases from pig pancreas or fungal cultures, trypsin, chymotrypsin, subtilisin, elastase, papain, ficin were from Sigma or other specialist suppliers). α–amylases and proteinases were extracted from homogenized material in water or tris-glicine buffer with or without 1mM $CaCl_2$. Single bands of amylases or proteinases have been isolated using micropreparative variants of IEF or affinity chromatography and used for detection of inhibitors. In most cases the amylases present in total extracts, when used in detection methods, showed clear pictures of interaction with their inhibitors comparable with those obtained with enzymes extracted from guts or partially purified.

2.3 Methods for protein fractionation

2.3.1 Flat-bed-PAGE in homogenous buffer system

This type of electrophoresis was chosen since it allowed the inclusion of α–amylase in the gel without effecting its concentrating during the separation of proteins which is in contrast to the discontinuous buffer system (see in 2.4.2). The solution containing 6% acrylamide, 0.26 % N,N'-methylenebisacrylamide, 7.4 mM Tris - 57 mM glycine buffer pH 8.3, riboflavin and TEMED was poured into a casting form consisting of two glass plates, one of which possessed juts for forming slots for samples. The 2 mm thick gel was polymerized in the light. The thickness of the gel could be reduced if a plastic supporting film GelFix for PAGE (Serva) was used providing a covalent binding to the gel layer. The length of the gel was in the range 10 to 20 cm and the position of the slots was determined depending on analysis. The gel was put on the cooling plate of the Multiphor II Electrophoresis unit (LKB) and connected to electrode buffer tanks by paper wicks. Electrode buffers were the same as in the gel. Protein samples were loaded in slots in a volume of between 0.5-20 mcl. Wheat seed albumins were separated in 120x200 mm gel with a path length for bromphenol blue of 9 cm for 1 h at voltage 1000 V, current 40 mA and 10° C. Wheat protein bands were visualized by transfer of the gel into 10 % TCA.

2.3.2 SDS-PAGE and mass-spectrometry (MS)

SDS-PAGE of proteins was carried out according to Laemmli (1970) with modifications. Proteins were heated in a standard SDS-PAGE sample buffer with 2-mercaptoethanol (for estimation of sizes of peptides) or without reductant (for following detection of intact inhibitors). For Edman sequencing proteins were transferred to PVDF membrane using a semi-dry method and CAPS buffer.

For MS peptides were isolated from PAGE and digested by trypsin. Peptide digests were mixed 1:1 with matrix (2 mg/mL of R-cyano-4-hydroxycinnamic acid was dissolved in 49.5% (v/v) ethanol, 49.5% (v/v) acetonitrile, and 1% (v/v) of a 0.1% (v/v) TFA solution). 1 mcl of the mixture was spotted onto the target plate and the sizes of peptides were determined using a Micromass M@LDI-LR mass spectrometer (Waters, Manchester, UK) using a standard peptide mass fingerprinting method and mass acquisition between 800 and 3500 m/z. The laser firing rate was 5Hz, 40 random aims per spot, 10 shots per spectrum, 10 spectra per scan, 10 scans combined, 10% adaptive background subtracted, smoothed (Savitzky-Golay) and centroided. The MALDI-MS was tuned to 10,000 FWHM (Full Width at Half Height) and calibrated with a tryptic digest of alcohol dehydrogenase following the manufacturer's instructions.

For sequencing by MS peptides were concentrated and desalted using Zip-Tips (Millipore), dried, dissolved in 70% (v/v) methanol containing 1% (v/v) formic acid, sonicated for 3 min and then loaded into nanoflow tips (Waters). Electrospray ionization-MS was performed on a quadrupole time-of-flight (Q-TOF) I mass spectrometer (Micromass, Manchester, UK) equipped with a z-spray ion source. Instrument operation, data acquisition and analysis were performed using MassLynx/Biolynx 4.0 software. The sample cone voltage and collision energy were optimized for each sample. The Micro channel plate (MCP) detector voltage was set at 2800 V. Scanning was performed from m/z 100-3500. Prior to acquisition the mass spectrometer was calibrated using a solution of [Glu[1]]-fibrinopeptide B as described by the manufacturer.

2.3.3 Isoelectric focusing (IEF)

IEF of proteins was carried out on a variety of available IEF gel systems (Servalyt Precotes pH 3-10 (Serva) etc.) using LKB Multiphor II apparatus or a Phast System (Pharmacia). Cytochrome c (pI 10.65), horse myoglobin (pI 7.3) and whale myoglobin (pI 8.3) (Serva) were used as markers. Proteins were detected in gels by Coomassie R-250 in 10 % TCA after removal of ampholites by washing in 10 % TCA alone.

For micropreparative IEF, 0.5–1 mg of proteins (e.g. trypsin or chymotrypsin inhibitors eluted from affinity columns) were applied in 0.4 ml water onto a 10x60-mm filter-paper strip across the pH gradient of a Servalyt Precotes (Serva) gel, and IEF was carried out using a Multiphor with a distance of 10 cm between electrodes. The gel was then placed in 40% (w/v) ammonium sulphate for 5 min and the opalescent protein bands excised and placed in 0.5 ml of water for 1 h to elute the inhibitors. Micropreparative IEF in PAG or in Ultrodex was used also for the isolation of insect and fungal α–amylases and proteinases.

Micropreparative fractionation of proteinases or inhibitors with higher resolution included IEF in DryStrips (GE Healthcare). Protein fractions obtained by affinity chromatography (e.g. proteinase inhibitors from *Cycas siamensis* or *Veronica hederifolia* seeds) were loaded onto two Immobiline DryStrip pH 3–10 NL gels and inhibitor bands located using gelatin replicas of one gel. The second gel was fixed in 10% (v/v) TCA revealing proteins as opaque bands. Zones containing protein were excised and TCA was removed by washing with cold acetone. Proteins were then extracted from the gel with 20% (v/v) ethanol containing 0.1% (v/v) TFA, followed by 30% (v/v) acetonitrile in 0.03 M ammonium bicarbonate and 4% (v/v) formic acid (for 1 h each) and finally freeze dried. Transfer of proteins from the electrophoresis, IEF or TLGF gels to nitrocellulose (NC) or PVDF membranes was performed using equipment for electro transfer or simple diffusion by contact of NC with gel for 1-12 hours (if detection of inhibitors or microsequencing was to be carried out).

2.3.4 Gel-filtration

Preparative gel-filtration of proteins was carried out in 40-100x1.5 cm columns filled with Sephadex G-100 or G-50 (Fine) or in Ultrogel AcA 54 (LKB) with either 0.1 M phosphate buffer pH 7.4 containing 0.2 M NaCl or 0.1 M ammonium acetate pH 7 (the latter if proteins were to be freeze-dried).

Thin layer gel-filtration (TLGF) of plant proteins was performed in 0.4-1 mm layer of Sephadex G-50 or G-100 Superfine (LKB) attached to 250x125 mm a glass plate or to Gel-bond to agarose film (Serva) with 0.1 M NaCl or 0.03 M ammonium acetate as the liquid phase. The Sephadex gel plate was set in apparatus for horizontal electrophoresis and connected to upper and lower tanks containing the liquid phase by filter-paper wicks. Proteins (0.5-10 mcg/mcl) were applied on gel in a minimal volume (0.5-2 mcl). Coloured proteins cytochrome c (12.5 kDa), horse myoglobin (17.8 kDa) and bovine hemoglobin (64.5 kDa) were used as markers. The elution rate was adjusted by changing the angle of inclination of the plate (for the spot of cytochrome c it was about 2 cm/h). The fractionation of wheat seed albumins with MW 12-60 kDa lasted about 3 h.

2.3.5 Affinity chromatography of proteinases and proteinase inhibitors

Affinity matrices using proteinases (trypsin, chymotrypsin and subtilisin) or proteinase inhibitors immobilized on CNBr-activated Sepharose or acrylamide or agarose gels were used for purification of proteinase inhibitors from plants and proteinases from insects (Conners et al., 2007; Konarev 1986a, Konarev et al., 2000, 2002a, 2004, 2008, 2009, 2011). For analytical work, certain types of inhibitor or proteinase were pre-absorbed from protein mixtures by modification of the method of Hejgaard et al. (1981) to facilitate the identification of other types of inhibitors or proteinases. 6 mcl of a 50% suspension of affinity gel in water (v/v) was added to 30 mcl of protein mixture extracted from seeds with water (1:4) and mixed frequently during 30 min period. The affinity gel was removed by centrifugation.

In preparative work with inhibitors highly specific to certain proteinase (for e.g., trypsin inhibitors from wheat grain and immobilized trypsin), 5-30 ml of solution containing inhibitors was passed through a column with 0.5-3 ml of the affinity gel with immobilized proteinase. For higher volume sample (0, 5 - 1 L) 5-10 ml of suspension of affinity gel was added to the solution and after 15-30 min of shaking the gel was collected using a sintered glass filter. The gel was washed with 10-50 volumes of the same solvent as used in the sample for loading (water, 0.2 M NaCl, 2 M urea, 0.1 M ammonium acetate etc.) and then with 10-50 volumes of 0.1 M NaH_2PO_4 with 0.5 M NaCl pH 5.0 and finally with water. The gel was transferred to a column and inhibitors were eluted with 0.015 M HCl. The process was monitored at 280 nm using standard equipment. Fractions obtained were freeze dried and analyzed for the presence of proteinase inhibitors using IEF combined with the gelatin replicas method (see below). Selected fractions were then separated by reverse phase HPLC with a C18 RP Phenomenex column and a gradient of 15–45% (v/v) acetonitrile in 0.1% trifluoroacetic acid.

Similar procedures were used for isolation of proteinases, e.g. grain borer *Rhyzopertha dominica* (Fab.) gut serine proteinases on soybean TI linked to agarose (Sigma) (Konarev et Fomicheva, 1991a; Konarev, 1992b).

In instances where the interaction between proteinase and inhibitor was weak (for e.g. the gluten-specific proteinase of sunn pest *Eurygaster integriceps* Put. and chymotrypsin inhibitor I from potato, Calbiochem), after loading of the sample the affinity gel was washed just with the solvent used in the sample (0.01 M ethanolamine with 0.2 M NaCl and 0.01% Triton X-100 pH 10) and eluted with water/0.01% (v/v) Triton X-100 followed by 0.01M HCl/0.01% (v/v) Triton X-100 (Konarev et al., 2011). Both eluates contained almost pure proteinase.

2.4 Detection of α-amylases and their inhibitors

2.4.1 Detection of α-amylases after PAGE or IEF

For the estimation of heterogeneity and electrophoretic mobility of the α–amylases 0.1 % soluble starch was added to the solution, described in 2.3.1, before polymerization of the gel for PAGE. After separation the gel was placed in 0.1 M acetic pH 5.4 (for insect and germinating wheat grain α–amylases) or 0.1 M phosphate buffer pH 7.0 (for human and mammalian α–amylases) containing 0.1 M NaCl and 1 mM $CaCl_2$ and incubated for 30 min at 37° C. Areas of α–amylase activity were observed as transparent bands on dark blue background after immersion of the gel in I/KI solution (50 mg I and 1000 mg KI per liter of water). In case of IEF the gel with the same composition was placed on top of a separating gel and after incubation at 37° C for 10-30 min put in I/KI solution.

2.4.2 Detection of α-amylase inhibitors after PAGE, IEF or TLGF of plant proteins

In order to detect α–amylase inhibitors in complex seed protein extracts 0.1 % starch and the α–amylase of interest were included in separating gel for PAGE (see 2.3.1) before polymerisation. The volume of added α–amylase solution approximately corresponded to the enzyme activity required to hydrolyze all the starch in the gel after 1 h at optimal pH and at 37° C (which was determined in preliminary tests using a row of small gels containing different volumes of α–amylase). The optimal volume of seed protein sample loaded in a slot of the gel depended strongly on the type of α–amylase included in the gel and its sensitivity to certain type of inhibitors. Therefore, when inhibitors of low sensitive α–amylase from sunn pest *E. integriceps* gut were to be identified, 10-15 mcl of protein samples extracted from bread wheat seeds (with water or 2 M urea (1:4)) were loaded. In the case of the much more inhibitor-sensitive α–amylase from mealworm beetle *Tenebrio molitor* L. larvae the extract was diluted 500-1000 times before loading. The key features of this method are that (1) the enzyme and substrate are present in the separating gel simultaneously and untimely hydrolysis is prevented by non-optimal pH and a low temperature during separation, (2) α–amylase included in the gel, in most cases, moves slower than the inhibitors and after PAGE in homogenous buffer system inhibitors are evenly surrounded by enzyme and finally (3) the presence of an α-amylase in all layers of the gel (in contrast to methods, where the gel is incubated in a α-amylase solution) provides a very high sensitivity of detection.

In the described conditions of electrophoresis α–amylases from human saliva, pig pancreas, gut of sunn pest *E. integriceps* and germinating wheat seed had a mobility lower than that of wheat albumins including amylase inhibitors. In case of α–amylase of mealworm beetle *T. molitor* larvae the enzyme moves faster than the inhibitors, so that separating gel should be extended in part between sample slots and cathode in order to compensate for the difference in speed of run of enzyme and inhibitors. After separation gels were put in 0.1 M acetic pH 5.4 (for insect and wheat germinating grain α–amylases) or 0.1 M phosphate buffer pH 7.0 (for human and mammalian α–amylases) containing 0.1 M NaCl and 1 mM $CaCl_2$ and incubated for 30-50 min at 37° C. For initial runs (with an "unknown" α–amylase) hydrolysis of starch should be checked every 5-10 min on small pieces of gel cut from the sides of the separating gel and immersed in iodine solution. Incubation may be stopped when the gel pieces have acquired a violet-pink colour. The whole separating gel can then be transferred to the iodine solution. Inhibitors become visible as dark blue bands of undigested starch on the light transparent background.

For detection of α–amylase inhibitors after IEF or TLGF of plant proteins the gel with the same composition as was used for detection at PAGE (see above) was placed onto separating gel for 10-30 min and then put in buffer with pH optimal for α–amylase and incubated for 15-40 min at 37° C until test pieces of the gel have acquired a violet-pink colour. More details will be given at description of examples of method application.

2.5 Detection of proteinases and proteinase inhibitors

2.5.1 Detection of proteinases

Proteinases were detected in protein mixtures isolated from insect, fungal or plant samples following separation by IEF or PAGE using plastic films covered by protein substrates (gelatin or glutenin) or solutions of synthetic substrates.

The most sensitive and applicable to detection of many serine and cysteine proteinases with wide substrate specificity (trypsin-, chymotrypsin-, subtilisin-, papain- or ficin-like enzymes) method is based on the use of layer of gelatin present in photographic materials. This approach exists in many variants (Burger and Schroeder, 1976; Harsulkar et al., 1998). We used undeveloped opaque photographic film ("Photo 65", Russia, for example) both for analytical and micropreparative scales (Konarev and Fomicheva, 1991a; Konarev et al., 1999b). E.g., for isolation of extracellular proteinases of fungi *Sclerotinia sclerotiorum* and *Colletotrichum lindemuthianum* cultural filtrate was applied to the gel using wide filter-paper strips. After IEF three 1x40 mm strips were placed on the gel along the pH gradient and incubated for 5 min at 40° C. The zones of the gel corresponding to hydrolyzed gelatin were cut out, immersed in 200 mcl of 25 % sucrose for 1 h and the solution was used for detection of proteinase inhibitors using the gelatin-replicas method (see below) (Konarev et al., 1999b).

For the detection of glutenin-specific proteinase from the salivary glands of sunn pest *E. integriceps*, a layer of insoluble in acetic acid glutenin was attached to the plastic film as substrate (Konarev et al, 2011).

2.5.2 Detection of proteinase inhibitors

There are many methods for detection of proteinase inhibitors in protein mixtures following separation of the proteins by PAGE or IEF. Some are based on the use of a gelatin layer on photographic film as a substrate for proteinases (Pichare & Kachole, 1994; Mulimani et al., 2002). However, most of these methods include immersion of the separating gel in a proteinase solution followed by laying of photographic film on the gel and incubation. But because of diffusion of proteins from the gel in solution the sharpness of the bands can be decreased. Also, immersion in certain proteinase solutions prevents the detection of the inhibitors of other proteinases in the same gel. We developed the so-called "gelatin replicas" method which includes consecutive contact of the separating gel with one to four replicas which can be developed by four different proteinases. The method is based on ability of gelatin layer to absorb proteins from the separating gel (as with protein blotting onto nitrocellulose) (Konarev, 1986a; Konarev et al., 2002a, 2002b, 2004, 2008). This approach gives also the opportunity to detect α–amylase and proteinase inhibitors in the same separating gel (Konarev, 1986b, 1996).

One to four pieces of undeveloped non-transparent photographic film were applied sequentially to IEF gel (for 2, 5, 20 and 30 min, respectively). The "gelatin replicas" containing inhibitors absorbed from the IEF gel were then applied to 0.8% (w/v) agarose gels containing 0.1 M Na_2HPO_4 (pH 9) and one of serine proteinases (Sigma): trypsin (1 mcg/ml), chymotrypsin (10 mcg/ml), elastase (4 mcg/ml) and subtilisin (0.3 mcg/ml), or some insect or fungal serine proteinase. Incubation for 30 min at 45° C allowed the proteases to digest the gelatin on the photographic film with the positions of inhibitors being detected as dark ''islands'' of undigested gelatin on the photographic layer. The similar approach was used for detection of plant and insect cysteine proteinase inhibitors (Konarev, 1984; 1986a,1990a; Konarev et al, 1999a, 2002b), 0.1 M NaH_2PO_4 with 0.1 M DTT being included in agarose gels together with proteinase.

Besides, gelatin films could be effectively used for detection of inhibitors on NC replica from separating gel, soaked with proteinase solution and washed by buffer. This approach can be used, for example, for detection of inhibitors after SDS-PAGE or in "cross" method for improving of resolution of inhibitor or enzyme spectra.

2.5.3 Cross methods for detection of inhibitors of single α–amylase or proteinase IEF bands

In order to identify inhibitors of individual proteinase bands in complex proteinase mixtures without special purification of proteinases or inhibitors, the "cross" methods has been developed (Konarev, 1990a; Konarev et Fomicheva, 1991a). Proteinases and protein mixture containing inhibitors extracted from seeds were applied to the individual gels using wide paper strips (about 9 cm for 120x120 mm gel). At low contention in extract, inhibitors could be preliminary enriched by affinity chromatography. After separation a photographic film was placed for 20 min on the gel to obtain a replica of fractionated proteins and then a gelatin replica was applied to the gel with separated proteinases, at right angles to each other, and incubated for 0.5-1 h at 38° C. Narrowings or brakes in the proteinase bands corresponds to positions of their inhibitor bands. In our hands this method is applicable for the inhibitors from mature wheat grain and serine and cysteine proteinases from insects and germinating grains. The same approach was used for α–amylases and their inhibitors (see 3.1.).

3. The use of detection methods and affinity chromatography in study of hydrolases and their inhibitors in relation to the problems of plant diversity, evolution, pest resistance, food quality and medicine

3.1 Analysis of polymorphism of α–amylase and proteinase inhibitors in plants

The detection methods may be used for studying plant diversity and evolution by analysis of polymorphism of α–amylase and proteinase inhibitors. They may also find utility in the field of pest resistance and food quality. The figures that follow demonstrate some examples of application of the methods described.

It is well known that exogenous α–amylase inhibitors in wheat and related cereals are represented by three main fractions: 12 kDa (with monomeric molecules), 24 kDa (dimeric) and about 60-66 kDa (tetrameric) which can be easily obtained by gel-filtration (Buonocore et al., 1977; Franco et al., 2002). Fig. 1, I, shows results of PAGE of different fractions of wheat grain proteins followed by detection of inhibitors of insect, mammalian and plant amylases. α–amylases differ in sensitivity to various fractions of inhibitors, α–amylase from pig pancreas being intermediate between α–amylase from human saliva and insect amylases in their interaction with monomeric inhibitors. The most of studied insect α–amylases were inhibited by 12 kDa albumin fraction in contrast to human α–amylase. α–amylase from germinating wheat grain (c) and exogenous α–amylases are inhibited by different protein components. The method allows detection of inhibitors of both highly-sensitive (to inhibitors) and low-sensitive α–amylases. So, for obtaining comparable visible inhibitor bands, in case of highly sensitive α–amylase from *T. molitor* larvae, amount of seed proteins applied to PAGE was approximately in 1,000 times lower than that for α–amylase from sunn pest *E. integriceps* gut. With the same amount of seed proteins as was used for gut enzyme, α–amylase from sunn pest salivary glands was not inhibited at all (not shown) that indicated its practically full insensitivity. The reason for the variation in sensitivity of mentioned α–amylases to inhibitors may be hypothetised to be the co-evolution of the insects and plants on the level of digestive enzymes and their inhibitors. For example, sunn pest is a highly specialized phytophage (in contrast to *T. molitor*); during co-evolution of this bug with wheat its digestive amylases developed decreased sensitivity to inhibitors from wheat grains that weakened negative role of this proteins (Konarev, 1981, 1996). The same can be true for the sunn pest digestive proteinases, insensitive to inhibitors from wheat grain (Konarev et al., 2011; Konarev and Fomicheva, 1991a) and also for proteinases of some other insect pests (Gatehouse, 2011).

Bread wheat is a hexaploid (amphidiploid) plant combining three different genomes that determines the specific composition of wheat proteins (V.G. Konarev, 1996; Shewry et al., 2003) including enzyme inhibitors. Fig. 1, II, shows the variability of inhibitors of three α–amylases in grains of hexaploid bread wheat *Triticum aestivum* (genome composition A^uA^uBBDD), tetraploid *T. turgidum* (A^uA^uBB) and diploid goatgrass (*Aegilops*) species related to donors of genomes D and B for *T. aestivum*.

Methods of detection can be efficiently used in screening of plant collections for inhibitor composition. Presence or absence of main α–amylase inhibitor fractions in accessions can be easily estimated using thin layer gel filtration (TLGF) in combination with a PAG replica, containing starch and target enzyme (Fig1, III&IV). So, accessions of *Ae. speltoides* from the World collection of Vavilov Institute of Plant Industry (III, b & c) differ in presence of monomeric inhibitor of insect α–amylase. Fig.1, IV shows main insect amylase inhibitor fractions present in wheat, wild and cultivated barley species, maize, and oat grains (Konarev & Fomicheva, 1991b; Konarev, 1992b).

Using of TLGF with detection methods appeared to be also effective in search for novel low-molecular weight proteinase inhibitors (see in 3.2.2 and 3.2.3).

IEF in combination with PAG-amylase-starch replica provided much higher resolution of protein fractionation (Fig.2) although with less sensitivity of inhibitor detection.

This approach was most suitable for work with inhibitors (from cereals) of highly sensitive α–amylase of *T. molitor* larvae, and was also applicable to inhibitors of α–amylases from beetles lesser grain borer *Rhyzopertha dominica*, granary weevil (*Sitophilus granarius* L.) and human salivary α–amylase, but was not effective enough for sunn pest *E. integriceps* gut α–amylase inhibitors. These methods were used for analysis of hundreds of accessions of wheat and related cereals for α–amylase inhibitor composition in relation to problem of wheat diversity, evolution and pest resistance (Konarev, 1982a, 1986b,1992b), and also for analysis of variability of Mexican bean weevil (*Zabrotes subfasciatus)* and azuki and bean weevil *(Callosobruchus chinensis)* α–amylase inhibitors among seed proteins of *Phaseolus* and *Vigna* accessions (Konarev et al., 1999c). Clear evolutionary links between the α–amylase inhibitor systems in bread wheat and in other wheat and *Aegilops* species related to genome donors to *T. aestivum* were established during researches with use of mentioned method.

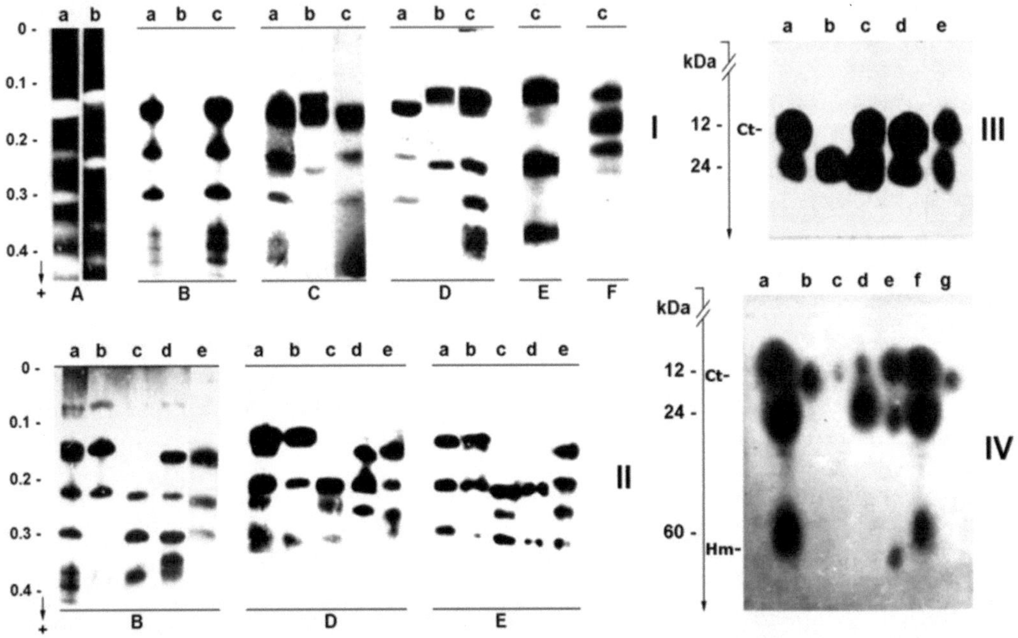

Panels I & II. 6% PAG containing 7.4 mM Tris - 57 mM glycine buffer pH 8.3 and 2 M urea (A), and additionally 0.1 % starch and α–amylases from: human saliva (B), pig pancreas (C), sunn pest (*Eurygaster integriceps* Put.) gut (D), mealworm beetle larvae (*Tenebrio molitor* L., E) and germinated wheat grain (F). I, panels A–F contain tracks of separated wheat proteins: a & b, bread wheat (*Triticum aestivum)* seed albumins 12 kDa (a) and 24 kDa (b), and total proteins extracted from wheat flour with 2 M urea (c). Panel II, B-E: tracks a-e, seed proteins extracted with tris-glycine buffer; a - T. *aestivum*; b, *Aegilops tauschii*; c, *Ae. longissima*; d, *T. turgidum*; e, *Ae. speltoides*. Proteins in separating gel A (I), were detected by fixation in 10 % TCA. α–amylase inhibitors were detected in separating gels (I, A–F and II, B,D & E) as described in 2.4.2. Panels III & IV. TLGF of seed proteins extracted by tris-glycine buffer in 0.4 mm layer of Sephadex G-100 Superfine. Ct & Hm, positions of coloured marker proteins cytochrome c and hemoglobin (spots not shown). 12, 24 and 60, approximate MW of inhibitor fractions. Panel III: a, *T. turgidum*; b & c, *Ae. speltoides* k-443 & k-1596; d, *Ae. longissima* k-194; e, *Ae. bicornis* k-904. Panel IV: a & f, *T. aestivum*; b & g, *Zea mays*; c, *Avena sativa*; d, *Hordeum vulgare*; e, *H. bulbosum*. The 6% PAG-replica containing 0.1 % starch, *T. molitor* α–amylase and tris-HCl buffer pH 8.3 has been placed on separating gel for 15 min and α–amylase inhibitors were detected (see in 2.4.2).

Fig. 1. Detection of α–amylase inhibitors from wheat and other cereals seed following either PAGE (I & II) or thin layer gel-filtration (TLGF, III & IV).

Fig. 2. demonstrates one of the fragments of analysis of the variability of insect (*T. molitor*) and human saliva α–amylase inhibitors (tracks bb`& ee`) in accessions of cultivated and wild wheat, *Aegilops* and rye species. The use of mentioned detection methods allowed us to find first that monomeric insect α–amylase inhibitors are controlled by chromosomes 6B and 6Dα of wheat (Konarev, 1982 &1996) and estimate the level of inter- and intraspecific variation of wheat by insect inhibitor composition and activity which impacts on the nature of pest resistance of wheat. Short reviews of these results on α–amylase inhibitors were published in English (Konarev, 1996, 1999a, 2000). Here we describe just examples of application of detection methods.

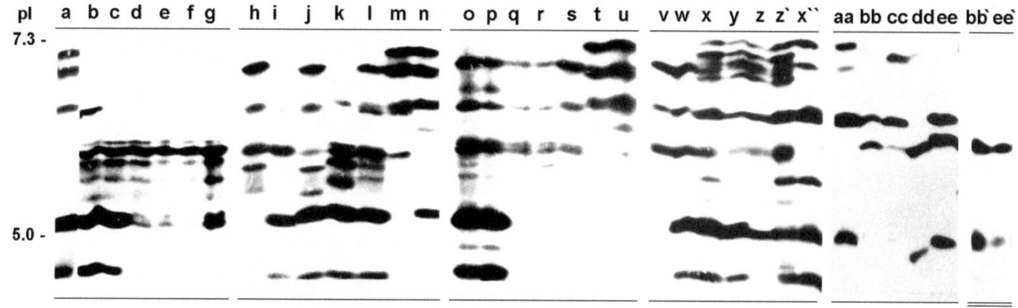

a & n, *Aegilops speltoides* k-1596; b, *Triticum militinae*; c & k, *T. timopheevii* k-29553; d, e & g, *T. araraticum*, different accessions; f, *T. zhukovskyi*; h, s & v, *T. persicum*; i, *Ae. longissima* k-194; j & w, *T. turgidum* k-42267; l, o & p, *T. timonovum*; m, t & u, *Ae. speltoides* k-452, k-443 & k-1316; q & r, *T. fungicidum*; x, y, z &z`, *T. aestivum* (x & y, var. Bezostaya 1; z &z`, var. Chinese Spring; x`` & aa, *Ae. tauschii*; bb & bb`, rye *Secale cereale* var. Vyatka 2; cc, *S. segetale*; dd, S. montanum; ee and ee`, S. ancestrale. Replicas containing 0.1% starch and *Tenebrio molitor* α–amylase (a-g, h-n, o-u, v-x`` & aa–ee) and human salivary α–amylase (bb`-ee`) were placed on separating gels and developed as described in 2.4.2.

Fig. 2. Detection of α–amylase inhibitors after IEF of wheat and other cereals seed proteins.

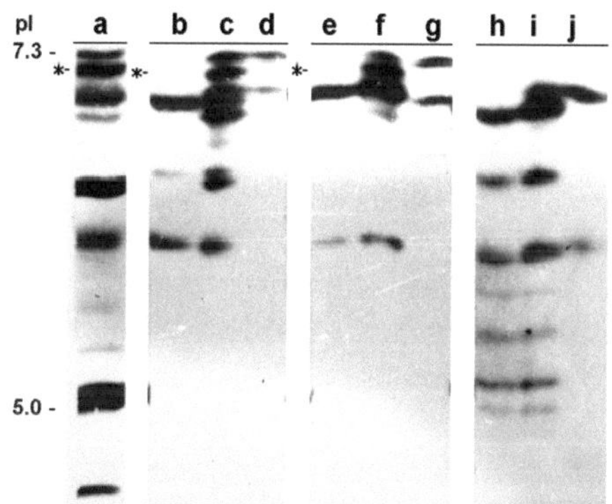

IEF of endosperm proteins and detection of inhibitors as described in 2.4.2. a, *Triticum aestivum*; b & h, *T. persicum*; d & g, *Aegilops tauschii* k-80; e & j, *T. boeoticum* (accession No 21); c, mixture of extracts b & d; f, mixture of extracts e & g; i, mixture of extracts h & j.

Fig. 3. Analysis of formation of hybrid molecules ("*") of dimeric *Tenebrio molitor* α–amylase inhibitors (24 kDa) *in vitro*.

The inhibitor set in hexaploid *T. aestivum* (genome formula A^uA^uBBDD) combines sets of inhibitors from *T. turgidum* (A^uA^uBB) and Ae. tauschii (DD), but additional components are visible (Fig.2 & 3). Fig 3, a-d, indicates that these inhibitor bands correspond to novel bands emerging in the mixture of proteins from *T. persicum* accession (species possessing the same genome composition as *T. turgidum,* but deprived of monomeric α–amylase inhibitors) and *Ae. tauschii* accession k-80 (rare form deprived of monomeric inhibitors). Presumably novel bands have hybrid nature and arose as a result of interchange of subunits of dimeric inhibitors controlled by different genomes. Another model mixture of proteins of *Ae. tauschii*

(DD) and diploid *T. boeoticum* (A^bA^b, rare accession possessing dimeric amylase inhibitor) also gave hybrid bands (f) in contrast to mixture of *T. persicum* and *T. boeoticum* proteins (i). Results indicate that hybrid bands emerge in combination of subunits controlled by B and D genomes or A and D genomes, and are absent in combination of subunits controlled by A and B genomes, that can be used as criteria for estimation of affinity of subunits.

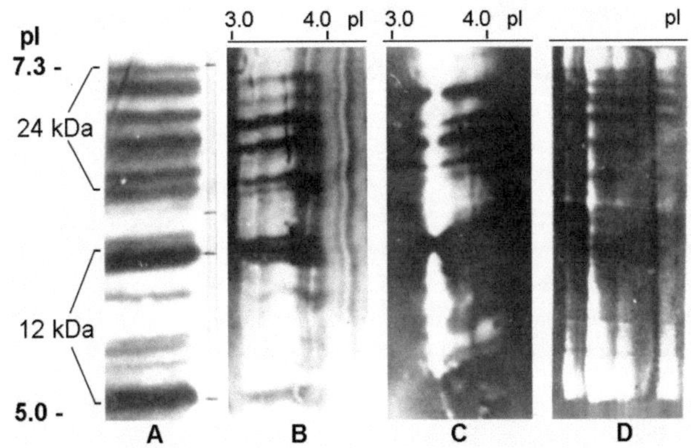

Fig. 4. Detection of inhibitors of α–amylase components using "cross"-method.

Protein extracts containing inhibitors and α–amylases were loaded onto detached gels with wide filter paper strips and separated by IEF. PAG replicas B-D containing 0.4 % starch were placed first onto the gel with separated inhibitors. After 15 min they were transferred onto the gel with separated α–amylases with a rotation of 90° relatively to the initial direction of separation. After a further 15 min, replicas were placed in a buffer with pI optimal for α–amylase and finally after 20 min moved to an iodine solution for staining. Replica from gel A contained *Tenebrio molitor* α–amylase and 0.4% starch and was used to demonstrate the spectrum of insect amylase inhibitors. Vertical direction on figure: IEF of water-soluble proteins from bread wheat var. Bezostaya 1 endosperm. Extracts were loaded on the gel using wide filter paper strips (A, B, C and D, loaded 2, 20, 20 and 10 mcg per mm of strip respectively). Horizontal direction: IEF of α–amylases from lesser grain borer (*Rhyzopertha dominica* F.) adults (B), granary weevil (*Sitophilus granarius* L.) adults (C) and human saliva. 12 and 24 kDa - approximate areas of protein spectrum with mainly monomeric and dimeric forms of inhibitors.

The "cross" variant of detection method can be useful for the preliminary characterisation of interaction of enzymes and inhibitors in complex mixtures. Fig. 4 demonstrates the uses of this approach by analyzing the interaction of individual bands from IEF inhibitor spectra with individual bands of heterogeneous α–amylase samples without special purification both of inhibitors and enzymes. In this case all IEF bands of each separated amylase are inhibited by similar inhibitor bands that indicate homogeneity of revealed enzyme bands in each analysed sample by sensitivity to inhibitors. The same approach was used for various serine and cysteine proteinases from insects and plants and their inhibitors (see method description in part 2.5.3). It allowed to reveal differences of bands of certain IEF proteinase spectrum in sensitivity, for example, to trypsin or chymotrypsin inhibitor bands and, correspondingly, to classify proteinase bands. In case of low concentration, inhibitors can be preliminary enriched by affinity chromatography (Konarev, 1990a; Konarev and Fomicheva, 1991a).

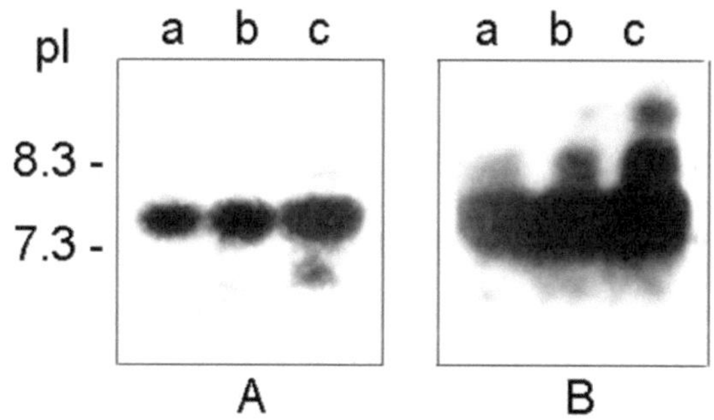

a, b c, 1, 3 & 4 mcl of protein solution extracted from *Hordeum bulbosum* endosperm with water (w/v 1:4) were loaded and separated by IEF. A PAG replica containing 0.4 % starch and *Tenebrio molitor* α–amylase (A) and gelatin replica (B) were put on the same gel sequentially and developed for either amylase or trypsin inhibitory activity.

Fig. 5. Detection of bifunctional α–amylase/trypsin inhibitor in protein spectra.

PAG-amylase-starch and gelatin replicas can be used in parallel for the detection of bifunctional α–amylase/proteinase inhibitors. Fig. 5 shows the results of such an approach for the revealing of the bifunctional insect α–amylase/trypsin inhibitor in IEF spectra of proteins from grains of wild barley species *Hordeum bulbosum* (Konarev & Fomicheva, 1991b; Konarev, 1992b). Bifunctional α–amylase from germinating wheat grain/subtilisin inhibitor and insect α–amylase/trypsin inhibitor have been detected in wheat and maize grains by the same methods (Konarev 1985, 1986b, 1992b). Revealing bifunctional nature of certain inhibitors opens up wide opportunities for their purification by affinity chromatography.

Sometimes the composition of inhibitors in the sample appears to be very complicated but the using of a combination of the gelatin replica method with a simplified version of affinity chromatography, affinity adsorption, can be useful. IEF spectra of proteins isolated from wheat and related cereals grains contain chymotrypsin, subtilisin (Fig.6., A&B), trypsin and amylase inhibitors (replicas not shown). The gelatin replicas A & B were obtained from the same separating gel sequentially and developed by chymotrypsin and subtilisin respectively. This showed that most of the bands revealed in replicas A and B coincided and corresponded to chymotrypsin/subtilisin inhibitors. Other bands have been revealed just in one of the replicas. A third replica from the same IEF gel developed by trypsin (not shown) allowed us to identify trypsin/chymotrypsin inhibitors (indicated on replica A with "*") in protein spectra of diploid wheat species *T. monococcum* (f), *T. boeoticum* (l) and *T. urartu* (m) and also in one of two accessions of *Ae. longissima* (g). In order to identify more clearly subtilisin inhibitors, the chymotrypsin/subtilisin inhibitors were removed from the protein extract by selective affinity adsorption using Sepharose gel with immobilized chymotrypsin. Treated extracts were separated by IEF and a gelatin replica from the gel was developed by subtilisin (replica C). The use of PAG-amylase-starch replica from the same gel showed that major bands visible on replica C correspond to germinating wheat grain α–amylase/subtilisin inhibitor (**). Most of the analyzed accessions of diploid, tetraploid and hexaploid wheat species and *Ae. tauschii* have similar major band of this inhibitor (except one of the accessions of *T. dicoccoides* in which it is absent). In their turn such inhibitors in accessions of *Aegilops* species from section *Sitopsis* differ from each other and from wheat inhibitors by pI.

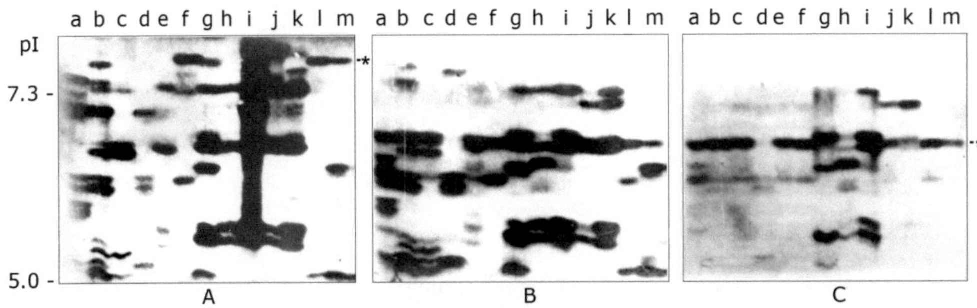

Proteins were extracted with water and separated by IEF in pI range 5 to 9 and gelatin replicas A & B were obtained from the same gel sequentially. Chymotrypsin inhibitors were removed from the extracts by affinity adsorption, the remaining proteins were separated by IEF and replica C obtained. Replica A was developed by chymotrypsin and replicas B and C by subtilisin (see in 2.5.2). a, *Ae. tauschii* κ-133; b, *T. aestivum* var. Chinese Spring; c, *T. dicoccum* κ-81; d, *T. dicoccoides* k-5201; e, *T. durum* var. Novomichurinka; f, *T. monococcum* k-18140; g & h, *Ae. longissima* k-194 and k-178; i- *Ae. bicornis* (k-904); j & k, *Ae. speltoides* k-1316 & k-198); l, *T. boeoticum* k-40117; m, *T. urartu* k-33871. Samples were loaded in volume 20 mcl. *, position of trypsin/chymotrypsin inhibitor; **, position of bifunctional germinating wheat grain α–amylase/subtilisin inhibitor.

Fig. 6. Analysis of component composition of chymotrypsin and subtilisin inhibitors from endosperm of various wheat (*Triticum*) and goatgrass (*Aegilops*) species.

Bifunctional α–amylase/subtilisin inhibitors from cereal seeds are considered to be involved in fundamental mechanisms providing plant defence and the regulation of endogenous α–amylase activity. So, screening of plant collections using proposed detection methods can assist in the search for novel forms of inhibitor of this type which can be then purified by affinity chromatography.

The gelatin replicas method has been used for studying the proteinase inhibitor polymorphism in seeds and leaves of hundreds of accessions of cultivated and wild wheats and related species. Evolutionary links of proteinase inhibitor systems of diploid and polyploid wheat and *Aegilops* species have been established and the possibility of using of this approach for wheat variety identification was shown (Konarev, 1986b, 1987a, 1987b, 1988, 1989, 1992b, 1993, 1996, 2000). Similar approaches have been used for analysis of changes in inhibitor composition in potato leaves and tubers after mechanical damage or infection of leaves by *Phytophthora infestans* (Konarev & Fasulati, 1996; Konarev, 2000b; Konarev & Zoteeva, 2006).

Use of gelatin and PAG-amylase-starch replicas in combination with IEF allows the study of polymorphism of insect α–amylase and serine and cysteine proteinases inhibitors in some legumes. The results obtained were generally in agreement with data of other researches on DNA–markers and allowed the taxonomy in section *Ceratotropis* of vigna to be clarified (Konarev et al, 1999c, 2002b) and later assisted in the description of novel vigna species.

Gelatin replicas combined with IEF, TLGF and affinity chromatography were used for analysis polymorphism and characterisation of serine proteinase inhibitors in seeds of several hundred species of Compositae and other representatives of wide angiosperm group, asterids, which never before were studied for such inhibitors (Konarev et al., 1999a, 2000, 2002a, 2004). The distribution and variability of various inhibitor types have been first studied and novel inhibitor forms have been discovered (see below).

The same combination of methods have been used for revealing and characterisation of serine proteinase inhibitors in seeds of various gymnosperms (Konarev et al., 2008 & 2009).

As addition to methods for detection of α–amylase and proteinase inhibitors after analytical separation in gels, the techniques for identification of other plant proteins with presumably protective role, lectins, have been also developed (Konarev, 1990b).

3.2 The search for novel forms of proteinases and proteinase inhibitors and their purification using combination of detection methods and affinity chromatography

3.2.1 Serine proteinase inhibitor from Cycas seeds

Proteinase inhibitors (PIs) play an important role in the molecular interaction and co-evolution of plants with phytophagous organisms. Serine PIs have been well studied in angiosperms but until recently not identified in gymnosperms. Among the latter, the Cycadales are of particular interest since they represent the most primitive living seed plants, related to extinct seed ferns, and are sometimes considered a "missing link" between vascular non-seed plants and the more advanced seed plants. With use of gelatin replicas method serine proteinase inhibitors were found in several representatives of two of the four major groups of gymnosperms, the Cycadales and the economically important Coniferales. Inhibitors of subtilisin, a typical enzyme of fungi and bacteria, were identified in members of both orders, being particularly active in the Cycadales. In two Cycas species these inhibitors were also active against trypsin and chymotrypsin, proteinases typical of both fungi and animals. Using combination of IEF, gelatin replicas method and affinity chromatography several inhibitor forms from C. *siamensis* seeds have been purified. They appeared to be highly heterogeneous. A small portion of the analytical research, an example of using the gelatin replicas method to monitor the affinity chromatography and determine the degree of heterogeneity and similarity of inhibitor fractions eluted directly from trypsin- and chymotrypsin-Sepharose or consequently eluted from one media after another, is shown in Fig. 7.

For the purification of inhibitor isofom, IEF in DryStrip NL pH 3-10 of fraction obtained by affinity chromatography on chymotrypsin-Sepharose combined with gelatin replicas method was used. Partial sequencing of an isoform showed its similarity to Kunitz-type inhibitors from angiosperms. Analysis of expressed sequence tag (EST) databases confirmed the presence of mRNAs encoding Kunitz-type inhibitors in the Cycadales and Coniferales and also demonstrated their presence in a third major group of gymnosperms, the Ginkgoales. The results show that gymnosperms and angiosperms contain similar type of serine PIs which may provide protection against microbial pathogens or limit the activity of symbiotic microorganisms (Konarev et al., 2008, 2009).

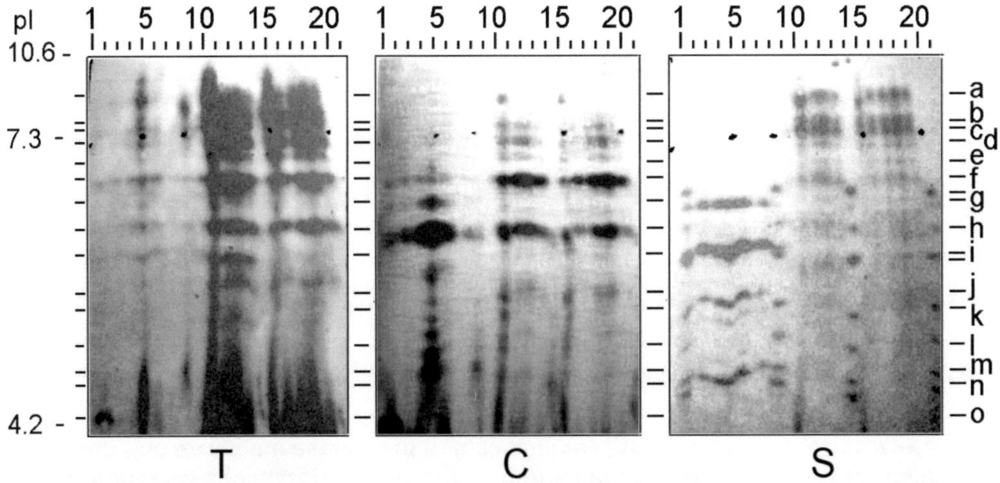

Proteins were extracted with water and loaded onto affinity chromatography columns, eluted by 0.1 M HCl, and the eluates separated by IEF in Servalyt Precotes pH 3-10. The inhibitors were detected in gelatin replicas by trypsin (T), chymotrypsin (C) and subtilisin (S). 1, 5, 8, 15 and 20, markers pI, position of horse myoglobin (pI 7.3) on replicas marked by ink dots; 2, 8, 14 and 20, proteins extracted from seeds with water; 3-7, eluates from chymotrypsin-Sepharose column; 9-13, eluates from trypsin-Sepharose column on which eluate from chymotrypsin-Sepharose column was loaded; 15-19, eluate from trypsin-Sepharose column. a-o, designations of positions of inhibitor bands.

Fig. 7. Monitoring affinity chromatography and analysis heterogeneity of serine proteinase inhibitors from seeds of *Cycas siamensis*.

3.2.2 Trypsin inhibitor from seeds of veronica with an unusual helix-turn-helix proteinase inhibitory motif

Joint use of IEF, TLGF and gelatin replicas method for screening of seed proteins from numerous representatives of asterids revealed some of them to contain several unusually low-molecular weight trypsin inhibitors. Trypsin inhibitor from ivyleaf speedwell (*Veronica hederifolia* L., Lamiales, Plantaginaceae) seeds (VhTI) has MW near 4 kDa, which is much lower than that of majority of known inhibitors in plants. It was purified using affinity chromatography followed by isoelectric focusing combined with gelatin replica method. The single bands of detected inhibitor were additionally fractionated by reverse phase HPLC (Konarev et al., 2004). At that time no homology with known inhibitors was found. Both the native inhibitor and corresponding especially synthesized peptide have been crystallized and their three-dimensional structures were determined (Conners et al., 2007). It was found that this inhibitor contains an unusual for inhibitors helix-turn-helix proteinase inhibitory motif. The subsequent analysis allowed us to reveal that VhTI is a member of wide group of plant peptides with antimicrobial and inhibitory activities found by other authors (Nolde et al., 2011; Park et al., 1997), first characterized by spatial structure. Their amino acid sequences can differ greatly from VhTI but also contain four cysteine residues in configuration nX-CXXXC-nX-CXXXC-nX. Further analysis of data clarified that some forms of plant thionins (e.g. α–purothionins) and neurotoxins from scorpions (Chagot et al., 2005) and marine snails (Möller et al., 2005) also possess a similar cysteine configuration and two alpha-helices connected by a loop and stabilized by two disulphide bridges. Today it is clear that effective enzyme inhibitors with some additional features (e.g. antimicrobial activity) could be constructed from simple helical motifs and VhTI and VhTI-like peptides provide a new scaffold on which to base the design of novel serine protease inhibitors e.g. as anticancer drugs for using in medicine.

3.2.3 Unique low-molecular weight cyclic trypsin inhibitor from sunflower seeds

Polymorphism of serine proteinase inhibitors was first studied in sunflower and other Compositae seeds using IEF, TLGF and gelatin replicas methods. Highly active trypsin inhibitors slightly active also to chymotrypsin with pI around 10 and trypsin/subtilisin (TSI) inhibitors with lower pI were detected in seeds of sunflower (*Helianthus annuus* L.) and related species (Konarev, 1995). Polymorphic TSI was found (using gelatin replicas method) to inhibit extracellular proteinase of pathogenic for sunflower fungi *Sclerotinia sclerotiorum* (Konarev et al., 1999b) and probably belong to potato inhibitor I family (Konarev et al., 2000, 2002a). The trypsin inhibitor was difficult to analyze because it could be only detected by the gelatin replica method but was not visible on SDS-PAGE until it was linked to PAG by glutaraldehyde. The inhibitor was purified by affinity chromatography on trypsin-Separose followed by revere phase HPLC. SDS-PAGE and mass-spectrometry showed that it was extremely low-molecular weight peptide (1, 4 kDa) (Konarev et al., 1998, 1999a, 2000) whose amino acid sequence could not be determined for some time. It turned to be a cyclic peptide, with no free N-terminus (Luckett et al., 1999). The analysis revealed that this inhibitor called SFTI-1 corresponds by sequence to a loop (containing a reactive centre) of well known and widely distributed Bowman-Birk inhibitors which in sunflower exists independently of the rest part of the inhibitor molecule being in the cyclic form (Luckett et al., 1999) that was later confirmed by other researchers (Korsinczky et al., 2001 & 2004; Mylne et al., 2011). This similarity can also be result of convergence. SFTI-1 was found only in *Helianthus* and related *Tithonia* species and not in any studied asterids including Compositae (Konarev et al., 2000, 2002a, 2004). SFTI-1 is considered as the most potent naturally occurring plant Bowman-Birk inhibitor known so far (Korsinczky et al., 2004). More than one hundred publications on SFTI-1 have appeared since 1999. It was revealed that the inhibitor was extremely robust, resistant to proteolysis (no N- or C-ends) which, coupled to its high potency of inhibition for such proteinases as matripatse, thrombin, kallikrein 4 etc. made it an excellent candidate for use as a template for further development of drugs for therapy against cancer, thromboembolism and other proteinase-related human pathologies (Li, et al., 2007; Luckett et al., 1999; Swedberg et al., 2009; Białas & Kafarski, 2009) or agents for plant protection against pests and pathogens.

3.2.4 Glutenin-specific proteinase of sunn pest *Eurygaster integriceps* Put. responsible for wheat gluten degradation

Sunn pest *E. integriceps* and related wheat bugs cause huge losses in grain quality in Russia, South and South-East Europe, Middle East and Central Asia. The main damaging agent is the proteinase of the salivary glands which is injected into maturating wheat grains for extraintestinal digestion. The traces of enzyme remaining in the grains after the bug`s attack show activity in dough mixing before baking even several years later. The proteinase degrades the wheat gluten proteins, leading to a loss of gluten and dough viscoelasticity and poor processing properties. The sunn pest proteinase was studied during the last decades in many laboratories but attempts to purify it failed probably due to narrow substrate specificity, low sensitivity to known proteinase inhibitors (which prevented the use of affinity chromatography), instability in purified form and lack of suitable methods for its detection at purification. We have developed method applicable for detection of sunn pest proteinase in protein extracts both from salivary glands and damaged seeds based on using of IEF in combination with thin layer of insoluble in acetic acid glutenin (protein determining quality of gluten and substrate for bug's proteinase) attached to supporting plastic film. In order to make it possible to use affinity chromatography on immobilized potato chymotrypsin inhibitor I at its low affinity to the enzyme, conditions of fractionation were modified so that proteinase left linked to inhibitor until elution. It was found that enzyme binds to the inhibitor at high pH (about 10) that made it possible washing column after loading just by solvent used for sample. The common washing by buffers with lower pH was omitted and proteinase began to elute after replacement of solvent by water. The elution continued after changing water to 0.01M HCl. Both fractions gave in SDS-PAGE the same almost pure proteinase band with traces of low-molecular proteins which can be easily removed by gel-filtration. The absence of defined elution peak can be explained by low affinity of interaction. IEF followed by detection using glutenin film and SDS-sedimentation methods confirmed proteinase nature of isolated fraction. The addition to all solutions of 0.01% non-ionic detergent Triton X-100 improved stability of proteinase after purification. This 28 kDa protein was partially sequenced by mass spectrometry and Edman degradation which showed homology to serine proteases from various insects. Three full length clones were obtained from cDNA isolated from sunn pest salivary glands using degenerate PCR based on the sequences obtained. The cleavage site of the protease was determined using recombinant and synthetic peptides and shown to be between the consensus hexapeptide and nonapeptide repeat motifs present in the high molecular weight subunits of wheat glutenin. Homology models were generated for one of the proteinase isoforms identified in this study. The novel specificity of this protease and data obtained may find various applications in both fundamental and applied studies, e.g. in design of effective inhibitors for improving wheat resistance to pests and limiting proteinase activity in food technologies. Besides, the proteinase cleaves one of epitopes of glutenin defined by Wal et al (1999) as minimal epitope for the HLA–DQ8 form of celiac disease, so probably it can be used for gluten proteins modification to decrease their toxicity and autoimmune activity for gluten-sensitive people (Konarev et al., 2011).

4. Conclusion

The proposed set of universal and sensitive methods for detection among plant proteins of insect, mammalian, fungal, bacterial and plant α–amylase and proteinase inhibitors and also these hydrolases themselves significantly simplify the analysis of inhibitor polymorphism and the search for novel forms of enzymes and their inhibitors. The detection methods enhance the capacity of affinity chromatography, mass-spectrometry and other techniques of protein fractionation and characterisation. They may be used in a wide variety of fields to answer many questions from plant systematics to the identification of proteins with potential uses in plant resistance as well as medicine. The further improvement in resolution, sensitivity and specificity of these techniques will help increase the efficiency of mentioned research.

5. Acknowledgment

These above described studies were carried out according to the Program of Fundamental Investigations on Plant Protection of the Russian Academy of Agricultural Sciences and were supported by the following grants from: The Russian Foundation for Basic Research (93-04-21290-a); Japan International Research Center for Agricultural Sciences (1997-1998); The Royal Society Joint Project "Characterization of Sunflower Seed Proteins" (683072.P810); an "Ex-Agreement Visit" Grant from The Royal Society of London (2002); a Fellowship Grants from Rothamsted International (2005) and the BBSRC Underwood fund (2006). Rothamsted Research receives grant-aided support from the Biotechnology and Biological Sciences Research Council of the United Kingdom. Authors are very grateful to Prof. Peter Shewry for continuing support and encouragement over the years of these projects. The authors would also like to thank all their colleagues who participated in the research.

6. References

Białas, A. & Kafarski, P. (2009). Proteases as anti-cancer targets - molecular and biological basis for development of inhibitor-like drugs against cancer. *Anticancer Agents Med. Chem.*, Vol.9, No.7, pp. 728-762, ISSN 1871-5206

Buonocore, V.; Petrucci, T. & Silano V. (1977). Wheat protein inhibitors of α–amylase. *Phytochemistry*, Vol.16, No.7, pp. 811-820, ISSN 0031-9422

Buonocore, V.; Poerio, E.; Gramenzi, F. & Silano, V. (1975). Affinity column purification of amylases on protein inhibitors from wheat kernel. *Journal of Chromatography*, Vol.114, No1., pp.109-114, ISSN 0021-9673

Buonocore, V.; Poerio, E.; Gramenzi, F. &; Silano, V. (1975). Affinity column purification of amylases on protein inhibitors from wheat kernel. *Journal of Chromatography*, Vol.114, No1. pp. 109-114, ISSN 0021-9673

Burger, W. C. & Schroeder, R. L. (1976). A sensitive method for detecting endopeptidases in electrofocused thin-layer gels. *Analytical Biochemistry*, Vol.71, No.2, pp. 384-392, ISSN 0003-2697

Chagot, B.; Pimentel, C.; Dai, L.; Pil, J.; Tytgat, J.; Nakajima, T.; Corzo, G.; Darbon, H. & Ferrat, G. (2005) An unusual fold for potassium channel blockers: NMR structure of three toxins from the scorpion *Opisthacanthus madagascariensis*. *Biochem J.* Vol. 388, No.1, pp. 263-71, ISSN 0264-6021

Conners, R.; Konarev, Al.V.; Forsyth, J.; Lovegrove, A.; Marsh, J.; Joseph-Horne, T.; Shewry, P. & Brady, R. L. (2007). An unusual helix-turn-helix protease inhibitory motif in a novel trypsin inhibitor from seeds of veronica (*Veronica hederifolia* L.). *The Journal of Biological Chemistry*, Vol.282, No.38, pp. 27760-27768, ISSN 0021-9258

Dunaevsky, Y. A.; Elpidina, E.; Vinokurov, K.; Belozersky, M. (2005). Protease inhibitors in improvement of plant resistance to pathogens and insects. *Molecular Biology*, Vol.39, No.4, pp. 608–613, ISSN 0026-8933

Fontanini, D.; Capocchi, A.; Saviozzi, F. & Galleschi, L. (2007). Simplified electrophoretic assay for human salivary alpha–amylase inhibitor detection in cereal seed flours. *J Agric Food Chem.*, Vol.55, No.11, pp. 4334-4339, ISSN 0021-8561

Franco, O.L.: Rigden, D.J.; Melo, F.R. and Grossi-De-Sá, M.F. (2002). Plant alpha-amylase inhibitors and their interaction with insect alpha-amylases. *Eur. J. Biochem.*, (January), Vol.269, No.2, pp. 397-412, ISSN 0014-2956

Gatehouse, J.A. (2011). Prospects for using proteinase inhibitors to protect transgenic plants against attack by herbivorous insects. Curr. Protein. Pept. Sci., (Aug 1), Vol.12, No.5, pp. 409-16, ISSN 1389-2037

Giri, A.P. & Kachole, M. S. (1996). Detection of electrophoretically separated amylase inhibitors in starch-polyacrylamide gels. *Journal of Chromatography,* Vol.752, No.1-2, pp. 261-264, ISSN 0021-9673

Habib, H. & Fazili, K. M. (2007). Plant protease inhibitors; a defence strategy in plants. *Biotechnology and Molecular Biology Review,* Vol.2, No.3, (August 2007), pp. 68-85, ISSN 1538-2273

Harsulkar, A.M.; Giri, A.P.; Gupta, V.S.; Sainani, M.N.; Deshpande, V.V.; Patankar, A.G. & Ranjekar, P.K. (1998). Characterization of *Helicoverpa armigera* gut proteinases and their interaction with proteinase inhibitors using gel X-ray film contact print technique. *Electrophoresis,* Vol.19, No.8-9, (Jun), pp. 1397-402, ISSN 1522-2683.

Hejgaard, J. (1981). Isoelectric focusing of subtilisin inhibitors. Detection and partial characterisation of cereal inhibitors of chymotrypsin and microbial proteases. *Analytical Biochemistry,* Vol.116, No.2. P.444-449, ISSN 0003-2697

Konarev, Al.V. & Fomicheva, Yu.V. (1991a). Cross analysis of the interaction of alpha-amylase and proteinase components of insects with protein inhibitors from wheat endosperm. *Biokhimiya* (*Biochemistry,* Moskow). Vol.56. No.4, pp. 628-638, ISSN 0006-2979

Konarev, Al.V. & Fomicheva, Yu.V. (1991b). Hydrolase inhibitors and coevolution of cereals and harmful insects. *Proceedings of IX All-Union Conference on Plant immunity to diseases and pests,* pp. 270-271, Minsk, USSR (In Russian)

Konarev, Al.V. & Fasulati, S.R. (1996). Proteinases of colorado potato beetle and their plant proteinaceous inhibitors. *Proceedings of XX Int. Congr.of Entomology,* Firenze, Italy, August 25-31, 1996

Konarev, Al.V. & Zoteeva, N.M. (2006). Induced by phytophthora infection accumulation of proteinase inhibitors in potato varieties. Proceedings of All-Russia scientific-practical conference " Induced immunity of agricultural crops as important direction in plant protection", p.45-47, Bolshie Vyazemy, Moskowskaya oblast., 15-16 Nov, 2006. (in Russian)

Konarev, Al.V. (1981). Specificity of wheat grain albumins (alpha-amylase inhibitors) towards enzymes of sunn pest (*Eurygaster integriceps* Put.). *Bulletin VIR,* Vol. 114., pp.30-32, ISSN 0202-5361 (In Russian)

Konarev, Al.V. (1982a). Component composition and genetic control of insect alpha-amylase inhibitors from wheat and aegilops grain. *Doklady VASKhNIL,* No.6. P. 42-44, ISSN 0042-9244 (In Russian)

Konarev, Al.V. (1982b). Identification of inhibitors of endogenous and exogenous alpha-amylases among wheat proteins. *Bulletin VIR.* V. 118.P. 11-12, ISSN 0202-5361 (In Russian)

Konarev, Al.V. (1984). Identification of inhibitors of thiol proteinases of insects and grain in wheat and other cereals. *Doklady VASKhNIL*, No.10,pp. 13-15, ISSN 0042-9244 (In Russian)

Konarev, Al.V. (1985) Methods for analyzing the component composition of cereal alpha-amylase and proteinase inhibitors. *Prikladnaya Biokhimiya i Microbiologia (Applied Biochemistry and Microbiology*, Moscow), Vol.21, No.1, pp. 92-100, ISSN, 0555-1099

Konarev, Al.V. (1986a). Analysis of proteinase inhibitors from wheat grain by gelatin replicas method. Biokhimiya (*Biochemistry*, Moskow), Vol.51, No. 2, pp. 195-201, ISSN 0006-2979

Konarev, Al.V. (1986b). Relationship and variability of proteinase and amylase inhibitors in wheat and related cereals. *Agricultural Biology*, No.3., pp. 46-52, ISSN 0131-6397 (In Russian)

Konarev, Al.V. (1987a). Component composition of trypsin inhibitors from grain and leaves of wheat. *Doklady VASKhNIL*, No.12, pp. 6-9, ISSN 0042-9244 (In Russian)

Konarev, Al.V. (1987b). Variability of trypsin-like proteinase inhibitors in wheat and related cereals. *Agricultural Biology*, No.5, pp. 17-24, ISSN 0131-6397 (In Russian)

Konarev, Al.V. (1988). Polymorphism of chymotrypsin-subtilisin inhibitors and possibilities of its use in wheat variety identification. In: *Biochemical identification of varieties. Materials of III International symposium ISTA*, V. Konarev & I. Gavriljuk (Ed.), pp. 176-181, VIR, Leningrad, USSR

Konarev, Al.V. (1989). The evolutionary variability and genetic control of chymotrypsin-subtilisin inhibitors in wheat. *Doklady VASKhNIL*, No.5, pp. 8-10, ISSN 0042-9244 (In Russian)

Konarev, Al.V. (1990a). Cross analysis of interaction of endogenous alpha-amylase and proteinase components with protein inhibitors in wheat. *Doklady VASKhNIL*. No.1, pp. 10-13, ISSN 0042-9244 (In Russian)

Konarev, Al.V. (1990b). Detection of lectins in analytical isoelectrofocusing of wheat proteins. *Biokhimiya* (*Biochemistry*, Moskow), Vol. 55, No.11, pp. 1975-1983, ISSN 0006-2979

Konarev, Al.V. (1992a). Analysis of native trypsin inhibitors from endosperm of wheats with various genome composition. *Agricultural Biology*, No.5, pp. 10-16, ISSN 0131-6397 (In Russian)

Konarev, Al.V. (1992b). Systems of hydrolase inhibitors in cereals: organisation, functions and evolutionary variability. *Doctor of Biological Sciences Dissertation*, Scientific Council in the Russian Academy of Sciences A.N.Bach Institute of Biochemistry, 356 p., Moscow, Russia (In Russian)

Konarev, Al.V. (1993). Component composition, genome and chromosome control of trypsin inhibitors from bread wheat leaves, roots and germs. *Agricultural Biology*, No.1., pp. 43-52, ISSN 0131-6397 (In Russian)

Konarev, Al.V. (1995). Proteinaceous hydrolase inhibitors in mechanisms of plant defense. *Proceedings of 1st All-Russia Plant Protection Congress*, pp. 205-206, VIZR, St. Petersburg, Russia (December 4-9, 1995) (In Russian)

Konarev, Al.V. (1996). Interaction of insect digestive enzymes with protein inhibitors from plants and host-parasite coevolution. *Euphytica*, Vol.92, No.1-2, pp. 89-94, ISSN 0014-2336

Konarev, Al.V. (2000a). Insect and fungal enzyme inhibitors in study of variability, evolution and resistance of wheat and other Triticeae Dum. cereals. In: *Wheat Gluten*, P.R. Shewry & A.Tatham, (Ed.), 526-530, ISBN 0-85404-865-0, Royal Society of Chemistry, Cambridge, UK

Konarev, Al.V. (2000b). Proteinase inhibitors and resistance of potato to Colorado potato beetle. In "*Modern systems of plant protection and novel approaches to improving of potato resistance to Colorado potato beetle*", K.G. Skryabin & K.V.Novozhilov (Ed.), pp.35-40, Nauka, Moscow, Russia, ISBN 5-02-005214-0 (In Russian)

Konarev, Al.V.; Anisimova, I.N.; Gavrilova, V.A. & Shewry, P.R. (1999a). Polymorphism of inhibitors of hydrolytic enzymes present in cereal and sunflower seeds. In: *Genetics and breeding for crop quality and resistance*, G.T.S. Mugnozza, E.Porceddu & M.A.Pagnotta, (Ed.), 135-144, ISBN 0792358449, Kluwer Academic Publishers, Dordrecht, NL

Konarev, Al.V.; Kochetkov, V.V.; Bailey, J.A. & Shewry, P.R. (1999b). The detection of inhibitors of the *Sclerotinia sclerotiorum* (Lib) de Bary extracellular proteinases in sunflower, *J. Phytopathology*, Vol.147, No.2, pp. 105-108, ISSN 0931-1785

Konarev, Al.V.; Tomooka, N.; Ishimoto M. & Vaughan, D.A. (1999c). Variability of the inhibitors of serine, cysteine proteinases and insect alpha-amylases in *Vigna* and *Phaseolus*. In: *Genetics and breeding for crop quality and resistance*, G.T.S.Mugnozza, E.Porceddu & M.A.Pagnotta (Ed.), 173-181, ISBN 0792358449, Kluwer Academic Publishers, Dordrecht, NL

Konarev, Al.V.; Anisimova, I.N.; Gavrilova, V.A.; Rozhkova, V.T.; Fido, R.; Tatham, A.S. & Shewry, P.R. (2000). Novel proteinase inhibitors in seeds of sunflower (*Helianthus annuus* L.): polymorphism, inheritance and properties. *Theoretical and Applied Genetics*, Vol.100, No.1, pp. 82-88, ISSN 0040-5752

Konarev, Al.V.; Anisimova, I.N.; Gavrilova, V.A.; Vachrusheva, T.E.; Konechnaya, G.Yu.; Lewis, M. & Shewry, P.R. (2002a). Serine proteinase inhibitors in Compositae: distribution, polymorphism and properties. *Phytochemistry*, Vol.59, No. 3, pp. 279-291, ISSN 0031-9422

Konarev, Al.V.; Tomooka, N. & Vaughan, D.A. (2002b). Proteinase inhibitor polymorphism in the genus *Vigna* Savi subgenus *Ceratotropis* and its biosystematic implications. *Euphytica*, Vol., 123, No.2, pp. 165-177, ISSN 0014-2336

Konarev, Al.V.; J.Griffin; Konechnaya, G.Yu.& Shewry, P.R. (2004). The distribution of serine proteinase inhibitors in seeds of the Asteridae. *Phytochemistry*, Vol.65, No.2, pp. 3003-3020, ISSN 0031-9422

Konarev, Al.V.;.Lovegrove, A. & P.R.Shewry. (2008). Serine proteinase inhibitors in seeds of *Cycas siamensis* and other gymnosperms. *Phytochemistry*, Vol.69, No. 13, pp.2482-2489, ISSN 0031-9422

Konarev, Al.V.; Lovegrove, A. & Shewry P. R. (2009). Inhibitors of microbial serine proteinases in cycads and other gymnosperms. *Proceedings of XIV International Congress on Molecular Plant-Microbe Interactions (IS-MPMI)*, p.20, Toronto, Canada, July 19-23

Konarev, Al.V.; Beaudoin F.; Marsh, J.; Vilkova, N.A.; Nefedova, L.I.; Sivri, D.; Koksel, H.; Shewry, P.R. & Lovegrove, A. (2011). Characterization of a glutenin-specific serine proteinase of sunn bug *Eurygaster integricepts* Put. *Journal of Agricultural and Food Chemistry*. Vol.59, No.6, pp. 2462-2470, ISSN 0021-8561

Konarev, V.G. (1996). Problems of species and genome in plant breeding. In: Molecular biological aspects of applied botany, genetics and plant breeding. Theoretical bases of plant breeding, A.V.Konarev (Ed.), pp. 14-24, VIR, St. Petersburg, Russia (In English)

Korsinczky, M.L.; J.; Schirra, H.J.; Rosengren, K.J.; West, J.; Condie, B.A.; Otvos, L.; Anderson, M.A. & Craik, D.J. (2001). Solution structures by 1H NMR of the novel cyclic trypsin inhibitor SFTI-1 from sunflower seeds and an acyclic permutant. *Journal of Molecular Biology*, Vol.311, No. 3, pp. 579-591, ISSN 0022-2836

Korsinczky, M.L.; Schirra, H.J, & Craik, D.J. (2004). Sunflower trypsin inhibitor-1. *Curr Protein Pept Sci.* Vol.5, No.5, pp. 351-364, ISSN 1389-2037

Li, P.; Jiang, S.; Lee, S.-L.; Lin, C.Y.; Johnson, M.D.; Dickson, R.B.; Michejda, C.J. & Roller, P.P. (2007). Design and synthesis of novel and potent inhibitors of the type II transmembrane serine protease, matriptase, based upon the sunflower trypsin inhibitor-1. *J. Med. Chem.*, Vol.50, No.24, pp. 5976–5983, ISSN 0022-2623

Luckett, S.; Garcia, R.S.; Barker, J.J.; Konarev, Al.V.; Shewry, P.R.; Clarke, A.R. & Brady R.L. (1999). High-resolution structure of a potent, cyclic proteinase inhibitor from sunflower seeds. *Journal of Molecular Biology, Vol.*290, No.2., (July), pp. 525-533, ISSN 0022-2836

Möller, C.; Rahmankhah, S.; Lauer-Fields, J.; Bubis, J.; Fields, G.B. & Marí, F. (2005). A novel conotoxin framework with a helix–loop–helix (Cs α/α) fold. *Biochemistry*, Vol.44, No.49, (November 16), pp. 15986–15996, ISSN 0006-2960

Mosolov, V.V. & Valueva, T.A. (2005) Proteinase inhibitors and their function in plants: A review. Applied Biochemistry and Microbiology, Vol.41, No.3, pp. 227-246, ISSN 0003-6838

Mulimani, V.H.; Sudheendra, K. & Giri, A.P. (2002). Detection of legume protease inhibitors by the Gel-X-ray film contact print technique. *Biochemistry and Molecular Biology Education*, Vol.30, No.1, pp. 40–44, ISSN 1539-3429

Mylne, J.S.; Colgrave, M.L.; Daly, N.L.; Chanson, A.H.; Elliott, A.G.; McCallum, E.J.; Jones A. & Craik D.J. (2011). Albumins and their processing machinery are hijacked for cyclic peptides in sunflower. *Nat Chem Biol.* , Vol.7, No.5, (May), pp.257-259, ISSN 1552-4450

Nagaraj, R.H. & Pattabiraman, T.M. (1985). Isolation of an amylase inhibitor from *Setaria italica* grains by affinity chromatography on blue-sepharose and its characterization. *J. Agric. Food Chem.*, Vol.33, No.4, pp. 646–650, ISSN 0021-8561

Nolde, S..B.; Vassilevski, A.A.; Rogozhin, E.A.; Barinov, N.A.; Balashova, T.A.; Samsonova, O.V.; Baranov, Y.V.; Feofanov, A.V.; Egorov, T.A.; Arseniev, A.S. & Grishin, E.V. (2011). Disulfide-stabilized helical hairpin structure and activity of a novel antifungal peptide EcAMP1 from seeds of barnyard grass (*Echinochloa crus-galli*). *J Biol Chem*; Vol.286, No.28, (July 15), pp. 25145-25153, ISSN 0021-9258

Park, S.S.; Abe, K.; Kimura, M.; Urisu, A. & Yamasaki, N. (1997). Primary structure and allergenic activity of trypsin inhibitors from the seeds of buckwheat (*Fagopyrum esculentum* Moench). FEBS Letters, Vol.400, No.1, pp. 103-107, ISSN 0014-5793

Pichare, M.M. & Kachole M.S. (1994). Detection of electrophoretically separated protease inhibitors using X-ray film. *Journal of Biochemical and Biophysical Methods*, Vol.28, No.3, (April 1994), pp. 215-224, ISSN 0165-022X

Saxena, L.; Iyer B.K. & Ananthanarayan, L. (2010). Purification of a bifunctional amylase/protease inhibitor from ragi (*Eleusine coracana*) by chromatography and its use as an affinity ligand. *J. Chromatogr. B Analyt. Technol. Biomed. Life. Sci.*, Vol.878, No.19. pp. 1549-1554, ISSN 1570-0232

Shewry, P. R. & Lucas, J.A. (1997) Plant proteins that confer resistance to pests and pathogens. Advances in Botanical Research, Vol.26, pp. 135–192, ISSN 0065-2296

Shewry, P.R.; Halford, N.G. & Lafiandra, D.(2003). Genetics of wheat gluten proteins. Advances in *Genetics*, Vol.49, pp. 111-184, ISSN 0065-2660

Swedberg, J.E.; Nigon, L.V.; Reid, J.C.; de Veer, S.J.; Walpole, C.M.; Stephens, C.R.; Walsh, T.P.; Takayama, T.K.; Hooper, J.D.; Clements, J.A.; Buckle, A.M. & Harris, J.M. (2009). Substrate-guided design of a potent and selective kallikrein-related peptidase inhibitor for kallikrein 4. *Chemistry & Biology*, Vol.16, No.6, pp. 633–643, ISSN 1074-5521

Van de Wal, Y.; Kooy, Y. M. C.; van Veelen, P.; Vader, W.; August, S. A.; Drijfhout, J. W.; Pena, S. A. & Koning, F. (1996). Glutenin is involved in the gluten-driven mucosal T cell response. *Eur. J. Immunol.*, Vol. 29, No.10, pp. 3133–3139, ISSN 0014-2980

12

Affinity Chromatography as a Key Tool to Purify Protein Protease Inhibitors from Plants

Elizeu Antunes dos Santos, Adeliana Silva de Oliveira, Luciana Maria Araújo Rabêlo, Adriana Ferreira Uchôa and Ana Heloneida Araújo Morais
Universidade Federal do Rio Grande do Norte
Brazil

1. Introduction

Several and distinct physiological processes in all the life forms are dependent on proteases, as processing and turnover of endogenous proteins, digestion of food proteins, regulation of formation and lysis of the clots, activation of apoptosis pathways, plant germination, sporulation, hormone activation, translocation through membranes, fertilization, control of immune response, cell differentiation and growing (Bode & Huber, 2000; Chou & Cai, 2006; Turk et al., 2000). Proteases are also involved in replication and propagation of infectious diseases, and the imbalance of their activity can cause important pathological disorders as inflammation, stroke, cancer and parasite infection (Chou & Cai, 2006; Johansson et al., 2002; Powers et al., 2002).

The principal naturally occurring control way of the proteases activity is achieved by the action of inhibitors of protein nature, which bind specifically and block proteases. Protease inhibitors (PIs) are found in all living organisms and are among the most intensively studied proteins. In plants they are widely distributed among different botanical families and have been found in reproductive organs, storage organs and vegetative tissues. They are synthesized constitutively in seeds or can be induced in tissue, as leaves, by the attack of herbivore or abiotic stress (Fan & Wu, 2005; Laskowski & Kato, 1980; Xavier-Filho, 1992). At least four PI families are known and can be distinguished based in their interaction with the protease class that they inhibit (Fan & Wu, 2005; Koiwa et al., 1997; Xavier-Filho, 1992). Plants PIs has received special attention because of their roles and potential biotechnological applications in agriculture as bioinsecticide, nematicidal, acaricidal, antifungal and antibacterial and in the biomedical field they are remarkable candidates in the production of drugs for human disease healing.

Purification of these inhibitors is a necessary and critical step in order to define their structural characteristics and binding specificity to the proteases. Isolate these from all other proteins that are present in the same biological source is a difficult task, since the PIs have a large molecular diversity. Nevertheless, due their specific and reversible binding capacity to the enzymes (without undergoing chemical change), PI purification can be greatly enhanced by the use of affinity chromatography techniques, where the binding agent are particular proteases (Gomes et al., 2005; Araújo et al., 2005; Oliveira et al., 2007a, 2007b).

2. Proteases

Proteases (EC[1] 3.4), also called peptidases, peptide hydrolase, and proteolytic enzymes, constitute a great group of enzymes that hydrolyze peptide bonds (Barrett, 1997; Rawlings et al., 2010). These enzymes are subdivided into two groups or class: exopeptidases (EC 3.4.11-19) and endopeptidases (EC 3.4.21-24 and EC 3.4.99). The first ones show capacity to hydrolyze amino acids from the N- or C-terminus while the second group cleaves the internal peptide bonds of polypeptides and, thus, are frequently assigned of proteinases[2] (Fan & Wu, 2005; Rawlings et al., 2010).

According to their catalytic mechanism and specificity, endopeptidases are classified in four major groups[3]: metalloproteases, aspartic proteases, cysteine and serine proteases (Fan & Wu, 2005; Rawlings et al., 2010), being the latter two classes better characterized (table 1).

Class	*Proteinases*	*Amino acid residue or metal in active site*	*Optimum pH*
Serine proteinase (EC 3.4.21)	trypsin, chymotrypsin, elastase, thrombin, cathepsins[4] A and G	Ser, His, Asp	7 - 9
Cysteine proteinase (EC 3.4.22)	papain, bromelain, cathepsins B, C, H, K, L, O, S and W	Cys	4 - 7
Aspartic proteinase (EC 3.4.23)	pepsin, cathepsins D and E, renin	Asp, Try	Below 5
Metalloproteinase (EC 3.4.24)	caboxipeptidases A and B, aminopeptidases, thermolysin	Metal ion (usually Zn)	7 - 9

Table 1. Proteinases classification (Modified from Fan & Wu, 2005).

2.1 Serine proteases

Serine proteases comprise a large group of peptidases characterized by presenting a catalytic serine residue. In fact, three residues (serine, histidine and aspartic acid) are essential in catalytic process. They work together to cleave the peptide bond of the substrate (Hedstrom, 2002) (see figure 1).

---

[1] The Enzyme Commission number (EC number) proposed by the Nomenclature Committee of the IUBMB (International Union of Biochemistry and Molecular Biology), is a numerical classification system for enzymes according to the nature of the chemical reactions they catalyze.

[2] In spite of the term "proteinase" be recommended as a synonymous with "endopeptidase", the name "protease" is also traditionally accepted to nominate all known endopeptidases or proteinases.

[3] For further details about classification of proteases see MEROPS (http://merops.sanger.ac.uk/), a database which constitutes a comprehensive and integrated information resource about peptidases and their protein inhibitors.

[4] Cathepsins are a group of proteases called by alphabetic letters in ascending order according to their order of discovery and can be grouped by to their mechanism of catalysis in serine, cysteine or aspartic proteases.

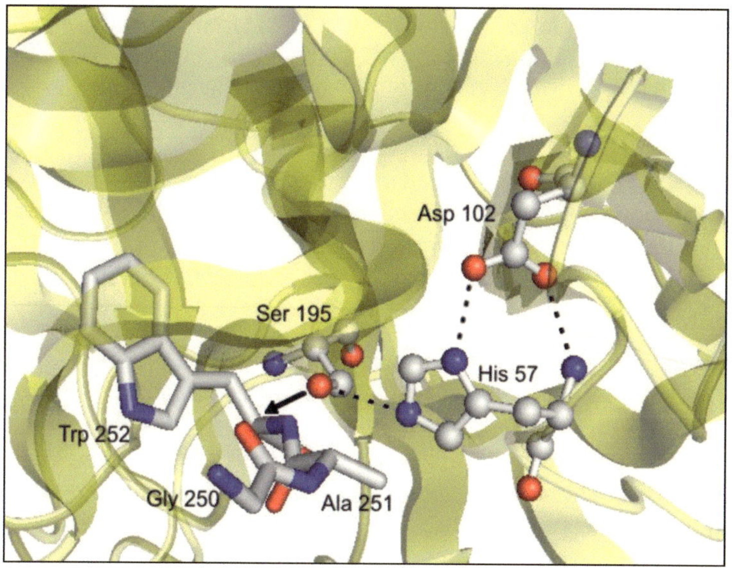

Fig. 1. The catalytic triad of the serine proteases.
Three-dimensional structure of bovine chymotrypsin (PDB code 8GCH) represented by transparent yellow ribbons. Catalytic triad (His 57, Asp 102 and Ser 195) is shown in ball-and-stick representation. The substrate (Gly-Ala-Trp tripeptide) is represented by sticks (note that the tryptophan lateral chain of the substrate is positioned in the recognize cleft of the chymotrypsin). Dotted line, hydrogen bonds. Arrow, cleavage site on the substrate. Figure made with PyMOL.

Trypsin, chymotrypsin and elastase are the largest and best studied serine proteases and they are involved in the protein digestion of the diet in animals, including human. They have the same three-dimensional structures and active site architecture, but differ in the substrate specificity: while trypsin cleaves peptide bonds on the C-terminal side of a positively-charged residue (Lys or Arg), except when it is followed by proline, chymotrypsin prefers large hydrophobic residues (Phe, Trp, Tyr) and elastase acts on small neutral amino acid residues (Ala, Gly, Val) (Hedstrom, 2002; Powers et al., 1977).

Other related serine proteases are implicated with blood coagulation process: plasmin, plasma kallikrein and clotting factors (X, XI, XII and thrombin) (Levi et al., 2006; 2010). Serine proteases such as proteinase 3, cathepsin G and, particularly, elastase from human leukocytes, play an important role in several inflammatory and pathologic processes (Liou & Campbell, 1995; Finlay, 1999; Shapiro, 2000). A distinct family of serine proteases includes subtilisin, the main protease secreted in the beginning of sporulation by the gram-positive bacterium, *Bacillus subtilis* (Power et al., 1986).

Serine proteases similar to those from mammals, specially with respect to the optimum pH, are found as predominant digestive enzymes in a wide variety of insects, as Thysanura (Zinkler & Polzer, 1992), Orthoptera (Lam et al., 1999, 2000), Hymenoptera (Schumaker et al., 1993), Diptera, (Silva et al., 2006), Lepidoptera (Bernardi et al., 1996; Gatehouse et al., 1999; Novillo et al., 1997) and Hemiptera (Colebatch et al., 2001). Trypsin-like proteases were also found in Coleopteran insects, although be known that these insects have an acid intestinal fluid (Alarcon et al., 2002; Franco et al., 2004; Girard et al., 1998; Oliveira-Neto et al., 2004; Purcell et al., 1992; Zhu & Baker, 1999, 2000).

2.2 Cysteine proteases

Cysteine proteases present a catalytic dyad composed by cysteine and histidine, which work the same way as serine and histidine of the serine proteases to cleave peptide bonds (Fan & Wu, 2005). In all the live organisms, most cysteine proteases show catalytic activity in an optimum range slightly acid of pH (4.0-6.5) and are represented by proteins with molecular mass around 21-30 kDa (Rawlings & Barret, 1994; Turk et al., 2000).

This class of proteinases comprises several plant proteases, such as papain (the most studied), bromelain, actinidin, chymopapain, ficin and caricain. Plant proteases are involved in protein processing (activation, maturation, degradation) in virtually all aspect of the physiology and development of plant (Brzin & Kidric, 1995; Grudkowska & Zagdanska, 2004; Salas et al., 2008). Other important cysteine proteases include most of the lysosomal cathepsins, the calpains (a cytosolic "calcium-activated neutral protease"), caspases (essential for apoptosis process) (Fan et al., 2005; Xu & Chye, 1999) and several viral and parasite proteases (Fan & Wu, 2005; Otto & Schirmeister, 1997; Rawlings et al., 2010; Turk et al., 2000).

Cysteine proteases represent the main larval digestive enzymes of several pest insects of the coleoptera order, such as the bruchid bean weevil (*Acanthoscelides obtectus*), mexican bean weevil (*Zabrotes subfasciatus*) and cowpea bruchid (*Callosobruchus maculatus*) (Lemos et al., 1990; Silva et al., 1991, 2001; Xavier-Filho et al., 1989), but also species of chrysomelidae (Liu et al., 2004, Cristofoletti et al., 2005) and curculionidae (Cristofoletti et al., 2005). Some hemipterans also present cysteine proteases (Rahbé et al., 2003; Cristofoletti et al., 2005)

2.3 Aspartic proteases

Aspartic (or aspartyl) proteases were so named because their catalytic mechanism involves residues of aspartic acid from their active site. These enzymes work at acidic (or neutral) pH and their specificity is typically for peptide bonds between two hydrophobic amino acid residues (Simões & Faro, 2004). Aspartic proteases are widely distributed in animals, yeast, virus and plants, performing various functions (Fan & Wu, 2005; Pearl, 1987). In vertebrates, these enzymes act in the digestion of dietary protein (pepsin and chymosin) and lysosomal protein (cathepsins D and E) and regulation of blood pressure (renin). HIV1-protease, essential for the life-cycle of HIV, is another example of aspartic protease (Brik & Wong, 2003).

2.4 Metalloproteases

Metalloproteases constitute a large family of proteases occurring in bacteria, fungi and animals (including man), which require metal ion for catalysis (Fan & Wu, 2005). Most metalloproteases posses zinc in the catalytic site, coordinated via three amino acids residues (among histidin, glutamate, aspartate, lysine and arginine) (Gomis-Ruth et al., 1994; Stocker & Bode, 1995). Many important biologic events are exercised by metalloproteases. Caboxipeptidase A and aminopeptidases are typical metalloproteases involved in peptide digestion (Rosenberg et al., 1975). The group of the matrix metalloproteases (MMPs) is responsible for differentiation and remodeling of the extracellular matrix (ECM), in addition to the cleavage of protein receptors and ligands. Because of their ability to hydrolyze articular cartilage, these enzymes are associated with arthritis and rheumatism (Zhen et al., 2008). Membrane metalloproteases are also related to cancer and inflammation (Seals & Courtneidge, 2003).

3. Protease inhibitors

3.1 Definition, occurrence and distribution

Protease inhibitors (PIs) are proteins naturally occurring in living organisms and able to inhibit and so, control the activity of proteases. They are ubiquitous proteins, occurring in animals, microorganisms and plants. In this former, they constitute one of the more abundant classes of proteins, being founded in reproductive, vegetative and storage organs. Seeds and tubercles contain about 10% of their total protein as PI (Brzin & Kidrič, 1995; Mandal et al., 2002; Ussuf et al., 2001).

Protease inhibitors were identified, isolated and purified from different monocotyledon and dicotyledonous plant species. Among monocots, investigations were directed particularly to plants of the grass (Poaceae) family, as rice, barley, corn, wheat, rye and sorghum as main representants. Among dicotyledons, the Solanaceae family, represented by tomato, potato and tobacco, and legume (Fabaceae) family, represented by beans, soybeans and peas, have received special attention. However, other families were studied on a minor scale, for example, Moraceae, Araceae and Caricaceae families (Brzin & Kidrič, 1995; Schuler et al., 1998).

In different plant tissues and organs were detected, isolated and purified proteinase inhibitors, as in fruit pulp (Araújo et al., 2004), pollen (Rogers et al., 1993), floral buds, seedlings (Lim et al., 1996), apples peel (Ryan et al., 1998), string (Misaka et al., 1996), latex (Monti et al., 2004), roots, stems, leaves, fruits (Brzin & Kidrič, 1995) and particularly in tubers (Huang et al., 2008; Valueva et al., 1997, 1998, 1999) and seeds (Araújo et al., 2005; Bhattacharyya et al., 2007, 2009; Cavalcanti et al., 2002; Gomes et al., 2005; Macedo et al., 2004; Macedo et al., 2007; Oliveira et al., 2007a, b).

The expression of these inhibitors varies according to the maturation stage and tissue location. The levels of these inhibitors in plants are variable and depend on the stage of maturation, tissue location, time of harvest and storage, as also the variety of plant, with possible co-existence of different classes of inhibitors as well as a variety of isoforms in a single tissue or organ (Bhattacharyya et al., 2007, 2009; Brzin & Kidrič, 1995; Ryan, 1990).

3.2 Classification of protease inhibitors

PIs are primarily classified based on the class of protease that they inhibit, thus four main inhibitor families have been established: serine protease inhibitors, cysteine protease inhibitors, metalloprotease inhibitors and aspartyl protease inhibitors (Koiwa et al., 1997; Ryan, 1990). Compared with the serine and cysteine protease inhibitors, studies directed for purification, characterization and biotechnological use of aspartic protease and metalloprotease inhibitors from plants are still very few. For this reason and due to the limited space it will be discussed only the two first families.

3.2.1 Serine protease inhibitors

Inhibitors of serine protease in plants are grouped into subfamilies based on their molecular weight, structural similarity, presence of cysteine residues and disulfide content (Brzin & Kidrič, 1995; Koiwa et al., 1997; Ryan, 1990). Thus, it was established at least eight inhibitor subfamilies: Bowman–Birk, Kunitz, Potato I, Potato II, Cucurbit, Cereal, Thaumatin-like and Ragi A1 (Koiwa et al., 1997; Ryan, 1990). Despite the variety of the subfamily of serine protease inhibitors, the most studied inhibitors are Kunitz and Bowman-Birk groups (Koiwa et al., 1997; Lawrence & Koundal, 2002; Ussuf et al., 2001). Studies are mainly directed to those found in seeds of legume subfamilies (Norioka et al., 1988).

Inhibitors of the Bowman-Birk subfamily found in monocotyledons and dicotyledons were grouped into three classes of proteins based on molecular weight, number of reactive sites[5], cysteine residues and disulfide bridges. In monocots, Bowman-Birk inhibitors are divided into two classes: the first consisting of inhibitors of approximately 8 kDa possessing a single reactive site ("single-headed") and five disulfide bridges and the second consisting of inhibitors of about 16 kDa, composed of about 180 amino acid residues, with two reactive sites (dual head inhibitors or "double-headed") and ten disulfide bonds. Inhibitors of 16 kDa are composed of two domains, each of 8 kDa with high identity, similar to the 8 kDa inhibitors. The presence of these two domains of 8 kDa was explained due to gene duplication events, leading to this group of inhibitors (Qi et al., 2005). Mello and colleagues (2003) reported the presence of Bowman-Birk inhibitors consisting of about 250 amino acids residues, distributed in three areas of 8 kDa, with high identity. In this same study, these authors, analyzing the amino acid sequences of various Bowman-Birk inhibitors from monocotyledons, consisting of 8 kDa, observed that some members of this family of inhibitors found in corn and cane sugar are glycoproteins. In dicotyledons, Bowman-Birk inhibitors have a molecular mass of 8 kDa. They are generally composed of about 104 amino acid residues including 14 cysteine residues involved in seven disulfide bridges, having two reactive sites (Mello et al., 2003; Qi et al., 2005). The comparison between different members of this family showed that the first reactive site is located in the N-terminal region and is more conserved than the second reactive site, located in C-terminal region of the molecule (Prakash et al., 1996; Wu & Whitaker, 1991). These inhibitors can interact simultaneously and independently with two serine proteases, not necessarily identical, i.e., with two molecules of trypsin or a trypsin and a chymotrypsin molecule (Mello et al., 2003; Qi et al., 2005). The first inhibitor that gave rise to this protein subfamily was purified from soybean seeds and consisted of a single chain protein with 71 amino acid residues and two kinetically independent reactive sites, one for trypsin (Lysine16-Serine17) and the other for chymotrypsin (Leucine 44-Serine 45) (Odani & Ikenaka, 1973a, b).

Kunitz inhibitors subfamily are proteins with molecular weights ranging from 18 to 26 kDa, constituted by approximately 180 amino acid residues, having a low content of cysteine residues and involved in one or two disulfide bridges (Krauchenco et al., 2004; Pando et al., 2001). Members of this subfamily have been found in a variety of botanical families, however most research has been directed to inhibitors of the three legume subfamilies (Fabaceae family). Inhibitors of the Papilionoideae and Caesalpinioideae subfamilies usually have a polypeptide chain while inhibitors from the Mimosoideae

---

[5] The reactive (inhibitory) site corresponds to the region of the inhibitor that binds with the enzyme.

subfamily consist of two polypeptide chains joined by disulfide bridges, being characterized as dimeric proteins (Batista et al., 1996; Norioka et al., 1988). The presence of carbohydrates is a common feature for inhibitors of the Kunitz family; however, there are few reports on the structural aspects. The inhibitor purified from jackfruit seeds (*Artocarpus integrifolia*) was characterized as a glycoprotein member of the Kunitz subfamily presenting in its structure units of galactose, glucose, mannose, fucose, xylose, glucosamine and uronic acid (Bhat & Pattabiraman, 1989). The presence of mannose, xylose, fucose and other sugars were also detected in the structure of Kunitz inhibitor purified from the papaya latex (*Carica papaya*), called PPI (Odani et al., 1996). The presence of sugars was also detected as a constituent in structure of other Kunitz inhibitors, including the inhibitor purified from *Bauhinia rufa* (Sumikawa et al., 2006) and from seeds of *Swartzia pickelli* (Cavalcanti et al., 2002). However, the role of carbohydrates in inhibitors structure has not been clarified. In general, these inhibitors have only one reactive site. Because of this structural feature, they are known as one head inhibitors ("single headed") (Ryan, 1990). On the other hand, few representants of this family were characterized as inhibitors that have two sites for two different enzymes (Migliolo et al., 2010; Valueva et al., 1999; Bösterling & Quast, 1981). Arginine and Lysine residues are part of reactive sites of the inhibitors (Iwanaga et al., 2005). This inhibitor subfamily has the ability to differentially inhibit serine proteases from various sources as trypsin and chymotrypsin responsible for digestion in mammalian and insects, enzymes of the blood clotting, among others (Oliva et al., 2000; Batista et al.,1996) (Table 2).

3.2.2 Cysteine proteases inhibitors

Cysteine protease inhibitors from plants are grouped into a single well-characterized protein family, commonly called plant cystatins[6] or phytocystatins (Brzin & Kidrik, 1995; Margis et al., 1998). These proteins exhibit the ability to suppress catalytic activity of cysteine proteases members, including papain family, calpains, clostripains, streptococcal cysteine protease and viral cysteine proteases or caspases, also called apopains (Abe et al., 1994; Brzin

Source	Family	M	Aa	Db	pI	Pc	Specificity	Reference
Alocasia macrorrhiza	A	19.7	184	2			T, Q	Argall et al.,1994
Brassica juncea	B	20	178	2			T	Mandal et al., 2002
*Carica papaya**	C	24	184	2			T, Q	Odani et al., 1996
Terminalia arjuna	Cb	21.5					T, Q	Rai et al., 2008
Putranjiva roxburghii	Eu	34				1	T	Chaudhary et al., 2008
Bauhinia bauhinoides	L-Ce	20	154	0	6.9	1	T, Q, K	Oliva et al., 2001
*B. rufa**	L-Ce	20	144	1	4.6	1	PE	Sumikawa et al., 2006
B. variegata	L-Ce	18.5	167	2	4.8	1	T	Di Ciero et al., 1998
Caesalpinia bonduc	L-Ce	20		2		2	T, Q	Bhattacharyya et al., 2007

[6] Cystatin from cysteine protease inhibitor.

Source	Family	M	Aa	Db	pI	Pc	Specificity	Reference
C. echinata	L-Ce	20				1	T, K, Pl, XII, X	Cruz-Silva et al., 2004
Delonix regia	L-Ce	22	185	2		1	T, K	Pando et al., 2001
Schizolobium parahyba	L-Ce	20	169	2	4.5	1	Q	Souza et al., 1995, 2000
*Swartizia pickelli**	L-Ce	20	174	1		1	T, Q, K, Pl	Cavalcanti et al., 2002
Acacia confusa	L-M	19.4	176	2		2	T, Q	Hung et al, 1994
Adenanthera pavonina	L-M	20	176	2		2	T, Q, P	Macedo et al., 2004
Albizzia kalkora	L-M	19.7		2		2	T	Zhou et al., 2008
Calliandra selloi	L-M	20		1	4.0	2	T, Q, K	Yoshizaki et al., 2007
Dimorphandra mollis	L-M	23				1	T, K, Pl	Mello et al., 2001
Entada scandens	L-M	19.7		1	7.4	1	T	Lingaraju & Gowda, 2008
Enterolobium contortisiliquum	L-M	22	174	2		2	T, Q, XII, K, Pl	Batista et al., 1996
I. laurina	L-M	20	180	1		1	T	Macedo et al., 2007
Leucaena leucocephala	L-M	20	174	2		2	T, Q, K, Pl, XII	Oliva et al., 2000
Plathymenia foliolosa	L-M	19					T, Q	Ramos et al., 2008, 2009
Crotalaria pallida	L-P	32.5				2	T, Q, PE, P	Gomes et al., 2005
C. paulina	L-P	20	177		4.0	1	T, Q	Pando et al., 1999
Erythrina acanthocarpa	L-P	17.4	163	2			T, Q	Joubert, 1982
Glycine max	L-P	21	181	2		1	T, Q, X, K	Kim et al., 1985
Psophocarpus tetragonolobus	L-P	20.2	183	2		1	Q	Kortt, 1980
Artocarpus integrifolia	M	26				1	T, Q	Bhat and Pattabiraman, 1989
Murraya koenigii	R	27				1	T	Shee and Sharma, 2007
Solanum tuberosum	S	21	186	2	6.3	2	T, Q, NE	Valueva et al., 1997, 1998
S. tuberosum	S	21	186	2	5.2	2	T, Q	Valueva et al., 1997, 1998

Table 2. Serine proteases inhibitors (Kunitz)
M, Moraceae; R, Rutaceae; B, Brassicaceae; A, Araceae; C, Caricaceae; S, Solanaceae; L-Ce, Leguminosae-Caesalpinioideae; L-MI, Leguminosae-Mimosoideae; L-P, Leguminosae-Papilionoideae; Cb, Combretaceae; Eu, Euphorbiaceae, S, Sapindaceae; MM, molecular mass (kDa); Aa, amino acid; Db, disulfide bonds; pI, isoeletric point; Pc, polypeptide chain number; T, pancreatic trypsin; Q, pancreatic quimotrypsin; PE, pancreatic elastase; P, papain; Pl, plasmin; K, human plasma kallikrein; XII, factor XIIa; X, factor Xa; NE, human neutrofil elastase; (*) glycosylated inhibitors.

& Kidrik, 1995; Brzin et al., 1998; Fernandes et al., 1993; Gaddour et al., 2001; Margis et al., 1998; Pernas et al., 1998) with different affinity degrees. Phytocystatins comprise a polypeptide chain devoid of disulfide bridges. In general, they are small molecules that have molecular weights ranging from 7.5 kDa, as the phytocystatin purified from pumpkin seeds (Levleva et al., 1997), or the inhibitor purified from soybean seeds (*Glycine max*) with approximately 26 kDa (Misaka et al., 1996) (Table 3). However, some representants of this family have higher molecular masses, like the phytocystatins purified from potato tuber (*Solanum tuberosum*), with approximately 85 kDa (Waldron et al., 1993), tomato leaves (*Lycopersicon esculentum*), with similar molecular mass (Jacinto et al., 1998; Wu & Haard, 2000) and sunflower seeds (*Helianthus annuus*) with a molecular mass of 32 kDa (Kouzuma et al., 2000).

Phytocystatins possess three regions quite conserved, interacting with their target proteases: a central motif consisting of Gln-X-Val-X-Gly, where X represents any amino acid residue, a dipeptide usually formed by a proline and a tryptophan near C-terminus and a glycine

Source	Family	MM (kDa)	Aa	pI	Specificity	Reference
Helianthus annus	A	9	83	5.6	P, Fi, CH, CB, CL	Kouzuma et al., 1996
Helianthus annus	A	11	100	9.5	P, Fi, CH, CB, CL	Kouzuma et al., 1996
Daucus carota	Ap	18	133		P	Ojima et al., 1997
Carica papaya	Cp	11.3	99	5.1	P, Qp, Car, PIV	Song et al., 1995
Curcubita maxima	Cu	7.5		6.0	P, F	Levleva et al., 1997
Castanea sativa	F	11.2	102		P, Fi, Qp, CB,	Pernas et al., 1998
Oryza sativa	G	11.8	102	5.3	P, CH, CB, CL	Abe et al., 1987
Oryza sativa	G	11.9	107		P, CH, CB, CL	Kondo et al., 1990
Zea mays	G	18	135		P, CB, CH, CL	Abe et al., 1992; 1994
Zea mays	G	9.2	85	5.2	P, Fi, Qp, CB, CH	Abe & Whitaker, 1988
Persea americana	L	11.3	100		P	Kimura et al., 1995
Vigna unguiculata	Le	10.7	97		P	Fernandes et al., 1993
Phaseolus lunatus	Le	14		5.5	P, CB, CH, CL	Brzin et al., 1998
Glycine max	Le	26	245		P	Misaka et al., 1996
Glycine max	Le	12	100		P	Zhao et al., 1996
Glycine max	Le	11	92		P	Zhao et al., 1996
Hordeum vulgare	Le				P, Fi, Qp	Gaddour et al., 2001
Chelidonium majus	Pa	10	90	9.3	P, CL, CH	Rogelj et al., 1998
Malus domestica	R	11.3	83		P, Fi, B	Ryan et al., 1998
Fragaria x ananassa	R	23.1	206		P, CH, CB	Martinez et al., 2005

Table 3. Biochemical characterization of Phytocystatins
MM, molecular mass; Aa, amino acid number; pI, Isoelectric point; A, Asteraceae; Ap, Apiaceae; G, Gramineae (Poaceae); L, Lauraceae; L-Ce, Leguminosae-Caesalpinioideae (Fabaceae); S, Solanaceae; F, Fagaceae; Ca, Caryophyllaceae; Cp, Caricaceae; Cu, Curcubitaceae; Pa, Papaveraceae; R, Rosaceae; S, Solanaceae; An, ananain; Ac, actinidin; Cr, cruzipain; CG, Cathepsin G; CB, Cathepsin B; CH, Cathepsin H; PPE, porcine pancreatic elastase; HNE, human neutrofil elastase; B, Bromelain; Qp, Quimopapain; Fi, Ficin; Car, Caricain; PIV, papaya proteinase IV; PIII, papaya proteinase III; P, papain.

residue located near the N-terminal region, known as flexible region of phytocystatins (Kondo et al., 1990; Stubbs et al., 1990; Margis et al., 1998). The three interaction regions can be found as tandem domains along protein structure. Thus, phytocystatins can be grouped into two distinct classes: one comprising phytocystatins of low molecular weight, with a single cystatin domain (Abe et al., 1987, 1992; Fernandes et al., 1993; Pernas et al., 1998); other class comprising phytocystatins of high molecular mass, with multiple cystatin domains, that is, with three interaction regions occurring repeatedly in primary structure of the inhibitor (Bolter, 1993; Diop et al., 2004; Kouzuma et al., 2000), and in this latter case, they are called multicystatins. Multicystatins from potato tuber and tomato leaves possess eight domains that can interact simultaneously with eight cysteine proteases (Jacinto et al., 1998; Waldron et al., 1993; Wu & Haard, 2000), however, sunflower multicystatin has three domains (Kouzuma et al., 2000) and the one from bean-to-string (*V. unguiculata*) has two cystatin domains (Diop et al., 2004).

3.3 PI inhibition mechanisms

In general terms, all the PIs bind to their specific protease preventing access of the substrate to the active site. For some PIs, the docking occurs directly in the protease active site, while for others the binding takes place in a neighborhood of the catalytic centre but leading to its steric hindrance (Krowarsch et al., 2003).

The majority of the known PIs, specially the serine protease inhibitors, interact with the enzyme catalytic sites in a "canonical" manner, similar to the enzyme-substrate interaction, via an exposed reactive site loop of conserved conformation (Bode & Huber, 2000) (figure 2). A well known member of this group of classic inhibitors is the Kunitz-type trypsin inhibitor

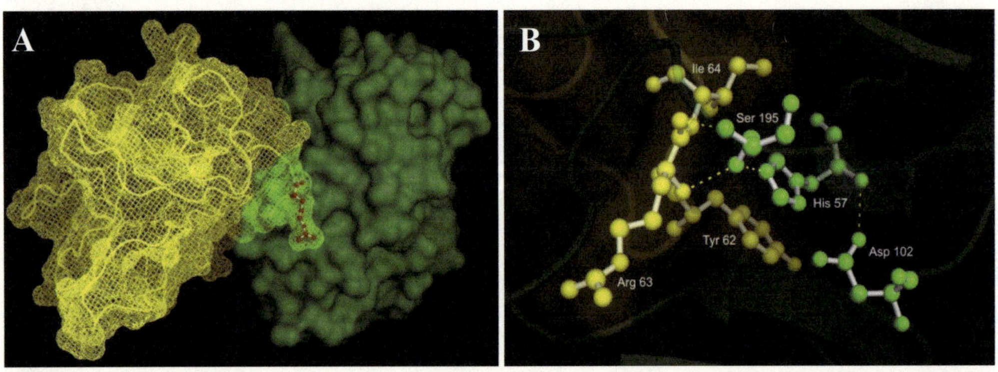

Fig. 2. Stereo diagram showing the interaction between porcine pancreatic trypsin and the soybean trypsin inhibitor (SKTI).
(A) SKTI is shown in yellow ribbon and mesh representation and trypsin in transparent dark green surface. The light green intersection represents the interface between the reactive site and the catalytic site, with the Arg 63 side chain (red ball and stick representation) of the inhibitor protruding into the specificity cleft of the trypsin. (B) Detail of the interaction interface of the trypsin catalytic triad (green ball and stick) and the reactive site loop residues of the SKTI (yellow ball and stick). Dotted line, hydrogen bonds. Figure made with PyMOL (PDB code 1AVW).

from soybean (SKTI, SBTI or STI), a potent trypsin inhibitor but that also inhibits in a lesser extent the chymotrypsin (De Vonis Bidlingmeyer et al., 1972) and plasmin (Nanninga & Guest, 1964). The SKTI is a 21.5 kDa non-glycosylated protein containing 181 amino acid residues in a single polypeptide chain crosslinked by two disulfide bridges. Its reactive site loop possess an arginine residue (Arg 63) whose side chain fits into the specificity pocket of the trypsin, while its carbonyl carbon makes contact with the serine (Ser 195) of the active site without, however, suffering catalysis and thus it blocks the enzymatic action (Song & Suh,1998; Macedo et al., 2007, 2011; Krowarsch et al., 2003) (figure 2B). One network of hydrogen bonds between inhibitor and enzyme is also formed outside of the reactive site to stabilize the complex (Song & Suh, 1998).

This canonical conformation of the reactive loop characterizes a mechanism of competitive inhibition and is also found in the trypsin inhibitors from *Erythrina caffra* seeds and *Psophocarpus tetragonolobus* chymotrypsin inhibitor (Song & Suh, 1998; Krauchenco et al., 2003). Trypsin inhibitors, such as from *Swartzia pickelii*, can present also glutamine residue in the reactive site (Cavalcanti et al., 2002). Specific inhibitor for chymotrypsin, in general, possess leucine residue in its reactive site (Kimura et al., 1993; Dattagrupta et al., 1996) and in the *Bauhinia rufa* elastase inhibitor one valine was identified in the reactive site (Sumikawa et al., 2006).

It would be interesting to note that there are Kunitz-type inhibitors able to inhibit proteases belonging to different mechanistic classes. Protease inhibitor purified from *Prosopis juliflora* seeds (PjTKI) presents a competitive inhibition mechanism directly interacting between its reactive site (Arg 64) and the catalytic site in target trypsin (Ser 195). Moreover, PjTKI also possesses an inhibitory activity against papain and a cysteine protease present in the digestive system of several phytophagous insect-pests (Oliveira et al., 2002; Franco et al., 2002). This bifunctional property was also observed for the Kunitz inhibitor from *Adenanthera pavonina* seeds (ApTKI). ApTKI was a strong non-competitive inhibitor of trypsin and moderate noncompetitive inhibitor to papain. Different from PjTKI, that was incapable of simultaneous inhibition of trypsin and papain, the interaction sites of the ApTKI did not overlap, and it formed a ternary complex that was observed through *in vitro* and *in silico* methods (Macedo et al., 2004; Prabhu & Pattabiraman, 1980; Migliolo et al., 2010).

Non-canonical mechanisms of inhibition are common to other classes of PI. A typical case is the inhibition of papain-like cysteine protease by the cystatins, which interact with the enzyme surface subsites adjacent to the active site, blocking it without direct contact with the catalytic groups (Stubbs et al., 1990; Bode & Huber, 2000; Turk et al., 2000).

3.4 Biological function of protease inhibitors

Protease inhibitors of plant origin are known for many years and their participation in a variety of endogenous and exogenous events continue to be the subject of much research. Regulation of endogenous proteases, storage function and defensive role against predators and pathogens are the principal proposed biological functions for plant PIs (Lawrence & Koundal, 2002). The proposed role for protease inhibitors as storage proteins was first suggested by Pusztai (1972). Plant storage proteins function as a store of nitrogen, carbon and sulfur and they are so considered when are deposited in tissues of reserves in concentrations higher than 5% of the total protein content (Shewry, 2003). In seeds, tubers and other plant tissue reserves, PIs are found from 1 to 10%, being one of the major storage proteins (Mandal et al., 2002; Brik & Wong, 2003; Shewry, 2003). Besides their storage function, by providing nitrogen and sulphur required during germination, other functional roles have been assigned for the trypsin inhibitors, such as regulating endogenous plant proteases to prevent precocious germination, inhibiting trypsin during passage through the animal's gut, thus helping in seed dispersal, and protecting plants against pests and diseases (Derbyshire et al 1976; Laskowski & Kato, 1980; Shewry, 2003). Confirmation of the role of PIs in plants is primarily based on *in vitro* experiments using insect gut proteases as well as the efficiency of these inhibitors on a variety of proteases from pathogens such as fungi, viruses, mites and nematodes (Fosket, 1994; Dunaevsky et al., 2005; Fan & Wu, 2005). The second line of evidence that strengthens the role of these molecules as defense compounds of plants results from the fact that these inhibitors interfere with normal growth and development of microorganisms when added to artificial diet system (Araújo et al., 2005; Gomes et al., 2005). Another line of evidence is the induction level of these molecules following mechanical injury or another biotic or abiotic stress. And finally, the increased resistance to insects and pathogens in plants transformed with genes expressing PIs (Kiggundu et al. 2010; Masoud, 1993).

3.5 Biotecnological/pharmacological applications of PI

Despite the mechanisms to control proteases, loss of proteolytic control is observed in a wide range of diseases. Indeed, increased proteolysis has been shown to underpin various pathological processes and, as a result, PIs have emerged as a class of highly promising chemotherapeutic agents (Scott & Taggart, 2010). Thus, many inhibitors can be strong therapeutic candidates for treating diseases, such as cancer, fungical, parasitic and neurological disorders, inflammatory, immune, respiratory and cardiovascular diseases (Leung et al., 2000).

There is evidence that inhibitors suppress various stages of carcinogenesis, including initiation, promotion and progression. Although many PIs have the ability to prevent carcinogenic processes, the most potent are those with activity antichymotrypsin (Kennedy, 1998; Zhang et al., 2007) and it is not yet known how these inhibitors work suppressing carcinogenesis. It is known that there may be multiple inhibition pathways of carcinogenesis, e.g., preventing the release of superoxide radicals and hydrogen peroxide by polymorphonuclear leukocytes or other cell types, stimulated by tumor promoting agents (Kennedy, 1998). Recent studies indicate that Bowman-Birk type inhibitors are potential candidates for inhibition of carcinogenic activity (Sessa & Wolf, 2001).

Because PIs naturally inhibit a diversity of proteases from plants pathogens, their genes have been used for the construction of transgenic crop plants to be incorporated in integrated pest management programmes (Lawrence & Koundal, 2002).

4. Purification of PIs using affinity chromatography (on protease-matrix)

4.1 General advantages of affinity chromatography on protease-matrix

Affinity chromatography is a very efficient technique capable of purifying proteins based on reversible interactions between a protein and a specific ligand coupled to a chromatographic matrix. The affinity chromatography is usually used as one of the last steps in the purification process of PIs and combines advantages as great time-saving and high capacity of selection and concentration of the target protein from a complex mixture of contaminating substances in a large sample volume (Cuatrecasas, 1970). The binding property of the PIs with enzymes (figure 2) has been exploited for the affinity purification of inhibitors from various sources, especially for those of plant origin.

4.2 Coupling

The first step to carry out the purification of the inhibitor of interest by affinity chromatography is to immobilize its corresponding enzyme in a coupling gel. The gel substance (commonly based on a polysaccharide) should exhibit mechanical and chemical stability to the coupling and elution conditions, minimal nonspecific interaction with proteins and form a loose porous network which allows the free flow of large molecules. Cross-linked dextran (Sephadex[7]) and, better yet, derivatives of agarose (Sepharose[8]) are polysaccharides that have many of these features and therefore are widely used as coupling gel (Cuatrecasas, 1970). Many studies have reported the purification of PIs by using a specific protease immobilized by covalent bonding to agarose resins (CNBR-activated Sepharose 4B) and chitosan (Xi et al., 2005), that have low non-specific adsorption of proteins. The coupling of protease using cyanogen halides (CNBr-activated Sepharose) is well described and consists in linking primary amino groups directly in pre-activated matrices (figure 3). The establishment of multipoint connections provides greater stability to the immobilized enzyme. However, it is important that these multipoint connections do not interfere with the enzyme binding site and thus compromise the effectiveness of the interaction of the inhibitor with the protease. The resins of CNBr-activated Sepharose 4B type are stable in a wide range of pH (2-11) and have a good range of ligand coupling.

4.3 Sample application and washing

After coupling the protease of choice with the gel and before applying the mixture sample containing target PIs, the matrix must be packaged in a column and pre-equilibrated in binding buffer. Usually, Sepharose gels containing the immobilized enzyme are equilibrated with 2-3 volumes buffers containing NaCl at concentrations ranging from 0.1-0.5 M and pH is in the same range of optimal pH activity of the enzyme.

It is not possible to predict a single optimum flow rate for loading the sample in an affinity chromatography because the degree of interaction between ligands is widely variable. The binding is favored by the use of a very low flow rates in the sample loading, especially for weak affinity systems. In extreme cases of very weak affinity may be useful to stop the flow after loading the sample or to re-apply it. It is therefore interesting to test the best flow rate to each case, and it is not recommended flow rates higher than 10 $mL/cm^2/h$.

---

[7] Sephadex (from separation Pharmacia dextran) is a trademark of GE Healthcare (formerly: Pharmacia).
[8] Sepharose (from separation Pharmacia agarose) is a trademark of GE Healthcare.

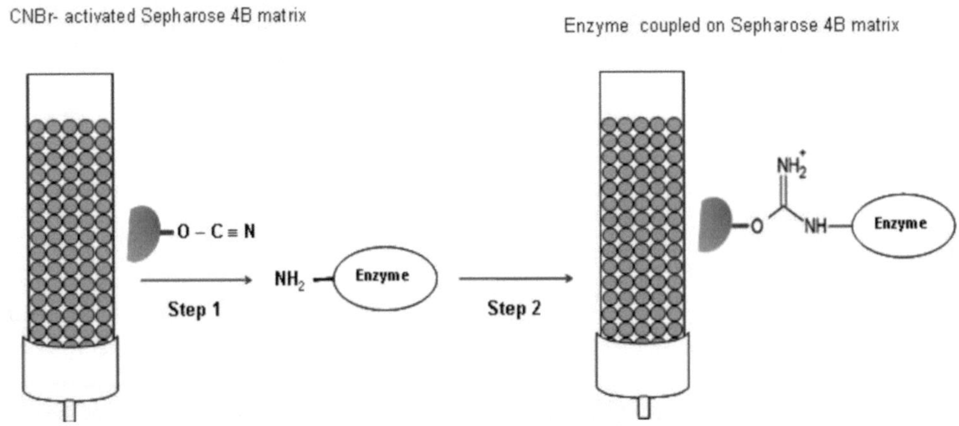

Fig. 3. Coupling the enzyme to CNBr-activated Sepharose 4B.

The volume of the sample loaded is not critical, since the principle of the technique is the affinity, however for interections with weak affinity it recommended to apply it in a small volume (about 5% of bed volume). In order to remove the unbound substances (and prior to the elution), the column should be washed with about 10 volumes of the starting buffer. The figure 4 presents a schematic view of the steps of an affinity chromatography.

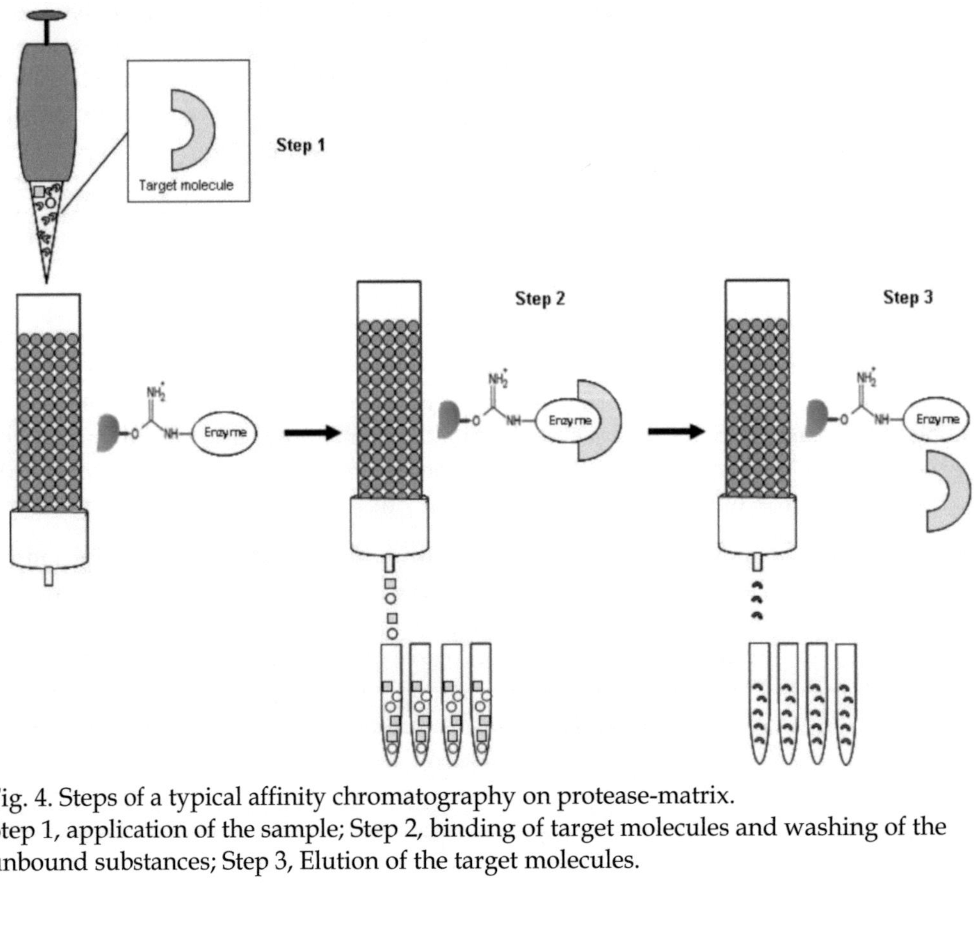

Fig. 4. Steps of a typical affinity chromatography on protease-matrix.

Step 1, application of the sample; Step 2, binding of target molecules and washing of the unbound substances; Step 3, Elution of the target molecules.

4.4 Elution

Methods of elution of target protein from the column can be selective (using a competitive ligand) and non-selective (changing the pH, ionic strength or polarity).

Some cysteine proteinase inhibitor has been isolated by affinity chromatography using immobilized papain. Systemin-inducible papain inhibitor from *Lycopersicon esculentum* leaves (Jacinto et al., 1998) was eluted as using 50 mM K_3PO_4 buffer, pH 11.5, containing 0.5 M NaCl and 10% glycerol (Anastasi et al., 1983). A cysteine proteinase inhibitor from chestnut seeds, *Castanea sativa*, named Cystatin CsC, retained in the affinity column on carboxymethylated-papain NHS-Superose was eluted of the column using 0.2 M trisodium phosphate buffer, pH 11.5, containing 0.5 M NaCl, then immediately neutralized using 2 M tris-HCl, pH 7.5, and desalted (Pernas et al., 1998). Solution of 10 mM NaOH was used for the recovery of chelidocystatin, cystatin from mature *Chelidonium majus* plants of the papain Sepharose affinity chromatography (Rogelj et al., 1998). Papain inhibitor of the immature fruit from *Malus domestica* was eluted of papain affinity column with solution of 50 mM HCl and immediately neutralised with 3 M Tris-HCl buffer, pH 10 (Ryan et al., 1998).

Protein inhibitors of serine proteases retained in protease trypsin-and chymotrypsin-Sepharose affinity column are generally recovered from resin by the use of fixed concentration of HCl. Trypsin is the enzyme most often used as a ligand for the purification of PIs by affinity. As example, Macedo et al. (2011) purified a trypsn inhibitor from *Sapindus saponaria* seeds (SSTI) using a trypsin-Sepharose column, where the inhibitor was eluted with 0.01 M HCl. Another inhibitor purified by the method of elution with fixed concentration of HCl was the inhibitor purified from *Adenanthera pavonina* seeds (APTI), using 0.1 M HCl (Macedo et al., 2004, 2011). Trypsin inhibitor of *Cocculus hirsutus* leaf (ChTI) was recouped of the affinity column with 0.2 N HCl, pH 3.0 (Bhattacharyya et al., 2009). HCl concentration of 0.015 M was efficient for removal from *Cycas siamensis* seeds inhibitor (Konarev et al., 2008). The mixture of 0.2 M Gly–HCl buffer in pH 3.0 containing 0.5 M NaCl was effective for the recovery of a Kunitz trypsin inhibitor of *Entada scandens* seeds (ESTI) bound the affinity chromatography on trypsin-Sepharose (Lingaraju & Gowda, 2008). Solution of 1 mM HCl was necessary to remove the trypsin inhibitors of *Pithecellobium dumosum* (Oliveira et al., 2007a, b; 2009) and *Crotalaria pallida* seeds (Gomes et al., 2005).

Purification of inhibitors from different classes of proteolytic enzymes using few purification steps and enzyme immobilization techniques has been used intensively for decades. Four competitive Kunitz-type trypsin inhibitors (JB1, JB2, JB3 and JB4) were purified from *Pithecellobium dumosum* by TCA precipitation, affinity chromatography on immobilized trypsin-Sepharose and reverse phase HPLC using Vydac C-18 column seeds with Ki values of 3.56, 1.65, 2.88 x 10^{-8} M and 5.70 x 10^{-10} M (Oliveira et al., 2007). The percentage inhibition of JB1, JB2 and JB3 on papain varied between 32.93 to 48.82% and was indicative of its bifunctionality with exception of JB4 that inhibited this activity in 9.9%. The papain inhibition by JB1 and JB2 were noncompetitive type and the Ki-values were 7.6 x 10^{-7} and 5.1 x 10^{-7} M, respectively. Among these a highly purified Kunitz-type inhibitors denominated PdKI (JB1) was isolated by affinity chromatography on trypsin-Sepharose column and HPLC (Figure 5) (Oliveira et al., 2007a).

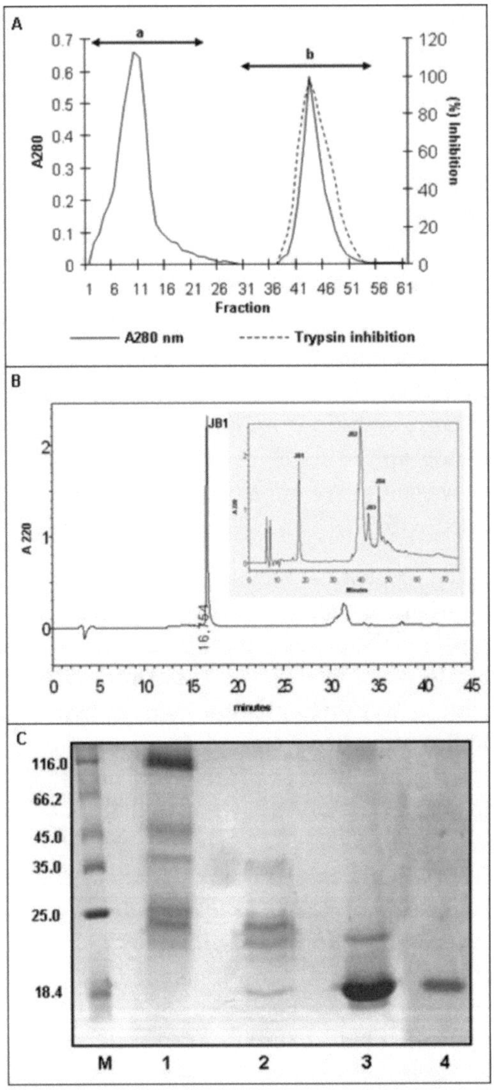

Fig. 5. (A) Chromatographic profile of Inhibitor of *Pithecellobium dumosum* seeds on Trypsin-Sepharose. Column (10 cm X 1.5 cm) was equilibrated with 50 mM Tris-HCl, pH 7.5 buffer and the retained proteins were eluted with 1 mM HCl. Fractions obtained were assayed against trypsin. (B) Elution profile on HPLC (Vydac C-18) column. The fractions obtained from Trypsin-Sepharose column were separated by semi-preparative reverse-phase HPLC column at a flow rate of 9 ml/min. Insets: the purified protein was then again subjected to analytical reverse-phase HPLC column at a flow rate of 1 ml/min. Both were eluted using a gradient of solvent B (60% acetonitrile in 0.1% TFA) in solvent A (0.1% TFA/H_2O), and monitored at 220 nm. (C) SDS-PAGE (15%) of PdKI from *P. dumosum* seeds, stained with silver nitrate. (M) Protein molecular weight markers: α-galactosidase (116 kDa), bovine serum albumin (66 kDa), ovalbumin (45 kDa), lactate degydrogenase (35 kDa), restriction endonuclease Bsp981 (25 kDa), â-lactoglobulin (18.4 kDa), and lysozyme (14.4 kDa). (1) Crude extract; (2) fraction treated with TCA; (3) Trypsin-Sepharose retained peak; (4) PdKI.

Although most studies describe elution of proteins retained on affinity column using a fixed molarity of HCl, however, it is interesting to test the elution of adsorbed proteins with solutions containing different molar concentrations of HCl, similar to stepwise (step by step). This procediment is important because high molar HCl concentrations may be unnecessary. For example, Figure 6A shows the elution profile of chymotrypsin-Sepharose column of a protein sample fractionated with ammonium sulfate at 30-60% saturation, which exhibited inhibitory activity to chymotrypsin. The column was equilibrated with Tris-HCl 0.05 M, pH 7.5. Non-adsorbed proteins were eluted with the same equilibration buffer and proteins adsorbed to matrix were eluted with 5 mM HCl. This experiment also used 10 and 100 mM HCl. However, only the concentration of 5 mM HCl was necessary for elution of total active proteins. In addition, inhibitors bind to immobilized enzyme linked to Sepharose affinity in different degrees. For example, Figure 6B shows elution profile of another sample containing high inhibitory activity against chymotrypsin. Three different molar concentrations of HCl were tested for elution of adsorbed proteins. The adsorbed proteins were differentially eluted with two concentrations of HCl. Proteins presenting lower affinity for matrix were eluted with 5 mM HCl, while proteins with higher affinity for matrix were only eluted with 100 mM HCl. No elution was observed when 10 mM HCl was used.

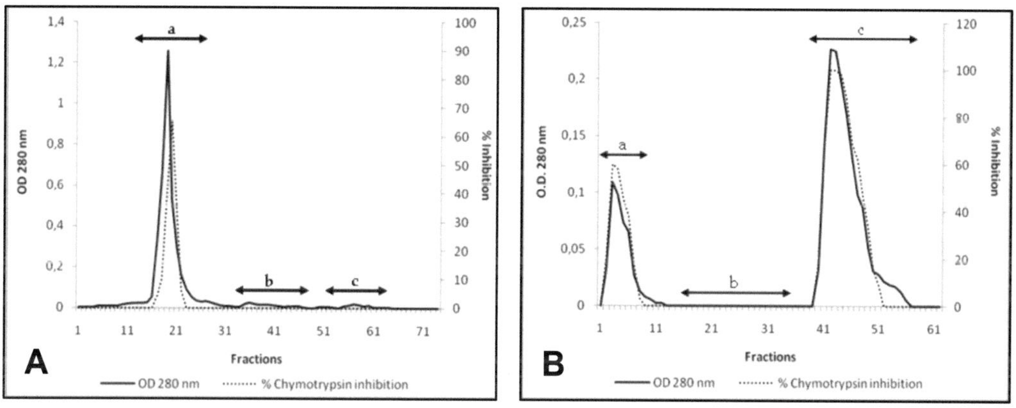

Fig. 6. Chromatographic profiles of different samples of inhibitors (A, B) from legume seeds on Chymotrypsin-Sepharose affinity column.
Eluition Conditions: a: 5 mM; b, 10 mM and c, 100 mM HCl.

Finally, affinity column are also useful to exclude contaminants from protein samples or delete unwanted activities. To retrieve specific inhibition activity for chymotrypsin in a sample showing predominant trypsin inhibitory activity against chymotrypsin, it was necessary to test binding ability on a trypsin-Sepharose column to exclude trypsin inhibitory activity in the sample (Figure 7). Thus, different concentrations of the sample were applied and inhibitory activity was monitored. Figure 7A shows a chromatographic profile in affinity column when 35 mg of protein was applied on a trypsin-Sepharose column after balanced with equilibration buffer. Inhibition tests showed that both fractions, non-adsorbed and adsorbed, inhibitory activity toward trypsin was 100% and inhibitory activity for chymotrypsin was determined only for non-adsorbed fractions. Following this procedure, 21 mg protein from the same sample was applied in trypsin-Sepharose column (Figure 7B). Inhibition assays showed that trypsin inhibition for non-adsorbed fractions were much lower than before (about 49% inhibition), compared to that obtained in adsorbed fractions and when compared to Figure 7A. However, the use of about 7 mg protein resulted in adsorption of total protein responsible for inhibitory activity when linked to trypsin immobilized on chromatographic matrix. Inhibitory activity against chymotrypsin was determined only in non-adsorbed fractions (Figure 7C). Fractions containing only inhibitory activity against chymotrypsin were pooled and applied to chymotrypsin-Sepharose affinity column, which all inhibitory activity against chymotrypsin was recovered free of any activity against trypsin (Figure 7D). This procedure allowed the elimination of total inhibitory activity for trypsin and total recovery of inhibitory activity to chymotrypsin present in sample.

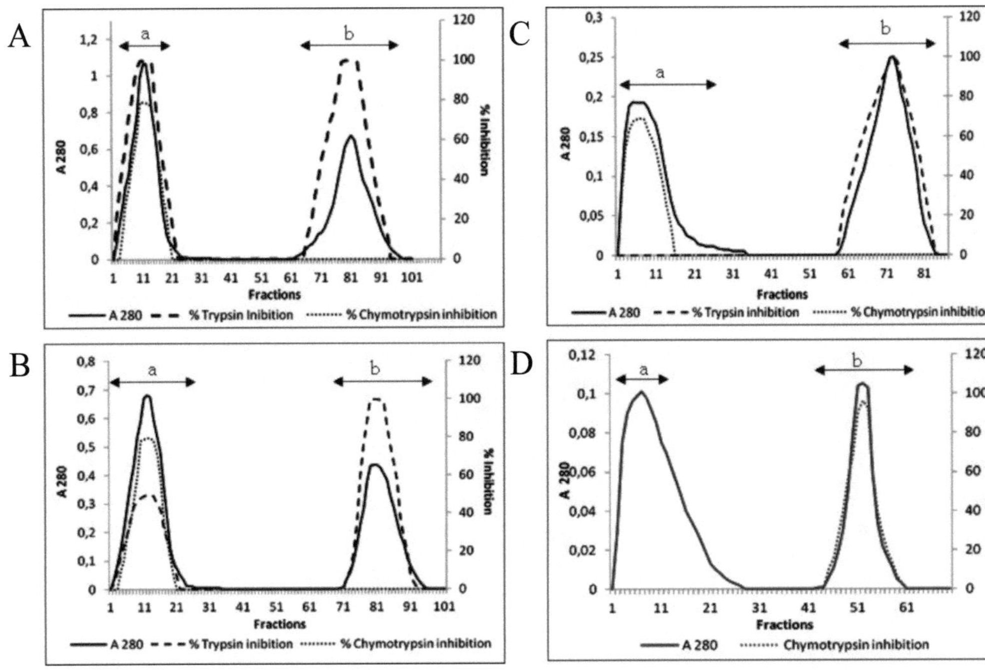

Fig. 7. Chromatographic profile of trypsin and chymotrypsin inhibitors in affinity columns. Seed extracts were loaded onto trypsin-Sepharose 4B (A-C) or chymotrypsin-Sepharose 4B (D) columns equilibrated with 50 mM Tris-HCl buffer, pH 8.0. Adsorbed proteins were eluted with 5 mM HCl: (3A) load 35 mg proteins, (3B) 21 mg (3C) 7 mg (3D) 3 mg, respectively. It was collected 2 mL/fraction and monitored at A = $\lambda_{280\ nm}$.

4.5 Detection (inhibitory activity)

Trypsin and chymotrypsin from bovine pancreas are serine protease more used *in vitro* assays for determination of the presence of inhibitory activity by the crude extracts of several origins, by accompaniment of inhibitory activity during all the isolation process, as well as to characterize inhibitor purified, including determination of dissociation constant (*Ki*), formation of inhibitor-serine protease complex and studies about stability of the inhibitory activity (Mello et al., 2001; Macedo et al., 2002, 2003, 2007; Pando et al., 1999; Gomes et al., 2005; Araujo et al., 2005; Oliveira et al., 2002, 2007a,b, 2009). Mature trypsin is composed for 223 amino acid residues with His57, Asp102 and Ser195 residues forming its catalytic triad (figure 2). Trypsin is 24 kDa distributed in a single chain polypeptide cross-linked by 6 disulfide bridges. Trypsin hydrolyzes specifically peptides on the carboxyl side of Lys and Arg amino acid residues (Walsh, 1970) in substrate protein such as azocasein (Xavier-Filho et al., 1989) or synthetic substrate as BAEE, benzoyl L-arginine ethyl ester (Avineri-Goldman et al., 1967; Delaage & Ladunski, 1968); TAME, p-toluenesulfonyl-L-arginine methyl ester (Bhattacharyy et al., 2006, 2007, 2009); tosyl-L-arginine methyl ester and BApNA, Na-benzoyl-L-arginine p-nitroanilide (Erlanger et al., 1961; Gomes et al., 2005; Araujo et al., 20005; Oliveira et al., 2007, 2009; Migliolo et al., 2010).

Chymotrypsin from bovine pancreas is a protein of 25 kDa, pI of 8.7, consisting of 241 amino acid residues composed by three peptide chains (A, B and C chain with 13, 131 and 97 residues, respectively), joined by disulfide bridges that shows capacity to hydrolyze peptide bonds on the C-terminal side of tyrosine, phenylalanine, tryptophan, leucine as well as methionine, isoleucine, serine, threonine, valine, histidine, glycine, and alanine (Appel, 1986; Ui, 1971). Hydrolyze of azocasein (Xavier-Filho et al., 1989) and release of p-nitroanilide from amide synthetic substrate (BTpNA, N-Benzoyl-L-tyrosine p-nitroanilide; N-Succinyl-L-phenylalanine-*p*-nitroanilide) are commonly used to assay chymotrypsin inhibitory activity (Nakahata et al., 2006; Macedo et al., 2007; Mello et al., 2001; Pando et al., 1999) as well as the hydrolyze of ester linkages of BTEE, N-Benzoyl-L-tyrosine ethyl ester (Bhattacharyy et al., 2006, 2007, 2009).

Both enzymes (Trypsin and chymotrypsin) are soluble, stabilized and can be stored in 1 mM HCl solution for 1 year at -20°C. The addition of calcium into of enzyme solution beyond stabilizing enzymes prevents the process of autolysis (Sipos & Merkel, 1970). The formation of p-nitroaniline (bright yellow) from amide substrate (BApNA, BTpNA) by trypsin and chymotrypsin is monitored at 405-410 nm (Nakahata et al., 2006; Macedo et al., 2007; Mello et al., 2001; Pando et al., 1999; Oliveira et al., 2007, 2009; Oliva et al., 1996).

The hydrolysis of azocasein (casein with 23.6 kDa conjugated to an azo-dye) is a procedure very used for the determination of proteolytic enzymatic activities that act in pH above of 5.0 as serine and cysteine proteases. Azocasein solution precipitates in pH below of 4.5. In general, in azocasein hydrolysis assays, reaction is stopped by the addition of trichloroacetic acid solution resulting in the formation of colored (red-orange) soluble components which are measured at (absorption maximum) 440 nm (Charney & Tomarelli, 1947).

5. Conclusions

The process of purification of biomolecules includes a combination of separation techniques, such as: extraction, fractionation by precipitation and chromatographies addressing different properties. The type and amount of technique will depend on the nature and characteristics of the molecule of interest, as well as the degree of purity desired in the final product. Among the chromatographic techniques, the affinity chromatography, by which specific biological properties can be exploited, stands out for its high purification capacity. For example, affinity chromatography on columns containing immobilized enzymes provides an efficient and rapid process of isolation and purification of protease inhibitors from different sources. This procedure provides advantages as high enrichment of inhibitor fraction, reduction of purification steps due to high binding specificity of the protein immobilized on a chromatographic matrix and purification of different protease inhibitors in the same fraction by differential elution of material retained on chromatographic matrix using different conditions to destabilize the adsorption between the complex enzyme-inhibitor. Protease inhibitors purified in this way can be evaluated for their use in agriculture and pharmaceutical industry. To analyze the application of protease inhibitors in these sectors are needed bioassays that require a lot of protein for development of dose-response curves. So the affinity chromatography represents a powerful tool for the enrichment of these proteins and enables the rapid achievement of these for analysis of activities like: bioinsecticide, bionematicide, bactericidal, anti-inflammatory, anticoagulant, and antitumor among others. The use of affinity chromatography on an industrial scale could also facilitate the achievement of these molecules for their use.

6. References

Abe, K.; Kondo, H. & Arai, S. (1987). Purification and characterization of a rice cysteine proteinase inhibitor. *Agricultural and Biological Chemistry*, Vol.51, No.10, (October 1987), pp. 2763–2768, ISSN 0002-1369

Abe, M. & Whitaker, J. R. (1988). Purification and characterization of a cysteine proteinase inhibitor from the endosperm of corn. *Agricultural and Biological Chemistry*, Vol.52, No.6 , 1583-1584, ISSN 0002-1369

Abe, M.; Abe, K.; Iwabuchi, K.; Domoto, C. & Arai, S. (1994). Corn cystatin I expressed in *Escherichia coli*: investigation of its inhibitory profile and occurrence in corn kernels. *Journal of biochemistry*, Vol.116, No.3, (Septembre 1994), pp. 488–492, ISSN 1756-2651

Abe, M.; Abe, K.; Kuroda, M. & Arai, S. (1992). Corn kernel cysteine proteinase inhibitor as a novel cystatin superfamily member of plant origin. Molecular cloning and expression studies. *European Journal of Biochemistry*, Vol.209, No.3, (November 1992), pp. 933–937, ISSN 0014-2956

Alarcon, F. J.; Martinez, T. F.; Barranco, P.; Cabello, T.; Diaz, M. & Moyano, F. J. (2002). Digestive proteases during development of larvae of red palm weevil, *Rhynchophorus ferrugineus* (Olivier, 1790) (Coleoptera: Curculionoidae). *Insect Biochemistry and Molecular Biology*, Vol.32, No.3, (March 2002), pp. 265-274, ISSN 0965-1748

Anastasi, A.; Brown, M. A.; Kembhavi, A. A.; Nicklin, M. J.; Sayers, C. A.; Sunter, D. C. & Barrett, A J. (1983). Cystatin, a protein inhibitor of cysteine proteinases. Improved purification from egg white, characterization, and detection in chicken serum. *The Biochemical journal*, Vol.211, No.1, (April 1983), pp. 129-138, ISSN 0264-6021

Araújo, C. L.; Bezerra, I. W. L.; Dantas, I. C.; Lima, T. V. S.; Oliveira, A. S.; Miranda, M. R. A., Leite, E. L. & Sales, M. P. (2004). Biological activity of proteins from pupls of tropical fruits. *Food Chemistry*, Vol.85, No.1, (March 2004), pp. 107-110, ISSN 0308-8146

Araújo, C. L.; Bezerra, I. W. L.; Oliveira, A. S.; Moura, F. T.; Macedo, L. L. P.; Gomes, C. E. M.; Barbosa, A. E. A. D.; Macedo, F. P.; Souza, T. M. S.; Franco, O. L.; Bloch-Jr, C. & Sales, M. P. (2005). In vivo bioinsecticidal activity toward *Ceratitis capitata* (fruitfly) and *Callosobruchus maculatus* (cowpea weevil) and in vitro bioinsecticidal activity toward different orders of insect pests of a typsin inhibitor purified from tamarind tree (*Tamarindus indica*) seeds. *Journal Agricultural and Food Chemistry*, Vol.53, No.11, (April 2005), pp. 4381-4387, ISSN 0021-8561

Argall, M. E.; Bradbury, J. H. & Shaw, D. C. (1994). Amino-acid sequence of a trypsin/chymotrypsin inhibitor from giant taro (*Alocasia macrorrhiza*). *Biochimica et Biophysica Acta*, Vol.1204, No.2, (February 1994), pp.189-194, ISSN 0006-3002

Avineri-Goldman, R.; Snir, I; Blauer, G. & Rigbi, M. (1969). Studies on the Basic Trypsin Inhibitor of Bovine Pancreas and Its Interaction with Trypsin. *Archives of Biochemistry and Biophysics*, Vol.121, No.1, (July 1969), pp. 107-116, ISSN 0003-9861.

Barrett, A J. (1997). Nomenclature Committee of the International Union of Biochemistry and Molecular Biology (NC-IUBMB). Enzyme Nomenclature. Recommendations 1992. Supplement 4: corrections and additions (1997). *European journal of biochemistry / FEBS*, *250*(1), 1-6.

Batista, I. F. C.; Oliva, M. L. V.; Araujo, M. S.; Sampaio, U. M.; Richardson, M.; Fritz, H. & Sampaio, C. A. M. (1996). Primary structure of a kunitz-type trypsin inhibitor from *Enterolobium contortisiliquum* seeds. *Phytochemistry*, Vol.41, No.4, (March 1996), pp. 1017-1022, ISSN 0031-9422

Bernardi, R.; Tedeschi, G.; Ronchi, S. & Palmieri, S. (1996). Isolation and some molecular properties of a trypsin-like enzyme from larvae of European corn borer *Ostrinia nubilalis* Hubner (Lepidoptera: Pyralidae). *Insect Biochemistry and Molecular Biology*, Vol.26, No.8-9, (September 1996), pp. 883-889, ISSN 0965-1748

Bhat, A. V. & Pattabiraman, T. N. (1989). Protease inhibitors from jackfruit seed (*Artocarpus integrifolia*). *Journal of Bioscience*, Vol.14, No.4, (December 1989), 351-365, ISSN 0250-5991

Bhattacharyya, A. & Cherukuri R. Babu, C. (2009). Purification and biochemical characterization of a serine proteinase inhibitor from *Derris trifoliata* Lour. seeds: Insight into structural and antimalarial features. *Phytochemistry*, Vol.70, No.6, (April 2009), pp. 703 712, ISSN 1873 3700

Bhattacharyya, A.; Mzumdar, S.; Leighton, S. M. & Babu, C. R. (2006). A Kunitz proteinase inhibitor from *Archidendron ellipticum* seeds: Purification, characterization, and kinetic properties. *Phytochemistry*, No.67, ISSN 0031-9422

Bhattacharyya, A.; Rai, S. & Babu, C. R. (2007). A trypsin and chymotrypsin inhibitor from *Caesalpinia bonduc* seeds: Isolation, partial characterization and insecticidal properties. *Plant Physiology and Biochemistry*, Vol.45, No.3-4, (March-April 2007), pp. 169-177, ISSN 0981-9428

Bode, W. & Huber, R. (2000). Structural basis of the endoproteinases-protein inhibitor interaction. *Biochimica et Biophysica Acta*, Vol. 1477, No. 1-2, (Mar 1999), pp. 241-252, ISSN 0006-3002

Bolter, C. J. (1993). Methyl jasmonate induces papain inhibitor(s) in tomato leaves. *Plant Physiology*, Vol.103, No.4, (December 1993), pp.1347–1353, ISSN 1532-2548

Bösterling, B. & Quast, U. (1981). Soybean trypsin inhibitor (Kunitz) is doubleheaded. Kinetics of the interaction of alpha-chymotrypsin with each side. *Biochimica et Biophysica Acta*, Vol.657, No.1, (January 1981), pp. 58-72, ISSN 0006-3002

Brik, A. & Wong, C. H. (January 2003). HIV-1 protease: mechanism and drug discovery. Organic and Biomolecular Chemistry, Vol.1, No.1, (January 2003), pp. 5–14, ISSN 1477-0539

Brzin, J. & Kidrič, M. (1995). Proteinases and their inhibitors in plants: role in normal growth and in response to various stress conditions. *Biotechnology and Genetic Engineering Reviews*, Vol. 13, 421–467, (December 1995), ISSN 0264-8725

Brzin, J.; Popovic, T.; Ritonja, A.; Puizdar, V. & Kidric, M. (1998). Related cystatin inhibitors from leaf and from seed of *Phaseolus vulgaris* L. *Plant Science*, Vol.138, No.1, (November 1998), pp. 17–26, ISSN 0168-9452

Cavalcanti, M. S. M.; Oliva, M L V; Fritz, H.; Jochum, M.; Mentele, R.; Sampaio, M.; Coelho, L. C.; Batista, I. F., Sampaio, C. A. (2002). Characterization of a Kunitz trypsin inhibitor with one disulfide bridge purified from *Swartzia pickellii*. *Biochemical and Biophysical Research Communications*, Vol.291, No.3, (March 2002), pp. 635-639.

Charney, J. & Tomarelli, R. M. (1947). A colorimetric method for the determination of the proteolytic activity of duodenal juice. The Journal of Biological Chemistry, Vol.171, No.2, (December 1947), pp. 501-505, ISSN 1083-351X

Chaudhary, N. S.; Shee, C.; Islam, A.; Ahmad, F.; Yernool, D.; Kumar, P. & Sharma, A. K. (2008). Purification and characterization of a trypsin inhibitor from *Putranjiva roxburghii* seeds. *Phytochemistry*. Vol.69, No.11, (August 2008), pp. 2120-2126, ISSN 0031-9422

Chou, K. C. & Cai, Y. D. (2006). Prediction of protease types in a hybridization space. *Biochemical and Biophysical Research Communications*. Vol. 339, No. 3, (January 2006), pp. 1015-1020, ISSN 006-291X

Colebatch, C. M.; East, P. & Cooper, P. (2001). Preliminary characterization of digestive proteases of the green mirid *Creontides dilutus* (Hemiptera:Miridae). *Insect Biochemistry and Molecular Biology*, Vol.31, No.4-5, (March 2001), pp. 415-423, ISSN 0965-1748

Cristofoletti, P. T.; Ribeiro, A. F. & Terra, W. R. (2005). The cathepsin L-like proteinases from the midgut of *Tenebrio molitor* larvae: Sequence, properties, immunocytochemical localization and function. *Insect Biochemistry and Molecular Biology*, Vol.35, No.8, (August 2005), pp. 883-901, ISSN 0965-1748

Cruz-Silva, I; Gozzo, A. J; Nunes, V. A; Carmona, A. K; Faljoni-Alario, A; Oliva, M. L. V; Sampaio, U. M; Sampaio, C. A. M & Araújo, M. S. A (2004). proteinase inhibitor from *Caesalpinia echinata* (pau-Brazil) seeds for plasma kallikrein, plasmin and factor XIIa. *Biological Chemistry*, Vol.385, No.11, (November 2004), pp. 1083-1086, ISSN 1437-4315

Cuatrecasas, P. (1970). Protein Purification by Affinity Chromatography. *The Journal of Biological Chemistry*. Vol.245, No.12, (June 1970), pp. 3059-3065, ISSN 0021-9258

Dattagupta J. K.; Podder, A.; Chakrabarti, C.; Sen, U.; Dutta, S. K. & Singh, M. (1996). Structure of a Kunitz-type chymotrypsin inhibitor from winged bean seeds at 2.95 Å resolution, Acta crystallographica - Section D - Biological crystallography, Vol.52, No.3, (May 1996), pp. 521–528 ISSN 0907-4449

De Vonis Bidlingmeyer, U.; Leary, T. R. & Laskowski, M. Jr.(1972). Identity of the tryptic and alpha-chymotryptic reactive sites on soybean trypsin inhibitor (Kunitz). *Biochemistry*, Vol.11, No.17, (August 1972), pp. 3303-3310, ISSN 0006-2960

Delaage, M. & M. Lazdunski, M. (1968). Trypsinogen, Trypsin, Trypsin-Substrate and Trypsin-Inhibitor Complexes in Urea Solutions. *European Journal of Biochemistry*, Vol.4, No.3, (April 1968), pp. 378–384, ISSN 1432-1033

Derbyshire, E., Wright, D. J., & Boulter, D. (1976). Legumin and Vicilin, Storage Proteins of Legume Seeds. *Phytochemistry*, Vol.15, No.1, (March 2001), pp. 3-24, ISSN 0031-9422

Di Ciero, L., Oliva, M. L.; Torquato, R.; Kohler, R.; Weder, J. K.; Camillo Novello, J.; Sampaio, C. A.; Oliveira, B. & Marangoni, S. (1998). The Complete Amino Acid

Sequence of a Trypsin Inhibitor from *Bauhinia Variegata* Var. Candida Seeds. Journal of Protein Chemistry, Vol.17, No. 8 (November 1998), pp. 827-834, ISSN 0277-8033.

Diop, N. N.; Kidric, M.; Repellin, A.; Gareil, M.; D'arcy-Lameta, A.; Pham, T. A. T. & Zuily-Fodil, Y. A. (2004). multicystatin is induced by drought-stress in cowpea (*Vigna unguiculata* (L.) Walp.) leaves. *FEBS letters*, Vol.577, No.3, (November 2004), pp. 545-550, ISSN 1873-3468

Dunaevsky, Y. E.; Elpidina, E. N.; Vinokurov, K. S. & Belozersky, M. A. (2005). Protease Inhibitors in Improvement of Plant Resistance to Pathogens and Insects. *Molecular Biology*, Vol. 39, No.4, (July 2005), pp. 608-613, ISSN 00268933

Erlanger, B. F.; Kokowsky, N. & Cohen, W. (1961). The preparation and properties of two chromogenic substrates of trypsin. *Archives of biochemistry and biophysics*, Vol.95, No.2, (November 1961), pp. 271–278, ISSN 1096-0384

Fan, S. & Wu, G. (2005). Characteristics of plant proteinase inhibitors and their applications in combating phytophagous insects. *Botanical Bulletin of Academia Sinica*, Vol. 46, No. 4, (October 2005), pp. 273-292, ISSN 0006-8063

Fan, T.J.; Han, L. H.; Cong, R. S. & Liang, J. (2005). Caspase Family Proteases and Apoptosis. *Acta Biochimica et Biophysica Sinica*, Vol.37, No.11, (September 2005), pp. 719–727 ISSN 1672-9145

Fernandes, K. V. S.; Sabelli, P. A.; Barratt, D. H. P.; Richardson, M.; Xavier- Filho, J. & Shewry, P. R. (1993). The resistance of cowpea seeds to bruchid beetles is not related to levels of cysteine proteinase-inhibitors. *Plant Molecular Biology*, Vol.23, No.1, (October 1993), pp. 215–219, ISSN 1573-5028

Finlay, B. B. (1999). Bacterial disease in diverse hosts. *Cell*. Vol. 96, No. 3, (February 1999), pp. 315–318, ISSN 1097-4172

Franco, O. L.; Dias, S. C.; Magalhães, C. P.; Monteiro, A. C. S.; Bloch-Jr, C.; Melo, F. R.; Oliveira-Neto, O. B.; Monnerat, R. G. & Grossi-De-Sá, M. F. (2004). Effects of soybean kunitz trypsin inhibitor on the cotton boll weevil (*Anthonomus grandis*). *Phytochemistry*, Vol.65, No.1, (January 2004), pp. 81-89, ISSN 0031-9422

Franco, O. L.; Grossi de Sá, M. F.; Sales, M. P.; Mello, L. V.; Oliveira, A. S. & Rigden, D. J. (2002). Overlapping binding sites for trypsin and papain on a kunitz-type proteinase inhibitor from *Prosopis juliflora*. *Proteins*, Vol.49, No.3, (November 2002), pp. 335–341, ISSN 1097-0134

Gaddour, K.; Vicente-Carbajosac, C. J.; Larac, P.; Isabel-Lamoneda, I.; Diaz, I. & Carbonero, P. A. (2001). constitutive cystatin-encoding gene from barley (*Icy*) responds differentially to abiotic stimuli. *Plant Molecular Biology*, Vol.45, No.5, (March 2001), pp. 599-608, ISSN 0167-4412

Gatehouse, A. M.; Norton, E.; Davison, G. M.; Barbe, S. M.; Newell, C. A. & Gatehouse, J. A. (1999). Digestive proteolytic activity in larvae of tomato moth, *Lacanobia oleracea*; effects of plant protease inhibitors in vitro and in vivo. *Journal of Insect Physiology*, Vol.45, No.6, (June 1999), pp. 545-558, ISSN 0022-1910

Girard, C.; Le Métayer, M.; Bonadé-Bottino, M.; Pham-Delègue, M. H., & Jouanin, L. (1998). High level of resistance to proteinase inhibitors may be conferred by proteolytic cleavage in beetle larvae. *Insect biochemistry and molecular biology*, Vol.28, No.4, (April 1998), pp. 229-237, ISSN 0965-1748.

Gomes, A. P. G., Dias, S., Bloch, C, Melo, F R, Furtado, J. R., Monnerat, R G, Grossi-de-Sá, M F, et al. (2005). Toxicity to cotton boll weevil Anthonomus grandis of a trypsin inhibitor from chickpea seeds. *Comparative biochemistry and physiology. Part B, Biochemistry & molecular biology, 140*(2), 313-9. doi:10.1016/j.cbpc.2004.10.013

Gomes, C. E. M.; Barbosa, A. E. A. D.; Macedo, L. L. P.; Pitanga, J. C. M.; Moura, F. T.; Oliveira, A. S.; Moura, R. M.; Queiroz, A. F. S.; Macedo, F. P.; Andrade, L. B. S.; Vidal, M. S.; Sales, M. P. (2005). Effect of trypsin inhibitor from *Crotalaria pallida* seeds on *Callosobruchus maculatus* (cowpea weevil) and *Ceratitis capitata* (fruit fly). *Plant physiology and biochemistry*, Vol. 43, No.12, (December 2005), pp. 1095-1102, ISSN 0981-9428

Gomis-Ruth. F. X.; Grams, F.; Yiallouros, I.; Nar, H.; Kusthardt, U.; Zwilling, R.; Bode, W. & Stocker, W. (1994). Crystal structures, spectroscopic features, and catalytic properties of cobalt(II), copper(II), nickel(II), and mercury(II) derivatives of the zinc endopeptidase astacin. A correlation of structure and proteolytic activity," *Journal of Biological Chemistry*, Vol. 269, No. 25, (June 1994), pp. 17111–17117, ISSN 1083-351X

Grudkowska, M. & Zagdańska, B. (2004). Multifunctional role of plant cysteine proteinases. Acta Biochimica Polonica. Vol.51, No.3, (July 2004), pp. 609-624, ISSN 0001-527X

Hedstrom, L. (2002). Serine Protease Mechanism and Specificity. *Chemical Reviews*, Vol.102, No.12, (December 2002), pp. 4501-4524, ISSN 1520-6890

Huang, G. J.; Ho, Y. L.; Chen, H. J.; Chang, Y. S.; Huang, S. S.; Hung, H. J. & Lin, Y. H. (2008). Sweet potato storage root trypsin inhibitor and their peptic hydrolysates exhibited angiotensin converting enzyme inhibitory activity *in vitro*. *Botanical Studies*, Vol.49, No.1, (October 2008), pp. 101-108, ISSN 0006-8063

Hung, C. H.; Lee, M. C. &; Lin, J. Y. (1994). Inactivation of *Acacia confusa* trypsin inhibitor by site-specific mutagenesis. *FEBS Letters*, Vol.353, No.3, (October 1994), pp. 312-314, ISSN 1873-3468

Iwanaga, S.; Yamasaki, N.; Kimura, M. & Kouzuma, Y. (2005). Contribution of conserved Asn residues to the inhibitory activities of kunitz-type protease inhibitors from plants. *Bioscience Biotechnology and Biochemistry*, Vol.69, No.1, (January 2005), pp. 220-223, ISSN 1347-6947

Jacinto, T.; Fernandes, K. V. S.; Machado, O. L. T. & Siqueira-Junior, C. L.(1998). Leaves of transgenic tomato plants overexpressing prosystemin accumulate high levels of cystatin. *Plant Science*, Vol.138, No.1, (November 1998), pp. 35–42, ISSN 0168-9452

Johansson, S.; Göransson, U.; Luijendijk, T.; Backlund, A.; Claeson, P. & Bohlin L. (2002). Neutrophil multitarget functional bioassay to detect anti-inflammatory natural products. *Journal of Natural Products*. Vol. 65, No.1, (January 2002), pp. 32-41, ISSN 0163-3864

Joubert, F. J. (1982). Purification and some properties of two proteinase inhibitors from *Erythrina acanthocarpa* seed. *Journal of Natural Products*, Vol.45, No.4, (July-August 1982), pp. 427-433, ISSN 0163-3864

Kennedy, A. R. (1998). The Bowman-Birk inhibitor from soybeans as an anticarcinogenic agent. *American Journal of Clinical Nutrition*, Vol.68, No.6, (December 1998), pp. 1406-1412, ISSN 1938-3207

Kiggundu, A.; Muchwezi, J.; Van der Vyver, C.; Viljoen, A.; Vorster, J.; Schlüter, U.; Kunert, K. & Michaud, D. (2010). Deleterious effects of plant cystatins against the banana weevil *Cosmopolites sordidus*. *Archives of insect biochemistry and physiology*, Vol.73, No.2, (February 2010), pp. 87-105, ISSN 1520-6327

Kim, S. H.; Hara, S.; Hase, S.; Ikenaka, T.; Toda, H.; Kitamura, K & Kaizuma, N. (1985). Comparative Study on Amino Acid Sequences of Kunitz-Type Soybean Trypsin Inhibitors, *Ti*[a], *Ti*[b], and *Ti*[c]. *The Journal of Biochemistry*, Vol.98, No.2, (August 1985), pp. 435-448, ISSN 1756-2651

Kimura, K.; Ikeda, T.; Fukumoto, D.; Yamasaki, N. & Yonekura, M. (1995). Primary structure of a cysteine proteinase inhibitor from the fruit of avocado (*Persea americana* Mill). *Bioscience Biotechnology and Biochemistry*, Vol.59, No.12, (December 1995), pp. 2328–2329, ISSN 0916-8451

Kimura, M.; Kouzuma, Y. & Yamasaki, N. (1993). Amino acid sequence of chymotrypsin inhibitor ECI from the seeds of *Erythrina variegata* (Linn.) var. Orientalis. *Bioscience, Biotechnology, and Biochemistry*, Vol.57, No.1, (January 1993), pp. 102-106, ISSN 0916-8451

Koiwa, H.; Bressan, R. A. & Hasegawa, P. M. (1997). Regulation of protease inhibitors and plant defense. *Trends in Plant Science*, Vol.2, No.10, (October 1997), pp. 379-384, ISSN 1360-1385

Konarev, A. V.; Lovegrove, A. & Shewry, P. R. (2008). Serine proteinase inhibitors in seeds of *Cycas siamensis* and other gymnosperms. *Phytochemistry*,Vol.69, No.13, (October 2008), pp. 2482-2489, ISSN 0031-9422

Kondo, H.; Abe, K.; Nishimura, I.; Watanabe, H.; Emori, Y. & Arai, S. (1990). Two distinct cystatin in rice seeds with different specificities against cysteine proteinases. *The Journal of Biological Chemistry*, Vol.265, No.26, (September 1990), pp. 15832–15837, ISSN 0021-9258

Kortt, A. A. (1980). Isolation and properties of a chymotrypsin inhibitor from winged bean seed (*Psophocarpus tetragonolobus*(L) Dc.). *Biochimica et Biophysica Acta*, Vol.624, No.1, (July 1980), pp. 237-248, ISSN 0005-2795

Kouzuma, Y.; Inanaga, H.; Doi-Kawano, K.; Yamasaki, N. & Kimura, M. (2000). Molecular cloning and functional expression of cDNA encoding the cysteine proteinase inhibitor with three cystatin domains from sunflower seeds. *Journal of Biochemistry*, Vol.128, No.2, (May 200), pp. 161-166, ISSN 0021-924X

Kouzuma, Y.; Kawano, K.; Kimura, M.; Yamasaki, N.; Kadowaki, T. & Yamamoto, K. (1996). Purification, characterization and sequencing of two cysteine proteinase inhibitors, *Sca* and *Scb*, from sunflower (*Helianthus annuus*) seeds. *Journal of Biochemistry*, Vol.119, No.6, (June 1996), pp. 1106–1113, ISSN 1756-2651

Krauchenco, S.; Nagem, R. A. P.; Silva, J. A.; Marangoni, S & Polikarpov, I. (2004). Three-dimensional structure of na unusual kunitz (STI) type trypsin inhibitor from *Copaifera langsdorfii*. *Biochimie*, Vol.86, No.3, (March 2004), pp. 167-172, ISSN 0300-9084

Krauchenco, S.; Pando, S. C.; Marangoni, S. & Polikarpov, I. (2003). Crystal structure of the kunitz (STI)-type inhibitor from *Delonix regia* seeds. *Biochemical and Biophysical Research Communications*, Vol.312, No.4, (December 2003), pp. 1303-1308, ISSN 1090-2104

Krowarsch, D.; Cierpicki, T.; Jelen, F. & Otlewski, J. (2003). Canonical protein inhibitors of serine proteases. *Cellular and molecular life sciences*, Vol.60, No.11, (November 2003), pp. 2427-2444, ISNN 1420-682X

Lam, W.; Coast, G. M. & Rayne, R. C. (1999). Isolation and characterization of two chymotrypsin from the midgut of *Lacusta migratoria. Insect Biochemistry and Molecular Biology,* Vol.29, No.7, (July 1999), pp. 653-660, ISSN 0965-1748

Lam, W.; Coast, G. M. & Rayne, R. C. (2000). Characterization of multiple trypsins from the midgut of *Lacusta migratoria. Insect Biochemistry and Molecular Biology,* Vol.30, No.1, (January 2000), pp. 85-94, ISSN 0965-1748

Laskowski, M. & Kato, I. (1980). Protein inhibitors of proteinases. *Annual review of biochemistry,* Vol. 49, No. 1, (July 1980), pp. 593-626, ISSN 0066-4154

Lawrence, P. L. & Koundal, K. R. (2002). Plant protease inhibitors in control of phytophagous insects. *Electronic Journal of Biotechnology,* Vol.5, No.1, (April 2002), pp. 93-109, ISSN 0717-3458

Lemos, F. J. A.; Campos, F. A. P. Silva, C. P. & Xavier-Filho, J. (1990). Proteinases and amylases of larval midgut of Zabrotes subfasciatus reared on cowpea (*Vigna unguiculata*) seeds. *Entomologia Experimentalis et Applicata,* Vol.56, No.3, (September 1990), pp. 219-227, ISSN 0013-8703

Leung, D.; Abbenante, G. & Fairlie, D. P. (2000). Protease Inhibitors: Current Status and Future Prospects. Journal of Medicinal Chemistry, Vol. 43, No.3, (February 2000), pp. 305-341, ISSN 1520-4804

Levi, M.; Levy, J. H.; Andersen, H. F. & Truloff, D. (2010). Safety of recombinant activated factor VII in randomized clinical trials. *The New England Journal of Medicine.* Vol. 363, No. 19, (November 2010), pp. 1791–800. ISSN 1533-4406

Levi, M.; Poll, T. V. D. & Cate, H. T. (2006). Tissue factor in infection and severe inflammation. *Seminars in thrombosis and hemostasis.* Vol. 132, No. 1, (February 2006), pp. 33–39, ISSN 1098-9064

Levleva, E. V.; Rudenskaya, Y. A.; Dunaevsky, Y. E. & Mosolov, V. V. (1997). A 7.5-kD inhibitor of cysteine proteinases from pumpkin seeds. *Biochemistry,* Vol.62, No.5, (Febrary 1997), pp. 551-556, ISSN 0006-2979

Lim, C. O., Lee, S. I., Chung, W. S., Park, S. H., Hwang, I. & Cho, M. J. (1996). Characterization of a cDNA encoding cysteine proteinase inhibitor from Chinese cabbage (*Brassica campestris* L. ssp. pekinensis) flower buds. *Plant Molecular Biology,* Vol.30, No.2, (January 1996), pp. 373-379, ISSN 0167-4412

Lingaraju, M. H. & Gowda, L. R. (2008). A Kunitz trypsin inhibitor of *Entada scandens* seeds: another member with single disulfide bridge. *Biochimica et Biophysica Acta,* Vol.1785, No.5, (May 2008), PP. 850-855, ISSN 0006-3002

Liou, T.G. & Campbell, E. J. (1995). Nonisotropic enzyme--inhibitor interactions: a novel nonoxidative mechanism for quantum proteolysis by human neutrophils. *Biochemistry,* Vol. 34, No. 49, (December 1995), pp. 16171-16177, ISSN 1520-4995

Liu, Y.;Salzman, R. A.; Pankiw, T. & Zhu-Salzman, K. Transcriptional regulation in southern corn rootworm larvae challenged by soyacystatin N. *Insect Biochemistry and Molecular Biology,* Vol.34, No.10, (October 2004), pp. 1069-1077, ISSN 0965-1748

Macedo, M. L. R.; de Sa, C. M.; Freire, M. G. M. & Parra, J. R. P. (2004). A kunitz-type inhibitor of Coleopteran proteases, isolated from *Adenanthera pavonina* L. seeds and its effect on *Callosobruchus maculatus. Journal of Agricultural and Food Chemistry,* Vol.52, No.9, (May 2004), pp. 2533-2540, ISSN 1520-5118

Macedo, M. L. R.; Filho, E. B. S. D.; Freire, M. G. M.; Oliva, M. L. V.; Sumikawa, J. T.; Toyama, M. H. & Marangoni, S. (2011). A trypsin inhibitor from *Sapindus saponaria*

L. seeds: purification, characterization, and activity towards pest insect digestive enzyme. *The protein journal*, Vol.30, No.1, (January 2011), pp. 9–19, ISSN 1875-8355

Macedo, M. L. R.; Garcia, V. A.; Freire, M. G. M. & Richardson, M. (2007). Characterization of a kunitz trypsin inhibitor with a single disulfide bridge from seeds of *Ingra laurina* (SW.) Willd. *Phytochemistry*, Vol.68, No.8, (April 2007), pp. 1104-1111, ISSN 1873-3700

Macedo, M. L. R.; Mello, G. C.; Freire, M. G. M.; Novello, J. C.; Marangoni, S. & Matos, D. G. G. (2002). Effect of a trypsin from *Dimorphandra mollis* seeds on the development of *Callosobruchus maculatus*. *Plant Physiology and Biochemistry*, Vol.40, No.10, (October 2002). pp. 891-898, ISSN 0981-9428

Macedo, M. L. R; Freire, M. G. M; Cabrini, E. C; Toyama, M. H; Novello, J. C. & Marangoni, S. (2003). A trypsin inhibitor from *Peltophorum dubium* seeds active against pest proteases and its effect on the survival of *Anagasta Kuehniella*. *Biochimica et Biophysica Acta*. Vol.1621, No. 2, (May 2003), pp. 170-182, ISSN 0006-3002

Mandal, S.; Kundu, P; Roy, B. & Mandal, R. K. (2002). Precursor of the inactive 2S seed storage protein from the Indian mustard *Brassica juncea* is a novel trypsin inhibitor. Charaterization, post-translational processing studies, and transgenic expression to develop insect-resistant plants. *The Journal of biological chemistry*, Vol.277, No.40, (October 2002), pp. 37161-37168, ISSN 1083-351X

Margis, R.; Reis, E. M. & Villet, V. (1998). Structural and phylogenetic relationships among plant and animal cystatins. *Archives of Biochemistry and Biophysics*, Vol.359, No.1, (November 1998), pp. 24-30, ISSN 1096-0384

Martinez, M.; Abraham, Z.; Gambardella, M.; Echaide, M.; Carbonero, P. & Diaz, I. (2005). The strawberry gene *Cyf1* encodes a phytocystatin with antifungal properties. *Journal of Experimental Botany*, Vol.56, No.417, (July 2005), pp. 1821-1829, ISSN 0022-0957

Masoud, S. A.; Johnson, L. B.; White, F. F. & Reeck, G. R. (1993). Expression of a Cysteine Proteinase Inhibitor (Oryzacystatin-I) in Transgenic Tobacco Plants. *Plant Molecular Biology*, Vol.21, No.4, (February1993), pp. 655-663, ISSN 0167-4412

Mello, G. C; Oliva, M. L. V; Sumikawa, J. T; Machado, O. L. T; Marangoni, S; Novello, J. C & Macedo, M. L. R (2001). Purification and characterization of a new trypsin inhibitor from *Dimorphandra mollis* seeds. *Journal of Protein Chemistry*, Vol.20, No.8, (November 2001), pp. 625-632, ISSN 0277-8033

Mello, M. O; Tanaka, A. S & Silva-Filho, M. C. (2003). Molecular evolution of Bowman-Birk type proteinase inhibitors in flowering plants. *Molecular Phylogenetics and Evolution*, Vol.27, No.1, (April 2003), pp. 103-112, ISSN 1095-9513

Migliolo, L.; Oliveira, S. O.; Santos, E. A.; Franco, O. L. & Sales, M. P. (2010). Structural and mechanistic insights into a novel non-competitive Kunitz trypsin inhibitor from Adenanthera pavonina L. seeds with double activity toward serine- and cysteine-proteinases. Journal of molecular graphics & modeling, Vol.29, No.2, (September 2010), pp. 148-156, ISSN 1093-3263

Misaka, T.; Kuroda, M.; Iwabuchi, K.; Abe, K. & Arai, S. (1996). Soyacystatin, a novel cysteine proteinase inhibitor in soybean, is distinct in protein structure and gene organization from other cystatins of animal and plant origin. *European Journal of Biochemistry*, Vol.240, No.3, (September 1996), pp. 609-614, ISSN 0014-2956

Monti, R.; Contiero, J. & Goulart, A. J. (2004). Isolation of natural inhibitors of papain obtained from *Carica papaya* Latex. *Brazilian Archives of Biology and Technology*. Vol.47, No. 5, (September 2004), pp. 747-754, ISSN 1516-8913

Nakahata, A. M.; Bueno, N. R.; Rocha, H. A. O., Franco, C. R. C.; Chammas, R.; Nakaie, C. R.; Josiutionis, M. G.; Nader, H. B.; Santana, L. A.; Sampaio, U. M. & Oliva, M. L. V. (2006). Structural and inhibitory properties of a plant proteinase inhibitor containing the RGD motif. *International Journal of Biological Macromolecules*, Vol.40, No.1, (Dezember 2006), pp. 22-29, ISSN 1879-0003

Nanninga, L. B & Guest, M. M. (1964). On the interaction of fibrinolysin (plasmin) with the inhibitors antifibrinolysin and soybean trypsin inhibitor. *Archives of biochemistry and biophysics*, Vol.108, (December 1964), pp. 542-551, ISSN 0003-9861

Norioka, N.; Hara, S.; lkenaka, T. & Abe, J. (1988). Distribution of the Kunitz and the Bowman-Birk family proteinase inhibitors in leguminous seeds. *Agricultural and Biological Chemistry*, Vol.52, No.5, (May 1988), pp. 1245-1252, ISSN 0002-1369

Novillo, C.; Castanero, P. & Ortego, F. (1997). Inhibition of digestive trypsin-like proteases from larvae of several Lepidopteran species by the diagnostic cysteine protease inhibitor E-64. *Insect Biochemistry and Molecular Biology*, Vol.27, No.3 (February 1997), pp. 247-254, ISSN 0965-1748

Odani, S. & Ikenaka, T. (1973a). Studies on soybean trypsin inhibitor. 8. Dissulfide bridges in soybean Bowman-Birk proteinase inhibitor. *Journal Biochemistry*, Vol.74, No.4, (October 1973), pp. 697-715, ISSN 1756-2651

Odani, S. & Ikenaka, T. (1973b). Scission of soybean Bowman-Birk proteinase inhibitor into two small fragments having either trypsin or chymotrypsin inhibitory activity. *Journal Biochemistry*, Vol.74, No.4, (October 1973), pp. 857-860, ISSN 1756-2651

Odani, S.; Yokokawa, Y.; Takeda, H.; Abe, S. & Odani, S. (1996). The primary structure and characterization of carbohydrate chains of the extracellular glycoproteinase inhibitor from latex of *Carica papaya*. *European Journal of Biochemistry*, Vol.241, No.1, (October 1996), pp. 70-82, ISSN 0014-2956

Ojima, A.; Shiota, H.; Higashi, K.; Kamada, H.; Shimma, Y.; Wada, M. & Satoh, S. (1997). An extracellular insoluble inhibitor of cysteine proteinases in cell culture and seeds of carrot. *Plant Molecular Biology*, Vol.34, No.1, (May 1997), PP.99–109, ISSN 0167-4412

Oliva, M. L. V; Souza-Pinto, J. C; Batista, I. F. C; Araújo, M. S; Silveira, V. F; Auerswald, E. A; Mentele, R; Eckerskorn, C; Sampaio, U. M & Sampaio, C. A. M. (2000). *Leucaena leucocephala* serine proteinase inhibitor: prymary structure and action on blood coagulation, kinin release and paw edema. *Biochimica et Biophysica Acta*, Vol.1477, No.1-2, (March 2000), pp. 64-74, ISSN 0006-3002

Oliva, M. L.; Mendes, C. R.; Santomauro-Vaz, E. M.; Juliano, M. A.; Mentele, R.; Auerswald, E. A.; Sampaio, M. U. & Sampaio, C. A. (2001). *Bauhinia bauhinioides* plasma kallikrein inhibitor: interaction with synthetic peptides and fluorogenic peptide substrates related to the reactive site sequence. *Current medicinal chemistry*, Vol.8, No.8, (July 2001), pp. 977-984, ISSN 0929-8673

Oliva, M. L;, Sallai, R. C.; Sampaio, C. A.; Fritz, H.; Auerswald, E. A.; Tanaka, A. S.; Torquato, R. J. & Sampaio, M. U. (1996). Bauhinia serine proteinase inhibitors: effect on factor X, factor XII and plasma kallikrein. *Immunopharmacology*, Vol. 32, No.1-3, (May 1996), pp. 85-87, ISSN 0162-3109

Oliveira, A. S.; Pereira, R. A.; Lima, L. M.; Morais, A. H. A.; Melo, F. R.; Franco, O. L.; Bloch Jr, C.; Grossi de Sá, M. F &, Sales, M. P. (2002). Activity toward bruchid pest of a kunitz-type inhibitor from seeds of the algaroba tree (*Prosopis juliflora* D.C.). *Pesticide Biochemistry and Physiology*. 2002, Vol.72, No.2, (February 2002), pp. 122–132, ISSN 0048-3575

Oliveira, A. S; Migliolo, L.; Aquino, R. O, Ribeiro, J. K. C.; Macedo, L. L. P.; Andrade, L. B. S.; Bemquerer, M. P.; Santos, E. A.; Kiyota, S. & Sales, M. P. (2007b). Identification of a Kunitz-type proteinase inhibitor from *Pithecellobium dumosum* seeds with insecticidal properties and double activity. *Journal of agricultural and food chemistry*, Vol. 55, No. 18, (September 2007), pp. 7342-7349. ISSN 1520-5118

Oliveira, A. S; Migliolo, L.; Aquino, R. O; Ribeiro, J. K. C; Macedo, L. L. P.; Andrade, L. B. S.; Bemquerer, M. P.; Santos, E. A.; Kiyota, S. & Sales, M. P. (2007a). Purification and characterization of a trypsin-papain inhibitor from *Pithecelobium dumosum* seeds and its in vitro effects towards digestive enzymes from insect pests. *Plant physiology and biochemistry*, Vol.45, No.10-11, (October-November 2007), pp. 858-865. ISSN 0981-9428

Oliveira, A. S; Migliolo, L; Aquino, R.O; Ribeiro, J. K.; Macedo, L. L.; Bemquerer, M. P.; Santos, E. A, Kiyota, S & Sales, M. P. (2009). Two Kunitz-type inhibitors with activity against trypsin and papain from Pithecellobium dumosum seeds: purification, characterization, and activity towards pest insect digestive enzyme. *Protein and peptide letters*, Vol.16, No.12, (December 2009), pp. 1526-32, ISSN 0929-8665

Oliveira-Neto, O. B.; Batista, J. A.; Rigden, D. J.; Fragoso, R. R.; Silva, R. O.; Gomes, E. A.; Franco, O. L.; Dias, S. C.; Cordeiro, C. M.; Monnerat, R. G. & Grossi-De-Sa, M. F. (2004).A diverse family of serine proteinase genes expressed in cotton boll weevil (*Anthonomus grandis*): implications for the design of pest-resistant transgenic cotton plants. *Insect Biochemistry and Molecular Biology*, Vol.34, No.9, (September 2004), pp. 903-918, ISNN 0965-1748

Otto, H. H. & Schirmeister, T. (1997). Cysteine proteases and their inhibitors. *Chemical Reviews*, Vol.97, No.1, (February 1997), pp. 133-171, ISSN 1520-6890

Pando, L. A.; Di Ciero, L.; Novello, J. C.; Oliveira, B.; Weder, J. K. & Marangoni, S. (1999). Isolation and characterization of a new trypsin inhibitor from *Crotalaria paulina* seeds. International Union of Biochemistry and Molecular Biology Life, Vol.48, No.5, (November 1999), pp. 519-523, ISSN 1521-6543

Pando, S. C; Oliva, M. L. V; Sampaio, C. A. M; Di Ciero, L; Novello, J. C (2001). Marangoni, S. (2001). Prymary sequence determination of a Kunitz inhibitor siolated from *Delonix regia* seeds. *Phytochemistry*, Vol.57, No.5, (July 2001), pp. 625-631, ISSN 0031-9422

Pearl, L. 1987. Sequence specificity of retroviral proteases. *Nature*, Vol.328, No.6130, (August 1987), pp. 351-354, ISSN 1476-4687

Pernas, M.; Sanchez-Monge, R.; Gómez, L. & Salcedo, G. A. (1998). Chestnut seed cystatin differentially effective against cysteine proteinases from closely related pests. *Plant Molecular Biology*, Vol.38, No.6, (December 1998), pp. 1235–1242, ISSN 0167-4412

Power, S. D.; Adams, R. M., & Wells, J. A. (1986). Secretion and autoproteolytic maturation of subtilisin. *Proceedings of the National Academy of Sciences of the United States of America*, Vol. *83*, No. 10, (May 1986), pp. 3096-100, ISSN 1091-6490

Powers, J. C.; Asgian, J. L.; Ekici, O. D. & James, K. E. (2002). Irreversible inhibitors of serine, cysteine, and threonine proteases. *Chemistry* Reviews, Vol. 102, No. 12, (November 8), pp. 4639-4750, ISSN 0009-2665

Powers, J. C.; Gupton, B. F.; Harley, A. D.; Nishino, N. & Whitley, R. J. (1977). Specificity of Porcine Pancreatic Elastase, Human Leukocyte Elastase and Cathepsin G. Inhibition with Peptide Chloromethyl Ketones. *Biochimica et Biophysica Acta Enzymology*, Vol.485, No.1, (November 1977), pp. 156-166, ISSN 0005-2744

Prabhu, K. S. & Pattabiraman, T. N. (1980). Natural plant enzyme inhibitors. Isolation and characterization of a trypsin/chymotrypsin inhibitor from Indian red wood (*Adenanthera pavonina*) seeds *Journal of the Science of Food and Agriculture*, Vol.31, No.10, (October 1980), pp. 967-980, ISSN 1097-0010

Prakash, B.; Selvaraj, S. S.; Murthy, M. R. N.; Sreerama, Y. N.; Rao, D. & Gowda, L. R. (1996). Analysis of the amino acid sequences of plant Bowman-Birk inhibitors. *Journal Molecular Evolution*, Vol.42, No.5, (May 1996), pp. 558-560, ISSN 0022-2844

Purcell, J. P.; Greenplate, J. T. & Sammons, R. D. (1992). Examination of midgut luminal proteinase activities in six economically important insects. *Insect Biochemistry and Molecular Biology*, Vol.22, No.1, (January 1992), pp. 41-47, ISSN 0965-1748

Pusztai, A. (1972). Metabolism of trypsin-inhibitory proteins in the germinating seeds of kidney bean (*Phaseolus vulgaris*), *Planta*, Vol.107, No.2, (May 1972), pp. 121-129, ISSN 1432-2048

Qi, R. F; Song, Z. W & Chi, C. W. (2005). Structural features and molecular evolution of Bowman-Birk protease inhibitors and their potential application. *Acta Biochimica et Biophysica Sinica*, Vol.37, No.5, (May 2005), pp. 283-292, ISSN 1745-7270

Rai, S., Aggarwal, K. K. & Babu, C. R. (2008). Isolation of a serine Kunitz trypsin inhibitor from leaves of *Terminalia arjuna*. *Current Science*, Vol.94, No.11, (June 2008), pp. 1509-1512, ISSN 0011-3891

Ramos, V. S.; Freire, M. G.; Parra, J. R. & Macedo, M. L. (2009). Regulatory effects of an inhibitor from *Plathymenia foliolosa* seeds on the larval development of *Anagasta kuehniella* (Lepidoptera). *Comparative biochemistry and physiology, Molecular & integrative physiology*, Vol.152, No.2, (Febrary 2009), pp. 255-261 ISSN 1095-6433

Ramos, V. S.; Silva, G. S.; Freire, M. G.; Machado, O. L.; Parra, J. R. & Macedo, M. L. (2008). Purification and characterization of a trypsin inhibitor from *Plathymenia foliolosa* seeds. *Journal of Agricultural and Food Chemistry*, Vol.56, No.23, (December 2008), pp. 11348-11355, ISSN 0021-8561

Rawlings, N D, & Barret, A. J. (1994). Families of serine peptidases. *Methods in Enzymology*, *244*, 19-61.

Rawlings, N. D.; Barrett, A. J. & Bateman, A. (2010). Merops: the peptidase database. *Nucleic Acids Research*, Vol.38, No. 13 Database issue, (January 2010), pp. D227–D233, ISSN (http://merops.sanger.ac.uk)

Rogelj, B.; Popovic, T.; Ritonja, A.; Strukelj, B. & Brzin, J. (1998). Chelidocystatin, a novel phytocystatin from *Chelidonium majus*. *Phytochemistry*, Vol.49, No.6, (November 1998), pp. 1645-1649, ISSN 0031-9422

Rogers, B. L.; Pollock, J.; Klapper, D. G. & Griffith, I. J. (1993). Sequence of the proteinase-inhibitor cystatin homologue from the pollen of *Ambrosia artemisiifolia* (short ragweed). *Gene*, Vol.133, No.2, (November 1993), pp. 219-221, ISSN 0378-1119

Rosenberg, R. C.; Root, C. A.; Bernstein, P. K. & Gray, H. B. (1975). Spectral studies of copper(II) carboxypeptidase A and related model complexes, *Journal of the American Chemical Society*, Vol. 97, No. 8, (April 1975), pp. 2092–2096, ISSN 1520-5126

Ryan, C. A. (1990). Protease inhibitor in plants: Genes for Improving defenses against insects and pathogens. *Annual Review of Phytopathology*, Vol.28, No.1, (September 1990), pp. 425-449, ISSN 00664286

Ryan, S. N., Laing, W. a, & McManus, M. T. (1998). A cysteine proteinase inhibitor purified from apple fruit. *Phytochemistry*, Vol.49, No.4, (October 1998), pp. 957-963, ISSN 0031-9422

Salas, C. E.; Gomes, M. T.; Hernandez, M. & Lopes, M. T. (2008). Plant cysteine proteinases: evaluation of the pharmacological activity. Phytochemistry, Vol.69, No.12, (July 2008), PP. 2263-2269, ISNN 0031-9422

Schuler, T. H.; Poppy, G. M.; Kerry, B. R. & Denholm, I. (1998). Insect-resistant transgenic plants. *Trends in Biotechnology*, Vol.16, No.4, (April 1998), pp. 168-175, ISSN 0167-7799

Schumaker, T. T. S.; Cristofoletti, P. T. & Terra, W. R. (1993). Properties and compartmentalization of digestive carbohydrases and proteases in *Scaptotrigona bipunctata* (Apidae: Meliponinae) larvae. *Apidologie*, Vol.24, No.1, (September 1992), pp. 3-17, ISSN 1297-9678

Scott, C. J. & Taggart, C. C. (2010). Biologic protease inhibitors as novel therapeutic agents. *Biochimie*, Vol.92, No.11, (November 2010), pp. 1681-1688, ISSN 1638-6183

Seals, D. F. & Courtneidge, S. A. (2003). The ADAMs family of metalloproteases: multidomain proteins with multiple functions. *Genes & Development*, Vol.17, No., (January 2003), pp. 7–30, ISSN 0890-9369

Sessa, D. J. & Wolf, W. J. (2001). Bowman-Birk Inhibitors in Soybean Seed Coats. *Industrial crops and products*, Vol.4, No.1, (July 2001), pp. 73-78, ISSN 0926-6690

Shapiro, S. D. (2002). Neutrophil elastase path clearer, pathogen killer, or Just pathologic? *American* Journal of Respiratory *Cell* and *Molecular Biology*. Vol. 26, No. 3, (March 2002), pp. 266–268, ISSN 1535-4989

Shee, C & Sharma, A. K. (2007). Purification and characterization of a trypsin inhibitor from seeds of Murraya koenigii. *Journal of Enzyme Inhibition and Medicinal Chemistry*, Vol.22, No.1, (Febrary 2007), pp. 115-120, ISSN 1475-6374

Shewry, P.R. (2003). Tuber storage proteins. Annals of Botany, Vol.91, No.7, (April 2003), pp. 755-769, ISSN 1095-8290

Silva, C. B. L. F.; Alcazar, A. A.; Macedo, L. L. P.; Oliveira, A. S.; Macedo, F. P.; Abreu, L. R. D.; Santos, E. A. & Sales, M. P. (2006). Digestive enzymes during development of *Ceratitis capitata* (Díptera:Tephritidae) and effects of SBTI on its digestive serine proteinase targets. *Insect Biochemistry and Molecular Biology*, Vol.36, No.7, (July 2006), pp. 561-569, ISSN 0965-1748

Silva, C. P. & Xavier-Filho, J. (1991). Comparasion between the levels of aspartic and cysteine proteinases of the larval midguts of *Callosobruchus maculatus* (F.) and *Zabrotes subfasciatus* (Boh.)(Coleoptera: Bruchidae). *Comparative Biochemistry and Physiology - Biochemistry & Molecular Biology*, Vol.99, No.3, (December 1990), pp. 529-533, ISSN 1096-4959

Silva, C. P.; Terra, W. R, & Lima, R. M. (2001). Differences in midgut serine proteinases from larvae of the bruchid beetles *Callosobruchus maculatus* and *Zabrotes subfasciatus*.

Archives of Insect Biochemistry and Physiology, Vol.47, No.1, (May 2001), pp. 18-28, ISSN 0739-4462

Simões, I. & Faro, C. (2004). Structure and function of plant aspartic proteinases. European journal of biochemistry, Vol. 271, N.11, (June 2004), pp. 2067-2075, ISSN 1432-1033

Sipos, T. & Merkel, J. R., (1970). An effect of calcium ions on the activity, heat stability, and structure of trypsin. *Biochemistry*, Vol.9, No.14, (July 1970), pp. 2766-2775, ISNN 1520-4995

Song, H. K. & Suh, S. W.(1998). Kunitz type soybean trypsin inhibitor revisited: refined structure of its complex with porcine trypsin reveals an insight into the interaction between a homologous inhibitor from *Erythrina caffra* and tissue type plasminogen activator. *Journal of Molecular Biology*, Vol.275, No.2, (January 1998), pp. 347-363, ISSN 0022-2836

Song, H. K.; Lee, K. N.; Kwon, K. S.; Yu, M. H. & Suh, S. W. (1995). Crystal Structure of an Uncleaved Alpha 1-Antitrypsin Reveals the Conformation of Its Inhibitory Reactive Loop, *FEBS Lett*, Vol.377, No.2, (December 1995), pp. 150-154, ISSN 0014-5793

Souza, E. M. T.; Mizuta, K.; Sampaio, M. U. & Sampaio, C. M. (1995). Purification and partial characterization of a *Schizolobium parahyba* chymotrypsin inhibitor. *Phytochemistry*, Vol. 39, No.3, (June 1995), pp. 521-525, ISSN 1873-3700

Souza, E. M. T.; Teles, R. C. L.; Siqueira, E. M. A. & Freitas, S. M. (2000). Effects of denaturing and stabilizing agents on the inhibitory activity and conformational stability of *Schizolobium parahyba* chymotrypsin inhibitor. *Journal of Protein Chemistry*, Vol. 19, No.6, (August 2000), pp. 507-513, ISSN 0277-8033

Stocker, W. & W. Bode. (1995). Structural Features of a Superfamily of Zinc-Endopeptidases: The Metzincins. *Curr Opin Struct Biol*, Vol.5, No.3, (June 1995), pp. 383-390, ISSN 0959-440X

Stubbs, M.T.; Laber, B.; Bode, W.; Huber, R.; Jerala, R.; Lenarcic, B. & Turk, V. (1990). The refined 2.4 A X-ray crystal structure of recombinant human stefin B in complex with the cysteine proteinase papain: a novel type of proteinase inhibitor interaction. *The EMBO Journal*, Vol.9, No.6, (June 1990), pp. 1939–1947, ISSN *0261-4189*

Sumikawa, J. T.; Nakahata, A. M.; Fritz, H.; Mentele, R.; Sampaio, M. U. & Oliva, M. L. A. (2006). Kunitz-type glycosylated elastase inhibitor with one disulfide bridge. *Planta Medica*, Vol.72, No.5, (April 2006), pp. 393-397, ISSN 0032-0943

Turk, B., Turk, D. & Turk, V. (2000). Lysosomal cysteine proteases: more than scavengers. *Biochimica et Biophysica Acta*, Vol. 1477, No.1-2, (Mar 2000), pp. 98-111, ISSN 0006-3002

Ui, N., Isoelectric points and conformation of proteins. II. Isoelectric focusing of a-chymotrypsin and its inactive derivative. *Biochimica et Biophysica Acta - Protein Structure*, Vol.229, No.3, (March 1971), pp. 582-589 ISSN 0005-2795

Ussuf, K. K., Laxmi, N. H., & Mitra, R. (2001). Proteinase inhibitors: Plant-derived genes of insecticidal protein for developing insect-resistant transgenic plants. *Current Science*, Vol.80, No.7, (April 2001), pp. 847-853, ISSN 00113891

Valueva, T. A., Revina, T. A., & Mosolov, V. V. (1999). Reactive sites of the 21-kD protein inhibitor of serine proteinases from potato tubers. *Biochemistry*, Vol.64, No.9, (September 1999), pp. 1074-1078, ISSN 0006-2979

Valueva, T. A.; Revina, T. A & Mosolov, V. V. (1997). Potato tuber protein proteinase inhibitor belonging to the kunitz soybean inhibitor family. *Biochemistry*, Vol.62, No.12, (December 1997), pp. 1367-1600, ISSN 1608-3040

Valueva, T. A.; Revina, T. A.; Kladnitskaya, G. V. & Moslov, V. V. (1998). Kunitz-type proteinase inhibitors from intact and Phytophthora-infected potato tubers. *FEBS letters*, Vol.426, No.1, (April 1998), pp. 131-134, ISSN 1873-3468

Waldron, C.; Wegrich, L. M.; Merlo, P. A. O. & Walsch, T .A. (1993). Characterization of a genomic sequence coding for potato multicystatin, an eight domain cysteine proteinase inhibitor. *Plant Molecular Biology*. Vol.23, No.4, (November 1993), pp. 801–812, ISSN 0167-4412

Walsh, K. (1970). Trypsinogens and trypsins of various species. In G. Perlmann & L. Lorand (Eds.), Methods of Enzymology, 19, 41. New York: Academic Press.

Walsh, K.: Trypsinogens and Trypsins of Various Species, *Methods in Enzymology Vol. 19*,G. Perlmann and L. Lorand, Academic Press, NY, 41, 1970

Wu, C. & Whitaker, J. R. (1991). Homology among trypsin/chymotrypsin inhibitors from red kidney bean, Brazilian pink bean, lima bean and soybean; *Journal of Agricultural and Food Chemistry*, Vol.39, No.9, (September 1991), pp. 1583-1589, ISSN 0021-8561

Wu, J. & Haard, J. N. (2000). Purification and characterization of a cystatin from the leaves of methyl jasmonate treated tomato plants. *Comparative Biochemistry and Physiology. Toxicology & Pharmacology*, Vol.127, No.2, (Sptember 2000), pp. 209–220, ISSN 1532-0456

Xavier-Filho, J. (1992). The biological roles of serine and cysteine proteinase inhibitors in plants. *Revista Brasileira de Fisiologia Vegetal*, Vol. 4, No.1, (Junho 1992), pp. 1-6, ISSN 0103-3131

Xavier-Filho, J.; Campos, F. A. P.; Ary, M. B.; Silva, C. P.; Carvalho, M. M. M.; Macedo, M. L. R.; Lemos, F. J. A.; Grant, G. (1989). Poor correlation between the levels of proteinase inhibitors found in seeds of different cultivars of cowpea (*Vigna unguiculata*) and the resistance/susceptibility to predation by *Callosobruchus maculatus*. *Journal of Agricultural and Food Chemistry*, Vol.37, No.4, (July 1989), pp. 1139-1143, ISSN 0021-8561

Xi, F. N., Wu, J. M., & Luan, M. M. (2005). Silica-supported Macroporous Chitosan Bead for Affinity Purification of Trypsin Inhibitor. *Chinese Chemical Letters*, Vol.16. No.8, pp. 1089-1092, ISSN1001-8417

Xu, F. X. & Chye, M. L. (1999). Expression of cysteine proteinase during developmental events associated with programmed cell death in brinjal. *The Plant Journal*, Vol. 17, No. 3, (January 1999), pp. 321 -327, ISSN 1365-313X

Yoshizaki, L.; Troncoso, M. F.; Lopes, J. L.; Hellman, U.; Beltramini, L. M. & Wolfenstein-Todel, C. (2007). *Calliandra selloi* Macbride trypsin inhibitor: isolation, characterization, stability, spectroscopic analyses. *Phytochemistry.*, Vol.68, No.21, (November 2007), pp. 2625-2634. ISSN 0031-9422

Zhang, H.; Gu, Y. T. & Xue, Y. X. (2007). Bradykinin-induced blood-brain tumor barrier permeability increase is mediated by adenosine 5'-triphosphate-sensitive potassium channel. *Brain Research*, Vol.1144, No.4, (May 2007), pp. 34-41, ISSN 1872-6240

Zhao, Y.; Botella, M. A.; Subramanian, L.; Niu, X.; Nielsen, S. S.; Bressan, R. A. & Hasegawa, P. M. (1996). Two wound-inducible soybean cysteine proteinase have greater insect digestive proteinase inhibitory activities than a constitutive homolog. *Plant Physiology*, Vol.111, No.4, (August 1996), pp. 1299–1306, ISSN 0032-0889

Zhen, E. Y.; Brittain, I. J.; Laska, D. A.; Mitchell, P. G.; Sumer, E.U.; Karsdal, M. A.& Duffin, K. L. (2008). Characterization of metalloprotease cleavage products of human articular cartilage. *Arthritis and Rheumatism*, Vol.58, No.8, (August 2008), pp. 2420-2431, ISSN 1529-0131

Zhou, J. Y.; Liao, H.; Zhang, N. H.; Tang, L.; Xu, Y. & Chen, F. (2008). Identification of a Kunitz inhibitor from *Albizzia kalkora* and its inhibitory effect against pest midgut proteases. *Biotechnology letters*, Vol.30, No.8, (August 2008), pp. 1495–1499, ISSN 0141-5492

Zhu, Y. C. & Baker, J. E. (1999). Characterization of midgut trypsin-like enzymes and three trypsinogen cDNAs from the lesser grain borer, *Rhyzopertha dominica* (Coleoptera: Bostrichidae). *Insect Biochemistry and Molecular Biology*, Vol.29, No.12, (Dezember 1999), pp. 1053-1063, ISSN 0965-1748

Zhu, Y. C. & Baker, J. E. (2000). Molecular cloning and characterization of a midgut chymotrypsin-like enzyme from the lesser grain borer, *Rhyzopertha dominica*. *Archives of insect biochemistry and physiology*, Vol.43, No.4, (April 2000), pp. 173-184, ISSN 0739-4462

Zinkler, D. & Polzer, M. (1992). Identification and characterization of digestive proteinases from the firebrat *Thermobia domestica*. *Comparative Biochemistry and Physiology - Biochemistry & Molecular Biology*, Vol. 103, No. 3, (November 1992), pp. 669-673, ISSN 03050491

13

Identification of cGMP-Kinase Complexes by Affinity Chromatography

Salb Katharina, Schinner Elisabeth and Schlossmann Jens*
University Regensburg, Pharmacology and Toxicology, Institute of Pharmacy, Regensburg Germany

1. Introduction

During the last two decades it emerged that cGMP-dependent protein kinases (PKG) act as signalling molecules with pleiotropic physiological functions e.g. in the cardiovascular system, in the gastrointestinal tract, immune system and central nervous system (Hofmann et al., 2006). The cellular effects of PKG are transmitted by its phosphorylated substrates. The identification of different PKG substrates in the diverse organs and cells and the analysis of the interaction of PKG with its substrates are important issues which are strongly targeted by several research groups recently (summarized in (Schlossmann & Desch, 2009)). An important step for elucidating the identity and function of cGMP-kinase substrate proteins was the use of affinity chromatography methods utilizing immunoprecipitation or cyclic nucleotide based affinity columns. Thereby, a ternary complex of cGMP-kinase was purified which consists of the PKG1 isoform PKG1β, the inositol-trisphosphate receptor type 1 ($InsP_3R1$) and the inositol-trisphosphate receptor-associated cGMP-kinase substrate protein (IRAG). Functional analysis by genetic deletion of IRAG in mice revealed that IRAG is essential for relaxation of vascular smooth muscle and for inhibition of platelet aggregation. Meanwhile, it was elucidated that IRAG also interacts with further isoforms of $InsP_3R$, namely $InsP_3R2$ and $InsP_3R3$. Furthermore, the ternary PKG1 core complex associates with several other proteins which might tune the cellular function of PKG1 in various tissues. In comparison, the association of PKG1α with substrate proteins, MYPT1 or RGS2, was also identified by affinity chromatography including immunoprecipitation, GST-fusion proteins and/or His-tagged proteins. Therefore, the stable interaction of PKG with its substrate proteins might be a common theme and might lead to the view that direct interactions convert the intracellular function of PKG in analogy to the mechanisms which were identified for cAMP-dependent protein kinases. Therefore, affinity chromatography methods are essential tools for the identification of PKG substrate proteins and function in cells and tissues. The present review will give an overview of the structural and functional features of PKGs and of diverse PKG complexes which were identified by different affinity chromatographic techniques.

---

* Corresponding Author

2. cGMP-dependent protein kinases (PKG): Structure and function

PKG are stimulated by cGMP which is a key cellular second messenger. cGMP is synthesized by cytosolic soluble guanylyl cyclases (sGC) and membrane-bound particular guanylyl cyclases (pGC). sGC are activated by the gaseous molecule nitric oxide (NO). pGC are integral membrane proteins which are stimulated by the natriuretic peptides ANP, BNP, CNP or guanylin/uroguanylin. The different intracellular locations of these enzymes lead to the view that the localized synthesis of cGMP could differentiate the alternate function of NO and natriuretic peptides (Castro et al., 2010)

PKGs are Ser/Thr kinases which phosphorylate its substrates at the recognition site K/R K/R X S/T (one letter code of aa; X can be each aa). PKGs are expressed as three different proteins: PKG1α with 676 aa, PKG1β with 686 aa, PKG2 with 762 aa. PKG1 and PKG2 are encoded by two different genes, prkg1 and prkg2. The PKG1 isoforms are cytosolic enzymes, whereas PKG2 is plasma membrane-bound by an N-terminal myristoylation modification. The two PKG1 isoforms PKG1α and PKG1β are expressed from the prkg1 gene by differential splicing and thereby differ in the N-terminal first ~100 aa which reside isoleucine/leucine zipper regions (LZ) (Fig. 1). These LZ exhibit several isoform-specific functions: homodimerization, recognition of specific-substrates, intracellular targeting. Furthermore, the K_a for cGMP-activation of PKG1 isoforms varies more than 10-fold. The K_a (cGMP) of PKG1α is 0.059 μM and of PKG1β is 0.79 μM. Recently, the structural features of PKG1 could be more specified by the first crystal structures of the N-terminal leucine zipper of PKG1β and the cGMP-binding domains of PKG1β (Casteel et al., 2011; Kim et al., 2011). Homodimerization of PKG1β is mediated by eight Ile/Leu heptad repeats. Thereby, hydrophilic aa residues are coordinated in a "knobs into holes" structure which assists the dimer formation. NMR studies of the PKG1α N-terminal domain (aa 1-39) revealed an α-helical structure with a parallel monomeric association of the coiled-coil domain which warrants PKG1α homodimerization through fixed electrostatic interactions. The inhibitory sequence (IS) contains the substrate-like sequence PRT^{59}TR with autophosphorylation site (Thr^{59}) of PKG1α (Aitken et al., 1984) and the presumed pseudosubstrate KRQAISAE of PKG1β (Kemp & Pearson, 1990; Ruth et al., 1997). The IS inactivates the kinase when cGMP is absent.

Upon cGMP binding the kinase is activated. The holo-enzymes of PKG1α or PKG1β contain two different cyclic nucleotide-binding domains CNBD-A and CNBD-B (Fig. 1). The recently obtained cocrystals of PKG1β (aa92-227) containing the N-terminal cyclic nucleotide binding domain of PKG1β (CNBD-A) together with cGMP and cAMP revealed new insights into the cyclic nucleotide-binding mechanism (Kim et al., 2011). cGMP attachs only in the *syn* conformation to CNBD-A, whereas cAMP in either *syn* or *anti* configuration. Surprisingly, CNBD-A binds both cGMP and cAMP with high affinity. Only a 2-fold preference was found for cGMP. Therefore, a possible cooperative action of CNBD-A and CNBD-B in cyclic-nucleotide-binding is proposed which should ensure a more differential cyclic nucleotide binding profile of PKG1. The recently published crystal structure of the regulatory domain containing both cGMP-binding sites of PKG1α (aa78-355) revealed a switch helix which promotes the formation of a hydrophobic interface between both cGMP-binding sites (Osborne et al., 2011). The regulatory domain is followed by the catalytic domain comprising an ATP binding - and a substrate binding domain. The 3-dimensional structure of these domains was not reported so far. Therefore, the structural elucidation of these different PKG1-domains and the holo-enzyme are still new challenges which will be essential for the functional insight of PKG1.

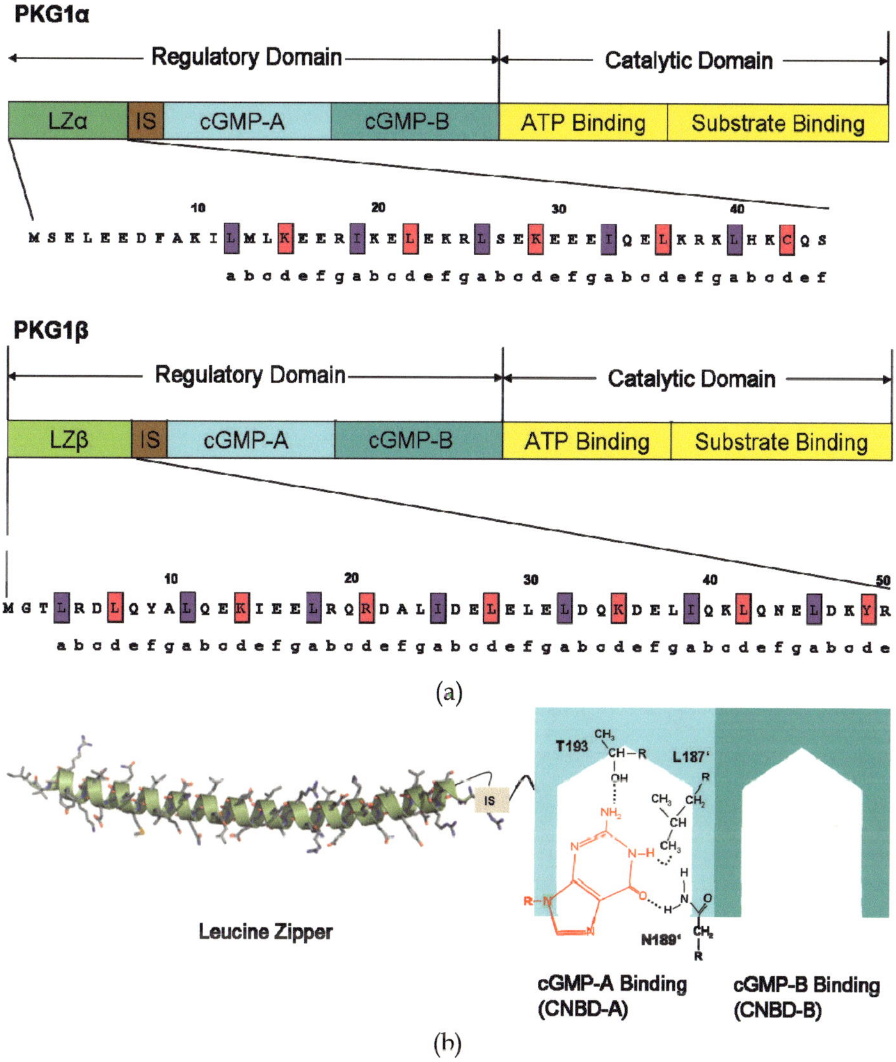

(A) The PKG1 is subdivided into the regulatory domain at the N terminus and the catalytic domain at the C terminus. The regulatory domain consists of the leucine zipper (LZ), the inhibitory sequence (IS) and the two cGMP binding pockets (cGMP-A/B = CNBD-A/B). The LZ is necessary for dimerization and docking to the substrates (Casteel et al., 2011). The primary sequences of the N-terminal 45 amino acids of the PKG1α LZ and of the N-terminal 50 amino acids of the PKG1β LZ are illustrated. The amino acids are grouped in a heptad repeat (a-g). The a-positions exhibit primary leucine or isoleucine residues (shaded in lilac) and the d-positions are mostly charged or hydrophilic residues (shaded in pink). (B) Scheme of PKG1β regulatory domain structure (Casteel et al., 2011). Leucine Zipper: The α-helical structure is depicted in green. Side chains in the helix are marked. The linker sequence between the leucine zipper and the autoinhibitory domain is variable between PKG1α and PKG1β, but is not important for the PKG1α or PKG1β phenotype (Ruth et al., 1997). IS: The inhibitory sequence is schematically drawn. It determines the high affinity PKG1α and the low affinity PKG1β phenotype (Ruth et al., 1997). cGMP-A binding pocket: cGMP binds in a *syn* conformation to CNBD-A (see 3.). The essential amino acids (T193, L187', N189') for cGMP-binding are shown.

Fig. 1. Structural requirements for interaction of PKG1α and PKG1β with their substrates

PKG1α

Substrate	MW(kDa)	Method for identification of interaction	Substrate function	Reference
MYPT1	130	Co-immunoprecipitation Fusion proteins	Enhanced myosin phosphatase activity mediating calcium desensitization	(Surks et al., 1999), (Wooldridge et al., 2004)
RGS2	24	Co-immunoprecipitation Fusion proteins	Inhibition of G_q mediated IP_3 synthesis	(Sun et al., 2005), (Tang et al., 2003)

PKG1β

Substrate	MW(kDa)	Method for identification of interaction	Substrate function	Reference
IRAG- $InsP_3R1$/ $InsP_3R2$/ $InsP_3R3$	125 230	Co-immunoprecipitation Fusion proteins cGMP-agarose	Inhibition of calcium release from IP_3 sensitive stores after phosphorylation of IRAG Smooth muscle relaxation Inhibition of platelet activation	(Haug et al., 1999), (Casteel et al., 2005), (Schlossmann et al., 2000), (Geiselhoringer et al., 2004), (Desch et al., 2010), (Schinner et al., 2011), (Masuda et al., 2010)
PDE5	100	Co-immunoprecipitation cGMP-agarose	Enhanced cGMP degradation	(Rybalkin et al., 2002), (Wilson et al., 2008)
Phospho-lamban	6	Co-immunoprecipitation cGMP-agarose	Increase of calcium uptake by SERCA (sarcoplasmic reticulum Ca^{2+}-ATPase)	(Koller et al., 2003), (Lalli et al., 1999)
TFII-1	120-150	Fusion proteins	Coordinating the activity of multiple transcriptions factors, signaling molecules and histone deacetylases	(Casteel et al., 2005), (Casteel et al., 2002) (Roy, 2001), (Wen et al., 2003)

PKG1

Substrate	MW(kDa)	Method for identification of interaction	Substrate function	Reference
CRP4	22.5	Fusion proteins	Nociceptive behaviour in spinal cord Enhanced transcriptional activation of smooth muscle specific proteins; complex with the serum response factor (SRF)	(Schmidtko et al., 2008) (Huber et al., 2000), (Zhang et al., 2007)

Table 1. Identification of PKG1 interaction with substrates.

The table gives an overview of substrates interacting with PKG1α, PKG1β or both PKG1 isoforms, their molecular weights, the methods for identification of interaction and their main function.

PKG1 exhibits various functions e.g. in the cardiovascular system, in the lung, in the intestine, in platelets and in the central nervous system. PKG1 is important for the relaxation of vascular and gastrointestinal smooth muscle. This enzyme inhibits platelet activation and thereby protects against arterial thrombosis. Furthermore, in the central nervous system PKG1 is involved in processes of learning and memory (Hofmann et al., 2009). In some tissues - e.g. smooth muscles - both PKG1 isoforms are expressed. However, there are tissues where only the α-isoform is present e.g. in the lung or in the cerebellum, whereas in other cells exclusively the β-isoform is found e.g. in the human platelets or in the hippocampus. These different locations lead to the view that both isoforms exhibit different functions in tissues and cells. Therefore, the identification of isoform-specific substrates was a breakthrough to dissect the individual functions of the PKG1α and PKG1β. Particularly, regarding smooth muscle contractility the identification of substrates of PKG1α (regulator of G-protein signalling 2 (RGS2) and myosin phosphatase subunit 1 (MYPT1)) and of PKG1β (inositol trisphosphate receptor-associated cGMP kinase substrate protein (IRAG)) substantiated a detailed view of functional regulation by these PKG1 isoforms. RGS2 phosphorylation regulates the intracellular activity of $G_{q/11}$ and thereby reduces the intracellular synthesis of $InsP_3$ by phospholipase C. MYPT1 phosphorylation leads to calcium desensitization of the cytoskeletal contractility (Schlossmann & Desch, 2009). Phosphorylation of IRAG inhibits the intracellular calcium release via $InsP_3R1$. Furthermore, IRAG is an important substrate for the NO/cGMP-dependent inhibition of platelet function and thereby prevents arterial thrombosis (Hofmann et al., 2006; Schlossmann & Desch, 2011). Various other substrates were identified in tissues and cells which lead to more precise understanding of physiological PKG functions (Hofmann et al., 2009; Schlossmann & Desch, 2009). Particularly, the identification of different PKG1-complexes together with its substrates directed to specific signalling pathways of the PKG1 isoforms. The elucidation of these complexes by affinity chromatography will be reviewed below in part 3 (PKG complexes by affinity chromatography).

3. PKG-complexes by affinity chromatography

3.1 Immunoprecipitation methods

Immunoprecipitation utilizes binding of a protein from a tissue or cell lysate to the respective primary protein-specific antibody which is linked to a Sepharose matrix followed by precipitation of the protein-antibody complex e.g. by centrifugation. If there are interacting proteins that build a stable complex with the antibody-bound protein, they can also be detected in the precipitate and hence are so called co-immunoprecipitated. The matrices used most commonly for co-immunoprecipitation experiments are protein A-Sepharose or protein G-Sepharose. Protein A and protein G are cell surface proteins from staphylococcus aureus or from streptococcus, respectively, which are able to efficiently bind immunoglobulins. Protein G-Sepharose binds goat, sheep and mouse primary antibodies with higher affinity than protein A-Sepharose (Kaboord & Perr, 2008).

For the PKG1, there are different protein complexes, which were identified by co-immunoprecipitation techniques. An identified PKG complex is the trimeric macrocomplex consisting of the PKG1β, IRAG and the $InsP_3R1$ (Fig. 2, Table 1). The previously unknown protein IRAG was detected as a PKG1 substrate that could be phosphorylated in the presence of cGMP in bovine tracheal smooth muscle membranes. The multimeric complex was identified by co-immunoprecipitation experiments with the specific antibodies against the three proteins (Schlossmann et al., 2000). In each co-immunoprecipitation test, all three components of the complex could be detected in the precipitate. Later, Masuda et al. investigated the interaction between all $InsP_3R$-subtypes and IRAG-GFP proteins by co-immunoprecipitation experiments with anti-GFP antibody (Masuda et al., 2010). Both the neuronal S^{2+}-$InsP_3R1$ and the peripheral S^{2-}-$InsP_3R1$ interacted with IRAG(GFP) after transient expression in COS7 cells. By expressing IRAGΔE12(GFP) which lacked the coiled-coil interaction site with the $InsP_3R$, the $InsP_3R1$ could not be detected in the immunoprecipitation pellet. In contrast, the PKG1β was precipitated with IRAG and with IRAGΔE12. In conclusion, there was a stable interaction between PKG1β and IRAG but not between PKG1β and the $InsP_3R1$. Additionally, it was shown, that $InsP_3R2$ and $InsP_3R3$ also interact with full-length IRAG but not with IRAGΔE12 in COS7 cells (Masuda et al., 2010).

After that, in an immunoprecipitation experiment with anti-PKG1 antibody phospholamban (PLB) was identified as an additional component of the above described signalling complex (Koller et al., 2003) (Fig. 2, Table 1). Upon co-expression of $InsP_3R1$, IRAGa, PKG1β and PLB in COS7 cells, $InsP_3R1$ and PKG1β could be precipitated with anti-PLB antibody. Moreover, solubilized bovine tracheal smooth muscle membranes were incubated with antibodies against IRAG, PLB and PKG1 and in each precipitate the trimeric complex could be detected. Thus it was demonstrated that PLB is part of the IRAG/PKG1β/$InsP_3R1$ signalling complex.

In 2008, the PDE5 could be verified as a further component of the macrocomplex in human platelets (Wilson et al., 2008) (Fig. 2, Table 1). Incubation of platelet lysates with the antisera against PDE5, PKG and $InsP_3R1$ allowed co-immunoprecipitation of all three proteins. IRAG was not directly detected in the complex because a primary anti-IRAG antibody was not available. However, Wilson et al. suggested that IRAG was also a component of this PDE5 signalling complex, because after ^{32}P-incorporation into the anti-PKG immunoprecipitate a band corresponding to the molecular weight of IRAG was detected in the autoradiogram. Furthermore, it was shown that only the PKG-associated form of PDE5 could be phosphorylated and activated after 8-Br-cGMP treatment and also only the $InsP_3R1$-associated PDE5 could be selectively activated by PKG1. This was demonstrated by measurements of PDE activities in different cellular fractions which contained or were devoid of $InsP_3R1$.

Additionally to the trimeric complex, the interaction between MYPT1 -the myosin-binding subunit (MBS) of myosin phosphatase- and PKG1α could be verified by immunoprecipitation experiments (Fig. 2, Table 1). Lysates from saphenous vein smooth muscle cells were incubated with anti-PKG1α and anti-MYPT1 antibodies and in the first case MYPT1 and in the second case PKG1α was detected in the protein A-Sepharose precipitate (Surks et al., 1999). Furthermore, in the PKG1α-immunoprecipitate, PP1 phosphatase activity was measured and also inhibited by the phosphatase inhibitor ocadaic acid. After that, to detect possible further PKG1α substrates in the PKG1α-PP1M complex, cGMP and PKG1α were added to anti-MYPT1 immunopellets. In the presence of [γ-^{32}P]ATP there was a significantly increased phosphorylation of MYPT1. In 2003, it was hypothesized that the leucine zipper domains of MYPT1 and PKG1α mediate the interaction between the two proteins (Surks & Mendelsohn, 2003). To further investigate the relevance of the MYPT1 leucine zipper for the interaction, co-immunoprecipitation experiments were performed with chicken smooth muscle cells and chicken smooth muscle tissue (Huang et al., 2004). Aorta from embryonic day 15 expressed a leucine zipper positive MYPT1 (LZ^+) and gizzard smooth muscle tissue from day 7 expressed a LZ^- MYPT1 isoform. Both LZ^+ and LZ^- MYPT1 isoforms associated with PKG1 and the presence of cGMP increased the binding of the MYPT1 to PKG1 in the embryonic aorta. In contrast, in cultured smooth muscle cells, neither LZ^+ nor LZ^- MYPT1 isoforms co-immunoprecipitated with PKG1α. These results suggested that the binding of MYPT1 to PKG1α is not mediated by a LZ-LZ interaction. Because PKG1α prefers binding to RR and RK motifs (Dostmann et al., 2000) and there is an RK motif in the aa 888–928 sequence of MYPT1, the relevance of this sequence was investigated (Given et al., 2007). Mutants were generated, which lack or contain this sequence and which also lack or contain the leucine zipper. These four MYPT1 mutant proteins were added to adult chicken gizzard lysate and chicken aorta lysate and co-immunoprecipitation experiments with anti-PKG1α antibody were performed. Only the fragments that contained the amino acids 888-928 interacted with PKG1α and the interaction was independent of the expression of the leucine zipper (Given et al., 2007). The RK motif within the aa 888-928 sequence is located at aa 916 and 917, that were mutated to alanine (MYPT1A) or glutamic acid (MYPT1E) to investigate whether these two amino acids are important for binding of PKG1α. It was shown that MYPT1A binds to PKG1α whereas the interaction between MYPT1E and PKG1α is abolished. Therefore the amino acids R^{916} K^{917} are important for the binding of PKG1α and the interaction depends on the charge of the amino acid residues and not on the size. At last, it was proved that an accessory protein may be necessary for interaction, because there is a loss of the MYPT1-PKG1α interaction if partially purified PKG1α is used in the co-immunoprecipitation experiment with the MYPT1 fragment (Given et al., 2007).

Additionally to the IRAG- and the MYPT1- PKG1 complexes, an interaction between PKG1α and the regulator of G-protein signalling-2 (RGS2) was demonstrated (Fig. 2, Table 1). When recombinant [^{35}S] RGS2 protein and recombinant PKG1α were combined, a co-immunoprecipitation of the RGS2 protein with PKG1α was detected (Tang et al., 2003).

3.2 Affinity chromatography by GST-fusion or His-tagged proteins

The interaction between two proteins can also be investigated by generation and analysis of GST- and His-tagged fusion proteins. A GST-tagged fusion protein contains the glutathione S-transferase enzyme and can be purified with glutathione beads. A Polyhistidine-tag consists of at least five histidine residues and His-tagged fusion proteins bind to Ni^{2+}-matrices. These methods were used for the analysis of the interaction of substrate proteins with PKG1α or PKG1β. The interactions of the different GST- and His-tagged proteins are summarized in table 2.

The interaction between RGS2 and PKG1α was also demonstrated by analysing GST-fusion proteins (Tang et al., 2003). A GST-PKG1α (1-59) fusion protein, which contained the PKG1α leucine zipper domain, interacted with the [^{35}S] RGS2 protein. The N-terminal region of

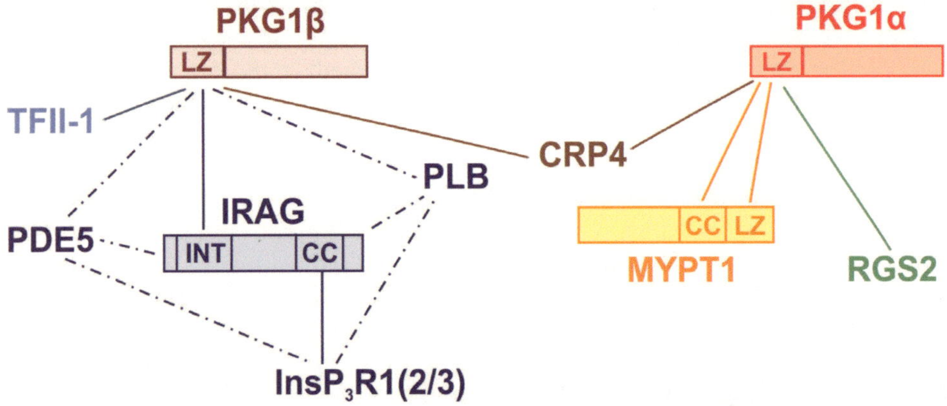

Fig. 2. PKG1α and PKG1β complexes identified by affinity chromatography methods
The filled lines indicate confirmed stable interactions between two complex proteins. When the respective interaction sites are known they are illustrated in the figure. The broken lines indicate putative interactions of complex proteins.
Complex proteins: PKG1α/β: cGMP-dependent protein kinase 1α/β; IRAG: inositol-trisphosphate receptor-associated cGMP-kinase substrate protein; $InsP_3R1/2/3$: inositol 1,4,5-trisphophate receptor subtype 1/2/3; PDE5: phosphodiesterase 5; TFII-1: general transcriptional regulator 1; PLB: phospholamban; CRP4: Cystein rich LIM protein 4; MYPT1: myosin targeting subunit of myosin phosphatase; RGS2: regulator of G-protein signalling 2
Interaction sites: LZ: leucine zipper; INT: IRAG-interaction site with PKG1β; CC: coiled-coil domain

RGS2 (GST-RGS2-N; residues 1-79) bound endogenous PKG1α from venous and arterial VSMC lysates. In contrast, the GST-RGS2 fusion protein of the C-terminus (GST-RGS2-C (80-211)) did not interact with the cGMP-dependent protein kinase. Summing up, the N-terminus of RGS2 binds the N-terminal leucine zipper domain of PKG1α in vascular smooth muscle cells (Fig. 2).

GST-fusion proteins were also generated to investigate the MYPT1-PKG1α interaction (Surks et al., 1999). A GST-MYPT1 protein (GST MYPT1 (850-1030) WT), a GST-tagged protein of the C-terminus of the Myosin-binding subunit interacted with PKG1α. After that, it was demonstrated with GST-fusion proteins of the N-termini of PKG1α (GST PKG1α (1-59)), PKG1β (GST PKG1β (1-92)) and PKG2 (GST PKG2 (1-272)) that only PKG1α interacts with MYPT1 (Fig. 2). Subsequently, the relevance of the leucine zipper of MYPT1 for the interaction with PKG1α was also investigated with GST-tagged proteins. GST-fusion proteins of MYPT1 mutants were generated in which two consecutive leucines were mutated to alanine (MYPT1 1007A1014A mutant and MYPT1 1021A1028A mutant). Only the wild-type GST-MYPT1 bound endogenous PKG1α from human aortic smooth muscle cell lysate, the MYPT1 mutants did not interact with the cGMP-dependent protein kinase (Surks & Mendelsohn, 2003). The MYPT1 leucine zipper was also relevant for homodimerization because these leucine zipper mutant GST-MYPT1 (GST MYPT1 CT_{181} L1021A/L1028A and GST MYPT1 CT_{181} L1007A/L1014A) did not bind MYPT1 from VSMC lysate (Surks & Mendelsohn, 2003). On the other hand the PKG1α leucine/isoleucine zipper

	fusion protein	interacting partner	interaction	reference
PKG1α/PKG1α	GST PKG1α (1-59) WT	VSMC lysate: PKG1α	yes	(Surks & Mendelsohn, 2003)
	GST PKG1α (1-59) L26P	VSMC lysate: PKG1α	yes	
	GST PKG1α (1-59) L12A/I19A	VSMC lysate: PKG1α	yes	
	GST PKG1α (1-59) I33A/L40A	VSMC lysate: PKG1α	yes	
MYPT1/PKG1α	GST MYPT1 CT_{181} (850-1030) WT	VSMC lysate: PKG1α	yes	(Surks et al., 1999)
	GST PKG1α (1-59)	VSMC lysate: MYPT1	yes	
	GST PKG1β (1-92)	VSMC lysate: MYPT1	no	
	GST PKG2 (1-272)	VSMC lysate: MYPT1	no	
	GST MYPT1 CT_{181} (850-1030) WT	VSMC lysate: PKG1α	yes	(Surks & Mendelsohn, 2003)
	GST MYPT1 CT_{181} (850-1030) L1021A/L1028A	VSMC lysate: PKG1α	no	
	GST MYPT1 CT_{181} (850-1030) L1007A/L1014A	VSMC lysate: PKG1α	yes	
	GST PKGIα (1-59) WT	His MYPT1 CT_{100} (930-1030)	yes	(Sharma et al., 2008)
MYPT1/MYPT1	GST MYPT1 CT_{181} WT	VSMC lysate: MYPT1	yes	(Surks & Mendelsohn, 2003)
	GST MYPT1 CT_{181} L1021A/L1028A	VSMC lysate: MYPT1	no	
	GST MYPT1 CT_{181} L1007A/L1014A	VSMC lysate: MYPT1	no	
PKG1β/PKG1β	GST PKG1β (1-110) E29K	PKG1β E29K	yes	(Casteel et al., 2005)
	GST PKG1β (1-110) D26K/E31R	PKG1β D26K/E31R	yes	
IRAG/PKG1β	GST PKG1α WT	Myc IRAG WT	no	(Casteel et al., 2005)
	GST PKG1β WT	Myc IRAG WT	yes	
	GST PKG1β (1-110) WT	His/Myc IRAG (1-415) WT	yes	
	GST PKG1β (1-110) E29K	His/Myc IRAG (1-415) WT	no	
	GST PKGIβ (1-110) D26K/E31R	His/Myc IRAG (1-415) WT	no	
	GST IRAG (1-415) WT	purified PKG1β	yes	
	GST IRAG (1-415) R120A/R124A	purified PKG1β	no	
	GST IRAG (1-415) R124A/R125A	purified PKG1β	no	
	GST PKG1β WT	Myc IRAG R124A/R125A	no	
TFII-I/PKGIβ	GST PKG1α WT	Myc TFII-I WT	no	(Casteel et al., 2005)
	GST PKG1β WT	Myc TFII-I WT	yes	
RGS2/PKG1α	GST RGS2-N (1-79)	VSMC lysate: PKG1α	yes	(Tang et al., 2003)
	GST RGS2-C (80-211)	VSMC lysate: PKG1α	no	
	GST PKG1α (1-59) WT	$[^{35}S]$RGS-2	yes	
CRP4/PKG1	Myc CRP4	PKG1α	yes	(Zhang et al., 2007)
	Myc CRP4	PKG1β	yes	

Table 2. Interaction of GST- and His-tagged fusion proteins (PKG1α, PKG1β and/or its substrate proteins). More details of these interactions are given in chapter 3.

was dispensable for formation of PKG dimers, which was shown by analysing the interaction between GST-tagged wild type and mutant amino-terminal PKG1α and endogenous PKG1α from VSMC lysate (Surks & Mendelsohn, 2003). Lee et al. (Lee et al., 2007) provided data that supported the importance of both the leucine zipper and the C-terminal CC domain of MYPT1 for the formation of the PKG1α-MYPT1 complex. Therefore, the interaction between the leucine zipper of PKG1α (PKG1α_{1-59}) and the C-terminal CC domain of the Myosin Binding Subunit was tested (Sharma et al., 2008). A C-terminal His_6-MYPT1 CT_{100} (residues 930-1030) protein which was bound to Ni^{2+}-NTA resin precipitated GST-PKG1α_{1-59}. In isothermal titration calorimetry studies and NMR experiments the importance of the residues 929-970 (MYPT1 CT_{42}) for the interaction with the leucine zipper of PKG1α was demonstrated (Sharma et al., 2008). In short, if the leucine-zipper motif of MYPT1 is absent, the PKG1α leucine-zipper binds to the coiled coil region and upstream segments of MYPT1 (Lee et al., 2007).

GST-fusion proteins were also generated to further investigate the interaction between IRAG and PKG1β. A GST-tagged full-length PKG1β interacted *in vivo* with a Myc-tagged IRAG (full-length IRAG WT) after expression of the proteins in COS7 cells (Casteel et al., 2005). Full-length PKG1α showed no binding of IRAG. Because both PKG1 isoforms bind via salt sensitive ionic interactions, the involvement of charged residues within the leucine zipper of PKG1β was tested. Acidic residues in the N-terminal sequence of PKG1β were mutated and the GST-tagged mutants were expressed in E.coli. These proteins were incubated with the Myc- and His-tagged N-terminal IRAG (His/Myc IRAG (1-415) WT). Two mutants of PKG1β (GST PKG1β (1-110) E29K and GST PKGIβ (1-110) D26K/E31R) showed no longer binding to IRAG and therefore it was concluded that acidic residues in PKG1β mediate the interaction with IRAG. On the other hand, basic residues in the IRAG interaction site (residues 100-132) (Ammendola et al., 2001) are necessary for binding of PKG1β. To determine the specific amino acids, pairs of basic residues were mutated to alanine and the GST-tagged IRAG mutants were tested for their ability to bind PKG1β in vitro. Some of the mutants showed completely disrupted binding (GST IRAG (1-415) R120A/R124A and GST IRAG (1-415) R124A/R125A) and others showed decreased binding to the cGMP-dependent protein kinase (Casteel et al., 2005). The E29K and D26K/E31R PKG1β mutants, which exhibited mutations of the acidic residues that are important for the interaction with IRAG, could form stable dimers after expression in COS7 cells. Therefore, these residues are not necessary for PKG1β homodimerization.

The importance of PKG1 homodimerization was demonstrated by the cardiovascular deficits exhibited by transgenic mice expressing a dimerization-deficient form of PKG1α (Michael et al., 2008). To further understand the molecular details of PKG1 dimerization and the association of PKG1 dimers with GKAPs, Casteel et al. solved a crystal structure of the PKG1β dimerization/docking domain by investigating a His-tagged PKG1β D/D domain (Casteel et al., 2011) (described in 2.).

Similarly to the IRAG-PKG1β interaction, the interaction between PKG1β and the general transcriptional regulator TFII-I was investigated (Casteel et al., 2005). Like IRAG, Myc-tagged TFII-I was only bound by GST-tagged PKG1β but not by PKG1α. It was shown that electrostatic interactions are important for binding of PKG1β to GST-TFII-I, because high NaCl concentrations (400 mM) disrupted binding. Acidic residues in the PKG1β leucine zipper are important for the interaction with basic amino acids of TFII-I which was proved by analysis of GST-tagged PKG1β and TFII-I mutants (Casteel et al., 2005). TFII-I is not only bound but also phosphorylated by PKG1 and as a consequence of this transcriptional activation of a serum response factor-dependent reporter gene is enhanced (Casteel et al., 2002).

Fusion proteins were also generated to analyze the interaction between the cysteine-rich LIM-only protein CRP4 and PKG1 in vascular smooth muscle cells (Zhang et al., 2007). After immunoprecipitation of endogenous CRP4 from PAC1 cells PKG1 was detected in the precipitate and Myc-epitope-tagged CRP4 interacted with PKG1α and PKG1β after expression in CV1 cells. Additionally, it was shown that cGMP/PKG1 mediated phosphorylation of CRP4 had no effect on CRP4-PKG1 association and that Ser104 is the major PKG phosporylation site.

3.3 Cyclic nucleotide affinity methods

As cGMP-dependent protein kinases are activated by cyclic GMP and contain cGMP binding domains they can be purified or bound by cGMP-agarose beads. The PKG1β/IRAG/$InsP_3R1$ macrocomplex was not only identified by co-immunoprecipitation experiments with the respective primary antibodies but also after expression of the three proteins in COS7 cells and then incubating the cell lysate with 8-AET-cGMP-agarose beads (Ammendola et al., 2001). The cGMP-agarose affinity method was also performed with colon smooth muscle tissues from IRAG-deficient and WT mice (Desch et al., 2010). Thereby it was demonstrated that the interaction of PKG1β and $InsP_3R1$ is destroyed by IRAG deletion and that the $InsP_3R1$ misses a stable interaction site for PKG1β.

In 2003, Phospholamban (PLB) was identified as a component of the PKG1β/IRAG/$InsP_3R1$ signalling complex after bovine tracheal microsomal membrane proteins were purified by cGMP agarose, phosphorylated by 8-pCPT-cGMP and separated by SDS-PAGE or Tricin-SDS-PAGE (Koller et al., 2003). Beside $InsP_3R1$, IRAG and PKG1β, Phospholamban and RhoA were phosphorylated in the presence of cGMP. Smooth muscle α-actin and smooth muscle calponin were also present in the cGMP-agarose complex and were identified by MALDI-TOF. Additionally, after pre-purification by cGMP-agarose and phosphorylation in presence of 8-pCPT-cGMP the microsomal membrane proteins were immunoprecipitated with anti-$InsP_3R1$, anti-IRAG, anti-PKG1 and anti-PLB antibody. In each case, the autoradiogram showed phosphorylated $InsP_3R1$, IRAG and PKG1β which is a further indication that PLB is a component of the macrocomplex and interacts with PKG1β.

Furthermore, PDE5 is present in the cGMP-agarose complex (Wilson et al., 2008) and an increased PDE activity was measured in the cGMP-agarose pellet after 8-Br-cGMP treatment.

Margarucci et al. investigated the rapid spatial responses of cAMP and cGMP signalling complexes induced by collagen stimulation of human platelets (Margarucci et al., 2011). The platelets were isolated from whole blood, activated by collagen related peptide (CRP) for 5 minutes and after platelet lysis the proteins were bound to a 1:1 mixture of 2-AHA-cGMP- and 8-AET-cGMP-agarose beads. After that, the proteins were eluted, digested and identified and quantified by LC-MS/MS. PKG1β, PKG2, PDE2A and PDE5A were identified as primary and IRAG, $InsP_3R1$ and $InsP_3R2$ as secondary interactors. After CRP stimulation, the secondary interactors showed increased enrichment versus control whereas binding of PKG1β was unaltered and the amount of PKG2 slightly decreased. Moreover, different phosphopeptides were identified and also quantified and it was shown, that phosphorylation of IRAG at Ser670 is reduced by 60% after incubation with collagen. Additionally, the autophosphorylation of PKG1β at Ser63 was reduced which presumably corresponded to a reduced kinase activity and lead to increased platelet activation after CRP stimulation.

4. Function of PKG complexes

4.1 IRAG-PKGIβ and the ternary PKG macrocomplex

IRAG is a 125 kDa substrate of PKG1β which is located at the endoplasmic reticulum membrane anchored by a C-terminal tail. IRAG is expressed in high amounts in smooth muscle tissues including the vasculature, the gastrointestinal tract and the lung. Furthermore, IRAG is found in high concentration in platelets. IRAG expression is also found in lower quantity in further tissues including heart, osteoclasts and thalamus. The interaction of IRAG was identified by various methods including co-immunoprecipitation, cGMP affinity chromatography, two-hybrid analysis and mutagenesis studies. Thereby, an N-terminal located 33 aa IRAG domain (aa152-184) was identified which interacts with the Ile/Leu-zipper domain of PKG1β but not with PKG1α. Two arginines of the IRAG interaction domain (R124 and R125) are essential for the electrostatic interaction with acidic aa of PKG1β (D26, E29, E31) (Casteel et al., 2005). PKG1β and IRAG form a ternary complex together with $InsP_3R1$ in smooth muscle, platelets and osteoclasts. IRAG is the core component of the complex which interacts with the $InsP_3R1$ by its central coiled-coil domain. Deletion of this coiled-coil domain prevent the $InsP_3R1$ interaction but does not alter IRAG-PKG1β binding. However, the deficiency of IRAG in IRAG-KO mice destructs the ternary complex showing that $InsP_3R1$ does not stably interact with PKG1β. The function of IRAG in the ternary complex was revealed by targeted deletion of the coiled-coil domain or total deletion of IRAG in mice. Thereby, it was elucidated that IRAG is essential for NO- and ANP-mediated smooth muscle relaxation (Desch et al., 2010). IRAG deletion did not alter basal blood pressure but prevented the blood pressure drop upon LPS-induced sepsis (Desch et al., 2010). In platelets, IRAG deletion lead to hyperaggregability and an enhanced amount of platelets in the blood. The NO/cGMP-mediated inhibition of platelet activation was prevented upon IRAG deletion (Schinner et al., 2011). Concludingly, IRAG prevented arterial thrombosis (Antl et al., 2007). Recently, it was reported that IRAG is one of seven proteins which exhibited a polymorphism which was associated with enhanced platelet aggregability known to be a major factor for the incidence of cardiovascular diseases (Johnson et al., 2010). A recent study showed that IRAG might be involved in the activation and attachment of osteoclasts (Yaroslavskiy et al., 2010). Furthermore, it was reported that IRAG is also able to interact with further $InsP_3R$ isoforms, namely $InsP_3R2$ and $InsP_3R3$ (Masuda et al., 2010). The exact mechanism of IRAG interaction with $InsP_3R$ and the inhibition of $InsP_3R$ gating was not established so far and remains a fundamental subject of IRAG function.

4.2 MYPT1-PKG1α

PKG1α interaction with MYPT1, a 130 kDa myosin-binding subunit of phosphatase 1, was revealed by co-immunopreciptation, two hybrid analysis and mutagenesis studies. The Ile/Leu-zipper domain of PKG1α (aa 1-59) interacts with MYPT1. There are two diverse developmentally regulated MYPT1 isoforms existing in vascular smooth muscle: an isoform containing a C-terminal leucine zipper (LZ+) which is sufficient to bind to PKG1α and thereby sensitizes cGMP-dependent relaxation. Nitric oxide tolerance was related to enhanced degradation of the LZ+-isoform. Moreover, a decreased expression of LZ+ was correlated with congestive heart failure. The leucine-zipper deficient isoform (LZ-) alternatively interacts with PKG1α through its N1-N2 coiled-coil domain but is cGMP-insensitive (Payne et al., 2006).

4.3 RGS2-PKG1α

The PKG1α-Ile/Leu-Zipper domain interacts with the N-terminal 79aa of RGS2. This RGS2-domain includes an amphipathic α-helical domain with Ser46 and Ser 64. Upon cGMP-dependent phosphorylation these Ser residues enable membrane targeting. Vice versa, localization of PKG1α at the plasma membrane was enhanced by RGS2 phosphorylation suggesting a thereby stabilized complex of these proteins. Thus, RGS2 is activated by PKG1α and hence G_q-triggered $InsP_3$-synthesis is attenuated (Tang et al., 2003). Accordingly, the deletion of RGS2 in mice impairs cGMP-dependent reduction of calcium transients in vascular smooth muscle cells. Furthermore, NO-dependent reduction of blood pressure is suppressed in RGS2-deficient mice (Sun et al., 2005). Interestingly, the hypertensive phenotype was correlated to nitric oxide function only in inactive phases of mice (during the day). In contrast, sympathetic activation and increased vascular adrenergic responsiveness correlated with enhanced blood pressure of RGS2-deficient mice in the active phases (during the night) (Obst et al., 2006). Recent reports revealed that ANP-signalling is also transduced by the PKG1α-RGS2 complex mediating the suppression of angiotensin II-induced cardiac hypertrophy (Klaiber et al., 2010).

4.4 Further PKG-substrate complexes

There are several further proteins which interact with the PKG1 isoforms. However, the elucidation of these PKG interacting proteins as PKG substrates was obtained only occasionally. The Cystein rich LIM protein CRP4 was identified as PKG1α- and PKG1β-interacting protein by two hybrid analysis using a smooth muscle cDNA library. CRP4 is phosphorylated as substrate by PKG1 (Huber et al., 2000). The CRP4 protein interacts via its third zinc finger domain with PKG1 associating together in a complex with serum response factor (SRF). Upon cGMP/PKG1 activation CRP4 enhances SRF/DNA association and thereby regulates the expression of smooth muscle-specific genes (Zhang et al., 2007).

A further PKG1β interacting protein is the general transcriptional regulator TFII-1 which resides as interface a R4 helix-loop-helix motif (aa 491-628). TFII-1 interacts with basic aa to the Leu/Ile domain of PKG1β similarly as reported for the IRAG protein. The TFII-1 interaction is cGMP-independent. TFII-1 transactivation of a serum-response element was enhanced by PKG1β (Casteel et al., 2005).

5. Conclusion

The identification of PKG-interacting proteins by affinity chromatographic methods clarified several new signalling pathways of PKG1. Furthermore, the dissection of PKG1 isoform specific functions of PKG1α and PKG1β was substantiated by the identification of specific substrate proteins in several tissues e.g. in smooth muscle, in platelets and in osteoclasts as shown above. The advanced elucidation of the function of the known PKG-interacting and substrate proteins will be a cue for the determination of pathophysiological consequences in cardiovascular, haematological and gastrointestinal diseases. Furthermore, the diverse signal transduction pathways of PKG1 isoforms and the detailed molecular analysis of their interactions and possible interferences will lead to new therapeutic horizons. However, the ubiquitous expression of PKG1 and the pleiotropic functions of the PKG1 isoforms require a very subtle intracellular signal regulation by probably a variety of further substrate proteins which were not elucidated so far. Particularly, in immunogical functions, in the renal system and in the central nervous system there are a variety of mechanisms which are not explained by the current knowledge about PKG signal transduction. The use of interaction assays including affinity chromatography methods is tempting to find new pathways which will aid in the identification of new (patho)physiological aspects of PKG signalling.

6. Acknowledgement

Work done in the authors laboratory was supported by DFG, GRK760 and SFB699. Tobias Holzammer aided in the art work.

7. References

Aitken, A., Hemmings, B.A. & Hofmann, F. (1984) Identification of the residues on cyclic GMP-dependent protein kinase that are autophosphorylated in the presence of cyclic AMP and cyclic GMP. *Biochim Biophys Acta*, 790, 219-225.

Ammendola, A., Geiselhoringer, A., Hofmann, F. & Schlossmann, J. (2001) Molecular determinants of the interaction between the inositol 1,4,5-trisphosphate receptor-associated cGMP kinase substrate (IRAG) and cGMP kinase Ibeta. *J Biol Chem*, 276, 24153-24159.

Antl, M., von Bruhl, M.L., Eiglsperger, C., Werner, M., Konrad, I., Kocher, T., Wilm, M., Hofmann, F., Massberg, S. & Schlossmann, J. (2007) IRAG mediates NO/cGMP-dependent inhibition of platelet aggregation and thrombus formation. *Blood*, 109, 552-559.

Casteel, D.E., Boss, G.R. & Pilz, R.B. (2005) Identification of the interface between cGMP-dependent protein kinase Ibeta and its interaction partners TFII-I and IRAG reveals a common interaction motif. *J Biol Chem*, 280, 38211-38218.

Casteel, D.E., Smith-Nguyen, E.V., Sankaran, B., Roh, S.H., Pilz, R.B. & Kim, C. (2011) A crystal structure of the cyclic GMP-dependent protein kinase I{beta} dimerization/docking domain reveals molecular details of isoform-specific anchoring. *J Biol Chem*, 285, 32684-32688.

Casteel, D.E., Zhuang, S., Gudi, T., Tang, J., Vuica, M., Desiderio, S. & Pilz, R.B. (2002) cGMP-dependent protein kinase I beta physically and functionally interacts with the transcriptional regulator TFII-I. *J Biol Chem*, 277, 32003-32014.

Castro, L.R., Schittl, J. & Fischmeister, R. (2010) Feedback control through cGMP-dependent protein kinase contributes to differential regulation and compartmentation of cGMP in rat cardiac myocytes. *Circ Res*, 107, 1232-1240.

Desch, M., Sigl, K., Hieke, B., Salb, K., Kees, F., Bernhard, D., Jochim, A., Spiessberger, B., Hocherl, K., Feil, R., Feil, S., Lukowski, R., Wegener, J.W., Hofmann, F. &

Schlossmann, J. (2010) IRAG determines nitric oxide- and atrial natriuretic peptide-mediated smooth muscle relaxation. *Cardiovasc Res*, 86, 496-505.

Dostmann, W.R., Taylor, M.S., Nickl, C.K., Brayden, J.E., Frank, R. & Tegge, W.J. (2000) Highly specific, membrane-permeant peptide blockers of cGMP-dependent protein kinase Ialpha inhibit NO-induced cerebral dilation. *Proc Natl Acad Sci U S A*, 97, 14772-14777.

Geiselhoringer, A., Werner, M., Sigl, K., Smital, P., Worner, R., Acheo, L., Stieber, J., Weinmeister, P., Feil, R., Feil, S., Wegener, J., Hofmann, F. & Schlossmann, J. (2004) IRAG is essential for relaxation of receptor-triggered smooth muscle contraction by cGMP kinase. *Embo J*, 23, 4222-4231.

Given, A.M., Ogut, O. & Brozovich, F.V. (2007) MYPT1 mutants demonstrate the importance of aa 888-928 for the interaction with PKGIalpha. *Am J Physiol Cell Physiol*, 292, C432-439.

Haug, L.S., Jensen, V., Hvalby, O., Walaas, S.I. & Ostvold, A.C. (1999) Phosphorylation of the inositol 1,4,5-trisphosphate receptor by cyclic nucleotide-dependent kinases in vitro and in rat cerebellar slices in situ. *J Biol Chem*, 274, 7467-7473.

Hofmann, F., Bernhard, D., Lukowski, R. & Weinmeister, P. (2009) cGMP regulated protein kinases (cGK). *Handb Exp Pharmacol*, 137-162.

Hofmann, F., Feil, R., Kleppisch, T. & Schlossmann, J. (2006) Function of cGMP-dependent protein kinases as revealed by gene deletion. *Physiol Rev*, 86, 1-23.

Huang, Q.Q., Fisher, S.A. & Brozovich, F.V. (2004) Unzipping the role of myosin light chain phosphatase in smooth muscle cell relaxation. *J Biol Chem*, 279, 597-603.

Huber, A., Neuhuber, W.L., Klugbauer, N., Ruth, P. & Allescher, H.D. (2000) Cysteine-rich protein 2, a novel substrate for cGMP kinase I in enteric neurons and intestinal smooth muscle. *J Biol Chem*, 275, 5504-5511.

Johnson, A.D., Yanek, L.R., Chen, M.H., Faraday, N., Larson, M.G., Tofler, G., Lin, S.J., Kraja, A.T., Province, M.A., Yang, Q., Becker, D.M., O'Donnell, C.J. & Becker, L.C. (2010) Genome-wide meta-analyses identifies seven loci associated with platelet aggregation in response to agonists. *Nat Genet*, 42, 608-613.

Kaboord, B. & Perr, M. (2008) Isolation of proteins and protein complexes by immunoprecipitation. *Methods Mol Biol*, 424, 349-364.

Kemp, B.E. & Pearson, R.B. (1990) Protein kinase recognition sequence motifs. *Trends Biochem Sci*, 15, 342-346.

Kim, J.J., Casteel, D.E., Huang, G., Kwon, T.H., Ren, R.K., Zwart, P., Headd, J.J., Brown, N.G., Chow, D.C., Palzkill, T. & Kim, C. (2011) Co-Crystal Structures of PKG Ibeta (92-227) with cGMP and cAMP Reveal the Molecular Details of Cyclic-Nucleotide Binding. *PLoS One*, 6, e18413.

Klaiber, M., Kruse, M., Volker, K., Schroter, J., Feil, R., Freichel, M., Gerling, A., Feil, S., Dietrich, A., Londono, J.E., Baba, H.A., Abramowitz, J., Birnbaumer, L., Penninger, J.M., Pongs, O. & Kuhn, M. (2010) Novel insights into the mechanisms mediating the local antihypertrophic effects of cardiac atrial natriuretic peptide: role of cGMP-dependent protein kinase and RGS2. *Basic Res Cardiol*, 105, 583-595.

Koller, A., Schlossmann, J., Ashman, K., Uttenweiler-Joseph, S., Ruth, P. & Hofmann, F. (2003) Association of phospholamban with a cGMP kinase signaling complex. *Biochem Biophys Res Commun*, 300, 155-160.

Lalli, M.J., Shimizu, S., Sutliff, R.L., Kranias, E.G. & Paul, R.J. (1999) [Ca2+]i homeostasis and cyclic nucleotide relaxation in aorta of phospholamban-deficient mice. *Am J Physiol*, 277, H963-970.

Lee, E., Hayes, D.B., Langsetmo, K., Sundberg, E.J. & Tao, T.C. (2007) Interactions between the leucine-zipper motif of cGMP-dependent protein kinase and the C-terminal region of the targeting subunit of myosin light chain phosphatase. *J Mol Biol*, 373, 1198-1212.

Margarucci, L., Roest, M., Preisinger, C., Bleijerveld, O.B., van Holten, T.C., Heck, A.J. & Scholten, A. (2011) Collagen stimulation of platelets induces a rapid spatial response of cAMP and cGMP signaling scaffolds. *Mol Biosyst*, DOI: 10.1039.

Masuda, W., Betzenhauser, M.J. & Yule, D.I. (2010) InsP3R-associated cGMP kinase substrate determines inositol 1,4,5-trisphosphate receptor susceptibility to phosphoregulation by cyclic nucleotide-dependent kinases. *J Biol Chem*, 285, 37927-37938.

Michael, S.K., Surks, H.K., Wang, Y., Zhu, Y., Blanton, R., Jamnongjit, M., Aronovitz, M., Baur, W., Ohtani, K., Wilkerson, M.K., Bonev, A.D., Nelson, M.T., Karas, R.H. & Mendelsohn, M.E. (2008) High blood pressure arising from a defect in vascular function. *Proc Natl Acad Sci U S A*, 105, 6702-6707.

Obst, M., Tank, J., Plehm, R., Blumer, K.J., Diedrich, A., Jordan, J., Luft, F.C. & Gross, V. (2006) NO-dependent blood pressure regulation in RGS2-deficient mice. *Am J Physiol Regul Integr Comp Physiol*, 290, R1012-1019.

Osborne, B.W., Wu, J., McFarland, C.J., Nickl, C.K., Sankaran, B., Casteel, D.E., Woods, V.L., Jr., Kornev, A.P., Taylor, S.S. & Dostmann, W.R. (2011) Crystal Structure of cGMP-Dependent Protein Kinase Reveals Novel Site of Interchain Communication. *Structure*, 19, 1317-1327.

Payne, M.C., Zhang, H.Y., Prosdocimo, T., Joyce, K.M., Koga, Y., Ikebe, M. & Fisher, S.A. (2006) Myosin phosphatase isoform switching in vascular smooth muscle development. *J Mol Cell Cardiol*, 40, 274-282.

Roy, A.L. (2001) Biochemistry and biology of the inducible multifunctional transcription factor TFII-I. *Gene*, 274, 1-13.

Ruth, P., Pfeifer, A., Kamm, S., Klatt, P., Dostmann, W.R. & Hofmann, F. (1997) Identification of the amino acid sequences responsible for high affinity activation of cGMP kinase Ialpha. *J Biol Chem*, 272, 10522-10528.

Rybalkin, S.D., Rybalkina, I.G., Feil, R., Hofmann, F. & Beavo, J.A. (2002) Regulation of cGMP-specific phosphodiesterase (PDE5) phosphorylation in smooth muscle cells. *J Biol Chem*, 277, 3310-3317.

Schinner, E., Salb, K. & Schlossmann, J. (2011) Signaling via IRAG is essential for NO/cGMP-dependent inhibition of platelet activation. *Platelets*.

Schlossmann, J., Ammendola, A., Ashman, K., Zong, X., Huber, A., Neubauer, G., Wang, G.X., Allescher, H.D., Korth, M., Wilm, M., Hofmann, F. & Ruth, P. (2000)

Regulation of intracellular calcium by a signalling complex of IRAG, IP3 receptor and cGMP kinase Ibeta. *Nature*, 404, 197-201.

Schlossmann, J. & Desch, M. (2009) cGK substrates. *Handb Exp Pharmacol*, 163-193.

Schlossmann, J. & Desch, M. (2011) IRAG and Novel PKG Targeting in the cardiovascular system. *Am J Physiol Heart Circ Physiol*, 301, H672-H682.

Schmidtko, A., Gao, W., Sausbier, M., Rauhmeier, I., Sausbier, U., Niederberger, E., Scholich, K., Huber, A., Neuhuber, W., Allescher, H.D., Hofmann, F., Tegeder, I., Ruth, P. & Geisslinger, G. (2008) Cysteine-rich protein 2, a novel downstream effector of cGMP/cGMP-dependent protein kinase I-mediated persistent inflammatory pain. *J Neurosci*, 28, 1320-1330.

Sharma, A.K., Zhou, G.P., Kupferman, J., Surks, H.K., Christensen, E.N., Chou, J.J., Mendelsohn, M.E. & Rigby, A.C. (2008) Probing the interaction between the coiled coil leucine zipper of cGMP-dependent protein kinase Ialpha and the C terminus of the myosin binding subunit of the myosin light chain phosphatase. *J Biol Chem*, 283, 32860-32869.

Sun, X., Kaltenbronn, K.M., Steinberg, T.H. & Blumer, K.J. (2005) RGS2 is a mediator of nitric oxide action on blood pressure and vasoconstrictor signaling. *Mol Pharmacol*, 67, 631-639.

Surks, H.K. & Mendelsohn, M.E. (2003) Dimerization of cGMP-dependent protein kinase 1alpha and the myosin-binding subunit of myosin phosphatase: role of leucine zipper domains. *Cell Signal*, 15, 937-944.

Surks, H.K., Mochizuki, N., Kasai, Y., Georgescu, S.P., Tang, K.M., Ito, M., Lincoln, T.M. & Mendelsohn, M.E. (1999) Regulation of myosin phosphatase by a specific interaction with cGMP- dependent protein kinase Ialpha. *Science*, 286, 1583-1587.

Tang, K.M., Wang, G.R., Lu, P., Karas, R.H., Aronovitz, M., Heximer, S.P., Kaltenbronn, K.M., Blumer, K.J., Siderovski, D.P., Zhu, Y. & Mendelsohn, M.E. (2003) Regulator of G-protein signaling-2 mediates vascular smooth muscle relaxation and blood pressure. *Nat Med*, 9, 1506-1512.

Wen, Y.D., Cress, W.D., Roy, A.L. & Seto, E. (2003) Histone deacetylase 3 binds to and regulates the multifunctional transcription factor TFII-I. *J Biol Chem*, 278, 1841-1847.

Wilson, L.S., Elbatarny, H.S., Crawley, S.W., Bennett, B.M. & Maurice, D.H. (2008) Compartmentation and compartment-specific regulation of PDE5 by protein kinase G allows selective cGMP-mediated regulation of platelet functions. *Proc Natl Acad Sci U S A*, 105, 13650-13655.

Wooldridge, A.A., MacDonald, J.A., Erdodi, F., Ma, C., Borman, M.A., Hartshorne, D.J. & Haystead, T.A. (2004) Smooth muscle phosphatase is regulated in vivo by exclusion of phosphorylation of threonine 696 of MYPT1 by phosphorylation of Serine 695 in response to cyclic nucleotides. *J Biol Chem*, 279, 34496-34504.

Yaroslavskiy, B.B., Turkova, I., Wang, Y., Robinson, L.J. & Blair, H.C. (2010) Functional osteoclast attachment requires inositol-1,4,5-trisphosphate receptor-associated cGMP-dependent kinase substrate. *Lab Invest*, 90, 1533-1542.

Zhang, T., Zhuang, S., Casteel, D.E., Looney, D.J., Boss, G.R. & Pilz, R.B. (2007) A cysteine-rich LIM-only protein mediates regulation of smooth muscle-specific gene expression by cGMP-dependent protein kinase. *J Biol Chem*, 282, 33367-33380.

Part 8

Affinity Chromatography as a Quantitative Tool

14

Affinity Chromatography as a Tool for Quantification of Interactions Between Drug Molecules and Their Protein Targets

Piotr Draczkowski, Dariusz Matosiuk and Krzysztof Jozwiak
Faculty of Pharmacy with Division of Medical Analytics, Medical University of Lublin
Lublin
Poland

1. Introduction

Separations in affinity chromatography are based on specific biological interactions between (bio)molecules and in many aspects resemble processes by which these species interact in a living organism. The technique is widely used in biomedical sciences to separate and detect certain molecules based on their defined specificity to other (bio)molecules immobilized on a stationary phase. Moreover,it can also be used for quantitative determination of affinity interactions and their physiological and pharmacological role in a living system. In both cases the same basic physico-chemical effects (hydrophobic interactions, electrostatic, hydrogen bonds etc.) lead to description of the equilibrium state between unbound, free-floating molecules and those forming a complex with a target. This allows the use of affinity chromatography system as a model for analyzing interactions that normally occur in human body. This approach was suggested by analytical chemists over 50 years ago, e.g. Soczewinski and Bieganowska stated in 1969:

"... *If the body is regarded as an extremely complex chromatographic system, in which the blood plays the role of the developing solvent, a certain parallelism can be expected between the behaviour of drugs and their chromatographic parameters in common "simple" partition systems.*"

Therefore, affinity chromatography is currently successfully employed in medicinal chemistry projects for detailed characterization of interactions between drug molecules and their protein targets. This type of liquid chromatography is referred to as analytical or quantitative affinity chromatography (QAC).

In most applications the assay is performed in HPLC systems using a column with a protein immobilized on the surface of the stationary phase. Relatively simple and rapid procedures (time of a single assay average between 5 to 15 min.), ability to multiple use of the same column without significant loss of properties of immobilized protein (Jozwiak et al., 2004) and the possibility to automate the analysis process, make this technique a promising method for screening and determination of relative affinities of a series of analyzed drug molecules. Loun & Hage (1996) reported that theirs column was stable even to 500-1000 injections of an analyte. Above mentioned advantages of this method additionally increase reproducibility of obtained results. Thus, affinity chromatography is considered by many medicinal chemists as good alternative to tedious *in vitro* tests using cell cultures (Jozwiak et al., 2004). It also allows to reduce time needed to perform the assays (Jozwiak et al., 2005).

Special procedures were developed in order to transform retention of a substance or it elution profiles to characterize the interaction between a drug and the target protein. Therefore, various protocols can be used to characterize binding equilibrium constants, the kinetics of drug-protein complex association/dissociation, relative amount of bound drug, number of binding sites in the system or forces responsible for complex formation. Simultaneous application of more than one active substance on the column allows defining interactions between them during binding to immobilized target. It is also possible to determine the affinity of various isomeric forms of the drug to immobilized protein, which is extremely important in a pharmaceutical research (Chen & Hage, 2006). Data obtained by QAC show high correlation with derived from reference methods and can be further used in studies of quantitative structure - activity relationships (QSAR) (Markuszewski & Kaliszan, 2002; Jozwiak et al., 2004).

This chapter reviews current aspects of application of QAC including basic issues of protein immobilization on the surface of the stationary phase, advantages and disadvantages of zonal and frontal elution techniques and vast information which can be provided by competition displacement in QAC studies. The last subsection provides short review of application of these techniques in medicinal chemistry investigations.

2. Immobilization of target proteins on the surface of the stationary phase

QAC assay requires a column where one of the partner of the drug-protein complex is immobilized on the surface of the stationary phase. Both, drug molecules and protein targets may be immobilized on the chromatographic bead particles. The latter option seems to be more versatile as it allows to investigate different types of active substances without having to change the column bead. It directly permits to compare properties of a series of substances on the basis of obtained chromatograms. The way in which the ligand is attached to its support is a key factor in any type of affinity chromatography. Immobilization methods for soluble cytosolic proteins are well established (Taylor, 1991; Turková, 1999; Kim & Hage, 2006; Scheil et al., 2006) and they are based mainly on chemical or physical mechanisms. Physical methods include protein adsorption (physical or ionic adsorption) or protein entrapment within insoluble gel matrix through which only small drug molecules can diffuse. The advantage of these methods is relatively small perturbation of the protein native structure, on the other hand immobilisation forces are weak and promotes the leak of adsorbed protein from the support during use, especially while temperature, pH or ionic strength is changed. Chemical immobilization methods mainly include protein attachment to the stationary phase by covalent bonds or cross-linking reactions. Covalent linkage may alter the native tertiary structure of the protein and cause a change in drug binding properties, but the target associates strongly with the support preventing the desorption phenomenon and increasing a column lifetime. The functional groups that usually take part in this binding are amino, epoxy, carboxyl, sulfhydryl, hydroxyl, diol and phenolic groups which according to the mode of linkage lead to a wide variety of binding reactions such as diazotization, amide bond formation, arylation, Schiff's base formation and amination

(Girelli & Mattei, 2005). Another type of immobilization method is biospecific adsorption. It uses the binding between the ligand of interest and a secondary ligand attached to the support. Although a variety of secondary ligands can be used for this purpose, two of the most common are avidin and streptavidin for the adsorption of biotin-containing compounds, and protein A or protein G for the adsorption of antibodies (Akerstrom & Bjorck, 1986; Wilchek & Bayeras, 1990; Bayer & Wilchek, 1996; Wilchek & Bayeras, 1998; Page & Thorpe, 2002, as cited in: Kim & Hage, 2006).

Various supports are commercially available or have been specifically developed for the immobilization processes, including silica based derivatized matrices (Narayanan et al., 1990; Ruhn et al., 1994; Mateo et al., 2000) and monoliths (Josic & Buchacher, 2001; Lebert, 2008). An important factor is a structure of the support used in preparation of affinity column, since it determines accessibility of protein active sites to substrates. The ideal support must be inert, stable and resistant to mechanical strength, so it can retain its tertiary structure, and this ensures the substrate accessibility to interact with the active sites. Other physical properties, such as porosity, pore size distribution and charge are also important, because they influence the kinetic process (Girelli & Mattei, 2005). Variety of available methods of protein connection with chromatographic beads, gives the opportunity to select and optimize right immobilization method for binding of our protein of interest. It should be noted that the immobilized ligand must as close as possible imitate the behaviour which it exhibits in natural conditions. The method chosen for protein binding should not disrupt the structure crucial for drug binding and provide proper orientation to eliminate any negative steric interactions. Proper immobilization allows retaining activity of the protein on the column, and even its conformational mobility (Beigi et al., 2004), which allows to study allosteric interactions (Chen et al., 2004).

In last years new techniques were developed to immobilize membrane proteins on the surface of chromatographic bead particles to describe the nature of interactions between drug molecules and target receptors. This is very valuable from pharmacological point of view considering that the membrane and transmembrane receptor proteins are targets for almost 75% of current pharmaceuticals (Landry & Gies, 2008). The first membrane protein that was analysed using QAC was glucose transporter present on the surface of red blood cells (GLUT1) (Yang and Lundahl, 1995). Two immobilization techniques were used in this system: in first proteoliposomes or cytoskeleton depleted membrane vesicles containing GLUT1 were immobilized in the pores of size exclusion chromatography beads (Superdex) (Gottschalk et al., 2000) by technique of repeated freezing - thawing (Lundqvist et al., 1998). In the second approach whole cells were immobilized on the surface of beads with positively charged groups (Zeng et al., 1997) or columns with wheat germ lectin agarose gel beads (Gottschalk & Lundahl, 2000). Currently, there are many reports in literature describing the columns with different immobilized receptors and membrane transporters. Most of them use silicon particles with immobilized phospholipids (IAM) (Pidgeon & Venkataram, 1989) which connect parts of the cell membranes of tissues showing high expression of the receptor or cultured cells transformed with receptor gene. It is known as cellular membrane affinity chromatography: CMAC, or CMC. Using this method columns with immobilized nicotinic receptor (nAChR) subtypes α3β4, α4β2, α3β2, α4β4, α7 and subunits β4 and α3 (Zhang et al., 1998; Wainer et al., 1999; Moaddel et al., 2005; Moaddel et al., 2008) purinergic receptors, both acting as ion channels (P2X family) (Trujillo et al., 2007) as well as G-protein coupled receptors (P2Y family)

(Moaddel et al., 2007), the P2Y-like receptor GRP17 (Temporini et al., 2009), β2 adrenergic receptor (Beigi et al., 2004) or the μ and κ opioid receptors (Beigi & Wainer, 2003) were constructed. Multi-receptor columns containing nAChRs, γ-amino-butyric acid receptors (GABA) and N-methyl D-aspartate receptors (NMDA) (Moaddel et al., 2002) were also developed. Kitabatake and co-workers (2008) constructed multi-receptor column and confirmed presence of two types of nAChRs: α7 nAChR and heteromeric nAChRs but also GABA, and NMDA receptors on the surface of 1321N1 and A172 astrocytoma cell lines. The results indicate that the columns can be used to characterize binding affinities of small molecules to each of the receptors, and that this approach can be used to probe the expression of endogenous membrane receptors. With similar immobilization technique beads with transmembrane transport proteins such as P-glycoprotein (Zhang, 2000), human organic anion transporter protein (hOAT1 and hOAT2) (Kimura et al., 2007) and human organic cationic transporter protein (hOCT1) were prepared (Moaddel et al., 2005).

Alternative method was used to immobilize α1 adrenergic receptor (Yu et al., 2005) or muscarinic receptor (Yuan et al., 2005). In this case, the membrane fragments with an interesting protein were subjected to adsorption on the surface of silica particles under vacuum and with use of ultrasounds. Since the phospholipid bilayer fragments show a spontaneous ability to connect, they enfold bead particles forming compact, durable coating of cell membrane. Same approach was used to construct multi-receptor column for the simultaneous determination of drug interactions with the purinergic, P2Y1 and histamine 1 receptors (Moaddel et al., 2010) and multi-receptor column prepared using the glioma cell membranes in order to identify the types of receptors on the surface of these cells by their specific ligands (Kitabatake et al., 2008).

Immobilization of membrane proteins on the inner surface of silica capillaries (open tubular column) is also applied. For this purpose, fragments of cell membrane containing membrane protein bind non-covalently with a capillary using the avidin-biotin pair. Capillaries with immobilized P-glycoprotein (Moaddel et al., 2004), and recently with immobilized cannabinoid receptors (CB1/CB2) (Moaddel et al., 2011) were developed using this approach.

When a native structure of a target protein cannot be obtained during immobilization a drug can be used as an immobilized ligand. However, in this situation any steric interactions should be eliminated by connecting drug molecule to the bead using a spacer with adequate length and hydrophobicity, in order to secure free access to the binding site on the protein molecule. It is important not to connect the drug using functional groups that participate in binding to the target protein. Application of QAC with immobilized drug molecules is particularly useful for identifying biomolecules targets for substances with known therapeutic effect. It is also used as a pre-clinical method of detection of undesired interactions with other system biomolecules (Guiffant et al., 2007).

3. Elution techniques

3.1 Zonal chromatographic studies of drug-protein binding

One method of QAC is a technique known as the zonal elution technique. It was first used by Dunn and Chaiken (1974) as modified low-pressure liquid chromatography method used to investigate the retention of *Staphylococcus* nuclease on the column with immobilized thymidine-5'-phosphate-3'-aminophenylphosphate. Zonal elution method can also be applied using a standard HPLC apparatus equipped with temperature control unit. This method involves injection on the column small volume of analyte solution and then isocratic elution with mobile phase, which usually has a composition and pH reflecting the physiological conditions. Compared to frontal elution (described in subsection 3.2) small amount of analyte is needed to perform the assay in zonal format. Detection is carried out on-line, however there are applications with off-line detection when chromatographic systems has low efficiency (Dunn & Chaiken, 1974).

In result of the analysis the retention factor also called capacity factor is determined. It is expressed by formula $k = (t_r/t_m) - 1$, where t_r is the retention time of the test substance, and t_m is the column dead time. Comparing the value of k for different substances, we can determine their relative affinity for the immobilized protein. Typical chromatogram obtained from zonal chromatography studies is shown in Figure 1. Studies on the affinity of benzodiazepines and related coumarins done by Noctor and colleagues using immobilized human serum albumin (HSA) (Noctor et al., 1993, as cited in Bertucci et al., 2003) have shown a strong correlation (r = 0.999) between the percentage of bounded drug measured by the standard method of ultrafiltration, and the data obtained from chromatographic studies expressed as $k/(k + 1)$.

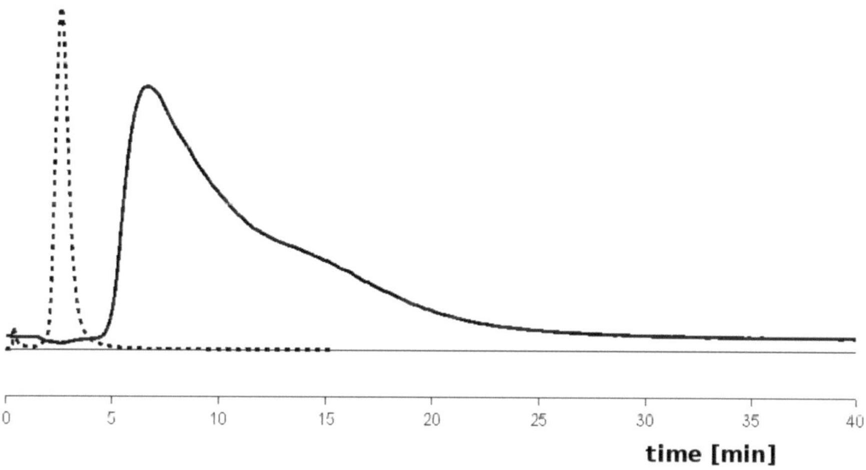

Fig. 1. Typical chromatogram obtained from zonal AC. Comparison of elution peak profiles of ketamine (solid) and negative control phenylbutazon (dashed line) on the column with immobilized nicotinic receptor (nAChR), subtype α3β4. Adapted with permission from Jozwiak, K., Ravichandran, S., Collins, J. R., & Wainer, I. W. (2004). Interaction of noncompetitive inhibitors with an immobilized α3β4 nicotinic acetylcholine receptor investigated by affinity chromatography, quantitative-structure activity relationship analysis, and molecular docking. *J Med Chem*, Vol. 47, No. 16, pp. 4008-4021. Copyright 2011 American Chemical Society.

Similar results were reported by Cheng et al. (2004). In this case, the working curve between literature values of % drug bounded to HSA (by either ultrafiltration or dialysis method) and $k/(k + 1)$ factor determined by chromatography method showed good linearity with the coefficient of 0.96, which is acceptable considering diversity of drugs tested and the variety of %binding data resources used.

According to the Equation 1 (Chaiken, 1987, as cited in Cheng et al., 2004):

$$k = \frac{K_a m_L}{V_M} \tag{1}$$

compound retention factor is closely related with equilibrium constant of binding reaction to immobilized protein target. This allows to define an order relative binding strength of different compounds by comparing their retention factors (Xuan & Hage, 2005) In Equation 1, V_M is the void volume of the column (i.e., the elution volume of a non-retained solute), m_L is the moles of active binding site and K_a is association equilibrium constant for the injected solute at this site .

Zonal elution technique is also used to determine the forces that play a fundamental role in the formation of drug-protein complex. Changing assay conditions (e.g., pH, ionic strength, content of organic modifier) allows to determine which factors affect the most binding reaction. For example, retention dependence on mobile phase pH indicates a considerable contribution of Coulomb interactions in the binding of a drug. In turn, addition of organic modifier can accelerate the elution of analyte by disturbing the hydrophobic interactions (Hage & Chen, 2006). It also allows to examine the change in binding of drugs when standard physiological system conditions will be changed as a result of pathological lesions (Basiaga & Hage, 2010).

Temperature studies allow to define changes in enthalpy and entropy of interactions between the drug and immobilized protein. It follows from the equilibrium constant depending on temperature, which can be described by the Equation 2:

$$\ln K_a = \frac{-\Delta H}{RT} + \frac{\Delta S}{R} \tag{2}$$

where ΔH express a change of enthalpy and ΔS entropy change in a place of interaction, R is the gas constant and T is absolute temperature. So if the system meets the assumption that the number of binding sites (m_L) does not change with temperature, and this is a single-site binding then $\ln K_a$ plotted against $1/T$ should be linear with a slope equal to $-\Delta H/R$, and intercept to ΔS. The total energy change can be calculated using Equation 3 (Kirkwood & Oppenheim, 1961, as cited in Yang & Hage, 1993):

$$\Delta G = -RT\ln(K_a) \tag{3}$$

The method, however, requires earlier determination of K_a at given temperature, for example, by conducting self-competition studies with test compound at different temperatures. It is also possible to designate the enthalpy and entropy of binding directly from the value of retention factor. In this case, if a binding has a single-site character, the plot of $\ln k$ against $1/T$ is linear with a slope equal to $-\Delta H/R$ and the intercept equal to $[\Delta S/R + \ln(m_L/V_M)]$. However, the calculation of the value of ΔH and ΔS requires the prior determination of concentration of binding sites for the analyte (m_L/V_M) (Yanda & Hage, 1993).

3.2 Frontal affinity chromatography (FAC) in drug-protein interaction investigations

Another commonly used method of determining the drug-target protein affinity is the frontal technique. Although this elution format is practically no longer used in analytical applications of chromatography, it is still successfully applied in QAC assays, and has some advantages over zonal elution methods. For the first time it was used in 1975 by Kasai and Ishii. In contrast to zonal elution, a test substance is applied continuously on the column as an addition to the mobile phase in specified concentration. The result is essentially a titration of active sites within the column. As the mobile phase flows through the column, the analyte saturates binding sites on the immobilized protein and we can observe a gradual increase of the amount of unbound analyte leaving the column. This produces a vertical rise in the chromatographic trace, called breakthrough curve which ends or plateaus when the immobilized target is fully saturated. Initial, relatively flat portion of the chromatographic traces represents the non-specific and specific binding of the tested compound to the cellular membranes and the target. Inflection points of breakthrough curves shift to shorter breakthrough times (volumes) as the ligand concentration increases (see Figure 2).

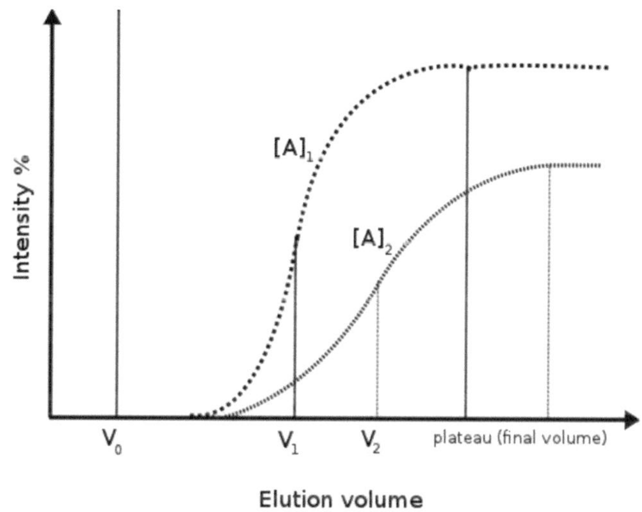

Fig. 2. Typical breakthrough curves of two analyte concentration $[A]_1$ (higher) and $[A]_2$ (lower). V_o represents breakthrough volume of the ligand in the absence of the binding event.

Extent and asymmetry of obtained chromatographic profile are related to the analyte-target protein binding kinetics. Measuring the breakthrough times for several concentrations and fitting the results to equations based on various reaction models allow to characterize a nature of binding affinity and the amount of immobilized target on a column. This is accomplished by plotting number of apparent moles of analyte required to reach the mean point of the breakthrough curve ($1/m_{Lapp}$) versus $1/[A]$ (where $[A]$ is applied analyte concentration). According to Equation 4 (Loun & Hage, 1992), in case of single site reaction, plot ($1/m_{Lapp}$) versus $1/[A]$ should give a linear response (see Figure 3) with a slope equal to $1/(K_A m_L)$ and an intercept of $1/m_L$. Dividing the intercept by the slope allows to obtain information about equilibrium binding constant of analysed interaction.

$$m_{Lapp} = \frac{m_L K_A[A]}{1 + K_A[A]} \text{ or } \frac{1}{m_{Lapp}} = \frac{1}{K_A m_L[A]} + \frac{1}{m_L} \quad (4)$$

A similar approach, but using a combination of both nonlinear and linear fits, can be used for more complex systems that involve multisite interactions (Jacobson et al., 1993; Tweed et al., 1997, as cited in Hage & Chen, 2006).

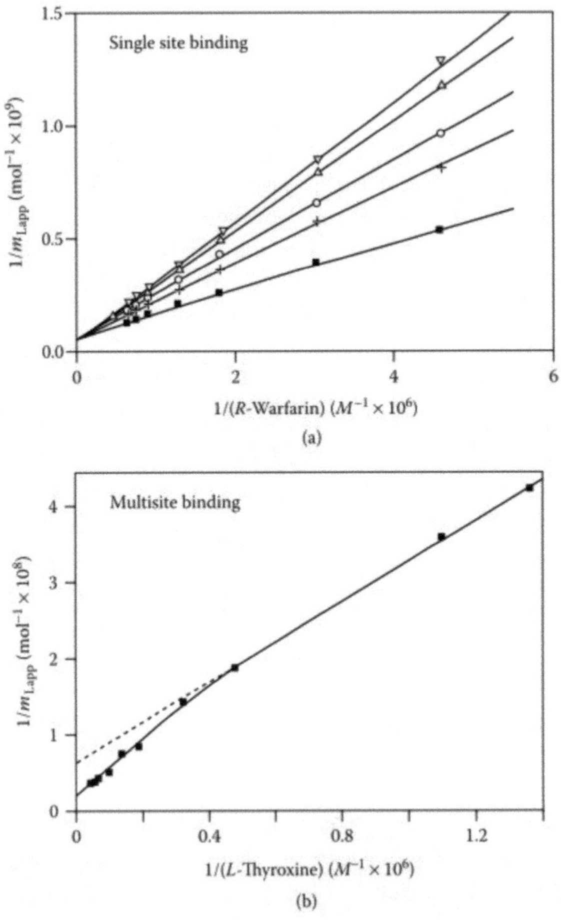

Fig. 3. Examples of double-reciprocal frontal analysis plots for systems with (a) single-site binding and (b) multisite binding. Reprinted with permission from: Loun, B., & Hage, D. S. (1994). Chiral separation mechanisms in protein-based HPLC columns. 1. Thermodynamic studies of (*R*)- and (*S*)-warfarin binding to immobilized human serum albumin. *Anal Chem*, Vol. 66, No. 21, pp. 3814-3822; Tweed, S. A., Loun, B., & Hage, D. S. (1997). Effects of ligand heterogeneity in the characterization of affinity columns by frontal analysis. *Anal Chem*, Vol. 69, No. 23, pp. 4790-4798. Copyright 2011 American Chemical Society.

The simplest binding event, involving the interaction of a ligand with a single type of binding site can also by described by Equation 5. There may be multiple, equivalent sites in a given target molecule but the model assumes their independence.

$$V - V_0 = \frac{B_t}{[A]_0 + K_d} \qquad (5)$$

In this basic FAC equation (5), two variables are present: $[A]_0$ (ligand infusion concentration) and $V\text{-}V_0$ (breakthrough volume V for the ligand corrected by the breakthrough volume of the ligand in the absence of the binding event V_0). This simple equation indicates that once B_t (corresponding to the dynamic capacity of the affinity column for the ligand) and the concentration of the analyte are known, the dissociation constant can be determined from a single measurement of its $V\text{-}V_0$. In order to determine B_t, various concentrations of the compound are pumped through the column and the corresponding $V\text{-}V_0$ values are measured. The analysis of changes in $V\text{-}V_0$ versus $[A]_0$ by means of Lineweaver–Burk type double reciprocal plot or standard nonlinear regression analysis, B_t (equal to the reciprocal of the y intercept) and dissociation constant K_d (in M, expressed as negative reciprocal of the x intercept) are obtained (Calleri et al., 2009). Reliable measurement of V_0 requires a suppression of specific binding or application of a saturating ligand concentration. In case of membrane proteins it can be done by measuring retention of marker using a column constructed with membranes from cells that do not express the target protein (Moaddel et al., 2005). Other approach that can be employed to differentiate between specific and non-specific interactions is parallel chromatography system in which analysed compounds are simultaneously applied on a column with immobilized membranes containing target protein (experimental) and on a column with immobilized membranes from the same cell line that does not express the target protein (control). The assumption is that all of the non-target interactions between a test compound or protein and the cellular membranes will be the same for the control and experimental cell lines and will be reflected in the chromatographic retention on the control column. Then the difference in compound retention between control and experimental column will reflect only the specific binding. This system may be applied in both, zonal and frontal chromatographic studies (Baynham et al., 2002, as cited in Moaddel & Wainer, 2006).

In frontal QAC connected with mass spectrometry (FAC-MS), ligand is sequentially infused at increasing concentrations, but with no washing steps between infusions. This allows to determine binding parameters in a single experiment. This is accomplished by infusing a FAC–MS column sequentially starting with the lowest of a series of concentrations of analyte. For this approach, referred to as a modified staircase, the summed concentrations ($[A]_0$+y) refer to initial concentration of the ligand for the first step of the staircase but for the second step of the staircase, it will be the sum of the initial concentration plus the concentration of the second step. Similarly, the concentration of the ligand for the third step of the staircase will be the sum of the initial, second and third steps, and so on for the remaining concentrations. A plot of $[A]_0$+y versus reciprocal breakthrough volume supports the determination of B_t and K_d by linear regression analysis. This type of assays referred to as direct measurements might not always be advantageous, especially if each tested ligand required unique mass spectrometric conditions. (Chan et al., 2003; Slon-Usakiewicz et al., 2005). Indirect methods will be described further (see subsection 3.3) since they require usage of known competitive marker ligand. Figure 4 illustrates an ideal chromatogram that would arise from the application of this procedure. Note that successively higher void marker (a compound that has no affinity for the immobilized protein target and gives the same elution front whether the target protein is present in the column or not) concentrations are applied as well, to ensure accurate measurement of V_0.

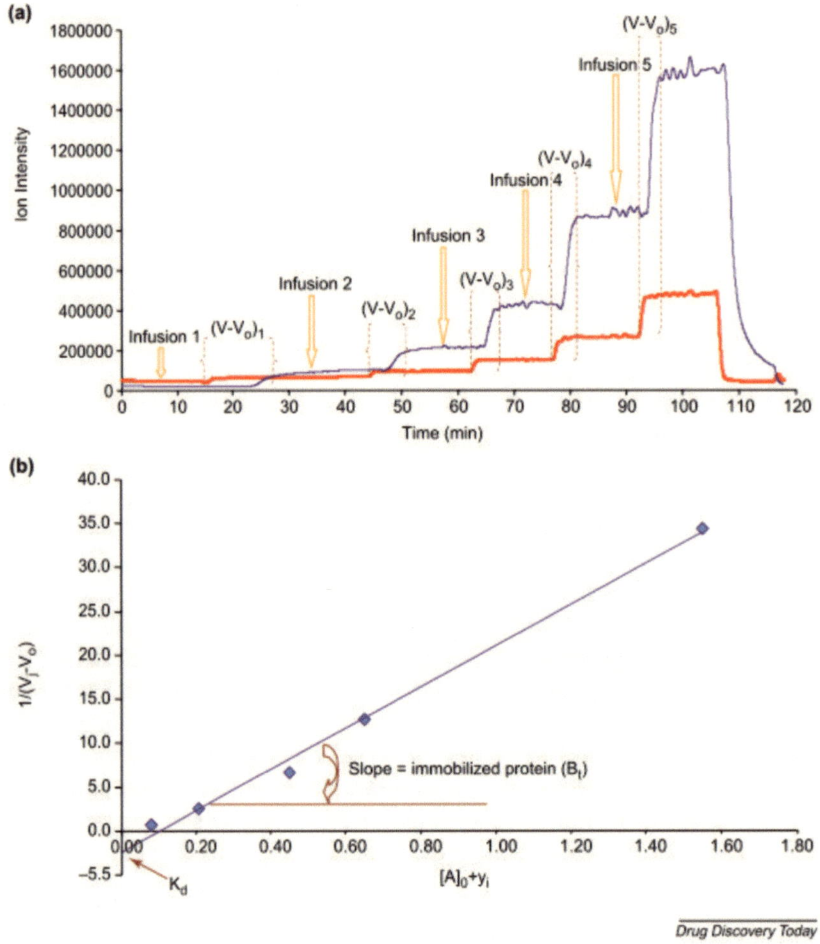

Fig. 4. Depiction of a typical 'modified staircase' experiment to evaluate K_d for a protein target (immobilized in the column) and a small molecule ligand. (a) The ligand (blue) is infused at increasing concentrations starting from an initial (i) to a final (j) concentration along with a void marker (red). (b) The reciprocal of the breakthrough volumes, $1/(V_j-V_0)$, are plotted against the summed ligand concentrations, $[A]_0+y_j$, to produce a linear correlation from which the K_d value can be determined from the y intercept. The total amount of immobilized protein (B_t) in the column is then obtained from the slope . Reprinted from *Drug Discovery Today*, Vol. 10, No. 6, Slon-Usakiewicz, J. J., Ng, W., Dai, J. R., Pasternak, A., & Redden, P. R. (2005). Frontal affinity chromatography with MS detection (FAC-MS) in drug discovery, pp. 409-416, Copyright 2011, with permission from Elsevier.

The association constant measured from frontal chromatography can be directly related to the retention factor obtained from zonal elution chromatography using the same column. Kim & Wainer (2008) reported a linear relationship (r^2 = 0.9993, n = 7) between the standard association constants from frontal analysis and retention factors from zonal elution using reference drugs analysed on a column with immobilized HSA. This standard plot was later used for rapid determination of association constants of various drugs which show low to medium binding affinity to HSA. Determination of association constants was as fast as 1.5 min and showed a high correlation to equilibrium dialysis or ultrafiltration. The combination of frontal and zonal chromatography for determining association constants showed several advantages, one being rapid determination of association constant of drug to HSA.

Other notable advantages include an ease of automation and simultaneous ability to distinguish association constants of chiral compounds at the same time. Authors suggested that the same approach could be used for studying interaction of other drugs and proteins and should further improve overall drug screening process.

There are several advantages of frontal over zonal chromatographic method. In the FAC technique, the dynamic capacity of the affinity column for the ligand and the dissociation constant for the interaction can be obtained from a single set of experiments. In case of zonal elution technique to determine number of binding sites a separate assay or self-competition studies are required This makes this approach valuable in characterizing the properties of a column and in obtaining accurate measurements of binding affinity and activity.

The direct methods described above are applicable for a broad K_d range from submicromolar to low millimolar. The lower limit to K_d determination is dictated by the detectability of a given ligand, and while this is both compound- and detector-dependent. Zonal chromatographic approach requires a protein stationary phase concentration much greater than the applied ligand concentration to ensure the experiment is conducted in the linear region of the binding isotherm. Thus consumption of target protein is increased and miniaturization capabilities are lower. Smaller column leads to lower amount of immobilized protein but also require lower ligand concentrations and this challenges the detectability of the ligand. From the same reasons application of zonal chromatography is limited to low and mid-micromolar K_d range because either strong (and thus difficult to remove from a column) or weak (and thus not well retained) interactions can lead to low concentrations of ligand at the detector (Chan et al., 2003).

Frontal elution technique is particularly useful in studies on solvent and temperature influence on drug interaction with target molecule. As it was mentioned earlier, analyte retention shifts may be due to alterations in either the affinity or number of binding sites. In frontal chromatography this is not a problem, since data on both affinity and activity are provided in the same experiment. However, frontal techniques need to use much larger amount of analyte (Hage & Chen 2006).

For more theoretical information and practical considerations about frontal and zonal elution techniques see comprehensive chapters about QAC in book "Handbook of Affinity Chromatography" (2006) edited by prof. David S. Hage.

3.3 Competition displacement studies

QAC allows to study relationships between different drugs interacting with the same target protein by observing the effect of the addition of one compound on the retention of the second. In zonal technique, the retention of a drug is measured in specially prepared mobile phase with addition of constant, known concentration of competitive agent. Consecutive injections of an analyte can be applied to a series of mobile phases with increasing concentration of competitive agent and changes in the retention as a function of competitor concentration allow describing its interaction with test analyte. In case of direct single site competitive interactions we should see direct, linear dependence of reciprocal of analyte retention factor ($1/k$) on concentration of competitor added to the mobile phase *[I]*. Increasing value of $1/k$ (drop in the retention) with an increase of additive concentration suggest positive competition of both molecules on the binding site. If this relationship is nonlinear, and the increase of competitive reagent in the mobile phase decreases the retention of the analyte (increasing $1/k$), this suggests that there is negative allosteric interaction between them or multisite competitive interaction. The nonlinear nature of the dependence $1/k$ on *[I]* characterized by an increase in analyte retention as higher concentrations of compound are added to the mobile phase, indicates a positive allosteric interaction (Hage & Chen, 2006). Examples of such plots representing interactions above mentioned are shown in Figure 5.

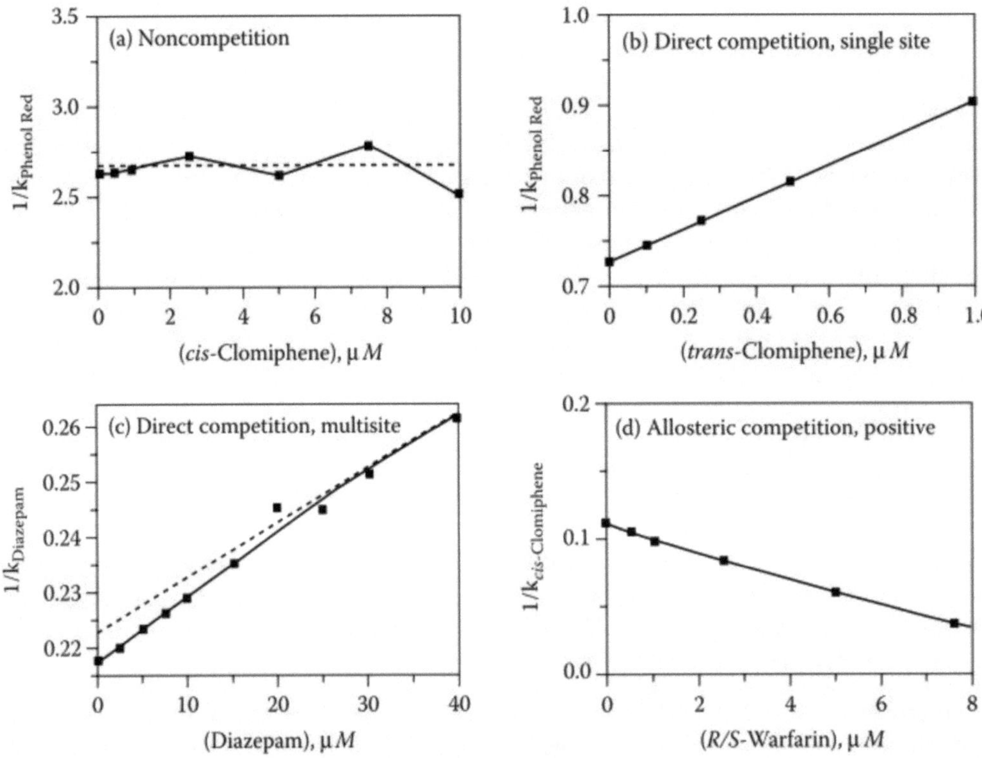

Fig. 5. Reciprocal plots prepared for analyte and competing agents with various types of competition on immobilized HSA columns . Reprinted from *Journal of Chromatography B*, Vol. 768, No. 1, Hage, D. S. (2002). High-performance affinity chromatography: a powerful tool for studying serum protein binding, pp. 3-30, Copyright 2011, with permission from Elsevier.

When the interaction is limited to a single binding site on the protein molecule, and analysed substances show no other interaction with the stationary phase, the Equation 6 describes the observed retention:

$$\frac{1}{k} = \frac{K_I V_M [I]}{K_a m_L} + \frac{V_M}{K_a m_L} \tag{6}$$

In this equation, the V_M determines the volume of eluted substances when they are not interacting with the ligand (the column dead volume) m_L describes the number of moles of binding sites for the analyte (A) and competitive agent (I), *[I]* the concentration of added competitive agent, and K_a and K_I binding equilibrium constants, respectively (Hage et al., 2009). In this case, the ratio of the slope of a graph of $1/k$ dependence on *[I]* to its intercept will determine the K_I. Determination of K_a requires separate measurement of the concentration of binding sites in the column (m_L/V_M). Similar approach can be used to define binding affinity of multisite and allosteric interaction by choosing the equation best describing the shape of the dependence $1/k$ on *[I]* (Hage & Chen, 2006; Joseph & David 2010). In case of simple allosteric interactions this Equation 7 will have the form (Chen & Hage, 2004):

$$\frac{k_0}{(k_0 - k)} = \frac{1}{(\beta_{I \to A} - 1)}\left(1 + \frac{1}{K_{IL}[I]}\right) \tag{7}$$

In this equation, k_0 defines the retention factor of an analyte without addition of compound I in the mobile phase, and the K_{IL} equilibrium binding constant of I with the immobilized protein. In the case of allosteric interactions between compound A and I, the presence of compound I leads to increase or decrease in the binding of substance A, which in turn changes the equilibrium binding constant (K_{AL} to K'_{AL}). This change is represented in above equation as $\beta_{A \to I}$, which is equivalent to the ratio of K'_{AI}/K_{AL}. If the analysed interaction fulfill the assumption described by above equation (7), the plot of $k_0/(k_0-k)$ against $1\ /\ [I]$ should be linear. Intercept in this case is equivalent to $1/(\beta_{A \to I-1})$, and the slope is $1/[(\beta_{A \to I}-1)K_{IL}]$. Based on these values it is possible to calculate the $\beta_{A \to I}$ and K_{IL}. The value of $\beta_{A \to I} > 1$ indicates a positive allosteric effect of compound I on binding of the analyte A, while $\beta_{A \to I} < 1$ indicates a negative allosteric interaction between these two substances. The value of $\beta_{A \to I}$ equal to zero suggests competitive interaction between I and A on the immobilized ligand, while the value of $\beta_{A \to I}$ equal to unity indicates the absence of any effect of compound I on the binding of the substance A (Chen & Hage, 2004). Therefore, it is important that the retention of injected analyte resulted solely from its interaction with the immobilized ligand, and not from the column overload. In order to confirm this, the retention time should remain unchanged at different initial concentration of injected sample. If not, the amount should be decreased until the retention time remains constant and/or increase the volume of the column. Also number of injections at different flow rates should be done to confirm that the processes of association/dissociation are fast enough (compared to how much time analyte spends in the column) to create inside the column local state of a binding equilibrium (Loun & Hage, 1995). Considerations on the factors that need to be pointed out using the zonal QAC are described in publications by prof. D. S. Hage (2002).

Competition and displacement technique allows to study drug-protein interactions occurring on a single binding site, which interacts with injected compound even though the drug is bound to several different sites on the protein. In this case, the analysed compound (drug) should be used as a competitive addition to the mobile phase, and as injected compound analyte with known, specific binding site on the protein. In addition, this approach allows to directly calculate the binding equilibrium constant of analysed drug from the ratio of slope to intercept in plot $1/k$ against *[I]* (Loun & Hage, 1995; Chen et. al., 2004; Yoo et al., 2009, Mallik et al., 2008).

Knowing the location of specific binding site for a substance on the target protein and applying an analyte on the column in presence of it, becomes possible to define binding site of analysed drug (Yoo et al., 2009; Mallik et al., 2008). Carrying out series of this type of experiments we can draw a map of allosteric protein binding sites (Chen et al., 2004).

It is also possible to use the same compound as an injected analyte and a competitive agent (self-competition studies). In this case the Equation 8 describing single site competitive interaction will take the form (Hage, 2002):

$$\frac{1}{k} = \frac{V_M[A]}{m_L} + \frac{V_m}{K_a m_L} \quad (8)$$

Plotting the dependence of $1/k$ on *[A]* we can obtain information about the number of binding sites for the analyte. In case of a single site interactions it should be a linear relationship with a slope equal to m_L/V_M, and the ratio of the slope to intercept will determine K_a (Xuan & Hage, 2005).

As the zonal elution method, frontal chromatography can also be used in combination with competition-displacement technique. Increasing concentrations of the competitive ligand in the presence of constant concentration of a marker are added to the mobile phase and the effect on the breakthrough volumes of marker is measured. Decreasing breakthrough time of marker with increasing concentration of competing agent suggests direct competition between them. If positive or negative allosteric effect of displacement agent on binding reaction is occurring, shift to lower or higher marker breakthrough times should be observed. Using Equation 9 the relationship between displacer concentration *[D]* and marker retention volume can be used to determine the K_d value of the displacer as well as the number of active binding sites (B_t).

$$[D](V - V_0) = \frac{B_t[D]}{(K_d + [D])} \quad (9)$$

In above equation V is the retention volume of marker and V_0 is the retention volume of marker when the specific interaction is completely suppressed. From the plot of $[D](V\text{-}V_0)$ versus *[D]*, dissociation constant values for displacer ligand can be obtained (Moaddel & Wainer, 2006).

Chan and co-workers (2003) reported an alternative "competitive" assay format for K_d determination of tested ligands. FAC column is equilibrated with increasing concentrations of test ligand, with no washing between infusions. Each equilibration is bracketed by an infusion of the indicator (marker ligand detectable by MS) and void marker, and the adjusted breakthrough volume ($V\text{-}V_0$) of the indicator is monitored by mass spectrometry. At high indicator dilution relative to K_d of the particular interaction ($[A]_0$ is negligible compared with K_d), the breakthrough value is insensitive to slight changes in its concentration and achieves its maximum value (V_{limit}). In this mode the indicator is not competing with test ligand, but merely quantitating uncomplexed immobilized protein (column capacity). This operational mode is very practical because a weak ligand (i.e., with high K_d) can be applied at modest concentrations and still function in the linear region of its binding isotherm. Weak ligands break through the column quickly and are easily washed out, thus providing a rapid probe of column capacity. This method is referred to as the indirect staircase approach for K_d determination (Chan te al., 2003).

The competition displacement methods are insensitive to non-specific binding since they are measuring a retention changes only due to allosteric or competitive interactions between drugs on a specific binding surface of target protein.

Aside from the utility of the FAC technique to provide accurate dissociation constants measurements for individual ligands, combination of frontal QAC with mass spectrometry allows rapid screening of mixtures of substances for their pharmacological activity, and the results show a high correlation with those obtained by traditional methods. FAC-MS screenings can be done with and without an indicator. In the second case detection is performed at selected *m/z* values to detect individual ligands and void marker. This allows to evaluate set of ligands in a single experiment (Ng et al., 2007).

In the indicator method, analysed compound is applied on the affinity column simultaneously with the indicator and the extent (or percentage) to which they shifts an indicator is determined. To compare the reductions in the breakthrough times for indicator in the presence of the ligands, the %shift is quantified from Equation 10. This FAC-MS readout can be used to rank the binding of a series of ligand or ligand mixtures: the greater the percentage shifts, the greater the degree of competition with the indicator. Mixtures in which a significant displacement (or shift) of the indicator is observed merit further investigation and deconvolution (Slon-Usakiewicz et al., 2004, as cited in Calleri et al., 2010).

$$\%shift = \frac{t_I - t}{t_I - t_{NSB}} \times 100 \qquad (10)$$

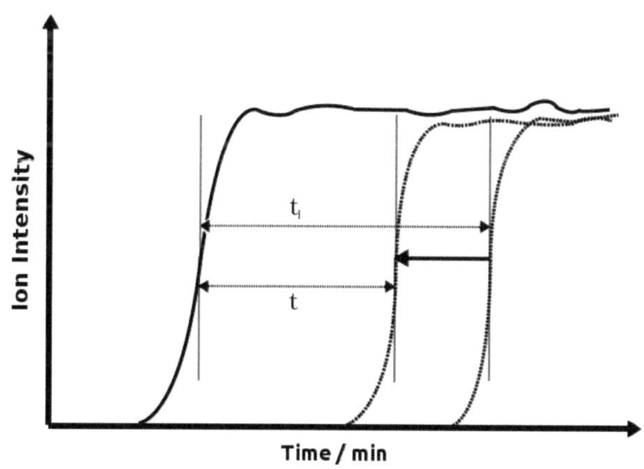

Fig. 6. Typical FAC curves obtained using the "indicator" screen method.

In above equation (10) t is the breakthrough time difference, measured at the inflection point, of the sigmoidal fronts between the indicator and a void marker in the presence of any competing ligand(s); t_{NSB} is the non-specific binding breakthrough time difference in the absence of immobilized protein (and is a constant for the indicator used); and t_I is the breakthrough time difference in the absence of any competing ligands. Typical FAC curves obtained using the indicator screen method is shown in Figure 6.

For more detailed information about FAC-MS technique the readers are referenced to reviews (Chan et al., 2003; Slon-Usakiewicz et al., 2005; Calleri et al., 2009).

3.4 Nonlinear chromatography for determination of kinetics parameters

Peaks obtained in zonal AC differ from Gaussian shape observed frequently in classical chromatographic analyses. Because we have to deal with column overload and slow kinetics of adsorption/desorption during assays, peaks exhibit a strong tailing, which increases with increased concentration of an analyte. Observed asymmetry can arise from a variety of other factors-including extra column effects, heterogeneity of the stationary phase, heterogeneous mass transfer or a non-linear isotherm (Wade et al., 1987, as cited in Moaddel & Wainer, 2006). The degree of deviation from a Gaussian distribution is a function of applied ligand concentration and the kinetics of ligand–receptor interactions occurring during the chromatographic process. An example of the effect of solute concentration on peak shape is presented in Figure 7, which shows the example of usage of nonlinear chromatography (NLC) for determination of kinetic parameters for the α3β4 nAChR allosteric inhibitors (Jozwiak et al., 2002).

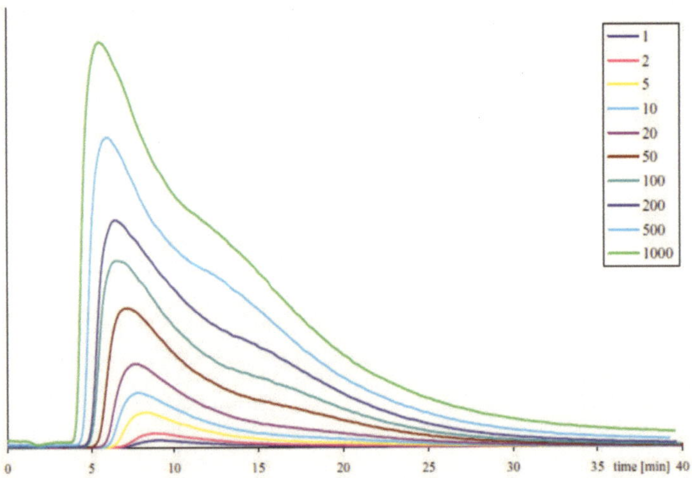

Fig. 7. The effect of increasing concentraions of mecamylamine, from 1 to 1,000 mM, on the chromatographic profiles of mecamylamine. For experimental details, see reference (Jozwiak et al., 2002).

While peak tailing (or fronting) is a problem in analytical separations, concentration-dependent asymmetry can be used with NLC techniques to characterize the separation processes occurring on the column. When the chromatographic process includes binding interactions between a ligand and an immobilized membrane bound receptor, the NLC approach can be used to calculate the association rate constant (k_{on}), dissociation rate constant (k_{off}) for the ligand-receptor complex and the equilibrium constant for complex formation (K). One approach to the analysis of NLC data is the Impulse Input Solution for the mass balance equation developed in 1987 by Wade and collaborators (Wade et al., 1987, as cited in Moaddel & Wainer, 2006). This approach is based upon the observation that when adsorption/desorption rates are slow, band broadening is insensitive to a moderate degree of column overload. In contrast to numerical integration methods this approach uses the analytical solution, which can be applied directly to fit experimental peak profiles. The Impulse Input Equation has been included in commercially available deconvolution software, and can be easily applied to NLC studies. The peak area parameter (a_0), peak center parameter (a_1), its width (a_2) and distortion (a_3) are the parameters used to describe the chromatographic traces. Thermodynamic and kinetics parameters of a drug-receptor complex formation are then calculated on the basis of the relationship:

$k' = a_1$ the real thermodynamic capacity factor

$k_d = 1/a_2 t_0$ desorption reaction rate constant

$K_a = a_3/C_0$ adsorption equilibrium constants, where C_0 is the concentration of solute injected multiplied by the width of the injection pulse (as a fraction of column dead volume)

$k_a = k_d K_a$ reaction rate constant of adsorption

4. Examples of application

QAC was successfully used to describe drug interactions with multiple different system proteins. In case of soluble proteins this approach was applied to determine binding of different active substances (drugs, hormones) with e.g. serum albumins (HSA), α1-acid glycoprotein (AGP) or nucleic estrogen receptors (hER).

Xuan & Hage (2005) conducted research on immobilized α1-acid glycoprotein (AGP) demonstrating good correlation (0.954) of observed retention factors of several compounds with their equilibrium binding constants to AGP designated by other methods. In the same work authors, using the self-competition technique, confirmed the literature data on existence of one binding site for propranolol enantiomers on this protein. Determined values of K_a were also included in the range of values known from other experiments. Performing the analysis at different temperatures, changes in enthalpy and entropy of propranolol enantiomers binding to α1-acid glycoprotein, and their contribution in the total energy change in the process of binding were determined. It was found, that this reaction depends mainly on the enthalpy, but entropy change also significantly affects the binding of propranolol. Linearity of the plot confirmed the single-side character of binding of this compound by the AGP.

Combination of zonal QAC with competition-displacement studies confirmed the negative allosteric nature of the interaction between verapamil and tamoxifen. It was determined that verapamil causes 41% decrease in binding constant of tamoxifen by immobilized HSA (Malik et al., 2008). As mentioned earlier, using compound of interest as a competitive addition to the mobile phase it is possible to define its interactions with target protein on a single binding site even though the drug is bound to several different sites on the protein.

Such solution has been used in the study of binding of both; hormones (Loun & Hage, 1995) and drugs (Chen et. al., 2004; Yoo et al., 2009; Mallik et al., 2008) to different places in the HSA molecule. Simultaneous injection of imipramine with L-tryptophan on the column with immobilized HSA, showed a competitive interaction between those two drugs, confirming that imipramine specifically connects to the indole-benzodiazepine binding site on the molecule of albumin (Sudlow site II). This analysis also allowed to determine the equilibrium constant of this interaction (Yoo et al., 2009). Analogously was defined binding site for the verapamil on this protein, demonstrating in this case, competitive interaction with warfarin, specifically connecting to Sudlow site I. This was confirmed by elution in the presence of tamoxifen. Nonlinear dependence of retention of this compound on verapamil concentration showed a high correlation to the equation describing the allosteric interaction, characteristic for interaction that is observed between the tamoxifen binding site and the warfarin binding site (Sudlow I) (Mallik et al., 2008). Team of Chen and colleagues (2004) determined the nature of the interaction of phenytoin by drawing a map of interactions of the drug with the major binding sites on the surface of HSA on the basis of the relationship between $1/k$ of the compounds with confirmed specific binding site on albumin molecule, and various concentrations of phenytoin as a mobile phase additive during elution. Thus suggested the potential impact of other concomitant medications on the efficacy of therapy with phenytoin.

The effect of different assay conditions on drug-protein interaction was used, for example to define the influence of long-chain fatty acids concentration in the plasma and glycation of plasma transport proteins on the binding of sulfonylurea drugs in diabetes. Experiments were carried out using columns with immobilized HSA (Basiaga & Hage, 2010).

Recently Sanghvi and co-workers (2010) successfully immobilized the ligand binding domains of the estrogen related receptors ERRα and ERRγ onto the aminopropyl silica liquid chromatography stationary phase, as well as the surface of the open tubular capillaries, creating the ERR-silica and ERR-OT columns. Both types were characterized using frontal chromatographic techniques with diethylstibesterol and the binding affinities, expressed as K_d values, to the immobilized receptors were consistent with the literature data. Biochanin A, the ERRα agonist, was also used to further characterization of properties of the ERRα-silica column, and obtained K_d value was consistent with the previously reported data. The ERRγ-silica column was characterized using nonlinear chromatographic techniques using a series of tamoxifen derivatives (tamoxifen, 4-hydroxy tamoxifen and N-desmethyl 4-hydroxytamoxifen, Endoxifen). The relative K_d values obtained for the derivatives were consistent with relative ability of the compounds to inhibit the cellular proliferation of the human-derived T98G glioma cell line, expressed as IC_{50} values. The results indicate that the relative retention of compounds on these columns reflects the magnitude of their inhibitory activity. Therefore columns containing immobilized ERRs can be used for a preliminary screen for anti-glioma agents, such as tamoxifen, which work as selective estrogen receptor modulator and that this method may replace current laborious and time consuming cellular uptake studies.

QAC has been applied to study drug-receptor interactions in the α1 adrenergic receptor system. The results showed a positive correlation of chromatographic data (k') with literature data defining the affinity of ligands on the basis of radio ligand studies with cell membrane homogenates containing α1 adrenergic receptor (Yu et al., 2005).

QAC was also used in determination of binding site structure of the α3β4 nAChR. Comparing the retention time of tritium labelled epibatidine ([^{3}H] EB) on four different columns: with immobilized either receptor subunit α3 or β4 subunit, both subunits immobilized on the same column and complete pentameric α3β4 receptor, it was found that the binding site for this compound is formed by several subunits of the complete receptor, and not on the individual subunits. This was confirmed by assays with nicotine as a competitive additive to the mobile phase. No impact of nicotine on the retention of [^{3}H] EB on columns with immobilized subunits was observed, while the epibatidine retention on the column with complete receptor decreased from k = 8.4 to k = 1.5, which indicates the interaction of nicotine on [^{3}H] EB binding site (Wainer et al., 1999). In the same work the effect of pH and ionic strength of mobile phase on the retention of [^{3}H] epibatidine in the α3β4 nAChR column was also examined using zonal elution technique. Analyses showed an increase in epibatidine retention with increasing mobile phase pH from pH 4 to pH 7 and a significant decrease in the retention of this compound with increasing ionic strength of mobile phase (5-200 mM ammonium acetate concentration). These results showed that the binding of competitive agonists is mainly electrostatic interaction between drug molecule and the binding site on the nAChR.

QAC in combination with displacement studies were used in case of α3β4 nAChR to confirm the location of binding sites of allosteric inhibitors of this receptor. Elution of bupropion, ketamine and dextromethorphan was conducted in the presence of mecamylamine (nAChR blocker) as an addition to the mobile phase. The linear dependence of eluted compounds $1/k'$ on the displacer concentration indicates that all tested inhibitors compete for the same binding site on the receptor molecule. Additional studies also showed no effect of mecamylamine in the mobile phase on epibatidine retention and no effect of nicotine in the mobile phase on the retention of ketamine and bupropion. It indicates that ketamine, bupropion, mecamylamine bind to other sites on the receptor surface than nicotine and epibatidine and do not compete for the binding site (Jozwiak et al., 2002). In the same work nonlinear chromatography (NLC) model was used to analyse interactions between the immobilized α3β4 nAChR, and its allosteric inhibitors. Peak profiles from zonal elution of allosteric inhibitors (ketamine, bupropion, mecamylamine, and dextrometorfan), recorded by the mass spectrometry were numerically fitted to Impulse Input Solution model using PeakFit v4.11 software (SPSS Inc., Chicago, IL). The results (k_{on}, k_{off}, K) were consistent with available literature data, thus confirming the effectiveness of the NLC in receptor drug binding kinetics study, in this case allosteric nAChR inhibitors. In further work enantioselectivity of the interaction of dextromethorphan (DM) and levomethorphan (LM) with an immobilized α3β4 nAChR subtype liquid chromatographic stationary phase has been compared to DM- and LM-induced non-competitive blockade of nicotine-stimulated ^{86}Rb+ efflux from cells expressing the α3β4 nAChR. Since DM and LM are enantiomers and have the same physicochemical properties, any chromatographic and pharmacological differences must be due to specific interactions with nAChR. Asymmetrical peaks were observed for both compounds and DM had significantly longer retention time than LM (Figure 8). Determined by NLC k_{on} values of both compounds did not significantly differ while k_{off} of DM was significantly lower than k_{off} value of LM. That was in good agreement with results of the functional inhibition studies which showed that DM and LM had equivalent potencies, i.e. the same IC_{50} values, but that DM inhibition lasted longer than the effect produced by LM. The effect of temperature on the chromatographic retention of DM and LM on the nAChR column was also determined using a sequence of temperature experiments ranging from 5 to 30°C. Respective *k*'s values of independently injected compound were determined and van't Hoff plots were constructed by plotting ln *k* versus 1/*T*. Results showed that the binding reaction is enthalpy driven. In addition, for DM and LM there was no significant difference in the $\Delta S°$ values, while the $\Delta H°$ value for the DM-nAChR interaction was significantly lower than respective value for LM-nAChR interaction showing that the enantioselectivity of α3β4 nAChR is mostly enthalpy-based (Jozwiak et al., 2003).The results of those studies demonstrated that non-linear chromatography approach of investigations of immobilized nAChRs can be useful in characterizing ligand–receptor binding interactions and predicting of properties of drugs and drug candidates. Additional NLC studies with this system were used to construct QSAR models of non-competitive inhibitors of the nAChR (Jozwiak et al., 2004), to develop a molecular model of these interactions (Jozwiak et al., 2004) and to predict IC_{50} values (Jozwiak et al., 2005)

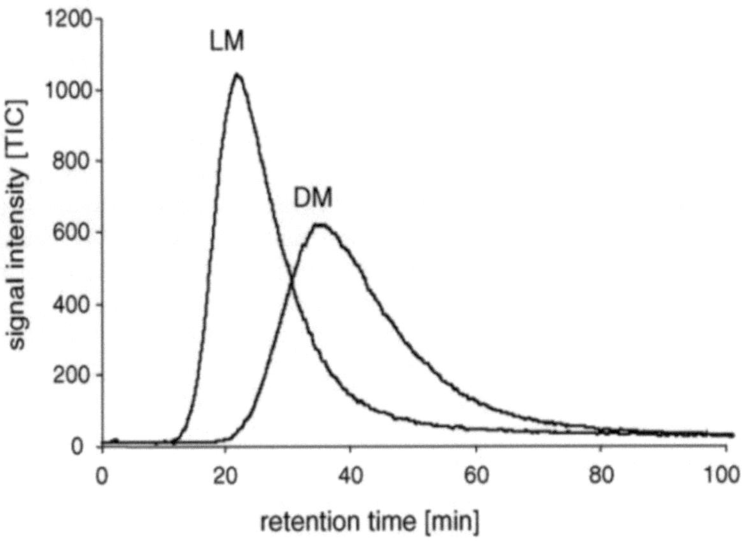

Fig. 8. The comparison of peak profiles of DM and LM obtained in independent experiments of consequent injections. Reprinted from *Journal of Chromatography B*, Vol. 797, No. 1-2, Jozwiak, K., Hernandez, S. C., Kellar, K. J., & Wainer, I. W. (2003). Enantioselective interactions of dextromethorphan and levomethorphan with the α3β4-nicotinic acetylcholine receptor: comparison of chromatographic and functional data, pp. 373-379, Copyright 2011, with permission from Elsevier

Competition displacement studies with [^{3}H]epibatidine and [^{125}I]α-bungarotoxin as marker radioligands specific for α3β4 and α7 nAChRs subtypes respectively were applied to determine K_i values and check subtype selectivity of newly synthesized derivatives of epiboxidine, synthetic epibatidine-related compounds (Rizzi et al., 2008).

Another interesting application of QAC is use of immobilized-enzyme reactors (IMERs). In medicinal chemistry research IMERs are applied to drug metabolism studies, enantioselective analyses and for the identification of substrates and inhibitors as potential drugs. Interestingly, the investigated enzymatic reaction took place directly on the column.

Attractive features of immobilized enzyme reactors are the increased enzyme stability and the reusability coupled to accuracy, automation and potential high throughput when they are inserted in a HPLC system (Bartolini et al., 2005). The approach requires neither highly purified enzyme nor use of labelled substrates (radio- or color labelled). In immobilized enzymes, inhibitors efficacies can be measured either as IC_{50} values or K_i values using numerical transformation. Enzyme kinetic parameters are determined by successive injection of a substrate at increasing concentrations and measuring the rate of enzymatic reaction (V) expressed as peak area of product formed after each injection. Fitting the data to Lineweaver–Burk double-reciprocal plot of $1/(V)$ against the substrate concentration $[S]$ (what is a linear transformation of the Michaelis–Menten plot), K_m and V_{max} values can be obtained. The y-intercept of such graph is equivalent to the inverse of $1/V_{max}$, the x-intercept of the graph represents $-1/K_m$. In order to obtain correct results the concentration of substrate should be normalized according to the formula 11.

$$[S] = \frac{C_{inj} \times V_{inj}}{BV} \tag{11}$$

where C_{inj} is the injected substrate concentration, V_{inj} is the injected volume and BV is the bed volume of the IMER.

To determine the inhibition constant (K_i) for a test drug a set of inhibitor injections in several different concentrations ($[I]$) at two or more concentration of a substrate should be performed. As noted by Dixon in 1953, if $1/V$ is plotted against inhibitor concentration $[I]$, a straight line plot is obtained for each substrate concentration $[S]$. The $[I]$ value of a intersect of those lines is equal to $-K_i$. If curves obtained for several different $[S]$ converge in the left upper quadrant of a chart, the inhibitor is competitive. If curves converge on the $[I]$ axis, the inhibitor is non-competitive. For not competitive inhibition, the lines are parallel.

Simultaneous injections of both a substrate at a fixed saturating concentration and increasing concentration of an inhibitor, result in increasing reduction of the peak area (A_i) in comparison to area obtained for a substrate alone (A_0). The percent inhibition ($100 - (A_i/A_0 \times 100)$) is then plotted against the inhibitor concentration to obtain the inhibition curves (Girelli & Mattei, 2005; Nie & Wang, 2009). Recently this technique has been used for the kinetic characterization of inhibitors specific to brain-targeted butyrylcholinesterase (BuChE) (Bartolini et al., 2009), acetylocholinesterase (AchE) (Bartolini et al., 2004; Bartolini et al., 2005; He et al., 2010) and β-secretase (human recombinant β-amyloid precursor protein cleaving enzyme, hrBACE1)(Mancini & Andrisano, 2010) as potential therapeutics for Alzheimer's disease. IMER was also used for rapid and cost-effective on-line chromatographic screening of matrix metalloproteinases (MMP-9 (Ma & Chun Yong Chan, 2010) and MMP-8 (Mazzini et al., 2011)) inhibitors that may be useful in cancer therapy and for determining the role of some derived plant products for treating NO-dependent smooth disorders using monolithic micro-IMER with covalently bounded arginase (André et al., 2011).

5. Conclusions

The drive to bring innovative drugs to market faster, without compromising quality and safety, induced need for new experimental techniques and methodologies. It is crucial to determine biological activity, drug-target interactions and physico-chemical properties of drug candidates as predictors of administration, distribution, metabolism, excretion (ADME) characteristics. In fact, development of rapid methodologies enabling to obtain those information is a key aspects of the drug discovery process. QAC as rapid, relatively simple technique with the possibility of automation proved to be useful alternative to conventional methods in the field of drug discovery and analysis. This method is facilitated by multiple use of the same column with the immobilized target and as a consequence the reproducibility of assays is increased. Thanks to the above features QAC becomes popular method of measuring binding affinity of the drug-protein interactions. Variety of data that we can obtain via this technique allows characterization of binding reactions as well as description of the binding site. The development of techniques of high-yielding synthesis increases demand for technology that would allow pharmaceutical companies for efficient and rapid biological screening of thousands of synthesized compounds (Renaud & Delsuc, 2009). Taking into account that approximately one drug is produced after 8000-10000 compounds were subjected to primary and secondary drug screens, the classical approach, one compound-one assay, becomes unsatisfactory (Caldwell, 2000; as cited in Nie & Wang, 2009) . Studies conducted by Ng and colleagues have shown that use of an automated high performance chromatographic system consisting of two affinity columns and a mass spectrometer as the detector, allow to analyse and rank activity of 10 000 compounds in just 24 hours (Ng et al., 2007). Thus QAC techniques seem to be an promising method of preliminary verification of drug candidates, which is an alternative to expensive and tedious in vitro assays.

6. References

Andre, C., Herlem, G., Gharbi, T., & Guillaume, Y. C. (2011). A new arginase enzymatic reactor: development and application for the research of plant-derived inhibitors. *J Pharm Biomed Anal,* Vol. 55, No. 1, pp. 48-53, ISSN 1873-264X (Electronic)0731-7085 (Linking)

Bartolini, M., Cavrini, V., & Andrisano, V. (2004). Monolithic micro-immobilized-enzyme reactor with human recombinant acetylcholinesterase for on-line inhibition studies. *J Chromatogr A,* Vol. 1031, No. 1-2, pp. 27-34, ISSN 0021-9673 (Print)0021-9673 (Linking)

Bartolini, M., Cavrini, V., & Andrisano, V. (2005). Choosing the right chromatographic support in making a new acetylcholinesterase-micro-immobilised enzyme reactor for drug discovery. *J Chromatogr A,* Vol. 1065, No. 1, pp. 135-144, ISSN 0021-9673 (Print)0021-9673 (Linking)

Bartolini, M., Greig, N. H., Yu, Q. S., & Andrisano, V. (2009). Immobilized butyrylcholinesterase in the characterization of new inhibitors that could ease Alzheimer's disease. *J Chromatogr A,* Vol. 1216, No. 13, pp. 2730-2738, ISSN 1873-3778 (Electronic)0021-9673 (Linking)

Basiaga, S. B., & Hage, D. S. (2010). Chromatographic studies of changes in binding of sulfonylurea drugs to human serum albumin due to glycation and fatty acids. *J Chromatogr B Analyt Technol Biomed Life Sci,* Vol. 878, No. 30, pp. 3193-3197, ISSN 1873-376X (Electronic)1570-0232 (Linking)

Beigi, F., & Wainer, I. W. (2003). Syntheses of immobilized G protein-coupled receptor chromatographic stationary phases: characterization of immobilized mu and kappa opioid receptors. *Anal Chem,* Vol. 75, No. 17, pp. 4480-4485, ISSN 0003-2700 (Print)0003-2700 (Linking)

Beigi, F., Chakir, K., Xiao, R. P., & Wainer, I. W. (2004). G-protein-coupled receptor chromatographic stationary phases. 2. Ligand-induced conformational mobility in an immobilized beta2-adrenergic receptor. *Anal Chem,* Vol. 76, No. 24, pp. 7187-7193, ISSN 0003-2700 (Print)0003-2700 (Linking)

Calleri, E., Temporini, C., & Massolini, G. (2011). Frontal affinity chromatography in characterizing immobilized receptors. *J Pharm Biomed Anal,* Vol. 54, No. 5, pp. 911-925, ISSN 1873-264X (Electronic)0731-7085 (Linking)

Calleri, E., Temporini, C., Caccialanza, G., & Massolini, G. (2009). Target-based drug discovery: the emerging success of frontal affinity chromatography coupled to mass spectrometry. *ChemMedChem,* Vol. 4, No. 6, pp. 905-916, ISSN 1860-7187 (Electronic)1860-7179 (Linking)

Chan, N. W., Lewis, D. F., Rosner, P. J., Kelly, M. A., & Schriemer, D. C. (2003). Frontal affinity chromatography-mass spectrometry assay technology for multiple stages of drug discovery: applications of a chromatographic biosensor. *Anal Biochem,* Vol. 319, No. 1, pp. 1-12, ISSN 0003-2697 (Print)0003-2697 (Linking)

Chen, J., & Hage, D. S. (2004). Quantitative analysis of allosteric drug-protein binding by biointeraction chromatography. *Nat Biotechnol,* Vol. 22, No. 11, pp. 1445-1448, ISSN 1087-0156 (Print)1087-0156 (Linking)

Chen, J., Ohnmacht, C., & Hage, D. S. (2004). Studies of phenytoin binding to human serum albumin by high-performance affinity chromatography. *J Chromatogr B Analyt Technol Biomed Life Sci,* Vol. 809, No. 1, pp. 137-145, ISSN 1570-0232 (Print)1570-0232 (Linking)

Cheng, Y., Ho, E., Subramanyam, B., & Tseng, J. L. (2004). Measurements of drug-protein binding by using immobilized human serum albumin liquid chromatography-mass spectrometry. *J Chromatogr B Analyt Technol Biomed Life Sci,* Vol. 809, No. 1, pp. 67-73, ISSN 1570-0232 (Print)1570-0232 (Linking)

Dixon, M. (1953). The determination of enzyme inhibitor constants. *Biochem J,* Vol. 55, No. 1, pp. 170-171, ISSN 0264-6021 (Print)0264-6021 (Linking)

Girelli, A. M., & Mattei, E. (2005). Application of immobilized enzyme reactor in on-line high performance liquid chromatography: a review. *J Chromatogr B Analyt Technol Biomed Life Sci,* Vol. 819, No. 1, pp. 3-16, ISSN 1570-0232 (Print)1570-0232 (Linking)

Gottschalk, I., Li, Y. M., & Lundahl, P. (2000). Chromatography on cells: analyses of solute interactions with the glucose transporter Glut1 in human red cells adsorbed on lectin-gel beads. *J Chromatogr B Biomed Sci Appl,* Vol. 739, No. 1, pp. 55-62, ISSN 1387-2273 (Print)1387-2273 (Linking)

Gottschalk, I., Lundqvist, A., Zeng, C. M., Hagglund, C. L., Zuo, S. S., Brekkan, E., Eaker, D., & Lundahl, P. (2000). Conversion between two cytochalasin B-binding states of the human GLUT1 glucose transporter. *Eur J Biochem,* Vol. 267, No. 23, pp. 6875-6882, ISSN 0014-2956 (Print)0014-2956 (Linking)

Gritti, F., & Guiochon, G. (2005). Critical contribution of nonlinear chromatography to the understanding of retention mechanism in reversed-phase liquid chromatography. *J Chromatogr A,* Vol. 1099, No. 1-2, pp. 1-42, ISSN 0021-9673 (Print)0021-9673 (Linking)

Guiffant, D., Tribouillard, D., Gug, F., Galons, H., Meijer, L., Blondel, M., & Bach, S. (2007). Identification of intracellular targets of small molecular weight chemical compounds using affinity chromatography. *Biotechnol J,* Vol. 2, No. 1, pp. 68-75, ISSN 1860-7314 (Electronic)1860-6768 (Linking)

Hage, D. S. (2002). High-performance affinity chromatography: a powerful tool for studying serum protein binding. *J Chromatogr B Analyt Technol Biomed Life Sci,* Vol. 768, No. 1, pp. 3-30, ISSN 1570-0232 (Print)1570-0232 (Linking)

Hage, D. S., Jackson, A., Sobansky, M. R., Schiel, J. E., Yoo, M. J., & Joseph, K. S. (2009). Characterization of drug-protein interactions in blood using high-performance affinity chromatography. *J Sep Sci,* Vol. 32, No. 5-6, pp. 835-853, ISSN 1615-9314 (Electronic)1615-9306 (Linking)

Hage, D. S., Ruhn, P. F. (2006). An Introduction to Affinity Chromatography, In: *Handbook of Affinity Chromatography,* Hage, D.S., pp. 3-15, CRC Press/Taylor&Francis, ISBN 978-0-8247-4057-3, Boca Raton

Hage, D.S. (Ed.). (2006). *Handbook of Affinity Chromatography,* CRC Press/Taylor&Francis, ISBN 978-0-8247-4057-3, Boca Raton, FL

Hage, D.S., Chen, J. (2006). Quantitative Affinity Chromatography: Practical Aspect, In: *Handbook of Affinity Chromatography,* Hage, D.S., pp. 595-628, CRC Press/Taylor&Francis, ISBN 978-0-8247-4057-3, Boca Raton

He, P., Davies, J., Greenway, G., & Haswell, S. J. (2010). Measurement of acetylcholinesterase inhibition using bienzymes immobilized monolith micro-reactor with integrated electrochemical detection. *Anal Chim Acta,* Vol. 659, No. 1-2, pp. 9-14, ISSN 1873-4324 (Electronic)0003-2670 (Linking)

Joseph, K. S., & Hage, D. S. (2010). Characterization of the binding of sulfonylurea drugs to HSA by high-performance affinity chromatography. *J Chromatogr B Analyt Technol Biomed Life Sci,* Vol. 878, No. 19, pp. 1590-1598, ISSN 1873-376X (Electronic)1570-0232 (Linking)

Josic, D., & Buchacher, A. (2001). Application of monoliths as supports for affinity chromatography and fast enzymatic conversion. *J Biochem Biophys Methods,* Vol. 49, No. 1-3, pp. 153-174, ISSN 0165-022X (Print)0165-022X (Linking)

Jozwiak, K., Haginaka, J., Moaddel, R., & Wainer, I. W. (2002). Displacement and nonlinear chromatographic techniques in the investigation of interaction of noncompetitive inhibitors with an immobilized alpha3beta4 nicotinic acetylcholine receptor liquid chromatographic stationary phase. *Anal Chem,* Vol. 74, No. 18, pp. 4618-4624, ISSN 0003-2700 (Print)0003-2700 (Linking)

Jozwiak, K., Hernandez, S. C., Kellar, K. J., & Wainer, I. W. (2003). Enantioselective interactions of dextromethorphan and levomethorphan with the alpha 3 beta 4-nicotinic acetylcholine receptor: comparison of chromatographic and functional data. *J Chromatogr B Analyt Technol Biomed Life Sci,* Vol. 797, No. 1-2, pp. 373-379, ISSN 1570-0232 (Print)1570-0232 (Linking)

Jozwiak, K., Moaddel, R., Yamaguchi, R., Ravichandran, S., Collins, J. R., & Wainer, I. W. (2005). Qualitative assessment of IC50 values of inhibitors of the neuronal nicotinic acetylcholine receptor using a single chromatographic experiment and multivariate cluster analysis. *J Chromatogr B Analyt Technol Biomed Life Sci,* Vol. 819, No. 1, pp. 169-174, ISSN 1570-0232 (Print)1570-0232 (Linking)

Jozwiak, K., Ravichandran, S., Collins, J. R., & Wainer, I. W. (2004). Interaction of noncompetitive inhibitors with an immobilized alpha3beta4 nicotinic acetylcholine receptor investigated by affinity chromatography, quantitative-structure activity relationship analysis, and molecular docking. *J Med Chem,* Vol. 47, No. 16, pp. 4008-4021, ISSN 0022-2623 (Print)0022-2623 (Linking)

Kasai, K., & Ishii, S. (1975). Quantitative analysis of affinity chromatography of trypsin. A new technique for investigation of protein-ligand interaction. *J Biochem,* Vol. 77, No. 1?, pp. 261-264, ISSN 0021-924X (Print)0021-924X (Linking)

Kim, H. S., & Wainer, I. W. (2008). Rapid analysis of the interactions between drugs and human serum albumin (HSA) using high-performance affinity chromatography (HPAC). *J Chromatogr B Analyt Technol Biomed Life Sci,* Vol. 870, No. 1, pp. 22-26, ISSN 1570-0232 (Print)1570-0232 (Linking)

Kim, H.S., Hage, D.S., Immobilization methods for affinity chromatography, In: *Handbook of Affinity Chromatography,* Hage, D.S., pp. 35-79, CRC Press/Taylor&Francis, ISBN 978-0-8247-4057-3, Boca Raton

Kimura, T., Perry, J., Anzai, N., Pritchard, J. B., & Moaddel, R. (2007). Development and characterization of immobilized human organic anion transporter-based liquid chromatographic stationary phase: hOAT1 and hOAT2. *J Chromatogr B Analyt Technol Biomed Life Sci,* Vol. 859, No. 2, pp. 267-271, ISSN 1570-0232 (Print)1570-0232 (Linking)

Kitabatake, T., Moaddel, R., Cole, R., Gandhari, M., Frazier, C., Hartenstein, J., Rosenberg, A., Bernier, M., & Wainer, I. W. (2008). Characterization of a multiple ligand-gated ion channel cellular membrane affinity chromatography column and identification of endogenously expressed receptors in astrocytoma cell lines. *Anal Chem,* Vol. 80, No. 22, pp. 8673-8680, ISSN 1520-6882 (Electronic)0003-2700 (Linking)

Landry, Y., & Gies, J. P. (2008). Drugs and their molecular targets: an updated overview. *Fundam Clin Pharmacol,* Vol. 22, No. 1, pp. 1-18, ISSN 1472-8206 (Electronic)0767-3981 (Linking)

Lebert, J. M., Forsberg, E. M., & Brennan, J. D. (2008). Solid-phase assays for small molecule screening using sol-gel entrapped proteins. *Biochem Cell Biol,* Vol. 86, No. 2, pp. 100-110, ISSN 0829-8211 (Print)0829-8211 (Linking)

Loun, B., & Hage, D. S. (1992). Characterization of thyroxine-albumin binding using high-performance affinity chromatography. I. Interactions at the warfarin and indole sites of albumin. *J Chromatogr,* Vol. 579, No. 2, pp. 225-235, ISSN 0021-9673 (Print)0021-9673 (Linking)

Loun, B., & Hage, D. S. (1994). Chiral separation mechanisms in protein-based HPLC columns. 1. Thermodynamic studies of (R)- and (S)-warfarin binding to immobilized human serum albumin. *Anal Chem,* Vol. 66, No. 21, pp. 3814-3822, ISSN 0003-2700 (Print)0003-2700 (Linking)

Loun, B., & Hage, D. S. (1995). Characterization of thyroxine-albumin binding using high-performance affinity chromatography. II. Comparison of the binding of thyroxine, triiodothyronines and related compounds at the warfarin and indole sites of human serum albumin. *J Chromatogr B Biomed Appl,* Vol. 665, No. 2, pp. 303-314, ISSN 1572-6495 (Print)1572-6495 (Linking)

Lundqvist, A., Ocklind, G., Haneskog, L., & Lundahl, P. (1998). Freeze-thaw immobilization of liposomes in chromatographic gel beads: evaluation by confocal microscopy and effects of freezing rate. *J Mol Recognit,* Vol. 11, No. 1-6, pp. 52-57, ISSN 0952-3499 (Print)0952-3499 (Linking)

Ma, X., & Chan, E. C. (2010). On-line chromatographic screening of matrix metalloproteinase inhibitors using immobilized MMP-9 enzyme reactor. *J Chromatogr B Analyt Technol Biomed Life Sci,* Vol. 878, No. 21, pp. 1777-1783, ISSN 1873-376X (Electronic)1570-0232 (Linking)

Mallik, R., Yoo, M. J., Chen, S., & Hage, D. S. (2008). Studies of verapamil binding to human serum albumin by high-performance affinity chromatography. *J Chromatogr B Analyt Technol Biomed Life Sci,* Vol. 876, No. 1, pp. 69-75, ISSN 1570-0232 (Print)1570-0232 (Linking)

Mancini, F., & Andrisano, V. (2010). Development of a liquid chromatographic system with fluorescent detection for beta-secretase immobilized enzyme reactor on-line enzymatic studies. *J Pharm Biomed Anal,* Vol. 52, No. 3, pp. 355-361, ISSN 1873-264X (Electronic)0731-7085 (Linking)

Markuszewski, M., & Kaliszan, R. (2002). Quantitative structure-retention relationships in affinity high-performance liquid chromatography. *J Chromatogr B Analyt Technol Biomed Life Sci,* Vol. 768, No. 1, pp. 55-66, ISSN 1570-0232 (Print)1570-0232 (Linking)

Mateo, C., Fernandez-Lorente, G., Abian, O., Fernandez-Lafuente, R., & Guisan, J. M. (2000). Multifunctional epoxy supports: a new tool to improve the covalent immobilization of proteins. The promotion of physical adsorptions of proteins on the supports before their covalent linkage. *Biomacromolecules,* Vol. 1, No. 4, pp. 739-745, ISSN 1525-7797 (Print)1525-7797 (Linking)

Mazzini, F., Nuti, E., Petri, A., & Rossello, A. (2011). Immobilization of matrix metalloproteinase 8 (MMP-8) for online drug screening. *J Chromatogr B Analyt Technol Biomed Life Sci,* Vol. 879, No. 11-12, pp. 756-762, ISSN 1873-376X (Electronic)1570-0232 (Linking)

Moaddel, R., & Wainer, I. W. (2006). Development of immobilized membrane-based affinity columns for use in the online characterization of membrane bound proteins and for targeted affinity isolations. *Anal Chim Acta,* Vol. 564, No. 1, pp. 97-105, ISSN 1873-4324 (Electronic)0003-2670 (Linking)

Moaddel, R., Bullock, P. L., & Wainer, I. W. (2004). Development and characterization of an open tubular column containing immobilized P-glycoprotein for rapid on-line screening for P-glycoprotein substrates. *J Chromatogr B Analyt Technol Biomed Life Sci,* Vol. 799, No. 2, pp. 255-263, ISSN 1570-0232 (Print)1570-0232 (Linking)

Moaddel, R., Calleri, E., Massolini, G., Frazier, C. R., & Wainer, I. W. (2007). The synthesis and initial characterization of an immobilized purinergic receptor (P2Y1) liquid chromatography stationary phase for online screening. *Anal Biochem,* Vol. 364, No. 2, pp. 216-218, ISSN 0003-2697 (Print)0003-2697 (Linking)

Moaddel, R., Cloix, J. F., Ertem, G., & Wainer, I. W. (2002). Multiple receptor liquid chromatographic stationary phases: the co-immobilization of nicotinic receptors, gamma-amino-butyric acid receptors, and N-methyl D-aspartate receptors. *Pharm Res,* Vol. 19, No. 1, pp. 104-107, ISSN 0724-8741 (Print)0724-8741 (Linking)

Moaddel, R., Jozwiak, K., Whittington, K., & Wainer, I. W. (2005). Conformational mobility of immobilized alpha3beta2, alpha3beta4, alpha4beta2, and alpha4beta4 nicotinic acetylcholine receptors. *Anal Chem,* Vol. 77, No. 3, pp. 895-901, ISSN 0003-2700 (Print)0003-2700 (Linking)

Moaddel, R., Lu, L., Baynham, M., & Wainer, I. W. (2002). Immobilized receptor- and transporter-based liquid chromatographic phases for on-line pharmacological and biochemical studies: a mini-review. *J Chromatogr B Analyt Technol Biomed Life Sci,* Vol. 768, No. 1, pp. 41-53, ISSN 1570-0232 (Print)1570-0232 (Linking)

Moaddel, R., Musyimi, H. K., Sanghvi, M., Bashore, C., Frazier, C. R., Khadeer, M., Bhatia, P., & Wainer, I. W. (2010). Synthesis and characterization of a cellular membrane affinity chromatography column containing histamine 1 and P2Y(1) receptors: a multiple G-protein coupled receptor column. *J Pharm Biomed Anal,* Vol. 52, No. 3, pp. 416-419, ISSN 1873-264X (Electronic)0731-7085 (Linking)

Moaddel, R., Oliveira, R. V., Kimura, T., Hyppolite, P., Juhaszova, M., Xiao, Y., Kellar, K. J., Bernier, M., & Wainer, I. W. (2008). Initial synthesis and characterization of an alpha7 nicotinic receptor cellular membrane affinity chromatography column: effect of receptor subtype and cell type. *Anal Chem,* Vol. 80, No. 1, pp. 48-54, ISSN 0003-2700 (Print)0003-2700 (Linking)

Moaddel, R., Rosenberg, A., Spelman, K., Frazier, J., Frazier, C., Nocerino, S., Brizzi, A., Mugnaini, C., & Wainer, I. W. (2011). Development and characterization of immobilized cannabinoid receptor (CB1/CB2) open tubular column for on-line screening. *Anal Biochem,* Vol. 412, No. 1, pp. 85-91, ISSN 1096-0309 (Electronic)0003-2697 (Linking)

Moaddel, R., Yamaguchi, R., Ho, P. C., Patel, S., Hsu, C. P., Subrahmanyam, V., & Wainer, I. W. (2005). Development and characterization of an immobilized human organic cation transporter based liquid chromatographic stationary phase. *J Chromatogr B Analyt Technol Biomed Life Sci,* Vol. 818, No. 2, pp. 263-268, ISSN 1570-0232 (Print)1570-0232 (Linking)

Narayanan, S. R., Kakodkar, S. V., & Crane, L. J. (1990). "Glutaraldehyde-P", a stable, reactive aldehyde matrix for affinity chromatography. *Anal Biochem,* Vol. 188, No. 2, pp. 278-284, ISSN 0003-2697 (Print)0003-2697 (Linking)

Ng, W., Dai, J. R., Slon-Usakiewicz, J. J., Redden, P. R., Pasternak, A., & Reid, N. (2007). Automated multiple ligand screening by frontal affinity chromatography-mass spectrometry (FAC-MS). *J Biomol Screen,* Vol. 12, No. 2, pp. 167-174, ISSN 1087-0571 (Print)1087-0571 (Linking)

Nie, Y.-l., Wang, W.-h. (2009) .Immobilized Enzyme Reactor in On-line LC and Its Application in Drug Screening. *Chromatographia,* Vol. 69, pp. S5–S12

Pidgeon, C., & Venkataram, U. V. (1989). Immobilized artificial membrane chromatography: supports composed of membrane lipids. *Anal Biochem,* Vol. 176, No. 1, pp. 36-47, ISSN 0003-2697 (Print)0003-2697 (Linking)

Renaud, J. P., & Delsuc, M. A. (2009). Biophysical techniques for ligand screening and drug design. *Curr Opin Pharmacol,* Vol. 9, No. 5, pp. 622-628, ISSN 1471-4973 (Electronic)1471-4892 (Linking)

Rizzi, L., Dallanoce, C., Matera, C., Magrone, P., Pucci, L., Gotti, C., Clementi, F., & De Amici, M. (2008). Epiboxidine and novel-related analogues: a convenient synthetic approach and estimation of their affinity at neuronal nicotinic acetylcholine receptor subtypes. *Bioorg Med Chem Lett,* Vol. 18, No. 16, pp. 4651-4654, ISSN 1464-3405 (Electronic)0960-894X (Linking)

Ruhn, P. F., Garver, S., & Hage, D. S. (1994). Development of dihydrazide-activated silica supports for high-performance affinity chromatography. *J Chromatogr A,* Vol. 669, No. 1-2, pp. 9-19, ISSN 0021-9673 (Print)0021-9673 (Linking)

Sanghvi, M., Moaddel, R., Frazier, C., & Wainer, I. W. (2010). Synthesis and characterization of liquid chromatographic columns containing the immobilized ligand binding domain of the estrogen related receptor alpha and estrogen related receptor gamma. *J Pharm Biomed Anal,* Vol. 53, No. 3, pp. 777-780, ISSN 1873-264X (Electronic)0731-7085 (Linking)

Slon-Usakiewicz, J. J., Ng, W., Dai, J. R., Pasternak, A., & Redden, P. R. (2005). Frontal affinity chromatography with MS detection (FAC-MS) in drug discovery. *Drug Discov Today,* Vol. 10, No. 6, pp. 409-416, ISSN 1359-6446 (Print)1359-6446 (Linking)

Soczewinski, E., Bieganowska, M. (1969). Investigations on the relationship between molecular structure and chromatogenic parameters: I. Two homologous series of pyridine derivatives suitable as reference compounds in liquid-liquid partition chromatography. *J Chromatogr A,* Vol. 40, pp. 431-439

Taylor, R.F. (Ed.). (1991). *Protein Immobilization: Fundamentals and Applications,* Marcel Dekker, ISBN10 0824782712, New York

Temporini, C., Ceruti, S., Calleri, E., Ferrario, S., Moaddel, R., Abbracchio, M. P., & Massolini, G. (2009). Development of an immobilized GPR17 receptor stationary phase for binding determination using frontal affinity chromatography coupled to mass spectrometry. *Anal Biochem,* Vol. 384, No. 1, pp. 123-129, ISSN 1096-0309 (Electronic)0003-2697 (Linking)

Trujillo, C. A., Majumder, P., Gonzalez, F. A., Moaddel, R., & Ulrich, H. (2007). Immobilized P2X2 purinergic receptor stationary phase for chromatographic determination of pharmacological properties and drug screening. *J Pharm Biomed Anal,* Vol. 44, No. 3, pp. 701-710, ISSN 0731-7085 (Print)0731-7085 (Linking)

Turkova, J. (1999). Oriented immobilization of biologically active proteins as a tool for revealing protein interactions and function. *J Chromatogr B Biomed Sci Appl,* Vol. 722, No. 1-2, pp. 11-31, ISSN 1387-2273 (Print)1387-2273 (Linking)

Tweed, S. A., Loun, B., & Hage, D. S. (1997). Effects of ligand heterogeneity in the characterization of affinity columns by frontal analysis. *Anal Chem,* Vol. 69, No. 23, pp. 4790-4798, ISSN 0003-2700 (Print)0003-2700 (Linking)

Wade, J.L., Bergold, A.F., Carr, P.W. (1987). Theoretical description of nonlinear chromatography, with applications to physicochemical measurements in affinity chromatography and implications for preparative-scale separations. *Anal. Chem.*, Vol. 59, No. 9, pp. 1286–1295

Wainer, I. W., Zhang, Y., Xiao, Y., & Kellar, K. J. (1999). Liquid chromatographic studies with immobilized neuronal nicotinic acetylcholine receptor stationary phases: effects of receptor subtypes, pH and ionic strength on drug-receptor interactions. *J Chromatogr B Biomed Sci Appl*, Vol. 724, No. 1, pp. 65-72, ISSN 1387-2273 (Print)1387-2273 (Linking)

Winzor, D. J. (2004). Determination of binding constants by affinity chromatography. *J Chromatogr A*, Vol. 1037, No. 1-2, pp. 351-367, ISSN 0021-9673 (Print)0021-9673 (Linking)

Xuan, H., & Hage, D. S. (2005). Immobilization of alpha(1)-acid glycoprotein for chromatographic studies of drug-protein binding. *Anal Biochem*, Vol. 346, No. 2, pp. 300-310, ISSN 0003-2697 (Print)0003-2697 (Linking)

Yang, J., & Hage, D. S. (1993). Characterization of the binding and chiral separation of D- and L-tryptophan on a high-performance immobilized human serum albumin column. *J Chromatogr*, Vol. 645, No. 2, pp. 241-250, ISSN 0021-9673 (Print)0021-9673 (Linking)

Yang, Q., & Lundahl, P. (1995). Immobilized proteoliposome affinity chromatography for quantitative analysis of specific interactions between solutes and membrane proteins. Interaction of cytochalasin B and D-glucose with the glucose transporter Glut1. *Biochemistry*, Vol. 34, No. 22, pp. 7289-7294, ISSN 0006-2960 (Print)0006-2960 (Linking)

Yoo, M. J., Smith, Q. R., & Hage, D. S. (2009). Studies of imipramine binding to human serum albumin by high-performance affinity chromatography. *J Chromatogr B Analyt Technol Biomed Life Sci*, Vol. 877, No. 11-12, pp. 1149-1154, ISSN 1873-376X (Electronic)1570-0232 (Linking)

Yu, W., Yuan, B., Deng, X., He, L., Youyi, Z., & Qide, H. (2005). The preparation of HEK293 alpha1A or HEK293 alpha1B cell membrane stationary phase and the chromatographic affinity study of ligands of alpha1 adrenoceptor. *Anal Biochem*, Vol. 339, No. 2, pp. 198-205, ISSN 0003-2697 (Print)0003-2697 (Linking)

Yuan, B. X., Hou, J., He, L. C., & Yang, G. D. (2005). Evaluation of drug-muscarinic receptor affinities using cell membrane chromatography and radioligand binding assay in guinea pig jejunum membrane. *Acta Pharmacol Sin*, Vol. 26, No. 1, pp. 113-116, ISSN 1671-4083 (Print)1671-4083 (Linking)

Zeng, C. M., Zhang, Y., Lu, L., Brekkan, E., Lundqvist, A., & Lundahl, P. (1997). Immobilization of human red cells in gel particles for chromatographic activity studies of the glucose transporter Glut1. *Biochim Biophys Acta*, Vol. 1325, No. 1, pp. 91-98, ISSN 0006-3002 (Print)0006-3002 (Linking)

Zhang, Y., Leonessa, F., Clarke, R., & Wainer, I. W. (2000). Development of an immobilized P-glycoprotein stationary phase for on-line liquid chromatographic determination of drug-binding affinities. *J Chromatogr B Biomed Sci Appl*, Vol. 739, No. 1, pp. 33-37, ISSN 1387-2273 (Print)1387-2273 (Linking)

Zhang, Y., Xiao, Y., Kellar, K. J., & Wainer, I. W. (1998). Immobilized nicotinic receptor stationary phase for on-line liquid chromatographic determination of drug-receptor affinities. *Anal Biochem,* Vol. 264, No. 1, pp. 22-25, ISSN 0003-2697 (Print)0003-2697 (Linking)

Zhu, L., Chen, L., Luo, H., & Xu, X. (2003). Frontal affinity chromatography combined on-line with mass spectrometry: a tool for the binding study of different epidermal growth factor receptor inhibitors. *Anal Chem,* Vol. 75, No. 23, pp. 6388-6393, ISSN 0003-2700 (Print)0003-2700 (Linking)

Part 9

Practical Application of Affinity Chromatography in Research Field

15

The Value of Fungal Protease Inhibitors in Affinity Chromatography

Jerica Sabotič[1], Katarina Koruza[1,2], Boštjan Gabor[2,3], Matjaž Peterka[2,3], Miloš Barut[2,3], Janko Kos[1] and Jože Brzin[1]
[1]*Department of Biotechnology, Jožef Stefan Institute,*
[2]*BIA Separations,*
[3]*Centre of Excellence for Biosensors, Instrumentation and Process Control,*
[1,2,3]*Slovenia*

1. Introduction

Proteolytic enzymes (also known as proteases, proteinases or peptidases) offer a wide range of applications. They are routinely used in detergent, leather, food and pharmaceutical industries, as well as in medical and basic research. Therefore, effective isolation procedures are of great importance. The chapter describes the use of recently discovered protease inhibitors from basidiomycetes as affinity chromatography ligands for isolating proteases. Affinity columns with serine and cysteine protease inhibitors immobilized to the natural polymer Sepharose have been prepared, the chromatography procedure optimized and used for isolating proteases from various bacterial, plant and animal sources. The cysteine protease inhibitor macrocypin showed superior characteristics as a ligand, so was selected for immobilization to CIM (Convective Interaction Media) monolithic disks. Different immobilization chemistries and process conditions were optimized to determine the best conditions for high capacity and selectivity. A very effective method for isolating cysteine proteases was developed using affinity chromatography with the fungal cysteine protease inhibitor macrocypin immobilized to a CIM monolithic disk.

1.1 Proteases and their applications

Proteases occur in all groups of organisms, including bacteria, archaea, protists, fungi, plants, animals and viruses. They are classified according to their catalytic type into aspartic, cysteine, glutamic, serine and threonine peptidases, based on the type of the amino acid residue at the active site, and metallopeptidases, that require a catalytic divalent metal ion within the active site. The MEROPS database (http://merops.sanger.ac.uk) further classifies peptidases according to their evolutionary relationships and encompasses information on all known proteases (Rawlings et al., 2010).

Proteases offer a wide range of biotechnological applications. Alkaline proteases are routinely used in the detergent industry. Proteases with elastolytic and keratinolytic activities have been used in the leather industry for dehairing and baiting skins and hides. They are used in the food industry in cheese making and baking, in preparing the various protein hydrolysates used as flavour enhancers, in meat tenderization and in manufacturing protein-rich diets. In the pharmaceutical industry proteases have found uses as therapeutic agents as well as additives in preparations of slow-release dosage forms. In addition to industrial and medical applications, proteases have found use also in basic research, for example proteases with very selective peptide bond cleavage are used in protein sequencing and proteome analyses. Furthermore, unselective proteases are also used, for example proteinase K is used in nucleic acid isolation, and trypsin is widely used in animal cell culture maintenance (Kumar & Takagi, 1999; Østergaard & Olsen, 2010; Rao et al., 1998).

1.2 Protease inhibitors

Proteases play essential metabolic and regulatory functions in many biological processes, therefore the regulation of their activity is essential. Interaction with protease inhibitors constitutes a very important mechanism of protease regulation (Lopez-Otin & Bond, 2008; Rawlings et al., 2010). Protease inhibitors are either small molecules or proteins. They can be classified according to the source organism (microbial, fungal, plant, animal), according to their structure (primary and three-dimensional), or according to their inhibitory profile (broad-range, specific) and reaction mechanism (competitive, non-competitive, uncompetitive as well as reversible or irreversible). Commonly they are classified according to the class of protease they inhibit (for example: aspartic, cysteine or serine protease inhibitors) but a detailed classification of protein protease inhibitors based on their evolutionary relationship is available in the MEROPS database, together with a list of small molecule peptidase inhibitors (http://merops.sanger.ac.uk/inhibitors/). There are two general mechanisms by which protein inhibitors inhibit peptidases - irreversible "trapping" reaction, involving a conformational change of the inhibitor, and reversible tight-binding reactions, where the inhibitor forms a high-affinity interaction with the peptidase, most often at the active site (Christeller, 2005; Rawlings, 2010; Rawlings et al., 2004). The reversible protease inhibitors are used as ligands in affinity chromatography (Cuatrecasas et al., 1968).

1.2.1 Protease inhibitors of fungal origin

The only routinely used small molecule inhibitor of fungal origin is the irreversible inhibitor of cysteine proteases E-64, first isolated from the filamentous fungus *Aspergillus japonicus* (Hanada et al., 1978). Several other small molecule inhibitors with a broad inhibitory spectrum that are routinely used (e.g. pepstatin A, chymostatin, leupeptin, antipain, phosphoramidon, bestatin) were originally isolated from bacteria (Rawlings, 2010).

Information on protein protease inhibitors of fungal origin is limited. No protein metalloprotease inhibitors has been described, and only one family of aspartic protease inhibitors (family I34), which includes the highly specific inhibitor of the yeast proteinase A or saccharopepsin (Phylip et al., 2001). There are four families of serine protease inhibitors, namely the inhibitors of serine carboxypeptidase Y (family I51), *Aspergillus* elastase inhibitor family (I78), inhibitors of subtilisin-like proteases homologous to the subtilisin propeptide (family I9) and trypsin-specific protease inhibitors (family I66). Representatives of the latter two have been identified from higher fungi or mushrooms as well as from filamentous fungi (Rawlings, 2010). Furthermore, only three families of cysteine protease inhibitors have been described, namely mycocypins (families I48 and I85), that are unique to higher fungi, and a cysteine protease inhibitor family (I79) with only one representative found in one plant pathogenic fungal species (Rawlings, 2010).

The serine protease inhibitor cnispin (family I66), identified in the mushroom *Clitocybe nebularis*, is a 16.4 kDa protein with acidic isoelectric point. It is a very stable protease inhibitor that resists short-term exposure to extremes of pH (between pH 2 and pH 11). It is a very strong inhibitor of trypsin (family S1) with K_i in the nanomolar range, and a weak inhibitor of chymotrypsin (family S1), with K_i in the micromolar range. Inhibition of kallikrein (family S1) and subtilisin (family S8) is very weak and other proteases are not inhibited (Avanzo et al., 2009).

Mycocypins are cysteine protease inhibitors unique to basidiomycete mushrooms and belong to two MEROPS families (I48 and I85). Family I48 is represented by clitocypin, identified in *Clitocybe nebularis* (Brzin et al., 2000; Renko et al., 2010; Sabotič et al., 2006) and family I85 by macrocypins identified from *Macrolepiota procera* (Renko et al., 2010; Sabotič et al., 2009b). These are small (16.8 to 20 kDa) and exceptionally stable proteins, exhibiting high thermal and broad pH stability (Galeša et al., 2004; Kidrič et al., 2002; Sabotič et al., 2009b). They have the β-trefoil fold, which is composed of a core six-stranded β-barrel surrounded by 11 loops (Fig. 1) that provide a versatile surface for the inhibition of several types of proteases (Renko et al., 2010). They are very strong inhibitors of papain-like peptidases (family C1), including papain, cathepsins L, V, S, and K in the low nanomolar range and cathepsins B and H with higher inhibition constants. A second inhibitory reactive site is involved in the inhibition of the cysteine protease asparaginyl endopeptidase (AEP, legumain), of family C13 by clitocypin, macrocypins 1 and 3, and in the inhibition of the serine protease trypsin (family S1) by macrocypin 4 (Renko et al., 2010; Sabotič et al., 2007a; Sabotič et al., 2009b).

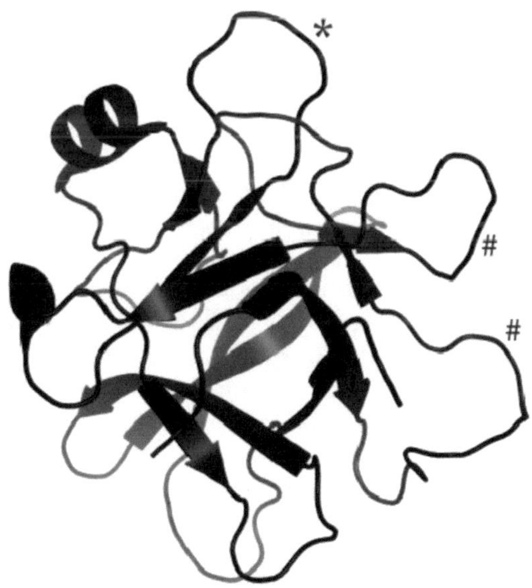

Fig. 1. Three-dimensional structure of macrocypin 1. The two loops involved in inhibiting papain-like proteases (family C1) are indicated with #, and the loop involved in legumain (family C13) inhibition with *.

1.3 Affinity chromatography

Affinity chromatography is a purification technique that exploits the unique biological properties of biomolecules to bind ligands specifically and reversibly (Cuatrecasas, 1970). It is exceptional in purification science as a tool for highly selective isolation of antigens (drugs, hormones, peptides, proteins, viruses, cell components), antibodies, enzymes, sugars, glycoproteins and glycolipids, immunoglobulins, nucleotide-binding and metal-binding peptides and proteins (Mallik & Hage, 2006). The use of specific interactions makes affinity chromatography a powerful tool in isolation of a particular substance from a complex mixture. In theory, it is capable of giving absolute purification from a complex mixture in a single process and it eliminates the need for ammonium sulphate precipitation, ion exchange, and gel filtration steps in an isolation protocol (Hermanson et al., 1992).

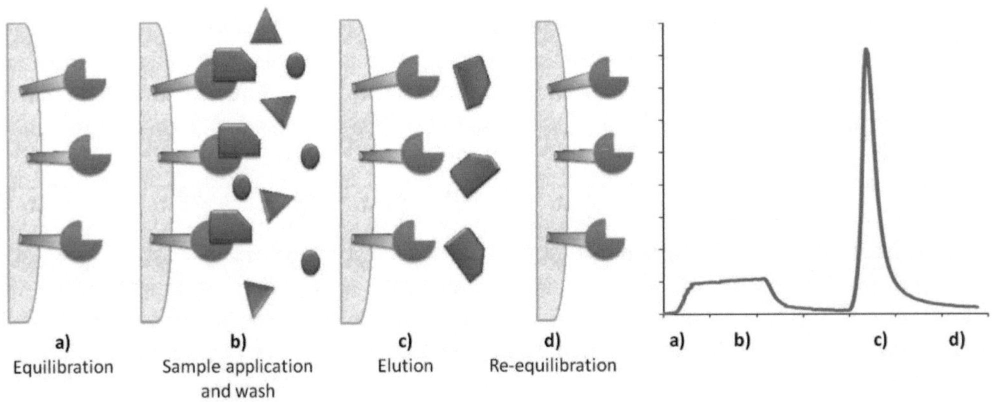

Fig. 2. Principle of affinity chromatography: a) Equilibration of the stationary phase in a binding buffer. b) Sample is applied to the column. Target molecules will bind specifically but reversibly to the ligand, non-specifically bound contaminants will pass through in the flow through during washing with binding buffer. c) The purified compound is released from the biospecific ligand by elution using specific competition or non-specific change in buffer composition. d) Stationary phase is re-equilibrated with a binding buffer. The different stages of affinity chromatography are represented in a typical chromatogram in the right panel.

In affinity chromatography one of the two interacting molecules, called the affinity ligand, is immobilized on a stationary phase, either by covalent immobilization or by physical adsorption (Tetala & van Beek, 2010). The ligand can consist of an immobilized sequence of DNA or RNA, a protein or enzyme, lectin, amino acids, immunoglobulin, a biomimetic dye, an enzyme substrate or inhibitor, or a small molecule (Hage, 2006; Healthcare, 2007). Target molecules to be purified are applied in a mobile phase known as the application buffer to the column containing insoluble polymer or gel. The buffer is generally chosen to allow optimal binding of the immobilized ligand to its target while other sample components pass through with little or no retention (Mallik & Hage, 2006). Interaction between the ligand and the target molecule in a sample can be the result of electrostatic or hydrophobic interactions, van der Waals' forces or hydrogen bonding. The retained target molecule is then eluted and several different methods are possible. If a ligand has a weak or moderate affinity for the target molecules, the elution can be performed under isocratic conditions, where composition of the mobile phase remains unchanged during washing and elution. This technique is known as weak affinity chromatography. When a ligand has strong affinity to the target molecule, it can be released by changing the mobile phase or column conditions. For the most part, elution buffer differs from the application buffer in pH, ionic strength, temperature or polarity. This approach is known as nonspecific elution. Another possibility is biospecific elution, where a buffer with the same pH, ionic strength and polarity as the application buffer is used for elution but with a competing agent that can bind to either the retained target or the immobilized ligand (Tetala & van Beek, 2010).

1.3.1 Supports in affinity chromatography

The role of stationary phase or support in affinity chromatography is ligand immobilization (Hage, 2006). A support is any material to which a biospecific ligand may be covalently attached. Typically, the material to be used as an affinity support is insoluble in the system where the target molecule is found (Hermanson et al., 1992). The support main characteristics are: large specific area, high rigidity and suitable form of the particles, hydrophilic character and high permeability. The particle size and porosity are designed to maximize the surface area available for immobilization of ligand and consequently binding of target molecule. Supports in classical or low-performance affinity chromatography are usually non-rigid particles with large diameters like agarose, but also organic polymers like polyurethane or inorganic materials as large diameter silica particles (Hage, 2006). In column chromatography the adsorption rate is limited by slow particle diffusion for larger beads or low axial velocities and high pressure for smaller beads. The consequence is limited access for the biomolecules to small pores in the case of classical chromatographic supports. Good flow properties are desirable for rapid separations as they save considerable time during column equilibration, regeneration and cleaning (Champagne, 2007). The use of affinity supports with more rigid and effective materials that can be used in high performance liquid chromatography systems (HPLC) gives a technique known as high-performance affinity chromatography (Hage, 2006). In comparison with low pressure chromatography, HPLC is basically an improved form of the technique, where a mobile phase is forced through the column under high pressure instead of being allowed to drip through under gravity. Another improvement is an extremely sensitive detection system and complete automation of the process. Development of more resistant stationary phases has resulted in faster and better resolution and explains why HPLC became the most powerful and versatile form of affinity chromatography (Wilson & Walker, 2005).

Moreover, a chromatographic support should provide appropriate surface chemistry for the immobilization of affinity ligand and chemical stability for immobilization, adsorption, desorption and regeneration of the support itself (Champagne, 2007). Ability to withstand a wide range of thermal, mechanical, chemical and physical conditions and resistance to microbial and enzymatic cleavage are also desired (Tetala & van Beek, 2010). Different substances have been described and employed as affinity matrices to improve the separation and to overcome the limitations of particulate stationary phase.

1.3.1.1 Natural supports

Natural polysaccharides, such as agarose and cellulose, possess highly ionic or carboxylate residues and they require processing before being suitable as affinity supports (Hermanson et al., 1992). The most commonly used agarose is commercially available as a beaded form, known under the trademark Sepharose (GE Healthcare) and it has been further modified and developed to enhance properties required for an effective chromatographic material. It is an uncharged, hydrophilic matrix with an abundance of hydroxyl groups on the sugar residues which can be easily derivatized for covalent attachment of ligand (Healthcare, 2007). The primary structure of agarose consists of alternating residues of D-galactose and 3-anhydrogalactose. The secondary and tertiary structure forms a fabric with large accessible pores. The knitted porous structure of agarose is knotted at the juncture of the pores with strong hydrogen bonds. The large accessible pore structure of Sepharose yields affinity supports with sufficient capacity when the target molecule or ligand is a large protein or polysaccharide (Hermanson et al., 1992).

1.3.1.2 Synthetic supports

Synthetic supports, such as acrylamide derivatives, methacrylate derivatives, polystyrene and its derivatives and different membranes are produced by polymerization of functional monomers, to give matrices suitable for affinity-based separations. Commercially available synthetic materials have superior physical and chemical durability and can withstand the process of separation better than the natural soft gels. Synthetic supports are made with monomers that contain primary or secondary hydroxyl groups that maintain hydrophilicity and allow compatibility with most coupling methods (Hermanson et al., 1992).

1.3.1.3 Monolithic supports

To overcome the difficulties of slow mass transfer in different synthetic supports, a new generation of monolithic solid supports was introduced in affinity chromatography. Monolith is a name for the chromatographic stationary phase consisting of a single piece of highly porous material (Barut et al., 2008). They are prepared in various dimensions with agglomeration-type or fibrous microstructures. In recent years, the polymeric macroporous material, based on radical co-polymerization of glycidyl methacrylate and ethylene glycol dimethacrylate (GMA/EDMA) has been introduced under the trademark CIM® (Convective Interaction Media). There are several reasons for the popularity of monolithic supports in HPLC and affinity chromatography. The main advantage of CIM monolithic systems versus traditional chromatographic supports are much better mass transfer properties, ease of use, the ability to be manufactured with a wide range of pore sizes and shapes, simple scaling up and scaling down, and low back pressure even at very high volumetric flow rates (up to 10 column volumes/min) without the loss of efficiency and capacity (Podgornik & Štrancar, 2005). Monolithic supports can carry many specific ligands for affinity chromatography and the vast potential of monoliths for immobilization of affinity ligands has been recognized, resulting in an increase in experience in applying this technology to bioanalytical and biotechnological activities (Brne et al., 2009; Champagne, 2007; Švec et al., 2003).

1.3.2 Support activation

Support activation is the first step of ligand immobilization to a chromatographic support. It is a process of chemically modifying the support in a way to form a covalent bond with the ligand of choice. The reaction conditions and the proportion of the reagents will determine the number of ligand molecules that can be attached to the support surface (Wilson & Walker, 2005). The most common activation method for polysaccharide supports is with

Cyanogen Bromide (CNBr). At high pH CNBr introduces cyanate esters and imido carbonates into the matrix by reacting with the endogenous hydroxyl groups (Fig. 3). This activation works also with other synthetic polymers containing hydroxyl groups (Hermanson et al., 1992).

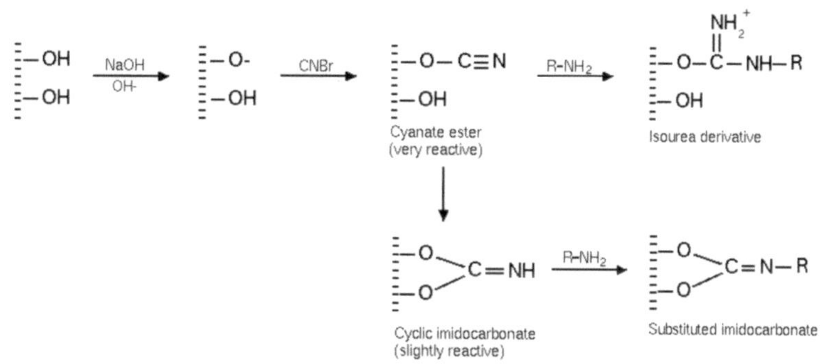

Fig. 3. Mechanism of activation of polysaccharide matrix by CNBr and subsequent coupling of amine-containing ligands (modified from Hermanson et al., 1992).

Alternative coupling procedures involve the use of bis-epoxides, N,N'-disubstituted carbodiimides, sulphonyl chloride, sodium periodate, N-hydroxysuccinimide esters and dichlorotriazines (Wilson & Walker, 2005). There are many different protocols for coupling a ligand to pre-activated supports and, as a consequence, it is easier to adjust the matrix and activation to suit the ligand than to adjust the ligand to suit the matrix and the activation. Moreover, many pre-activated matrices are commercially available.

1.3.3 Ligand and ligand immobilization

The ligand is a molecule that binds reversibly to a specific molecule, enabling separation by affinity chromatography. When a suitable ligand is available for the protein of interest, high selectivity, resolution and capacity of affinity chromatography are expected. Successful affinity separation requires a biospecific ligand that can be covalently attached to a solid support and retain its specific binding affinity for the target molecule after unbound material is washed away. Furthermore, the bond between the ligand and the target molecule must be reversible to allow the target molecule to be eluted in an active form. The most commonly used biological ligands are antibodies, antigens, inhibitors, substrates, cofactors and coenzymes, lectins, protein A and protein G, and, among non-biological ligands, tyrazine dyes and metal-chelates (Hermanson et al., 1992; Mallik & Hage, 2006; Turkova, 1993; Wilson & Walker, 2005).

The main criteria for the choice of an affinity ligand for target protein isolation are the functional groups of the ligand. To be immobilized, ligand must possess a functional group that will not be involved in the reversible binding of the ligand to the complementary target molecule, but which can be used to attach the ligand to the stationary support. The most common of such functional groups are $-NH_2$, -COOH, -SH and –OH (Wilson & Walker, 2005). In addition, the use of a long spacer arm is indispensable in the case of low-molecular weight affinity ligands, to provide accessibility to the binding site of the target molecule. For effective chromatography, an equilibrium dissociation constant, K_D, in the range of 10^{-4} to 10-

[8] M in free solution is required for successful separation (Healthcare, 2007; Turkova, 1993). When the dissociation constant is outside this range, altering elution methods may however help to execute successful affinity chromatography (Healthcare, 2007).

Ligands in biospecific chromatography can be classified into two types, monospecific and group specific. Monospecific ligands are those that bind only one molecule; an example is monoclonal antibody to a protein. In this case, a monospecific ligand is expected to bind the target molecules with large association constant, while binding weakly or not at all, to others (Miller, 2005). Group specific ligands have an affinity for a group of related substances rather than for a single type of molecule. The specificity derives from the selectivity of the ligand and the use of selective elution conditions. For example, a lectin can be used for affinity purification of glycoproteins, polysaccharides and glycolipids with the same glyco-signature (Healthcare, 2007; Miller, 2005). The plant lectin isolated from jack bean (*Canavalia ensiformis*) Concanavalin A (ConA) is a routinely used group specific ligand (Healthcare, 2007).

Several approaches have been reported for placing ligands within supports for chromatography. Examples include various covalent immobilization methods, as well as biospecific adsorption, entrapment and cross-linking. Covalent binding is preferred among the other immobilization methods, because it prevents leaking and combines the high selectivity of the reaction with the chemical and mechanical properties of the support (Benčina et al., 2004).

Introducing a spacer arm minimizes the risk of steric interference such as binding between support and target molecule or low accessibility of the ligand. Spacer arms are low-molecular-weight molecules interposed between the ligand and the solid support. They usually consist of linear hydrocarbon chains with functionalities on both ends for easy coupling to the support and ligand. One end of the spacer is immobilized to the matrix using traditional immobilization chemistries, while the other end is connected to the ligand, using a secondary coupling procedure. The result is immobilized ligand that sticks out from the matrix by a distance equal to the length of spacer arm chosen, where the optimal length is up to 10 carbon atoms (Hermanson et al., 1992). In addition, ligands involved in interaction must be sufficiently distant from the solid support to minimize steric interference with the binding processes (Cuatrecasas et al., 1968; Hermanson et al., 1992). With rigid support materials, a spacer molecule may also provide greater flexibility, allowing the immobilized ligand to move into position to establish the correct binding orientation with a protein. Sometimes the chemical structure of a spacer arm is critical for the success of separation. It is also important to consider the hydrophilicity of the spacer molecule, as some spacers are purely hydrophobic, for example methylene groups; others are hydrophilic, possessing carbonyl or imido groups (Wilson & Walker, 2005). Several ready to use supports of agarose, cellulose and polyacrylamide with a variety of spacer arms and pre-attached ligands are commercially available.

1.4 Affinity chromatography for purification of proteases

For the affinity chromatography of proteases different ligands from substrates or their analogues to various synthetic or naturally occurring inhibitors have been immobilized. Appropriate ligands for protease purification can be immobilized enzyme substrates themselves, but in practice, substrates are immediately converted to product after contact with the target enzyme (Healthcare, 2007; Polanowski et al., 2003). Therefore, the best results in protease purification may be gained by using highly specific inhibitors.

Synthetic inhibitors present effective ligands for many proteases. The synthetic inhibitor para-aminobenzamidine is used as affinity ligand for trypsin, trypsin-like serine proteases and zymogens. Benzamidine-Sepharose is commercially available and frequently used for removing molecules from cell culture supernatants, bacterial lysates or serum (Healthcare, 2007). Bovine basic pancreatic trypsin inhibitor (BPTI), or aprotinin, and inhibitors from legume seeds, such as soybean trypsin inhibitor (SBTI), are frequently employed protein ligands for serine protease isolation (Hewlett, 1990; Polanowski et al., 2003). Group specific inhibitors, that are selective and able to react with a protease from single class, are also effective ligands for specific classes of proteases. Egg white cystatin C coupled to Sepharose allows selective isolation of cysteine proteases from tissue or cell extracts, biological fluids and culture media (Tombaccini et al., 2001). Pepstatin A has been employed as a universal means of purifying aspartic proteases from a variety of sources. An example is isolation of an aspartic protease family from wild growing basidiomycete *Clitocybe nebularis* (Sabotič et al., 2009a).

2. Materials and methods

The protease inhibitors from mushrooms, cnispin and macrocypin, were selected as examples of application of fungal protease inhibitors in affinity chromatography for isolation of proteases. Similar protocols can be followed for other protease inhibitors.

2.1 Preparation of ligands

Recombinant protein protease inhibitors cnispin from *Clitocybe nebularis* and macrocypin from *Macrolepiota procera* were heterologously expressed in *Escherichia coli* BL21(DE3), refolded from inclusion bodies and purified as previously described (Avanzo et al., 2009; Sabotič et al., 2009b).

2.2 Immobilization of ligands to the Sepharose matrix

Recombinant cnispin and macrocypin were immobilized to the CNBr-activated Sepharose (GE Healthcare) according to the manufacturer's recommendations. For each ligand a 15 ml Sepharose column was prepared. The inhibitory activity of the Sepharose immobilized protease inhibitors was confirmed by a trypsin inhibition assay for cnispin and papain inhibition assay for macrocypin as described (Avanzo et al., 2009; Sabotič et al., 2009b).

2.3 Immobilization of ligands to monolith supports

Convective Interaction Media (CIM) epoxy and CIM carbonyldiimidazole (CDI) disks with a diameter of 12 mm and thickness of 3 mm, equipped with a special cartridge, were obtained from BIA Separations (Ljubljana, Slovenia). The CIM epoxy monoliths are synthesized from glycidyl methacrylate (GMA) and ethylene dimethacrylate (EDMA) monomers in the presence of pore forming solvents dodecanol and cyclohexanol. The resulting rigid monolith bears epoxy groups. CDI monoliths are prepared from epoxy monoliths hydrolyzed to obtain hydroxyl group, which are then treated with 1,1'carbonyldiimidazole (Benčina et al., 2004). Macrocypin was immobilized on CIM monoliths using these two different activation chemistries and, additionally, to a monolith that contained short functional spacers (see 2.3.3).

2.3.1 Immobilization on a CIM epoxy disk

Macrocypin was covalently immobilized to the original epoxy groups of the GMA/EDMA material. After conditioning the CIM epoxy disk with 0.5 M phosphate buffer pH 7.0 for 1 h, 2 ml of 2.5 mg/ml macrocypin in the same buffer was applied to the disk with a syringe. The monolithic disk was then incubated for 24 h at pH 7.0 at 45°C with gentle shaking. Afterwards, the monolith was washed for 1 h with water to remove unbound material, and then incubated for 1 h in 0.5 M H_2SO_4 at 50°C to inactivate the remaining epoxy groups. The disk was then washed with the mobile phase 0.1 M Tris-HCl, 0.6 M NaCl, pH 7.0 for 1 h at 1 ml/min flow, and stored in the same buffer at 4°C.

2.3.2 Immobilization on a CIM CDI disk

Macrocypin was covalently immobilized to a CIM CDI disk. The disk was first washed with 0.5 M phosphate buffer pH 7 for 1 h, and then permeated with 2 ml of macrocypin at 2.5 mg/ml in the same phosphate buffer. Afterwards, the disk was incubated with macrocypin solution for 24 h at pH 7.0 at 45°C under gentle shaking. The disk was then rinsed with distilled water for 1 h at 1 ml/min flow and stored in 0.5 M phosphate buffer pH 7 at 4°C.

2.3.3 Immobilization on CIM disk by a short spacer arm

A CIM ethylene diamine (EDA) disk containing (free) amine groups was first incubated in 50 mM phosphate buffer, pH 8.0 for 1 h and then derivatized with 10% (v/v) glutaraldehyde solution in the same buffer overnight in the dark at room temperature and under gentle shaking. The disk was washed with 50 mM phosphate buffer, pH 8.0 to remove the reagent and then with 0.5 mM phosphate buffer, pH 3.0. To immobilize macrocypin, the derivatized disk was permeated with 1 ml of macrocypin (2.5 mg/ml in 50 mM phosphate buffer pH 3.0) and allowed to react for 24 h in the dark, at room temperature and under gentle shaking. The disk was then washed with 0.5 M phosphate buffer, pH 3.0 for 1h. To reduce the formed Schiff's bases, the disk was additionally washed with 0.1 M sodium cyanoborohydride in 0.5 M phosphate buffer, pH 8.0, for 2 h. Free remaining groups were end-capped with 1 M monoethanolamine in phosphate buffer (0.5 M, pH 8.0) for 3 h. Finally, the disk was washed with distilled water for 1 h at 1 ml/min flow and stored in 0.01 M phosphate buffer at 4°C.

2.4 Protein extracts and enzymes

Crude protein extracts of papain-like cysteine proteases and legumain were prepared from plant (kiwi fruit and germinated bean seeds) and animal (pig kidney cortex) sources, and of the appropriate bacterial source for isolation of serine proteases. The above natural sources were selected based on previous experience and literature data as appropriate sources for target proteases. In addition, partially purified, commercially available bovine serine proteases and purified plant papain were used for characterization of the prepared affinity columns. The extracts were prepared in the appropriate binding buffers for direct application to the affinity chromatography column.

2.4.1 Kiwi fruit

Kiwi fruits (*Actinidia deliciosa*), previously stored at -20°C, were thawed and homogenized in 0.1 M Na acetate buffer pH 6.5 with 0.1 % $Na_2S_2O_3$ and 0.3 M NaCl. Before application to the affinity column the crude protein extract was cleared by centrifugation for 15 min at 16000 g and 4°C.

2.4.2 Germinated bean seeds

Bean seeds (*Phaseolus vulgaris* L.) were germinated aseptically at 28°C in the dark. After 3 days the cuticle was removed and the seeds homogenized at 4°C in 0.1 M Na acetate buffer, pH 6.5 supplemented with 0.5 M NaCl, 1.5 mM EDTA and 2 mM DTT using an ultraturax homogenizer (IKA-Labortechnik). The insoluble material was removed by centrifugation for 30 min at 12000 g and 4°C. Proteins in the supernatant were precipitated by ammonium sulphate at 80% saturation overnight at 4°C. The precipitate was collected by centrifugation (20 min, 16000 g), dissolved in 0.1 M Na acetate buffer, 0.5 M NaCl, pH 6.5, and dialyzed (7 kDa cut-off) against the same buffer. The extract was cleared by centrifugation (30 min, 16000 g) and applied to a Sephacryl S200 column (GE Healthcare) equilibrated with 0.1 M Na acetate, 0.5 M NaCl, pH 6.5 for size exclusion chromatography. Protein containing fractions (determined by measurement of absorbance at 280 nm) were pooled and concentrated by ultrafiltration (UM-10, Amicon).

2.4.3 Pig kidney cortex

Cortical tissue was dissected from pig kidney and homogenized in 0.1 M phosphate buffer, pH 6 supplemented with 0.3 M NaCl, 60 mM EDTA and 15 mM DTT using an ultraturax homogenizer (IKA-Labortechnik). The insoluble material was removed by centrifugation at 4 °C for 20 min at 8000 g and the supernatant cleared by centrifugation for 20 min at 16000 g and 4°C.

2.4.4 *Bacillus subtilis* culture supernatant

Bacillus subtilis was cultivated (0.5 % yeast extract, 1 % powdered milk) at 37°C and 220 rpm for 24 h. Bacteria were removed by centrifugation at 7000 g for 20 min at 4°C. Proteins in the culture supernatant were precipitated with ammonium sulphate at 80% saturation overnight at 4°C. After dialysis against 0.1 M Na acetate, 0.3 M NaCl, 1 mM EDTA, pH 6.5, the protein extract was cleared by centrifugation at 16000 g for 10 min at 4°C.

2.4.5 Enzymes

Partially purified bovine trypsin and chymotrypsin mixture (Fluka) was dissolved at 1 mg/ml in 0.1 M Na acetate, 0.3 M NaCl, 1 mM EDTA, pH 6.5. Papain from *Carica papaya* (Sigma) was dissolved at 1 mg/ml in 0.1 M Tris-HCl, 0.6 M NaCl, pH 7.0.

2.5 Affinity chromatography

2.5.1 Sepharose matrix affinity chromatography

A cnispin-affinity column was used to isolate subtilisin from the *B. subtilis* culture supernatant and trypsin from the partially purified bovine serine protease mixture.

Macrocypin-affinity chromatography was used for isolation of actinidin from kiwi fruit, legumain from germinated bean seeds and cysteine proteases from pig kidney cortex.

The Sepharose column was equilibrated with 10 volumes of the binding buffer appropriate for each protein extract (see 2.4). The protein sample was applied by gravity flow and the unbound proteins washed with the same binding buffer until the absorbance at 280 nm of the effluent approached zero. The bound proteins were eluted by a pH change using first 10 mM HCl and then 10 mM NaOH. Fractions were neutralized with 2 M Tris-HCl, pH 6.5, pooled and concentrated by ultrafiltration (UM-10, Amicon).

For isolation of cysteine proteases from germinated bean seeds and pig kidney cortex, an additional elution step was included before the pH change, namely the increased ionic strength using 0.7 M NaCl in a corresponding binding buffer.

2.5.2 CIM monoliths

Chromatographic experiments with papain were carried out with a high performance liquid chromatography system (Knauer). Papain (1 mg/ml in 0.1 M Tris-HCl, 0.6 M NaCl, pH 7.0) was pumped through the monolith coupled with macrocypin for 10 min at a flow rate of 1 ml/min. Bound proteins were eluted by a stepwise gradient, using 0.1 M glycine pH 2.0 at a flow rate of 1 ml/min. Fractions were collected in neutralization buffer 2 M Tris-HCl pH 8.0 to maintain papain stability and activity for further analyses.

The binding capacity of immobilized macrocypin was determined by pumping a 1 mg/ml papain solution in 0.1 M Tris-HCl, 0.6 M NaCl, pH 7.0 through the monolith for 10 min at a flow rate of 1 ml/min. During this experiment an early breakthrough of the papain was observed. However, when the experiment was continued and more sample was loaded onto the column, a substantial amount of papain was bound and eluted afterwards with 0.1 M glycine pH 2.0 and neutralized. The amount of papain bound was to a certain degree dependent on the amount of sample loaded. Based on this fact, we speculated that papain might exist in at least two different forms - one that is binding to the immobilized macrocypin and one that does not under the selected conditions. To obtain the papain form that binds to macrocypin, the eluted fraction was collected, desalted with PD-10 desalting columns (GE Healthcare) and re-applied to the monolithic disk. In this case, a typical breakthrough curve was obtained - the loaded papain bound to the column until a breakthrough was achieved. After that, the column was washed and the bound papain eluted in the same way as before. The dynamic binding capacity was calculated from this breakthrough curve at 50% breakthrough as previously described (Hage, 2006).

Legumain isolation experiments were performed on a fast protein liquid chromatography (FPLC) system (GE Healthcare). This technique is a type of high performance liquid chromatography, with lower operating pressure (1 - 2 MPa), adjusted for separation of amino acids, peptides and proteins (Wilson & Walker, 2005). The protein sample of germinated bean seeds (200 µl) was loaded onto a CIM epoxy disk, in a solution of 0.1 M Tris-HCl, 0.6 M NaCl, pH 7.0 - also used as mobile phase - at 0.2 ml/min. Bound proteins were eluted from disk with 0.1 M glycine, pH 2.0 and neutralized with 2 M Tris-HCl, pH 8.0 to prevent self-degradation.

2.6 Protein characterization

2.6.1 SDS-PAGE

Eluted proteins were separated on 12 % polyacrylamide gels using Low Molecular Weight (LMW) markers of 14.4 - 97 kDa (GE Healthcare). Proteins were visualized with Coomassie Brilliant Blue R-250 or silver staining, as appropriate, following the standard protocols.

2.6.2 Immunoblot analysis

Proteins separated by SDS-PAGE were transferred to a polyvinylidene difluoride membrane (Immobilon-P, Millipore) in a tank transfer system (Bio-Rad). The membrane was blocked for 1h in blocking buffer (50 mM Tris-HCl, pH 7.5, 150 mM NaCl and 5 % milk powder) and then incubated with anti-recombinant macrocypin serum (dilution 1:10000) overnight at 4°C. After washing at room temperature five times for 15 min in washing buffer (50 mM Tris-HCl, pH 7.5, 150 mM NaCl, 0.05 % Tween 20) the membrane was incubated with horseradish peroxidase conjugated goat anti-rabbit IgG (Dianova) secondary antibodies (dilution 1:20000) for 1 h at room temperature. The membrane was washed as above and the chemiluminescence detection performed with Lumi-Light^{PLUS} (Roche).

2.6.3 Protein sequencing

For N-terminal sequence analysis, proteins separated by SDS-PAGE were electro-transferred to an Immobilon-P membrane (Millipore) as described above, and visualized by Coomassie staining. Bands were excised and sequencing was performed on a Procise 492A Automated Sequencing System (Applied Biosystems). Alternatively, bands excised from a Coomassie Brilliant Blue stained SDS-PAGE were, after in-gel digestion with trypsin, analysed by mass spectrometry using electrospray ionization (LC-MS-MS) on MSD Trap XCT Ultra (Agilent).

2.6.4 Enzyme activity assays

The substrate Suc-Ala-Ala-Pro-Phe-7-amido-4-methylcoumarin (Suc-Ala-Ala-Pro-Phe-AMC) (Bachem) was used for the analysis of serine protease activity. Different amounts of samples were mixed with the buffer (0.1 M phosphate buffer, pH 8) and the reaction initiated by the addition of Suc-Ala-Ala-Pro-Phe-AMC substrate (30 μM). After 15 min incubation at 37°C the reaction was stopped by the addition of HCl and product formation monitored on a Luminescence Spectrometer LS30 (Perkin Elmer) using 370 nm for excitation and 460 nm for emission.

The substrate Z-Phe-Arg-AMC (Bachem) was used for measuring papain-like cysteine protease activity in a fluorimetric assay. Different amounts of samples were mixed with the buffer (0.1 M Na phosphate, pH 6.5, 3 mM DTT, 2 mM EDTA) and the reaction initiated with the addition of the substrate to a final concentration of 100 μM and incubated for 10 min at 37 °C. The reaction was stopped with iodoacetic acid and the release of the fluorescent product monitored at 370 nm excitation and 460 nm emission using Luminescence Spectrometer LS30 (Perkin Elmer).

The substrate Z-Ala-Ala-Asn-AMC (Bachem) was used for analysis of legumain activity. Different amounts of samples were mixed with the buffer (0.1 M phosphate-citrate buffer, pH 5.8, 0.1 % w/v CHAPS, 1 mM EDTA, 3 mM DTT), the reaction was initiated by the addition of the substrate (50 μM) and incubated for 15 min at 37°C. The reaction was stopped by the addition of iodoacetic acid and product formation (370/460 nm) monitored on a Luminescence Spectrometer LS30 (Perkin Elmer).

2.6.5 Zymogram analysis

Gelatin zymography was performed as described (Sabotič et al., 2007b) using developing solution at pH 8.0 (100 mM Tris-HCl, 200 mM NaCl, 5 mM $CaCl_2$) to analyse the general proteolytic activity of serine and cysteine proteases in samples. Following Coomassie Brilliant Blue staining the proteolytic activities are seen as white bands on a dark background.

Native PAGE, followed by detection of proteolytic activity using fluorescent substrate Z-Phe-Arg-AMC, was used for analysis of proteolytic activity of mammalian papain-like cysteine proteases as described (Budič et al., 2009).

3. Results and discussion

3.1 Cnispin-affinity chromatography for isolation of serine proteases

The serine protease inhibitor cnispin was used for the isolation of secreted proteases of *Bacillus subtilis*, which are mainly of the serine catalytic type. Bacterial secreted alkaline proteases are an indispensable detergent additive (Rao et al., 1998). Since their expression increases during the stationary growth phase (Priest, 1977), proteolytic activity in the culture supernatant was monitored for 24 h (Fig. 4a) to determine the optimal time of harvest. The culture supernatant after 24 h incubation that showed the highest proteolytic activity was used for precipitation of secreted proteins. They were applied to cnispin-affinity chromatography and unbound proteins washed off, followed by elution of the bound proteins by a change in pH (from pH 6.5 during application to pH 2 for elution). SDS-PAGE analysis revealed three bands in the pooled eluted fractions (23, 24 and 27 kDa), which also showed proteolytic activity on a gelatin zymogram (Fig. 4b). Furthermore, the specific activity against the chromogenic substrate Suc-Ala-Ala-Pro-Phe-AMC increased in the bound fractions relative to the unbound and applied sample (Fig. 4c). This activity was abolished by chymostatin, a broad-range inhibitor of serine proteases of families S1 (e.g. trypsin and chymotrypsin) and S8 (e.g. subtilisin).

Commercially obtained partially purified bovine trypsin mixture was further purified using the cnispin-affinity chromatography. Highly purified trypsin is for example used in protein digestion preceding mass spectrometry analyses. Cnispin is a very strong inhibitor of trypsin and a weak inhibitor of chymotrypsin, therefore trypsin purification was expected. The trypsin mixture (1 mg/ml) was applied to the column and bound proteins were eluted by lowering the pH. SDS-PAGE analysis revealed that a less contaminated trypsin was obtained with this additional step, seen as a predominant band at 23 kDa, which corresponds to the theoretical molecular mass of the bovine trypsin (Fig. 5).

Although cnispin affinity chromatography was successful in purifying serine proteases from the bacterial culture supernatant and the high trypsin specificity of cnispin enabled purification of trypsin from a mixture of similar enzymes, the amount of isolated proteases was very small. This indicates either a low capacity for protease binding to the immobilized inhibitor or ineffective elution of the protease from it and makes the described cnispin-affinity chromatography useful mainly for laboratory or analytical scale experiments.

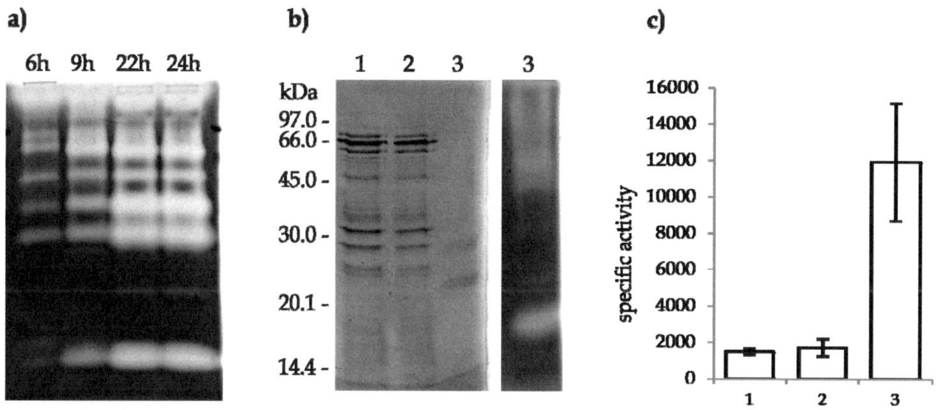

Fig. 4. Purification of secreted *B. subtilis* proteases by cnispin-affinity chromatography. (a) Gelatin zymogram of the *B. subtilis* culture supernatant collected at different times after inoculation. (b) SDS-PAGE (left) and gelatine zymogram (right) analysis of serine proteases isolated from the *B. subtilis* culture supernatant by cnispin-affinity chromatography. Lane 1, the applied precipitated secreted proteins; lane 2, unbound proteins; lane 3, eluted bound proteins. (c) Specific activity with standard deviation measured against the subtilisin substrate Suc-Ala-Ala-Pro-Phe-AMC. Sample numbers correspond to lanes in panel (b).

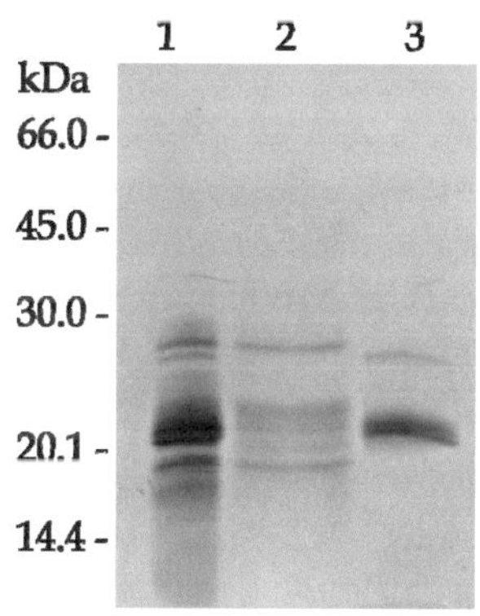

Fig. 5. Trypsin purification with cnispin-affinity chromatography. SDS-PAGE analysis of the trypsin mixture further purified by cnispin-affinity chromatography. Lane 1, the applied trypsin mixture; lane 2, unbound proteins; lane 3, eluted bound proteins.

3.2 Macrocypin-affinity chromatography for isolation of cysteine proteases

The cysteine protease inhibitor macrocypin 1 from *Macrolepiota procera* inhibits two families of cysteine proteases, utilizing different reactive loops (Fig. 1), namely papain-like proteases (family C1) and legumain (family C13). To determine whether macrocypin-affinity chromatography would be applicable for isolation of these proteases from plant and animal sources, we tested kiwi fruits as a source of papain-like protease actinidin, germinated bean seeds as a source of plant legumain and pig kidney cortex as a source of both papain-like cysteine proteases and legumain.

Actinidin (also called actinidain) is a papain-like cysteine protease abundant in kiwi fruits (*Actinidia deliciosa*). Several applications of actinidin have been considered, such as in meat tenderization (Aminlari et al., 2009), in cheese making (Lo Piero et al., 2011) and for isolation of various cell types from human and animal tissues (Mostafaie et al., 2008).

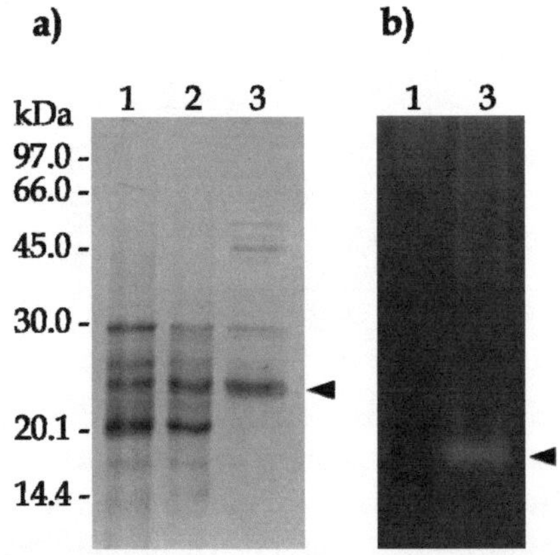

Fig. 6. Partial purification of actinidin from kiwi fruit using macrocypin-affinity chromatography. SDS-PAGE (a) and gelatine zymogram (b) analysis of actinidin purification by macrocypin affinity chromatography. Lane 1, kiwi fruit extract; lane 2, unbound proteins; lane 3, eluted bound proteins.

Macrocypin-affinity chromatography was used as a one-step purification procedure for actinidin from crude protein extracts of kiwi fruits, where a change in pH (from pH 6.5 during application to pH 11 for elution) was used for elution of bound proteins. SDS-PAGE analysis revealed a partial purification of actinidin, as a 24 kDa band corresponding to a mature actinidin was predominant (Fig. 6a). The presence of mature actinidin in the 24 kDa band was confirmed by N-terminal sequencing (H_2N-LPSYVDWRSA) and zymogram analysis confirmed its proteolytic activity (Fig. 6b).

Legumain-like proteases (also called vacuolar-processing enzymes) are abundant in seeds of leguminous plants and, in kidney bean seeds (*Phaseolus vulgaris*) legumain performs controlled hydrolysis of storage protein phaseolin during and after germination (Senyuk et al., 1998). A protein extract of germinating bean seeds was first purified by size exclusion chromatography and then applied to the macrocypin-affinity chromatography. Three different means of elution were performed to determine the optimal purification procedure, namely higher ionic strength (0.7 M NaCl in binding buffer), and elution by low pH (10 mM HCl) and high pH (10 mM NaOH). Purification of legumain was achieved and confirmed by mass spectrometry, where it was present in the 48 kDa band (Fig. 7a, arrow) that corresponds to the mature enzyme, in addition to phaseolin (47.5 kDa), which was probably retained on the column because of its abundance in the extract. The activity, measured against the legumain substrate Suc-Ala-Ala-Asn-AMC, in the applied sample, flow-through and eluted fractions confirmed the purification of legumain with elution by lowering the buffer pH (Fig. 7b). The change of ionic strength was not sufficient to elute the protease, while high pH causes its denaturation.

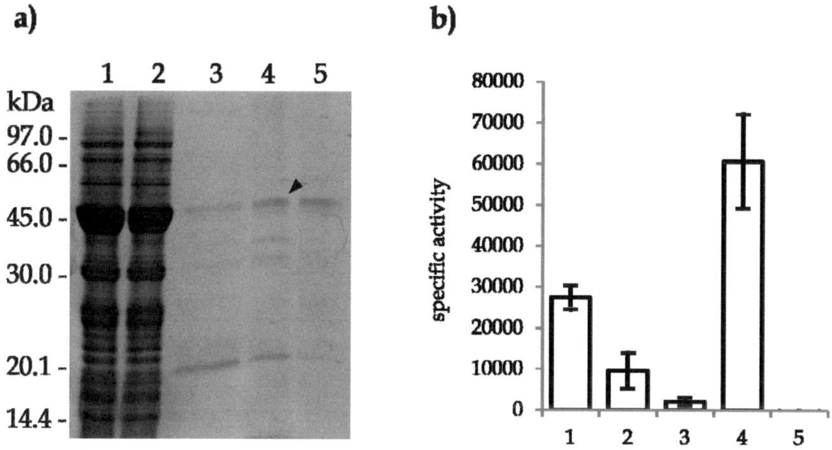

Fig. 7. Legumain purification from germinated bean seeds. (a) SDS-PAGE analysis of the macrocypin-affinity chromatography process. Lane 1, the applied crude protein extract; lane 2, unbound proteins; lane 3, bound proteins eluted by high ionic strength (0.7 M NaCl); lane 4, bound proteins eluted by low pH (10 mM HCl); lane 5 bound proteins eluted by high pH (10 mM NaOH). (b) Specific activity with standard deviation measured against the legumain specific substrate Z-Ala-Ala-Asn-AMC. Sample numbers correspond to lanes in panel (a).

Since isolation of plant legumain was successful with the macrocypin-affinity column, we wanted to confirm its applicability for animal legumain as well. Mammalian legumain is most abundant in kidneys (Chen et al., 1997), therefore, pig kidney cortex was used as a source material. Since many papain-like proteases are also present in mammalian tissues, their co-purification on the macrocypin-affinity column was expected. Two versions of the elution procedure were employed. First, for purification of papain-like proteases, the bound proteins were eluted (experiment A) by a change of pH only (from pH 6.0 in the application buffer to pH 12 in elution). Secondly, for purification of legumain and papain-like proteases (experiment B), elution with higher ionic strength (0.7 M NaCl in application buffer) was followed by low (10 mM HCl) and high pH (10 mM NaOH) elution. Papain-like proteases were isolated by both procedures, as seen by zymogram analysis (Fig. 8, lanes A3 and B4), however their specific activity was not increased (Fig. 8c, white bars). On the contrary, legumain isolation was, as with plant legumain, achieved optimally by elution with low pH, resulting in higher specific activity (Fig. 8c, black bars). However, as seen in the SDS-PAGE analysis (Fig. 8a), further purification steps would be required to obtain pure enzyme.

We have shown the applicability of macrocypin-affinity chromatography for isolation of cysteine proteases from plant and animal sources. The affinity chromatography purification procedure would have to be complemented by one other purification step to obtain a purified protein. Furthermore, the capacity of the macrocypin-affinity column was very low, possibly due to steric hindrance to protease binding to the immobilized inhibitor, making the protease inhibitor immobilized to the Sepharose matrix applicable only on the laboratory or analytical scale. Both of these issues were addressed by development of macrocypin-affinity chromatography on monolithic support.

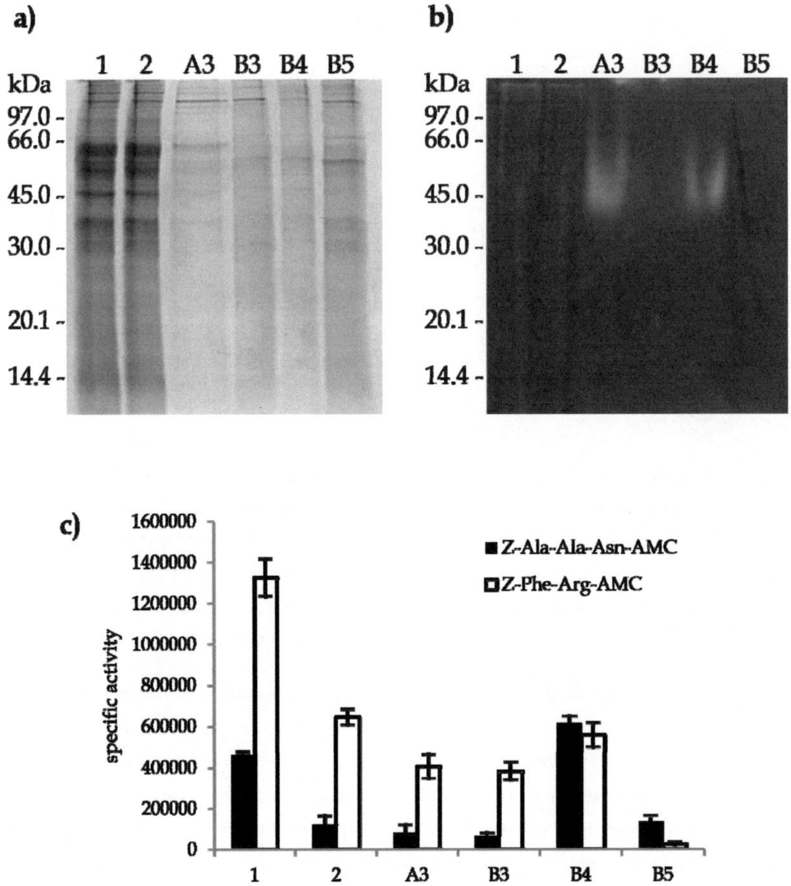

Fig. 8. Purification of cysteine proteases from pig kidney cortex using macrocypin-affinity chromatography. (a) SDS-PAGE analysis of applied, washed and eluted fractions: lane 1, the applied crude protein extract; lane 2, unbound proteins; lane A3, bound proteins eluted by high pH in experiment A; lane B3, bound proteins eluted by high ionic strength (0.7 M NaCl); lane B4, bound proteins eluted by low pH (10 mM HCl); and lane B5 bound proteins eluted by high pH (10 mM NaOH) in experiment B. (b) Zymogram analysis with the substrate Z-Phe-Arg-AMC for detection of papain-like cysteine protease activities in applied, washed and eluted fractions. Lane numbers correspond to those in panel (a). (c) Specific activity with standard deviation measured against the legumain specific substrate Z-Ala-Ala-Asn-AMC and papain-like protease substrate Z-Phe-Arg-AMC. Sample numbers correspond to lanes in panels (a) and (b).

3.3 Optimization of macrocypin-affinity chromatography on monolithic disk support

Three monolithic disks (CIM epoxy, CIM CDI and CIM with glutaraldehyde spacer arm) were used as supports for macrocypin immobilization (Fig. 9). Covalent binding of macrocypin to solid monolithic support was verified with immunoblot analysis (Fig. 10b). Results indicated that macrocypin immobilization was stable and macrocypin was not leaking from the support during experiments in all three versions of prepared disks.

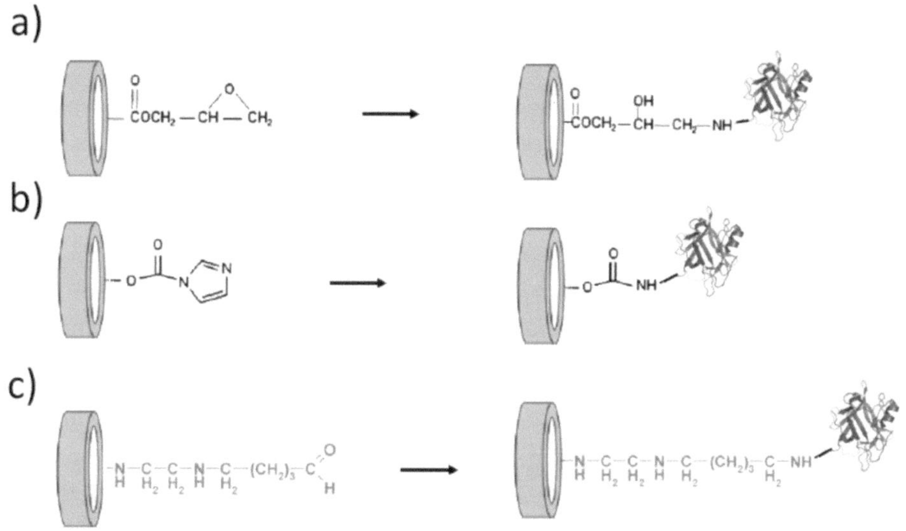

Fig. 9. Macrocypin immobilization on a: (a) CIM epoxy disk; (b) CIM CDI disk; (c) CIM with glutaraldehyde spacer arm.

3.3.1 Immobilization on CIM epoxy disk

The majority of the immobilizations on glycidyl methacrylate monolith supports were performed via epoxy groups, since they are formed *in situ* during the polymerization process and thus readily available for chemical modification. Macrocypin was immobilized on a CIM epoxy disk (Fig. 9a) as described in 2.3.1. Macrocypin solution was syringed through the disk to completely fill all the monolith pores. pH 7.0 was used as the optimal pH for coupling reaction between the epoxy groups and amino residues of macrocypin. A solution of 0.5 M H_2SO_4 at 50°C was employed for end-capping the remaining free epoxy groups and preventing side reactions. To analyze the CIM epoxy macrocypin affinity disk, 10 mg of crude papain solution was applied at 1 ml/min and eluted at lower pH. SDS-PAGE analysis (Fig. 10a) of eluted samples revealed more concentrated papain (23.4 kDa) in flow-through (Fig. 10a, lane C2) than from elution (Fig. 10a, lane C3).

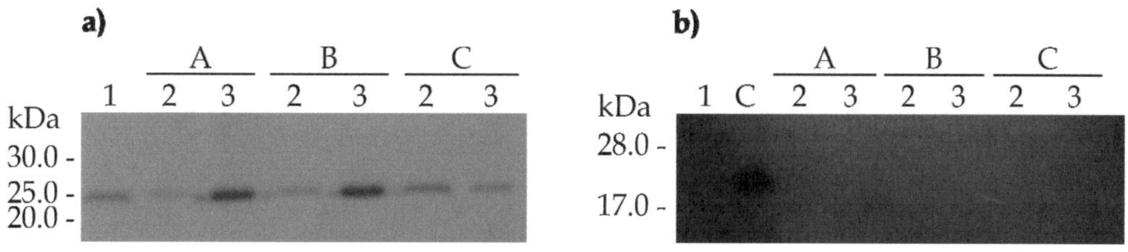

Fig. 10. Papain purification with monolith macrocypin-affinity chromatography. (a) SDS-PAGE analysis of papain purification on (A) CIM disk with glutaraldehyde spacer arm, (B) CIM CDI disk and (C) CIM epoxy disk; lane 1, the applied papain; lane 2, unbound papain; lane 3, eluted bound papain. (b) Immunoblot analysis of the same samples as in panel (a); polyclonal anti-macrocypin antibodies were used and immunoreactive bands visualized by chemiluminescent detection; lane C, macrocypin as positive control.

Although this method of immobilization was easily achieved, the active site of macrocypin was probably not accessible to the binding sites on papain, therefore macrocypin immobilization to CIM epoxy disk did not demonstrate the desired characteristics. Chemical modification of the epoxy groups to imidazole carbamate groups is one alternative to overcome the limitations of the epoxy method, so a macrocypin CIM CDI disk was prepared.

3.3.2 Immobilization on a CIM CDI disk

Imidazole carbamate groups that react with N-nucleophiles give an N-alkyl carbamate linkage, resulting from the reaction between hydroxyl groups obtained by hydrolysis of epoxy groups and 1,1′-carbonyldiimidazole (Hermanson et al., 1992). This activated support was used in macrocypin immobilization by means of nucleophilic substitution between the activated sites and primary amines on the protein, resulting in a stable amide linkage (Fig. 9b). This method is faster than the epoxy method and involves fewer steps (Mallik & Hage, 2006). In addition, the remaining free imidazole groups of the supports are rapidly self-deactivated after the immobilization process in aqueous solution, forming the original hydroxyl groups and releasing CO_2 and imidazole (Benčina et al., 2004; Nicoli et al., 2008).

To assess the performance of the CIM CDI macrocypin affinity disk, 10 mg of papain solution was loaded in 10 min to the disk and bound papain was eluted at low pH. SDS-PAGE analysis (Fig. 10a, lanes B) revealed better separation than with the CIM epoxy disk for the same amount of papain.

The CIM CDI disk demonstrated sufficient accessibility for protease to immobilized inhibitor, probably due to different steric orientation of the macrocypin on the support.

3.3.3 Immobilization on a CIM disk with spacer arm

Even better accessibility for the protease active site was achieved with the introduction of a spacer arm, which removes the inhibitor from the solid phase surface and minimizes steric interference during binding (Cuatrecasas, 1970). Polymer aldehyde groups formed after activation allow fast covalent binding of amino group bearing ligands under mild conditions, with elimination of water as the only side product (Ponomareva et al., 2010). The glutaraldehyde spacer (Fig. 9c) presumably provides greater flexibility, allowing the macrocypin to move into the right position to establish the correct binding orientation with protein. Affinity chromatography with 10 mg of papain solution was accomplished in 10 min and bound papain was eluted by lowering the pH. Compared with CDI and the epoxy immobilized disk, glutaraldehyde spacer revealed superior properties for papain separation, as seen on SDS-PAGE analysis (Fig. 10a, lanes A).

Macrocypin was also covalently immobilized on CIM disks with three different spacer arms: Ethylenediamine, 1,6-diaminohexane, and 1,4-butanediol diglycidyl ether. However, improved binding characteristics for papain were not observed (not shown).

3.4 Determination of binding capacities of CIM disks

The binding capacity of papain to macrocypin immobilized via glutaraldehyde and imidazole carbamate groups was found to be higher than to macrocypin immobilized via epoxy groups. Dynamic binding capacities were determined by measuring the breakthrough curve (Fig. 11). For CIM epoxy disk, the capacity was 0.34 mg/ml, for CIM CDI disk 5.1 mg/ml and for CIM disk with glutaraldehyde spacer arm 9.2 mg/ml.

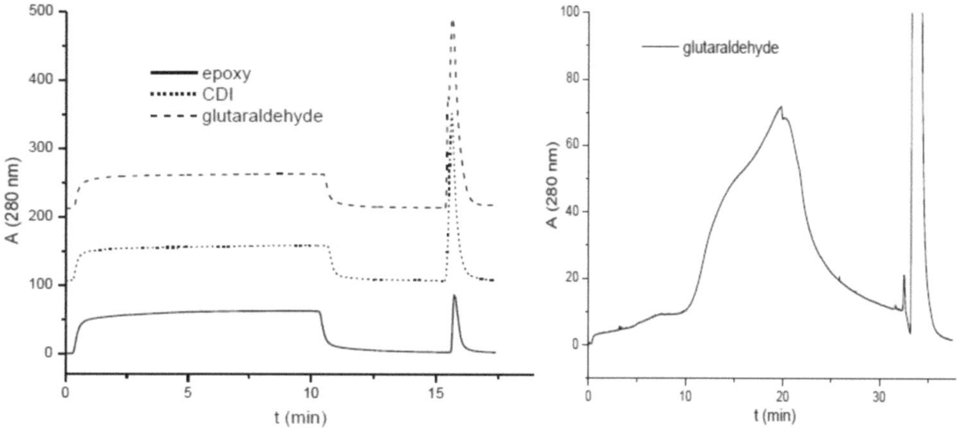

Fig. 11. Comparison of papain separations on CIM epoxy, CIM CDI and CIM with glutaraldehyde spacer arm (left). Chromatographic conditions: flow rate 1 ml/min; concentration of papain 1 mg/ml in 0.1 M Tris-HCl, 0.6 M NaCl, pH 7.0; elution with 0.1 M glycine pH 2.0; detection wavelength 280 nm. Eluted bound proteins were collected together and applied again to the disks to determine the 50% breakthrough. (right) An example of the breakthrough curve with papain bound to the CIM disk with macrocypin immobilized via a glutaraldehyde spacer arm.

3.5 Monolith macrocypin-affinity chromatography for isolation of plant cysteine protease

To explore the possibility of wider applications of macrocypin affinity disks, legumain from a crude protein extract of germinated bean seeds (*Phaseolus vulgaris*) was subjected to purification.

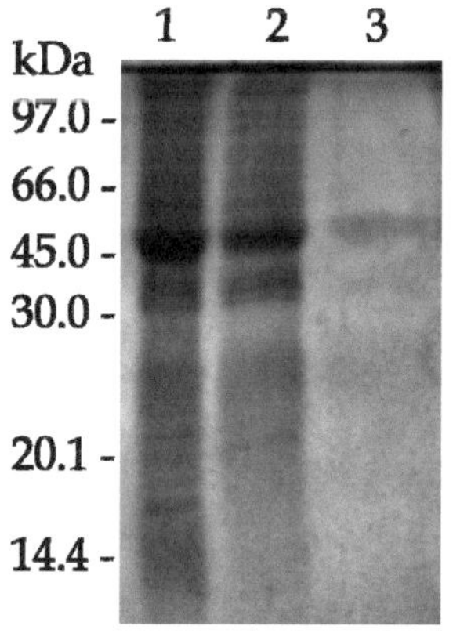

Fig. 12. SDS-PAGE analysis of legumain purification from germinated bean seeds with a CIM epoxy monolith disk. Lane 1, the applied crude protein extract; lane 2, unbound proteins; lane 3, bound proteins eluted with 0.1 M glycine, pH 2.0.

According to the experiences with plant and animal legumain isolation, a similar experiment was performed with a macrocypin CIM epoxy disk. The protein sample of germinated bean seeds was loaded onto a CIM epoxy disk and bound proteins were eluted by lowering the pH. Separation of legumain was confirmed by SDS-PAGE analysis, where a band at 48 kDa is visible (Fig. 12, lane 3). Isolation of proteolytically active legumain (determined by hydrolysis of the legumain specific substrate Z-Ala-Ala-Asn-AMC) using macrocypin immobilized to the monolith support confirms the latter's applicability for separation of cognate proteases from complex protein mixtures.

4. Conclusion

Protein protease inhibitors from mushrooms bound to solid matrices proved to be a useful tool for isolation of proteases from various natural sources. Protease inhibitors from higher fungi offer inhibitory patterns different from those from other sources, together with superior characteristics as affinity chromatography ligands, in terms of pH and temperature stability. They withstand the harsh conditions during immobilization procedures and retain their inhibitory activity through several elution cycles of extreme pH changes.

In most cases, the ligands of interest are immobilized onto conventional particle based chromatographic supports. This represents a widely used and well established technique used for selective isolation and purification of proteases. The Sepharose immobilized inhibitors were effective in isolating several different target proteases from various sources, however the method is applicable only on the analytical or laboratory scale. The well-known drawbacks of this type of matrix are their low intrinsic velocity of operation and mass transfer limitations. These can be effectively overcome by using monolith supports. Monolith supports are characterized by an open pore structure where the mass transfer between the mobile and the stationary phases is greatly enhanced by the convective flow. In this work, the monolith support with immobilized macrocypin provided a convenient approach for isolation of various target proteases. Selectivity of the columns was tested by SDS-PAGE and the best results were obtained with a glutaraldehyde spacer arm, indicating that steric hindrance was one of the reasons for low efficacy of the other systems tested. Thus the steric interference of analyte binding to the immobilized ligand is important to consider when designing an affinity column or disk. In conclusion, the CIM disk with macrocypin immobilized through the glutaraldehyde spacer arm could be used for rapid and effective purification of cysteine proteases from various sources.

5. Acknowledgements

This work was supported by the Slovenian Research Agency Grant No. P4-0127 (J.K.) and P4-0369 (A.P.). We are grateful to Dr Roger H. Pain for critical reading and language editing of the chapter. The authors gratefully acknowledge the help of Katarina Mustar and Manca Kocjančič with Sepharose-based affinity chromatography columns, dr. Adrijana Leonardi with protein N-terminus analysis and dr. Marko Fonović with mass spectrometry analysis.

6. References

Aminlari, M., Shekarforoush, S. S., Gheisari, H. R. & Golestan, L. (2009). Effect of actinidin on the protein solubility, water holding capacity, texture, electrophoretic pattern of beef, and on the quality attributes of a sausage product. *Journal of Food Science*, Vol. 74, No. 3, (April 2009), pp. C221-226, ISSN 1750-3841

Avanzo, P., Sabotič, J., Anžlovar, S., Popovič, T., Leonardi, A., Pain, R. H., Kos, J. & Brzin, J. (2009). Trypsin-specific inhibitors from the basidiomycete *Clitocybe nebularis* with regulatory and defensive functions. *Microbiology*, Vol. 155, No. 12, (December 2009), pp. 3971-3981, ISSN 1465-2080

Barut, M., Podgornik, A., Urbas, L., Gabor, B., Brne, P., Vidič, J., Plevčak, S. & Štrancar, A. (2008). Methacrylate-based short monolithic columns: enabling tools for rapid and efficient analyses of biomolecules and nanoparticles. *Journal of separation science*, Vol. 31, No. 11, (June 2008), pp. 1867-1880, ISSN 1615-9314

Benčina, K., Podgornik, A., Štrancar, A. & Benčina, M. (2004). Enzyme immobilization on epoxy- and 1,1'-carbonyldiimidazole-activated methacrylate-based monoliths. *Journal of separation science*, Vol. 27, No. 10-11, (July 2004), pp. 811-818, ISSN 1615-9306

Brne, P., Lim, Y. P., Podgornik, A., Barut, M., Pihlar, B. & Štrancar, A. (2009). Development and characterization of methacrylate-based hydrazide monoliths for oriented immobilization of antibodies. *Journal of Chromatography A*, Vol. 1216, No. 13, (March 2009), pp. 2658-2663, ISSN 1873-3778

Brzin, J., Rogelj, B., Popovič, T., Štrukelj, B. & Ritonja, A. (2000). Clitocypin, a new type of cysteine proteinase inhibitor from fruit bodies of mushroom *Clitocybe nebularis*. *The Journal of biological chemistry*, Vol. 275, No. 26, (June 2000), pp. 20104-20109, ISSN 0021-9258

Budič, M., Kidrič, M., Meglič, V. & Cigič, B. (2009). A quantitative technique for determining proteases and their substrate specificities and pH optima in crude enzyme extracts. *Analytical biochemistry*, Vol. 388, No. 1, (May 2009), pp. 56-62, ISSN 1096-0309

Champagne, J., Delattre, C., Shanthi, C., Satheesh, B., Duverneuil, L., Vijayalakshmi, M. A. (2007). Pseudoaffinity Chromatography Using a Convective Interaction media - Disk Monolithic Column. *Chromatographia*, Vol. 65, No. 11/12, (May 2007), pp. 639-648, ISSN 0009-5893

Chen, J. M., Dando, P. M., Rawlings, N. D., Brown, M. A., Young, N. E., Stevens, R. A., Hewitt, E., Watts, C. & Barrett, A. J. (1997). Cloning, isolation, and characterization of mammalian legumain, an asparaginyl endopeptidase. *Journal of Biological Chemistry*, Vol. 272, No. 12, (March 1997), pp. 8090-8098, ISSN 0021-9258

Christeller, J. T. (2005). Evolutionary mechanisms acting on proteinase inhibitor variability. *FEBS Journal*, Vol. 272, No. 22, (November 2005), pp. 5710-5722, ISSN 1742-464X

Cuatrecasas, P. (1970). Protein purification by affinity chromatography. Derivatizations of agarose and polyacrylamide beads. *The Journal of biological chemistry*, Vol. 245, No. 12, (June 1970), pp. 3059-3065, ISSN 0021-9258

Cuatrecasas, P., Wilchek, M. & Anfinsen, C. B. (1968). Selective enzyme purification by affinity chromatography. *Proceedings of the National Academy of Sciences of the United States of America*, Vol. 61, No. 2, (October 1968), pp. 636-643, ISSN 0027-8424

Galeša, K., Thomas, R. M., Kidrič, M. & Pain, R. H. (2004). Clitocypin, a new cysteine proteinase inhibitor, is monomeric: impact on the mechanism of folding. *Biochemical and Biophysical Research Communications*, Vol. 324, No. 2, (November 2004), pp. 576-578, ISSN 0006-291X

Hage, D. S. *Handbook of affinity chromatography* (2nd ed.), Taylor & Francis, ISBN 9780824740573, Boca Raton ; London

Hanada, K., Tamai, M., Yamagishi, M., Ohmura, S., Sawada, J. & Tanaka, I. (1978). Isolation and Characterization of E-64, a New Thiol Protease Inhibitor. *Agricultural and Biological Chemistry,* Vol. 42, No. 3, (August 1978), pp. 523-528, ISSN 0002-1369

GE Healthcare. (October 2007). *Affinity chromatography- principles and methods,*GE Healthcare, Retrieved from: <http://www.gelifesciences.com/aptrix/upp00919.nsf/content/LD_149605979-F640>

Hermanson, G. T., Mallia, A. K. & Smith, P. K. *Immobilized affinity ligand techniques* Academic Press, ISBN 0123423309 San Diego

Hewlett, G. (1990). Apropos aprotinin: a review. *Nature Biotechnology,* Vol. 8, No. 6, (June 1990), pp. 565-568, ISSN 0733-222X

Kidrič, M., Fabian, H., Brzin, J., Popovič, T. & Pain, R. H. (2002). Folding, stability, and secondary structure of a new dimeric cysteine proteinase inhibitor. *Biochemical and Biophysical Research Communications,* Vol. 297, No. 4, (October 2002), pp. 962-967, ISSN 0006-291X

Kumar, C. G. & Takagi, H. (1999). Microbial alkaline proteases: from a bioindustrial viewpoint. *Biotechnology advances,* Vol. 17, No. 7, (December 1999), pp. 561-594, ISSN 0734-9750

Lo Piero, A., Puglisi, I. & Petrone, G. (2011). Characterization of the purified actinidin as a plant coagulant of bovine milk. *European Food Research and Technology,* Vol. 233, No. 3, (May 2011), pp. 517-524, ISSN 1438-2377

Lopez-Otin, C. & Bond, J. S. (2008). Proteases: Multifunctional Enzymes in Life and Disease. *Journal of Biological Chemistry,* Vol. 283, No. 45, (November 2008), pp. 30433-30437, ISSN 0021-9258

Mallik, R. & Hage, D. S. (2006). Affinity monolith chromatography. *Journal of Separation Science,* Vol. 29, No. 12, (August 2006), pp. 1686-1704, ISSN 1615-9306

Miller, J. M. *Chromatography : concepts and contrasts* (2nd ed.), Wiley-Interscience, ISBN 0471472077, Hoboken

Mostafaie, A., Bidmeshkipour, A., Shirvani, Z., Mansouri, K. & Chalabi, M. (2008). Kiwifruit actinidin: a proper new collagenase for isolation of cells from different tissues. *Applied Biochemistry and Biotechnology,* Vol. 144, No. 2, (February 2008), pp. 123-131, ISSN 0273-2289

Nicoli, R., Gaud, N., Stella, C., Rudaz, S. & Veuthey, J. L. (2008). Trypsin immobilization on three monolithic disks for on-line protein digestion. *Journal of pharmaceutical and biomedical analysis,* Vol. 48, No. 2, (September 2008), pp. 398-407, ISSN 0731-7085

Østergaard, L. H. & Olsen, H. S., 2010. *Industrial Applications of Fungal Enzymes,* In: *The Mycota X Industrial Applications,* Hofrichter, M. (Ed.), pp. (269-290), Springer, ISBN 978-3-642-11458-8, Berlin Heidelberg

Phylip, L. H., Lees, W. E., Brownsey, B. G., Bur, D., Dunn, B. M., Winther, J. R., Gustchina, A., Li, M., Copeland, T., Wlodawer, A. & Kay, J. (2001). The potency and specificity of the interaction between the IA3 inhibitor and its target aspartic proteinase from *Saccharomyces cerevisiae. Journal of Biological Chemistry,* Vol. 276, No. 3, (Janary 2001), pp. 2023-2030, ISSN 0021-9258

Podgornik, A. & Štrancar, A. (2005). Convective Interaction Media (CIM)--short layer monolithic chromatographic stationary phases. *Biotechnology annual review*, Vol. 11, No. (October 2005), pp. 281-333, ISSN 1387-2656

Polanowski, A., Wilimowska-Pelc, A., Kowalska, J., Grybel, J., Zelazko, M. & Wilusz, T. (2003). Non-conventional affinity chromatography of serine proteinases and their inhibitors. *Acta biochimica Polonica*, Vol. 50, No. 3, (September 2003), pp. 765-773, ISSN 0001-527X

Ponomareva, E. A., Kartuzova, V. E., Vlakh, E. G. & Tennikova, T. B. (2010). Monolithic bioreactors: effect of chymotrypsin immobilization on its biocatalytic properties. *Journal of chromatography B*, Vol. 878, No. 5-6, (February 2010), pp. 567-574, ISSN 1873-376X

Priest, F. G. (1977). Extracellular enzyme synthesis in the genus *Bacillus*. *Bacteriological Reviews*, Vol. 41, No. 3, (September 1977), pp. 711-753, ISSN 0005-3678

Rao, M. B., Tanksale, A. M., Ghatge, M. S. & Deshpande, V. V. (1998). Molecular and biotechnological aspects of microbial proteases. *Microbiology and Molecular Biology Reviews*, Vol. 62, No. 3, (September 1998), pp. 597-635, ISSN 1092-2172

Rawlings, N. D. (2010). Peptidase inhibitors in the MEROPS database. *Biochimie*, No. (April 2010), ISSN 1638-6183

Rawlings, N. D., Barrett, A. J. & Bateman, A. (2010). MEROPS: the peptidase database. *Nucleic Acids Research*, Vol. 38, No. Database issue, (January 2009), pp. 227-233, ISSN 1362-4962

Rawlings, N. D., Tolle, D. P. & Barrett, A. J. (2004). Evolutionary families of peptidase inhibitors. *Biochemical Journal*, Vol. 378, No. 3, (March 2004), pp. 705-716, ISSN 1470-8728

Renko, M., Sabotič, J., Mihelič, M., Brzin, J., Kos, J. & Turk, D. (2010). Versatile loops in mycocypins inhibit three protease families. *The Journal of Biological Chemistry*, Vol. 285, No. 1, (Jan 1), pp. 308-316, 1083-351X

Sabotič, J., Galeša, K., Popovič, T., Leonardi, A. & Brzin, J. (2007a). Comparison of natural and recombinant clitocypins, the fungal cysteine protease inhibitors. *Protein Expression and Purification*, Vol. 53, No. 1, (May 2007), pp. 104-111, ISSN 1046-5928

Sabotič, J., Gaser, D., Rogelj, B., Gruden, K., Štrukelj, B. & Brzin, J. (2006). Heterogeneity in the cysteine protease inhibitor clitocypin gene family. *The Journal of Biological Chemistry*, Vol. 387, No. 12, (December 2006), pp. 1559-1566, ISSN 1431-6730

Sabotič, J., Popovič, T. & Brzin, J. (2009a). Aspartic Proteases from Basidiomycete *Clitocybe nebularis*. *Croatica chemica acta*, Vol. 82, No. 4, (January 2009), pp. 739-745, ISSN 0011-1643

Sabotič, J., Popovič, T., Puizdar, V. & Brzin, J. (2009b). Macrocypins, a family of cysteine protease inhibitors from the basidiomycete *Macrolepiota procera*. *FEBS Journal*, Vol. 276, No. 16, (August 2009), pp. 4334-4345, ISSN 1742-4658

Sabotič, J., Trček, T., Popovič, T. & Brzin, J. (2007b). Basidiomycetes harbour a hidden treasure of proteolytic diversity. *Journal of Biotechnology*, Vol. 128, No. 2, (February 2006), pp. 297-307, ISSN 0168-1656

Senyuk, V., Rotari, V., Becker, C., Zakharov, A., Horstmann, C., Muntz, K. & Vaintraub, I. (1998). Does an asparaginyl-specific cysteine endopeptidase trigger phaseolin degradation in cotyledons of kidney bean seedlings? *European Journal of Biochemistry*, Vol. 258, No. 2, (December 1999), pp. 546-558, ISSN 0014-2956

Švec, F., Tennikova, T. B. & Deyl, Z. *Monolithic materials : preparation, properties, and applications* (1st), Elsevier, ISBN 0444508791, Amsterdam ; Boston

Tetala, K. K. R. & van Beek, T. A. (2010). Bioaffinity chromatography on monolithic supports. *Journal of Separation Science*, Vol. 33, No. 3, (January 2010), pp. 422-438, ISSN 1615-9314

Tombaccini, D., Mocali, A., Weber, E. & Paoletti, F. (2001). A cystatin-based affinity procedure for the isolation and analysis of papain-like cysteine proteinases from tissue extracts. *Analytical biochemistry*, Vol. 289, No. 2, (February 2001), pp. 231-238, ISSN 0003-2697

Turkova, J. *Bioaffinity chromatography* (2nd ed.), Elsevier, ISBN 0444890300, Amsterdam, New York

Wilson, K. & Walker, J. M. *Principles and techniques of biochemistry and molecular biology* (6th ed.), Cambridge University Press, ISBN 978521828895, Cambridge, New York

16

Affinity-Based Methods for the Separation of Parasite Proteins

C.R. Alves, F.S. Silva, F.O. Oliveira Jr, B.A.S. Pereira, F.A. Pires and M.C.S. Pereira
Instituto Oswaldo Cruz – Fundação Oswaldo Cruz, Rio de Janeiro, RJ, Brasil

1. Introduction

Affinity chromatography-based techniques have been developed to purify parasite proteins and improve our understanding of the parasite life cycle. These advances can be translated into concrete proposals for new drugs, diagnostic methods and vaccines for parasite diseases and help to reduce social inequality.

Affinity chromatography has been demonstrated to be a powerful tool for the isolation and purification of parasite proteins and has potential applications for diagnosis and therapy. Many studies have focused on parasite proteins that modulate host cell defense, as gp63, a glycoprotein from *Leishmania spp.*, that is involved in the cleavage of the complement factor C3b to iC3b, which promotes adhesion of promastigotes to macrophages via complement receptor 3 (Brittgham et al., 1995). This route of internalization does not lead to production of oxygen radicals or NO and favors parasite subsistence within the host cell. Another example is the cysteine protease B (CPB), an important virulence factor of the *Leishmania (L.) mexicana* complex, that inhibits lymphocytes Th1 and/or promotes the Th2 response either through proteolytic activity or through epitopes derived from its COOH-terminal extension (Pereira et al., 2011).

Due to the important role of these molecules, many researchers seek to develop specific and potent inhibitors for therapeutic strategies. Aspartic protease, a potential target for antiparasitic therapies, has been isolated from *Trypanosoma cruzi* by affinity chromatography using a specific inhibitor of this enzyme (Pinho et al., 2009); this enzyme is target for treatment of infections caused by HIV (Wlodawer & Vondrasek, 1998) and Candida (Hoegl et al., 1999). This enzyme has also been reported in *Plasmodium* spp. and *Schistosoma mansoni*, where it plays an important role in host hemoglobin degradation (Klemba & Goldberg, 2002). Additionally, specific inhibitors of plasmepsins and renin are viable drugs for the treatment of patients with malaria and high blood pressure.

These parasite proteins, along with others, have been tested as new targets for chemo- and immunotherapies for parasite diseases. They have been assessed by lectins or protease inhibitor affinity chromatography. The separation of sugars based on lectin affinity is one of main procedure that has been used. This technique is based on the ability of lectins to bind specifically to certain oligosaccharide structures on glycoconjugates isolated from parasites. Parasite proteins are processed through a multi-lectin affinity column, and they bind to the immobilized lectins through their sugar chains. Certain glycoconjugates are important for the parasite life cycle, and lectin affinity chromatography can help to reveal their roles (Guha-Niyogi et al., 2001).

The use of protease inhibitors in affinity chromatography is another important approach for assessing parasite proteins. Proteases hydrolyze peptide bonds and can therefore degrade proteins and peptides that influence a broad range of biological functions, including the process of parasite infection (Mackeron et al., 2006). The specificity of the protease inhibitor used is an important aspect of this methodology; L- trans-epoxy-succinylleucylamido-(4-guanidino) butane (specific to cysteine-protease), pepstatin A (to aspartyl-protease) and aprotinin (to serine-protease) are frequently immobilized on a solid matrix for this technique.

Glycosaminoglycan (GAG) affinity is the only affinity chromatography method that is based on the sugar chains of lectin-like proteins. Some of these molecules (such as heparin sulfate, heparan sulfate, dermatan sulfate, keratan sulfate and chondroitin sulfate) contain complex oligosaccharide structures, which may be displayed on cell surfaces, incorporated into the extracellular matrix or attached to secreted glycoproteins, suggesting that they play structural roles (Dreyfuss et al., 2010). GAGs have been reported as potential candidates for therapeutic intervention against parasitic infections, such as leishmaniasis and Chagas diseases (Azevedo-Pereira et al., 2007; Oliveira-Jr et al., 2008).

According to the general principle of affinity chromatography (Fig. 1), a protein of interest is recovered based on its capacity to bind a specific functional group (ligand) that is immobilized on a bead material (matrix) that has been packed into a solid support (column). Although many ligands (enzymatic substrates, inhibitors of an enzyme, lectin, sugar residues, vitamins, enzyme cofactors, monoclonal antibodies) have been used to isolate proteins based on affinity, only lectin, an enzyme inhibitor and glycosaminoglycans have been used to obtain parasite proteins. The most commonly used matrix materials for the attachment of the ligand are polysaccharide derivatives (cellulose, dextran and agarose) and polyacrylamide.

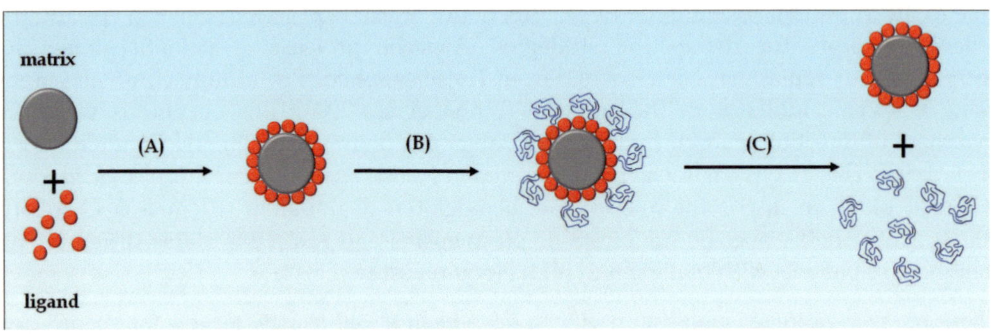

Fig. 1. The principle of affinity chromatography. The ligand is covalently bound to a matrix (A). The functionalized matrix is then able to bind to a target protein aided by a binding buffer (B). Afterwards, the bound proteins are eluted with a different buffer (C).

In these procedures, the soluble proteins are prepared from crude parasite lysates (or sub-cellular fractions) and loaded onto a column under chemical (buffer) and physical (temperature and pressure) conditions that promote the specific binding of the protein to the immobilized ligand (affinity) in what is known as the binding phase. Proteins that do not bind to the immobilized ligand under these conditions are removed from the solid phase by application of a constant liquid phase, which is referred to as the wash phase. Then, the bound protein can be recovered by changing the buffer conditions to favor desorption during the elution phase.

In this chapter, we describe the use of affinity chromatography to assess parasite proteins and the importance of these methods for public health. Several affinity chromatography protocols are considered. Additionally, we discuss our experience using affinity chromatography to obtain parasite proteins, and we include some unpublished results related to *Dermatobia hominis* third (L3) instar larvae proteases.

2. The use of affinity chromatography in parasite protein studies

2.1 Lectin affinity-based separation of parasite proteins

There are relatively few studies available in the current literature describing the use of lectins to affinity-purify glycosylated proteins from parasites. However, the reports on this subject demonstrate that this technique is useful for the retrieval of putative virulence factors or potential protective immunogens from a large array of parasites, including apicomplexan, trypanosomatids and nematodes (e.g., Fauquenoy et al., 2008, Gardiner et al., 1996, Smith et al., 2000). In addition to its utility in the isolation of parasite factors, lectin-based affinity chromatography is also a valuable resource for characterization of the structure of carbohydrates bound to proteins from these organisms due to the distinct specificities of the lectins that are available for this type of analysis.

Lectins are proteins that specifically bind to sugars, and they have been used for many types of studies, ranging from blood typing to immune regulation analysis (Rüdiger & Gabius, 2001). These proteins are generally isolated from plants (mostly legume seeds), where they can be found in abundance. Their usage is determined by the particular sugar structures that they are able to bind (Rüdiger & Gabius, 2001). The surveyed literature the use of six plant lectins [concanavalin A (Con A), ricin, jacalin, peanut agglutinin (PNA), wheat germ agglutinin (WGA) and Wisteria floribunda agglutinin (WFA)] in studies of parasites glycoproteins. Furthermore, one report described the use of Biomphalaria alexandrina lectin (BaSII), which in contrast to the others is a lectin obtained from an animal.

Con A is a lectin that can be extracted from jack beans of the species *Canavalia ensiformis* (family Fabaceae). It binds to mannose or glucose residues and is thus characterized as a mannose-binding lectin. This lectin presents a high affinity for the oligosaccharide GlcNAcβ2Manα6(GlcNAcβ2Manα3)-Manβ4GlcNAc. It is also known to be a potent mitogen (Beckert & Sharkey, 1970; Rüdiger & Gabius, 2001).

Ricin, along with jacalin and PNA, is a lectin that binds to galactose. Specifically, it binds with high affinity to the motif Galβ4GlcNAcβ2Manα6 (Galβ4-GlcNAcβ2Manα3) Manβ4GlcNAc. Ricin is highly toxic because it can impair ribosome activity through cleavage of the nucleobases of ribosomal RNA, and it has potential to be used as a biological weapon. This lectin is extracted from *Ricinus communis* (Family Euphorbiaceae) (Rüdiger & Gabius, 2001; Lord et al., 2003).

Jacalin binds to galactose and N-acetylgalactosamine, and presents a high affinity for the motif Galβ3GalNAcα. It is obtained from *Artocarpus integrifolia* (Family Moraceae). It is commonly used to isolate IgA from human plasma (Kabir, 1998, André et al., 2007).

Like Con A, PNA is a legume lectin and is isolated from plants that belong to the family Fabaceae. It is extracted from *Arachis hypogea* and binds specifically to the monosaccharide galactose and to the motif Galβ3GalNAcα, similarly to the binding motif of jacalin. PNA is used as a marker of T-cell subpopulations and to differentiate between the stages of the Leishmania parasites life cycle (Dumont & Nardelli, 1979, Wilson & Pearson, 1984, Rüdiger & Gabius, 2001).

WGA is obtained from the species *Triticum vulgare*. It presents a low affinity for N-acetylgalactosamine, but it binds to the sialic acid N-acetylneuraminic and to the motif GlcNAcβ4GlcNAcβ4GlcNAcβ4-GlcNAcβ4GlcNAc. This lectin has been shown to bind more avidly to activated human T lymphocytes (Hellström *et al.*, 1976, Rüdiger & Gabius, 2001).

WFA is isolated from *Wisteria floribunda*, a woody liana of the family Fabaceae. Although some uncertainty regarding its binding specificity remains, it seems that this agglutinin binds preferentially to the monosaccharide N-acetylgalactosamine and to the motif GalNAcα6GalNAc. WFA is used to fractionate lymphocyte populations, and although it is not mitogenic like Con A, it can induce lymphokine production in murine splenocytes (Jacobs & Poretz, 1980; Rüdiger & Gabius, 2001).

BaSII is a lectin that can be isolated from the snail *B. alexandrina*, an intermediate host of the trematoda parasite *Schistosoma mansoni*, the causative agent of schistosomiasis. It specifically binds to the motif Fucα1,2Galβ1,4Glc (Mansour, 1996).

2.1.1 General procedures for the isolation of parasite proteins by lectin affinity

The rational for lectin-based affinity chromatography is the same as for other types of affinity-based fractionation: a sample is exposed to a solid phase that has been coupled to an affinity separation molecule (a lectin, in this case) under conditions that are adequate for binding (Fig. 2A). The unbound material from the sample is washed away (generally using the same buffer applied to equilibrate the solid-phase), and in the final step, the affinity-bound fraction is recovered by altering the equilibrium conditions of the solid phase (by changing the system pH or salt concentration) or by adding a molecules that competes for the binding site of the ligand.

To provide several practical examples, a collection of lectin affinity-based methodologies used to isolate and/or characterize glycoproteins from distinct parasites is listed in the Table 1.

It is important to note that some techniques, such as metabolic radioactive labeling (by [3H]-myristic acid or [3H]-glucosamine, for example) and cell disruption (by Triton X-100, dioxane or hypotonic solution), must be applied prior to lectin chromatography to allow for the identification of molecules eluted from the column or the preparation of suitable samples for the chromatography column, respectively.

Table 1. Lectin affinity-based

Ligand	Organism	Isolated protein	First phase				Second phase	References
			Matrix	Bind	Wash	Elution	Methods	
WGA	*Trypanosoma cruzi*	85 kDa surface glycoprotein	Sepharose	10mM Tris-Hcl (pH 7.2), 150 mM NaCl	10mM Tris-Hcl (pH 7.2), 150 mM NaCl	0.1 M N-acetyl-D-glucosamine in 10mM Tris-Hcl (pH 7.2), 150 mM NaCl	None	Couto et al., 1990
Ricin	*Trypanosoma brucei rhodesiense*	Membrane Glycoprotein	Agarose	10 mM Mops buffer (pH 6.9), 1mM $MgSO_4$, 1 mM EGTA, 0.2% (v/v) Triton X-100, 10 µg/ml DNase I, 0.05 mM leupeptin, 2.5 mM PMSF, 5 mM iodoacetamide, 0.05 mM TPCK	10 mM Mops buffer (pH 6.9), 1mM $MgSO_4$, 1 mM EGTA, 0.2% (v/v) Triton X-100, 10 µg/ml DNase I, 0.05 mM leupeptin, 2.5 mM PMSF, 5 mM iodoacetamide, 0.05 mM TPCK	0.2 M acetic acid in 10 mM Mops buffer (pH 6.9), 1mM $MgSO_4$, 1 mM EGTA, 0.2% (v/v) Triton X-100, 10 µg/ml DNase I, 0.05 mM leupeptin, 2.5 mM PMSF, 5 mM iodoacetamide, 0.05 mM TPCK	None	Brickman & Balber, 1993
BaSII	*Schistosoma mansoni*	37 kDa glycoprotein	Sepharose	20 mM Tris-HCl (pH 8.0), 150 mM NaCl, 2 mM PMSF, 1% (v/v) Triton X-100, 1 mM PCMB, 1 mM *o*-phen, 1 mM iodoacetamide	20 mM Tris-HCl (pH 7.3), 0.1% (v/v) Triton X-100, 150 mM NaCl, 1mM $CaCl_2$, 1mM $MgCl_2$	300 mM L-fucose in 20 mM Tris-HCl (pH 7.3), 0.1% (v/v) Triton X-100, 150 mM NaCl, 1mM $CaCl_2$, 1mM $MgCl_2$	HPLC	Mansour, 1996
Concanavalin A or ricin	*Trypanosoma brucei*	Small Variable Surface Glycoprotein	Sepharose (Con A) or agarose (ricin)	10 mM Tris-HC (pH 7.4) ,150 mM NaCl, 1 mM $MgCl_2$, 5 mM $CaCl_2$, 2% NP-40, 100 µg/ml of antipain, leupeptin and E-64	10 mM Tris-HCl (pH 7.4), 150 mM NaCl, 1 mM $MgCl_2$, 5 mM $CaCl_2$, 2% NP-40, 100 µg/ml of antipain, leupeptin and E-64	0.5 M alpha-methyl mannoside (Con A) or 0.5 M galactose (ricin) in 10 mM Tris-HCl (pH 7.4), 150 mM NaCl, 0.1% NP-40, and 50 µg/ml antipain, leupeptin and E-64	Dylasis	Gardiner *et al.*, 1996
Concanavalin A	*Trypanosoma congolense*	Variant Surface Glycoprotein	Sepharose	10mM Tris-HCl (pH 7.5), 150 mMNaCl, 1mM $CaCl_2$, 1mM $MnCl_2$	10mM Tris-HCl (pH 7.5), 150 mMNaCl, 1mM $CaCl_2$, 1mM $MnCl_2$	Isoelectric focusing	Gel filtration (Biogel P 30)	Gerold *et al.*, 1996
WFA	*Trypanosoma congolense*	Variant Surface Glycoprotein	Agarose	50 mM Tris-HCl (pH 7.4), 0.02% sodium azide	50 mM Tris-HCl (pH 7.4), 0.02% sodium azide	50 mM Tris-HCl (pH 7.4), 0.02% sodium azide, 100 mM GalNAc	None	Gerold *et al.*, 1996
Concanavalin A	*Trypanosoma brucei*	Invariant surface glycoprotein heavily N-glycosylated	Sepharose	50 mM Tris buffer (pH 7.5), 150 mM NaCl, 1% (w/v) Triton X-100	10mM Tris buffer (pH 6.5), 0.1% (w/v) Triton X-100	0.5 M α-methylmannoside in 10mM Tris buffer (pH 6.5) containing 0.1% (w/v) Triton X-100	Ion exchange chromatograp hy (DEAE-52)	Nolan *et al.*, 1997

Table 1. (continued)

Ligand	Organism	Isolated protein	First phase				Second phase	References
			Matrix	Bind	Wash	Elution	Methods	
PNA or jacalin	*Haemonchus contortus*	Glycoprotein fractions	Agarose	10mM Tris-HCl (pH7.4, 0.5 M NaCl, 0.02% NaN3, 100 mM Ca^{2+}, 10 mM Mg^{2+}	10mM Tris-HCl (pH7.4, 0.5 M NaCl, 0.02% NaN3, 100 mM Ca_2, 10 mM Mg^{2+}	0.5 M galactose (PNA) or 0.8 M galactose (jacalin)	Gel filtration (Sephadex G-25); Anion Exchange chromatography (Mono Q – jacalin-binding material only)	Smith *et al.*, 2000
Concanavalin A	*Caenorhabditis elegans*	Glycoproteins	Agarose	10 mM Tris (pH 7.4) 0.5 M NaCl, 10 mM $CaCl_2$, 100 mM $MnCl_2$, 0.5% Triton X-100	10 mM Tris (pH 7.4) 0.5 M NaCl, 10 mM $CaCl_2$, 100 mM $MnCl_2$, 0.25% CHAPS	0.2M methylmannopyranoside and 0.2 M methylglucopyranoside in 10 mM Tris (pH 7.4) 0.5 M NaCl, 10 mM $CaCl_2$, 100 mM $MnCl_2$, 0.25% CHAPS	None	Redmond *et al.*, 2004
Ricin	*Trypanosoma brucei*	Glycoproteins presenting giant poly-N-acetyllactosamine carbohydrate chains	Agarose	50 mM Tris-HCl (pH 6.8), 400 mM NaCl, 0.8% Triton X-100, 0.1 M TLCK, 1 μg/ml leupeptin, 0.1% sodium azide	50 mM Tris-HCl (pH 6.8), 400 mM NaCl, 0.8% Triton X-100, 0.1 M TLCK, 1 μg/ml leupeptin, 0.1% sodium azide	30 mg/ml lactose and 30 mg/ml galactose in 12,5 mM Tris-HCl (pH 6.8), 100 mM NaCl, 0.2% Triton X-100, 25 mM TLCK, 0.25 μg/ml leupeptin, 0.025% sodium azide	None	Atrih *et al.*, 2005
Concanavalin A	*Toxoplasma gondii*	N-linked glycoproteins	Agarose	10 mM Tris-HCl (pH 7.5), 150 mM NaCl, 1 mM CaCl2, 1 mM MnCl2, 1% (v/v) Triton X-100, protease inhibitor mixture	10 mM Tris-HCl (pH 7.5), 150 mM NaCl, 1 mM CaCl2, 1 mM MnCl2, 1% (v/v) Triton X-100	1% (w/v) SDS in 100 mM Tris-HCl (pH 7.4) or 0.5 M α-methyl-	None	Fauquenoy et al., 2008
Concanavalin A	*Leishmania (V.) braziliensis*	Cysteine proteinases	Sheparose	20 mM Tris–HCl (pH 7.2), 5% (v/v) glycerol, 0.5% (v/v) CHAPS	20 mM Tris–HCl (pH 7.2), 5% (v/v) glycerol, 0.5% (v/v) CHAPS	50 mM α-Dmannose in 20 mM Tris–HCl (pH 7.2), 5% (v/v) glycerol, 0.5% (v/v) CHAPS	Anion exchange chromatography (DEAE-Sepahcel)	Rebello et al., 2009

During the affinity chromatography procedure, other methods, such as isoelectric focusing, may be used instead of the application of competing carbohydrates to elude the column-bound material. Furthermore, distinct affinity columns can be used in sequence to purify fractions with specific characteristics from a single sample.

As for the handling of the material that is eluted from an affinity column, many options for further purification are available, depending on the analysis method chosen for the study. Some of these options include: anion exchange chromatography, size exclusion chromatography and dialysis.

The combination of these accessible approaches allows for a vast array of study possibilities. Several examples of the results obtained by applying lectin-affinity chromatography in association with other techniques are described in the following paragraphs.

2.1.2 Parasite proteins isolated by lectin affinity chromatography

The structure of an N-linked oligosaccharide from a surface glycoprotein of *Trypanosoma cruzi*, an important human parasite that causes Chagas disease, was defined in a study using lectin chromatography (Couto et al., 1990). It was determined that the structure of this oligosaccharide is comprised of complex carbohydrate chains that possess a terminal sialic acid, α-L-fucose and a galactosyl(α1,3)galactose unit.

The cellular localization of glycoproteins of *Trypanosoma brucei rodhesiense*, a subspecies of the parasite responsible for the African sleeping sickness, was analyzed using ricin-based chromatrography (Brickman & Balber, 1993). It was observed that the ricin-binding proteins were primarily located in the vesicles of the lysosomal / endosomal system.

Gardiner et al., (1996) characterized small glycoproteins isolated from the surface of *Trypanosoma vivax*, which causes bovine trypanosomiasis. That study was the first to detail the characteristics of a *T. vivax* Variable Surface Glycoprotein (VSG). The isolated protein, designated ILDat 2.1 VSG, presented a molecular mass of 40 kDa and contained mannose (or a derivative sugar) in small quantities, and it was poorly retained by the lectin affinity column. It is possible that carbohydrates comprise only the C-terminal anchoring structure of this protein.

The characteristics of a fucosyllactose determinant of a *S. mansoni* glycoprotein were identified using affinity chromatography based on a lectin that was isolated from a host of this parasite, *B. alexandrina*. This determinant is expressed in the outer chain of a single unit of complex type N-linked oligosaccharides (Mansour, 1996).

Additionally, the VSG glycosyl-phosphatidylinositol membrane anchors of *Trypanosoma congolense*, another trypanosomatide species that causes bovine trypanosomiasis, were studied by lectin affinity (Gerold et al., 1996) using a modification of the technique in which the bound proteins are electrophoretically desorbed (Reinwald et al., 1981). This analysis allowed for description of the VSG GPI-anchor in this parasite: it contains a β1,6-linked galactose as the terminal hexose of the branch and an N-acetyl-glucosamine residue. Also, it was observed that *T. congolense* synthesizes two potential GPI-anchor precursors, one of which is insensitive to phospholipase C activity.

Nolan et al., (1997) identified a new invariant surface glycoprotein that is heavily N-glycosylated in the bloodstream forms of *Trypanosoma brucei* and designated it as ISG_{100}. This glycoprotein presents a large internal domain composed of a serine-rich repetitive motif, which was previously undescribed, and N-glycosylation sites on the N-terminal domain. Additionally, ISG_{100} is encoded by a single gene, whereas the trypanosomal plasma membrane proteins are commonly encoded by tandemly repeated genes that are part of a multigene family.

Potentially protective glycoprotein fractions from *Haemonchus contortus*, a parasitic nematode in ruminants, were also obtained by lectin chromatography (Smith *et al.*, 2000). The findings from that study confirmed the potential of the *H. contortus* PNA-binding glycoprotein fraction as an efficacious antigen against this parasite infection in sheep. Furthermore, this study identified another highly protective fraction that binds to jacalin. This second protective fraction presents sialyted versions of the same oligosaccharides that bound to the PNA column.

Another study on the protective properties of the glycoproteins of *H. contortus* was performed by the same group (Smith et al., 2003). The results showed that the four purified glycosylated zinc metalloproteinases from this parasite were such an efficacious antigen that, to an extent, they could account for most of the protection conferred by the urea-dissociated whole glycoproteins fraction. However, the role for the glycan moieties of these enzymes in the protection process was not clear.

The capacity of glycoproteins from *Caenorhabditis elegans*, a free living nematode, to induce protection from a challenge with *H. contortus* in sheep was assayed by Redmond et al. (2004). The lectin affinity methodology was able to identify glycoproteins with molecular masses between 25 and 200 kDa in extracts prepared from *C. elegans*, but the fractionated glycoproteins were not able to confer protection against an *H. contortus* challenge. These findings suggest that the conserved glycan moieties between these two species of worm are not solely responsible for the protections levels observed when native *H. contortus* antigens are used.

Trypanosoma brucei glycoproteins were shown to present distinctive structural features, such as the presence of giant poly-N-acetyllactosamine carbohydrate chains (Atrih et al., 2005). The recovered affinity-bound molecules were predominantly, but not exclusively, from the flagellar pocket. These glycoproteins carry massive glycans, representing the largest poly-LacNAc structures reported to that date, and they may produce a gel-like matrix in the lumen of the flagellar pocket and/or the endosomal/lysosomal system. Despite their remarkable size, these glycans present a very simple neutral structure, containing only mannose, galactose and N-acetylglycosamine.

Important glycoproteins from the apicomplexan parasite *Toxoplasma gondii* have also been analyzed by lectin affinity methods. It was shown that these components are pivotal factors for host invasion and intracellular development of parasites (Fauquenoy et al., *2008*).

Cysteine proteinases from promastigostes of *Leishmania (Viannia) braziliensis* were shown to be anchored to the membrane by glysoylphosphatidylinositol structures in an analysis of the hydrophobic fraction of promastigote forms. These enzymes are suggested to play a role in the process of parasite survival inside its hosts (Rebello et al., 2009).

2.1.3 Remarks on the isolation of proteins by lectin affinity chromatography

These reports provide examples of the uses of lectin affinity chromatography to identify potentially antigenic fractions of parasites that could be used for vaccine development. Also, they point to the potential of this method to characterize glyconjugates, such as the glycoproteins that are present on the parasite surface or secreted by these organisms. However, apart from these purely structural or clinically oriented applications, this method may also be relevant in other investigations, including studies of host-parasite interactions. This hypothesis is reinforced by reports indicating that lectin-glycan binding is important for the infection and virulence processes of some parasites, *e.g. Acanthamoeba castellanii* (Garate et al., 2006), *H. contortus* (Turner et al., 2008), *L. (V.) braziliensis* (Rebello et al., 2009) and *T. gondii* (Fauquenoy et al., 2008)

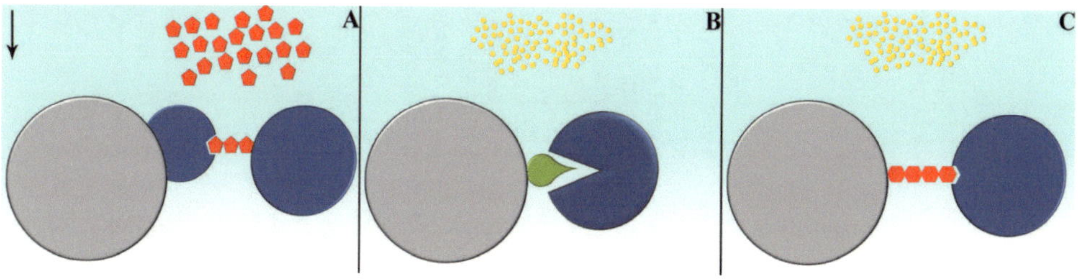

Fig. 2. Illustration of the affinity chromatography methodologies. The target molecules are bound to their ligands immobilized on a solid phase matrix. (A) Lectin affinity chromatography, (B) Protease inhibitor affinity chromatography and (C) Glycosaminoglycan affinity chromatography. Proteins = blue circle; carbohydrates = red pentagon and hexagon; protease inhibitors = green drop-like form; ions =yellow circles; and solid phase matrix beads = gray circle.

2.2 Protease inhibitors affinity-based separation of parasite proteins

Methodologies for the purification of parasite proteases have been applied in studies investigating the biological roles of these enzymes in parasite, including their participation in the infection process and in the survival of the parasites inside their hosts (McKerrow et al, 2006). Inhibitor affinity chromatography consists of the fractionation of parasite samples based on the reversible interactions between proteases and their specific inhibitors while the latter are covalently attached to a matrix (Fig. 2B). This technique can also be performed using irreversible inhibitors under particular conditions that will be described further in this section.

It is also interesting to note that, based on the specificity of the inhibitor used in the affinity chromatography, it is possible to suggest the enzyme class of the isolated protein. However, complementary analyses, such as characterization of the proteolytic activity, are often necessary to confirm these findings. Nevertheless, this purification strategy presents an initial advantage when compared to other methodologies.

In this section, fractionation approaches for serine-, aspartic acid- and cysteine proteases in specific parasites will be described. These approaches must take the class of the studied enzyme into consideration, as well as the inhibitor to be used and the characteristics of the mobile phase used for chromatography.

Table 2. Protease Inhibitors affinity-based

Ligand	Organism	Isolated protein	First phase				Second phase	References
			Matrix	Bind	Wash	Elution	Methods	
PNA or jacalin	*Haemonchus contortus*	Glycoprotein fractions	Agarose	10mM Tris-HCl (pH7.4, 0.5 M NaCl, 0.02% NaN3, 100 mM Ca^{2+}, 10 mM Mg^{2+}	10mM Tris-HCl (pH7.4, 0.5 M NaCl, 0.02% NaN3, 100 mM Ca_2, 10 mM Mg^{2+}	0.5 M galactose (PNA) or 0.8 M galactose (jacalin)	Gel filtration (Sephadex G-25); Anion Exchange chromatography (Mono Q – jacalin-binding material only)	Smith *et al.*, 2000
Concanavalin A	*Caenorhabditis elegans*	Glycoproteins	Agarose	10 mM Tris (pH 7.4) 0.5 M NaCl, 10 mM $CaCl_2$, 100 mM $MnCl_2$, 0.5% Triton X-100	10 mM Tris (pH 7.4) 0.5 M NaCl, 10 mM $CaCl_2$, 100 mM $MnCl_2$, 0.25% CHAPS	0.2M methylmannopyranoside and 0.2 M methylglucopyranoside in 10 mM Tris (pH 7.4) 0.5 M NaCl, 10 mM $CaCl_2$, 100 mM $MnCl_2$, 0.25% CHAPS	None	Redmond *et al.*, 2004
Ricin	*Trypanosoma brucei*	Glycoproteins presenting giant poly-N-acetyllactosamine carbohydrate chains	Agarose	50 mM Tris-HCl (pH 6.8), 400 mM NaCl, 0.8% Triton X-100, 0.1 M TLCK, 1 µg/ml leupeptin, 0.1% sodium azide	50 mM Tris-HCl (pH 6.8), 400 mM NaCl, 0.8% Triton X-100, 0.1 M TLCK, 1 µg/ml leupeptin, 0.1% sodium azide	30 mg/ml lactose and 30 mg/ml galactose in 12,5 mM Tris-HCl (pH 6.8), 100 mM NaCl, 0.2% Triton X-100, 25 mM TLCK, 0.25 µg/ml leupeptin, 0.025% sodium azide	None	Atrih *et al.*, 2005
Concanavalin A	*Toxoplasma gondii*	N-linked glycoproteins	Agarose	10 mM Tris-HCl (pH 7.5), 150 mM NaCl, 1 mM CaCl2, 1 mM MnCl2, 1% (v/v) Triton X-100, protease inhibitor mixture	10 mM Tris-HCl (pH 7.5), 150 mM NaCl, 1 mM CaCl2, 1 mM MnCl2, 1% (v/v) Triton X-100	1% (w/v) SDS in 100 mM Tris-HCl (pH 7.4) or 0.5 M α-methyl-	None	Fauquenoy et al., 2008
Concanavalin A	*Leishmania (V.) braziliensis*	Cysteine proteinases	Sheparose	20 mM Tris–HCl (pH 7.2), 5% (v/v) glycerol, 0.5% (v/v) CHAPS	20 mM Tris–HCl (pH 7.2), 5% (v/v) glycerol, 0.5% (v/v) CHAPS	50 mM α-Dmannose in 20 mM Tris–HCl (pH 7.2), 5% (v/v) glycerol, 0.5% (v/v) CHAPS	Anion exchange chromatography (DEAE-Sepahcel)	Rebello et al., 2009

Table 2. (continued)

Ligand	Organism	Isolated protein	First phase				Second phase	References
			Matrix	Bind	Wash	Elution	Methods	
Pepstatin A	*Neospora caninum*	52 kDa	Agarose	5 mM NaOAc, (pH 5.5)	5 mM NaOAc, (pH 5.5)	0.1 M Tris–HCl, 0.5 M NaCl, (pH 8.5)	None	Naguleswaran *et al*, 2005
Aprotinin	*Leishmania amazonensis*	115 and 56 kDa	Agarose	10 mM Tris-HCl, 5 mM $CaCl_2$, (pH7,5)	10 mM Tris-HCl, 5 mM $CaCl_2$, (pH7,5)	10 mM Tris-HCl, 1,5 mM NaCl, (pH7,5)	None	Silva-Lopez et al; 2005
Bacitracin	*Trichomonas vaginalis*	60 and 30 kDa	Agarose	20 mM NaOAc, (pH 4.0)	20 mM NaOAc, (pH 4.0)	0.1 M Tris-HCl, 1.0 M NaCl 25% 2-propanol, (pH 7.0)	Bio-Gel P-60	Sommer *et al*; 2005
Aprotinin	*Leishmania braziliensis*	60 and45 kDa	Agarose	10 mM Tris-HCl, 5 mM $CaCl_2$, (pH7,5)	10 mM Tris-HCl, 5 mM $CaCl_2$, (pH7,5)	10 mM Tris-HCl, 1,5 mM NaCl, (pH7,5)	None	Guedes HL *et al*; 2007
Aprotinin	*Trypanosoma cruzi*	75 kDa	Agarose	10 mM Tris-HCl, 5 mM $CaCl_2$, (pH7,5)	10 mM Tris-HCl, 5 mM $CaCl_2$, (pH7,5)	10 mM Tris-HCl, 1,5 mM NaCl, (pH7,5)	None	Silva-Lopez *et al*; 2008
Aprotinin	*Leishmania donovani*	115 kDa	Agarose	10 mM Tris-HCl, 5 mM $CaCl_2$, (pH 7.5)	10 mM Tris–HCl, 5 mM $CaCl_2$, (pH7,5)	10 mM Tris–HCl, 1.5 M NaCl, (pH7,5)	None	Choudhury *et al*; 2009
Pepstatin A	*Trypanosoma cruzi*	120, 48 and 56 kDa	Agarose	0.1 M NaOAc, 1.0 M NaCl, (pH 3.5)	0.1 M NaOAc, 1.0 M NaCl, (pH 3.5)	0.1 M Tris–HCl, 1.0 M NaCl, (pH 8.6)	HPLC gel filtration	Pinho *et al*; 2009
Aprotinin	*Leishmania donovani*	58 kDa	Agarose	10 mM Tris-HCl, 5 mM $CaCl_2$, (pH7,4)	10 mM Tris-HCl, 5 mM $CaCl_2$, (pH7,4)	10 mM Tris-HCl, 1 M NaCl, (pH7,5)	DEAE	Choudhury *et al*; 2010
Pepstatin A	*Rhipicephalus (B.) microplus*	42 kDa	Agarose	20 mM NaOAc, 1 M NaCl, (pH 5.3)	20 mM NaOAc, 1 M NaCl, (pH 5.3)	100 mM Tris-HCl, 1 M NaCl, (pH 8.6)	Mono-Q	Cruz *et al; 2010*
Benzamidine	*Plasmodium ookinete*	37 kDa	Sepharose	0.1% Triton X-100 in PBS	0.1% Triton X-100 in PBS	PBS	None	Li *et al;* 2010
Aprotinin	*Leishmania chagasi*	LCSII (105, 66, 60 kDa); LCSI (60, 58 kDa) and LCSIII (76, 68 kDa)	Agarose	10 mM Tris-HCl, 5 mM $CaCl_2$, (pH7,5)	10 mM Tris-HCl, 5 mM $CaCl_2$, (pH7,5)	10 mM Tris-HCl, 1,5 mM NaCl, (pH7,5)	HPLC	Silva-Lopez *et al*; 2010
E-64	*Trypanosoma cruzi*	60 kDa	Sepharose	20 mM Tris-HCl, 150 mM NaCl, 100 M PMSF, (pH 7.4)	20 mM Tris-HCl, 150 mM NaCl, 100 M PMSF, (pH 7.4)	20 mM Tris–HCl, 150 mM NaCl, 100 M PMSF, 2.0 M NaCl, (pH 7.4)	None	Bourguignon *et al* 2011

Aprotinin and pepstatin A are examples of inhibitors that are frequently used in the isolation of serine- and aspartic acid proteases, respectively, from many parasite species (Bond & Beynon). Other inhibitors that have been previously described in the isolation of serine proteases include soybean trypsin inhibitor (SBTI) and chloromethylketone (CMK). As for the purification of cysteine proteases, the use of three other inhibitors has been reported: L-transepoxysuccinyl-leucylamido-[4-guanidino]butane (E-64), bacitracin and glycyl-phenylalanyl-glycyl-semicarbazone (Table 2). It must be emphasized that these inhibitors cannot be used to isolate all of the proteases classes from parasites, as they present distinct affinities for members of different groups and families within these enzyme classes. Therefore, investigation of the possible variations present in the active site of these enzymes may prove useful.

The features of the buffer (temperature, pH and ionic strength) to be used may vary according to the ligand's physicochemical characteristics, the chemical environment of the parasite enzyme and the analyzed species of parasite. For example, distinct buffers were used for the purification of serine proteases from *S. mansoni* and *Trichinella spiralis* using benzamidine. It is also noteworthy that for each organism, a different matrix was used to immobilize the inhibitor, sepharose for *S. mansoni* and celite for *T. spiralis*. The use of distinct buffers in studies that are based on the same inhibitor is also noted in reports of SBTI, E-64, bacitracin and glycyl-phenyalanyl-glycyl-semicarbazone, all of which are cysteine protease inhibitors.

Affinity chromatography with an irreversible inhibitor has also been described previously; the cysteine-protease inhibitor is an example of this strong binding. In the interaction between E-64 and cysteine-protease, a covalent bond is established (Matsumoto, 1989). Therefore, a reaction between the epoxy groups of the inhibitor and the thiopropyl group of the sepharose matrix is necessary to bind E-64 to a solid support. This reaction prevents the reaction of E-64 with the cysteine residue at the protease catalytic center. However, this does not affect the bond between the inhibitor and cysteine-protease; instead, it only results in inhibition of the proteolytic activity (Govrin, 1999).

2.2.1 Parasite proteins isolated by cysteine-protease inhibitors affinity chromatography

There is only one published example of the use of E-64 affinity chromatography to assess cysteine-protease isolated from a parasite, and this study was conducted with the *T. cruzi* epimastigote. In this study, chromatography was useful for assessing the effects of β-Lapachone naphthoquinones on a 60 kDa cysteine-protease activity present in *T. cruzi*. The results demonstrated the potential of this protease inhibitor as a new antichagasic compound (Bourguignon et al., 2011). Another example of a cysteine-protease isolated by inhibitor affinity chromatography in parasites was described for *Plasmodium falciparum*. In this case, a glycyl-phenyalanyl-glycyl-semicarbazone-based column was used to isolate a protease with a molecular weight of 27 kDa, as determined by SDS-PAGE (Shenai et al, 2000).

2.2.2 Parasite proteins isolated by serine-protease inhibitors affinity chromatography

Aprotinin affinity-based chromatography was useful for the isolation of a serine-protease of 115 kDa (Silva-Lopez et al., 2005), a 68 kDa (Morgado- Diaz et al., 2004; Silva-Lopez et al., 2004) and a 56kDa (Silva-Lopez et al., 2004) from *L.(L.) amazonensis* compared to other purification procedures that were used to isolate parasite serine peptidase enzymes (Kong et al., 2000; Ribeiro de Andrade et al., 1998). In *Leishmania (V) braziliensis* promastigotes, 60 kDa and 45 kDa enzymes were purified using the aprotinin affinity-based and activity esterase assessed against N-alpha-benzoyl-L-arginine ethyl ester hydrochloride and Nalpha-p-tosyl-L-arginine methyl ester hydrochloride (Guedes et al., 2007). Furthermore, three protein profiles were isolated from *Leishmania chagasi* promastigotes, including LCSI (58 and 60 kDa), LCSII (60, 66, 105 and kDa) and LCSIII (68 and 76 kDa), which were characterized as serine-protease enzymes based on their activity toward α-N-ρ-tosyl-L-arginine methyl ester substrate (Silva-Lopez et al., 2010). Furthermore, serine proteases with molecular weights of 75 kDa (Silva-Lopez et al., 2008) and 115 kDa (Choudhury et al., 2009) were identified as excretory products of *T. cruzi* and components of the sub-cellular environment in *Leishmania donovani,* respectively, although the chromatography step was not able to produce a homogeneous fraction. Furthermore, a intracellular serine protease of 58 kDa was were purified from *Leishmania donovani* (Choudhury et al., 2010).

In addition, the aprotinin affinity-based chromatography was useful for the isolation of serine-proteases of 35 kDa and 26 kDa from *Anisakis simpZex* (Morris et al, 1994), 43 kDa from *Candida albicans* (Morrison et al, 1993), 15 kDa from *Schistosoma mansoni (*Salter et al, 2000*)*, 42 kDa from *Rhipicephalus (B.) microplus* (Cruz et al, *2010*), 60 kDa and 30 kDa from *Trichomonas vaginalis* (Sommer *et al*; 2005) and 35 to 52 from *Caenorhabditis elegans* (Geier et al; 1999).

Benzamidine-celite was applied in the isolation of serine proteases among the excreted or secreted proteins of *T. spirali.* The recovered proteases were not purified to homogeneity, and they showed molecular masses of 18 kDa, 40 kDa and 50 kDa (Todorova & Stoyanov). A similar finding was reported for the serine-proteases of *Chrysomya bezziana* larvae by using an SBTI-based column to purify four proteins with molecular masses of 13 kDa, 16 kDa, 26 kDa and 28 kDa (Muharsini et al., 2000).

Because it is possible to isolate heterogeneous products using inhibitors for affinity-based chromatography, we assessed a serine-protease from the third (L3) instar larvae of *D. hominis.* This ectoparasite causes dermatobiose in vertebrates, including humans, and it is particularly relevant in cattle, where it can cause a drop in production of meat and milk, leather as well depreciation (Maia & Guimarães, 1985).

Due to the association of DEAE-Sephacel with aprotinin agarose, it was possible to assess a serine protease from L3 larvae (Fig. 3). The fractions obtained by ion change chromatography containing estearasic activity were pooled and then fractionated on an aprotinin-agarose column. This fraction showed a profile with multiple bands by SDS-PAGE and silver staining, and only one band of enzyme activity (50 kDa) was detected by gelatin-SDS-PAGE at pH 7.5 (Fig. 3). Interestingly, this band of 50 kDa was not initially detected in the extracts from L3 by gelatin-SDS-PAGE. The expression of this enzyme is likely low in these larvae, and it can only be detected after concentration by chromatographic methods. The proposed strategy to isolate a serine protease allowed for the detection of a band of 50 kDa in extracts of L3 larvae, and this band had not been previously detected in the direct analysis of the total extract by gelatin-SDS-PAGE. Additionally, this fraction was found to have esterase activity (data not shown).

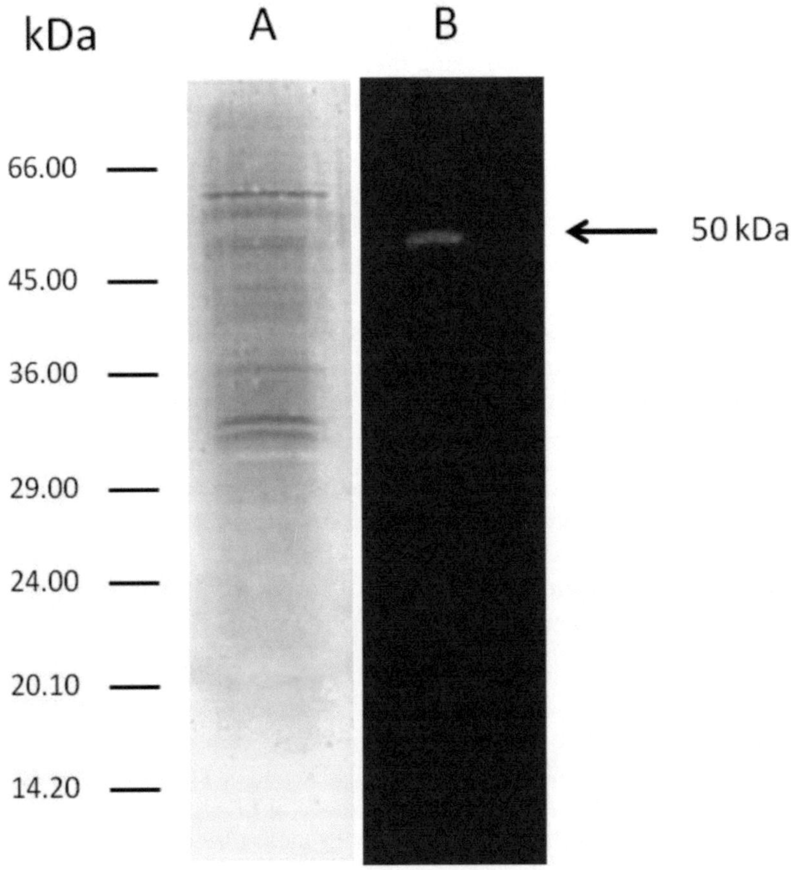

Fig. 3. Electrophoresis of proteins from L3 instar larvae of *Dermatobia hominis* eluted from a column of aprotinin-agarose. A total of 20 µg of protein from each fraction was resolved by SDS-PAGE (A) and gelatin-SDS-PAGE (B) and the bands were detected by silver staining and negative coloration, respectively. The arrow indicates a serine protease of 50 kDa. The molecular mass markers are indicated (kDa). These results are representative of two independent assays

2.2.3 Parasite proteins isolated by aspartyl-protease inhibitors affinity chromatography

Affinity-based chromatography based on pepstatine A was used to isolate a 52 kDa aspartyl protease from *Neospora caninum* tachyzoites (Naguleswaran et al., 2005) and a 45 kDa enzyme from *S.mansoni* (Valdivieso et al., 2003). In *Trypanosoma cruzi* epimastigotes, two aspartyl proteases were isolated (cruzipsin-I and cruzipsin-II). The molecular mass was estimated to be 120kDa by high performance liquid chromatography gel filtration, and the activities of these enzymes were detected in a doublet of bands (56 kDa and 48 kDa). These findings demonstrate that both proteases are novel *T. cruzi* acidic proteases. The physiological function of these enzymes in *T. cruzi* is not completely defined (Pinho et al., 2009).

An aspartyl protease with molecular mass of 37 kDa (plasmepsin) was isolated from the surface of *Plasmodium ookinete*, and its sequence was determined by mass spectrometry (Li et al., 2010). This protease was purified by using a benzamidine affinity-based column, which is typically used for the isolation of serine proteases. Structural similarity between the active site residues of the serine- and aspartyl proteases is possible, as some hydrogen-bonded residues can are arranged without any strain, such as in the formation of an oxyanion hole, in a manner that resembles the active site of a serine protease (Andreeva et al., 2004)

2.2.4 Remarks on the isolation of proteins by protease inhibitors affinity chromatography

Although the studies that have been conducted to isolate parasite proteases are of great medical interest, no parasiticide drug has been proposed thus far. In general, the chromatography methods involving inhibitor-based affinity-capture have been useful only to describe these enzymes in parasites and to establish their biochemical properties, their functions and their application in drugs tests.

Furthermore, the heterogeneous material obtained from affinity-based chromatography may require additional procedures for purification of the enzyme. Thus, other techniques must be applied to obtain proteases with greater purity, including molecular exclusion and ion exchange chromatography.

2.3 Glycosaminoglycans affinity-based separation of parasite proteins

Microbes have developed different strategies to gain access into mammalian cells (Bermúdez et al., 2010; Caradonna & Burleigh 2011; Soong et al., 2011). The first step involves the recognition of molecules at the surface of the target cell, which triggers the activation of signaling pathways that are implicated in the parasite internalization (Abban & Meneses 2010; Epting et al., 2011). Host cell surface sulfated proteoglycans have been implicated as key molecules at the host cell-parasite interface, mediating the adhesion and invasion of numerous parasitic microorganisms (Jacquet et al., 2001; Kobayashi et al., 2010; O'Donnell & Shukla 2010).

2.3.1 Structure of glycosaminoglycans

Proteoglycans (PGs) are composed of core proteins that are covalently linked to glycosaminoglycan (GAG) chains. As components of the extracellular matrix, the structural diversity of PGs depends on the identity of the core protein and the GAG composition. GAGs are linear polysaccharides comprised of disaccharide repeats containing uronic acid and hexosamine. GAGs vary in the type of hexosamine, hexose or hexuronic acid unit. The sulfated GAGs are classified as heparin [2-O-sulfo-β-D-glucuronic acid (GlcUA-2S) or 2-O-sulfo-α-L-iduronic acid (IdoUA-2S) and N-acetylglucosamine (GlcNAc) or N-sulfoglucosamine (GlcNS)], heparan sulfate [GlcUA, IdoUA or IdoUA-2S and GlcNAc or GlcNS], chondroitin sulfate [GlcUA and N-acetylgalactosamine (GalNAc)], dermatan sulfate [GlcUA or IdoUA and GalNAc] and keratan sulfate [galactose (Gal) and GlcNAc]. In fact, the structural diversity of PGs may provide sites of affinity for different ligands and, therefore, function as co-receptors or receptors for GAG-binding proteins (Dreyfuss et al., 2009; Ly et al., 2010). Although heparin is not found on the cell surface, this GAG has being commonly used as tool for pathogen-host cell interaction assays. Heparins are negativally charged structures and native heparin presents molecular weights in the range of 5 to 30 KDa, whereas commercial heparin preparations are in the range of 12 kDa to 15 kDa.

2.3.2 Role of heparin-binding proteins in pathogen-host cell

Many pathogens express surface proteins that interact with GAGs in different stages of their life cycle. Although some parasites can bind to multiple GAGs (Coppi et al., 2007; Fallgren et al., 2001), heparan sulfate proteoglycan (HSPG) has been implicated in the recognition and/or invasion process of a wide range of pathogens, including viruses, bacteria and protozoan parasites (Bambino-Medeiros et al., 2011; Dalrymple & Mackow 2011; Yan et al., 2006;). Despite the role of heparin-binding proteins in many physiological and pathological processes, the basis of the heparin-protein interaction at the molecular level is still unclear.

Thus, efforts have been concentrated to enhance methods for the isolation and characterization of heparin-binding proteins, and, in parallel, to determine the role of this GAG in pathogen-host cell interaction. Currently, heparin affinity chromatography has been applied to the purification of GAG-binding proteins from different pathogens (Table 3). In these chromatography assays, the heparin is covalently coupled to agarose or sepharose beads and its sulfates and carboxylates chains are able to bind many proteins by basic amino acids (Fig. 2C).

This technique has been used to isolate heparin-binding proteins without loss of their biological activity, leading to a better understanding of the mechanism involved in the parasite invasion process. For example, chlamydial outer membrane complex (OmcB), a 60 kDa cysteine-rich protein, displays a protein motif (50-70OmcB peptide) that acts as an acceptor molecule to bind heparan sulfate (HS) and promote Chlamydia invasion in eukaryotic cells (Stephens et al., 2001). Attachment of *Helicobacter pylori* to gastric epithelial cells also involves HS recognition. Two major proteins, one with a molecular mass of 71.5 kDa and pI 5.0 (HSBP50) and the other with a molecular mass of 66.2 kDa and pI 5.4 (HSBP54), have been identified on the surface of bacterial cells that are able to bind HS. The amino acid sequences of these proteins (HSBP50 - VPERAVRAHT; HSBP54 - VHLPADKTNV) are not homologous with bacterial adhesins or other HS-binding proteins (Ruzi-Bustos et al., 2001). Other proteins with the ability to bind heparin (66 and 60 kDa) have been identified in *Staphylococcus aureus*. The partial characterization of the amino acid sequences, which consist of DWTGWLAAA for the 66 kDa protein and MLVT for the 60 kDa protein, revealed no identity with HBPs from Chlamydia or *Helicobacter pylori*. HBPs from *S. aureus* have been demonstrated to be sensitive to heat and proteases, such as pronase E, proteinase K, pepsin and chymotrypsin (Liang et al., 1992). Interestingly, a 17-kDa heparin-binding protein with pI 4.6 has also been isolated from *S. epidermis* and *S. haemolyticus*, but the amino acid sequence similarity is low between these two organisms (MXTAHSYTXKYNGYTAN and MATQTKGYYYSYNGYV, respectively) and other bacterial HBPs (Fallgren et al., 2001).

Trypanosomatidaes also exploit HS for successful parasite attachment to and/or invasion of the mammalian and vector hosts. The adhesion of Leishmania amastigotes to macrophages is mediated by HS, but not other sulfated polysaccharides (Love et al., 1993). Two heparin-binding proteins, (65 and 54.5 kDa) from *L. (V.) braziliensis* promastigotes (HBP-Lb) recognize several molecules in the gut of *Lutzomyia intermedia* and *Lutzomyia whitmani* (Azevedo-Pereira et al., 2007). The biochemical characterization of these proteins revealed that only the 65-kDa HBP-Lb has metallo-proteinase activity, and this protein is primarily localized at the flagellar domain of the promastigotes. Surface plasmon resonance (spr) also demonstrated high-affinity binding at the flagellar domain, which forms a stable binding complex (Côrtes et al., 2011). In *T. cruzi*, HBPs also mediate parasite adhesion by recognition of PGHS on the surface of the target cells (Bambino-Medeiros et al., 2011; Calvet et al., 2003; Oliveira-Jr et al., 2008; Ortega-Barria & Pereira, 1991). Currently, three HBPs have been described in this parasite: a 60-kDa protein named penetrin (Ortega-Barria & Pereira, 1991) and two other proteins of 65.8 and 59 kDa that bind heparin, HS and chondroitin sulfate (CS). These proteins have been identified in both trypomastigotes and amastigotes (Oliveira-Jr et al., 2008). Interestingly, the HBP-HS binding is related to a specific region of the HS chain, the N-acetylated/N-sulfated HS domain, which promotes parasite attachment and invasion (Oliveira-Jr et al., 2008). Although only HS binding triggers *T. cruzi* invasion of mammalian cells (Ortega-Barria & Pereira, 1991; Calvet et al., 2003; Oliveira-Jr et al., 2008; Bambino-Medeiros et al., 2003), the multiple GAG recognition may provide an efficient association with other GAGs within the parasite life cycle. Recently, it has been demonstrated that sulfated proteoglycans are involved in the adhesion of epimastigotes to the luminal midgut epithelial cells of Rhodnius prolixus (Gonzalez et al., 2011).

2.3.3 Remarks on the isolation of proteins by glycosaminoglycans affinity chromatography

While the application of affinity chromatography has provided advances in our understanding of heparin-binding proteins, a large number of studies have focused on the parasite-host cell interface to improve our comprehension of the mechanisms that are activated by the receptor-ligand interaction (reviewed by Chen et al., 2008). The binding of Dengue virus to HS, for example, seems to result in the accumulation of virions at the surface of the human hepatoma cell line HuH-7 and elicit clathrin-dependent endocytosis (Hilgard & Stockert 2000). In addition to promote attachment and parasite invasion, HSPG also seems to be involved in the tropism of pathogen to specific tissues. The degree of HSPG sulfation guides the migration of Plasmodium sporozoites and the invasion of hepatocytes. Highly sulfated heparan sulfate at the surface of hepatocytes seems to regulate the proteolytic activity of the calcium-dependent protein kinase-6 on the CSP, which triggers the invasion of the parasite (Coppi et al., 2007).

Another interesting phenomenon is the release of syndecan-1, a transmembrane PGHS, as a mechanism of host defense inhibition. *Pseudomonas aeruginosa* induces syndecan-1 shedding through the enzymatic activity of LasA, leading to an enhancement of bacterial virulence (Park et al., 2001). A similar mechanism has been described for *Staphylococcus aureus* in which β-toxin, a secreted virulence factor, also induces syndecan-1 shedding by activating a metallo-proteinase involved in the host cell shedding mechanism, leading to enhancement of bacterial virulence due to the recruitment of inflammatory cells (Hayashida et al., 2009). Because heparan sulfate has been shown to be a receptor for a variety of pathogens, HS-binding polypeptides have been the subject of intense research and provide possibilities for drug intervention.

Table 3. Heparin affinity-based

Organism	Isolated proteins	First phase				Second phase	References
		Matrix	Bind	Wash	Elution	Methods	
Trypanosoma cruzi	Penetrin – 60 kDa	Sepharose	PBS, pH 7.2	PBS containing 0.05% Triton X-100, pH7.2	1.5M guanidine-HCl 2.0M potassium thiocyanate and NaCl gradient (0.25-3.0M)	None	Ortega-Barria and Pereira 1991
Staphylococcus aureus	*S. aureus*-HBP – 66 kDa and 60 kDa	Sepharose	PBS containing 100mM PMSF, pH6.0	PBS containing 100mM PMSF, pH6.0	NaCl gradient (0-1M), followed by 2M NaCl in 200mM Tris-HCl pH 8.0	None	Liang et al., 1992
Trypanosoma cruzi	Penetrin – 60 kDa	Sepharose	PBS, pH 7.2	PBS containing 0.05% Triton X-100, pH7.2	1.5M guanidine-HCl 2.0M potassium thiocyanate and NaCl gradient (0.25-3.0M)	None	Ortega-Barria and Pereira 1991
Chlamydial	Outer membrane complex (COMC) - 60 kDa	Agarose	50 mM DTT and 2% Triton X-100	PBS	2% SDS in PBS	None	Stephens et al., 2001
Helicobacter pylori	Outer membrane protein (OMP) - 71.5 kDa and 66.2 kDa	Sepharose	0.05M sodium acetate pH5.0	0.1M Sodium acetate, pH5.0	NaCl gradient (0-2M)	None	Ruzi-Bustos et al., 2001
Staphyloccocal	HBP – 17 kDa	Sepharose	PBS	None	NaCl gradient (0-2M)	None	Fallgren et al., 2001
Leishmania (V.) braziliensis	HBP-Lb – 65 kDa and 54.5 kDa	Sepharose	PBS containing 0.5% glycerol and 0.5% Chaps, pH7.2	PBS containing, 0.5% glycerol and 0.5% Chaps, pH7.2	PBS containing 0.5% glycerol, 0.5% Chaps and 2M NaCl, pH7.2	None	Azevedo-Pereira et al., 2007
Trypanosoma cruzi	HBP-Tc – 65.8 kDa and 59 kDa	Sepharose	PBS and 0.5% glycerol, pH 7.2	PBS, pH 7.2	NaCl gradient (0-1M)	None	Oliveira-Jr et al., 2008

3. Conclusion

The chromatographic procedures described here maintain the minimal amount of native folding necessary for proteins to retain their biological and biochemical activities. Thus, the materials used as supports for packed affinity columns, including agarose, sepharose and celite (from diatomaceous earth), to immobilize ligands, such as lectins, protease inhibitors and glycosaminoglycans, do not interfere with the functional properties of these proteins.

Furthermore, proteins obtained by affinity-based procedure have been useful in understanding the biological processes related to the life cycles of parasites and in the interaction with hosts. These studies are essential to developing strategies, such as the use of vaccines and drugs, to control the parasite diseases.

4. Acknowledgements

We acknowledge the financial support Brazilian funding agencies, including CAPES, CNPq, FAPERJ and PAPES (CNPq/Fiocruz). Dr. Carlos Roberto Alves and Dr. Mirian Claudia de Souza Pereira are research fellows of CNPq

5. References

Abban CY, & Meneses PI. 2010. Usage of heparan sulfate, integrins, and FAK in HPV16 infection. Virology. 403(1):1-16.

André S, Sanchez-Ruderisch H, Nakagawa H, Buchholz M, Kopitz J, Forberich P, Kemmner W, Böck C, Deguchi K, Detjen KM, Wiedenmann B, von Knebel Doeberitz M, Gress TM, Nishimura S, Rosewicz S, & Gabius HJ. 2007. Tumor suppressor p16INK4a--modulator of glycomic profile and galectin-1 expression to increase susceptibility to carbohydrate-dependent induction of anoikis in pancreatic carcinoma cells. FEBS J. 274(13):3233-56.

Andreeva N, Bogdanovich P, Kashparov I, Popov M, & Stengach M. 2004. Is Histoaspartic Protease A Serine Protease With A Pepsin-Like Fold? Proteins. 55(3):705-10.

Atrih A, Richardson JM, Prescott AR, & Ferguson MA. 2005. *Trypanosoma brucei* glycoproteins contain novel giant poly-N-acetyllactosamine carbohydrate chains. J Biol Chem. 280(2):865-71.

Azevedo-Pereira RL, Pereira MC, Oliveria-Junior FO, Brazil RP, Côrtes LM, Madeira MF, Santos AL, Toma L, & Alves CR. Vet Parasitol. 2007. Heparin binding proteins from *Leishmania (Viannia) braziliensis* promastigotes. 145(3-4):234-9.

Bambino-Medeiros R, Oliveira FO, Calvet CM, Vicente D, Toma L, Krieger MA, Meirelles MN, & Pereira MC. 2011. Involvement of host cell heparan sulfate proteoglycan in *Trypanosoma cruzi* amastigote attachment and invasion. Parasitology. 138(5):593-601.

Beckert WH, & Sharkey MM. 1970. Mitogenic activity of the jack bean (*Canavalia ensiformis*) with rabbit peripheral blood lymphocytes. Int Arch Allergy Appl Immunol. (4):337-41.

Bermúdez A, Alba MP, Vanegas M, & Patarroyo ME. 2010. 3D structure determination of STARP peptides implicated in *P. falciparum* invasion of hepatic cells. Vaccine. 28(31):4989-96.

Bond JS, & Beynon RJ., eds. 1989. Proteolytic enzymes: A practical approach. IRL Press/Oxford University Press, Oxford, United Kingdom

Bourguignon SC, Cavalcanti DF, de Souza AM, Castro HC, Rodrigues CR, Albuquerque MG, Santos DO, da Silva GG, da Silva FC, Ferreira VF, de Pinho RT, & Alves CR. 2011. *Trypanosoma cruzi*: insights into naphthoquinone effects on growth and protease activity. Exp Parasitol. 127(1):160-6.

Brickman MJ, & Balber AE. 1993. *Trypanosoma brucei rhodesiense*: membrane glycoproteins localized primarily in endosomes and lysosomes of bloodstream forms. Exp Parasitol. 76(4):329-44.

Brittingham, A, Morrison, CJ, McMaster, WR, McGwire, BS, Chang, KP, & Mosser, DM. (1995). Role of the Leishmania surface protease gp63 in complement fixation, cell adhesion, and resistance to complement-mediated lysis. J of Immunol 155 (6) :3102-11.

Calvet CM, Toma L, Souza FR, Meirelles MNL, & Pereira, MCP. 2003. Heparan Sulfate Proteoglycans mediate the invasion of cardiomyocytes by *Trypanosoma cruzi*. J Eukariot Microbiol. 50: 97-103.

Caradonna KL, & Burleigh BA. 2011. Mechanisms of Host Cell Invasion by Trypanosoma cruzi. Adv Parasitol. 76:33-61.

Chen Y, Götte M, Liu J, & Park PW. 2008. Microbial subversion of heparan sulfate proteoglycans. Mol Cells. 26(5):415-26.

Choudhury R, Bhaumik Sk, De T, Chakraborti, T. 2009. Identification, Purification & Characterization of a secretory serine protease in an indian strain of *Leishmania donovani*. Mol Cell Biochem. 320(1-2):1-14

Choudhury R, Das P, De T, Chakraborti T. 2010. Immunolocalization and characterization of two novel proteases in *Leishmania donovani*: putative roles in host invasion and parasite development. Biochimie. 92(10):1274-86.

Coppi A, Tewari R, Bishop JR, Bennett BL, Lawrence R, Esko JD, Billker O, & Sinnis P. 2007. Heparan sulfate proteoglycans provide a signal to *Plasmodium* sporozoites to stop migrating and productively invade host cells. Cell Host Microbe. 2(5):316-27.

Côrtes LMC, Pereira MCS, Olivera-Jr FOR, Suzana Corte-Real, Silva FS, Pereira BAS, Madeira MF, Moraes MTB, Brazil RP, & Alves RA. 2011. *Leishmania (Viannia) braziliensis*: insights on subcellular distribution and biochemical properties of heparin-binding proteins. Parasitology. 7: 1-8.

Couto AS, Gonçalves MF, Colli W, & de Lederkremer RM. 1990. The N-linked carbohydrate chain of the 85-kilodalton glycoprotein from *Trypanosoma cruzi* trypomastigotes contains sialyl, fucosyl and galactosyl (alpha 1-3)galactose units. Mol Biochem Parasitol. 39(1):101-7.

Dalrymple N, & Mackow ER. 2011. Productive dengue virus infection of human endothelial cells is directed by heparan sulfate-containing proteoglycan receptors. J Virol. 85(18):9478-85.

Ribeiro de Andrade AS, Santoro MM, De Melo MN, & Mares-Guia M. 1998. *Leishmania (Leishmania) amazonensis*: purification and enzymatic characterization of a soluble serine oligopeptidase from promastigotes. Exp Parasitol. 89(2):153-60.

Dreyfuss JL, Regatieri CV, Jarrouge TR, Cavalheiro RP, Sampaio LO, & Nader HB. 2009. Heparan sulfate proteoglycans: structure, protein interactions and cell signaling. An Acad Bras Cienc. 81(3):409-29.

Dreyfuss JL, Veiga SS, Coulson-Thomas VJ, Santos IA, Toma L, Coletta RD, & Nader HB. 2010. Differences in the expression of glycosaminoglycans in human fibroblasts derived from gingival overgrowths is related to TGF-beta up-regulation. Growth Factors. 28(1):24-33.

Dumont F, & Nardelli J. 1979. Peanut agglutinin (PNA)-binding properties of murine thymocyte subpopulation. Immunology. 37(1):217-24.

Epting CL, Coates BM, & Engman DM. 2011. Molecular mechanisms of host cell invasion by *Trypanosoma cruzi*. Exp Parasitol. 127(2):607.

Fallgren C, Utt M, & Ljungh A. 2001. Isolation and characterisation of a 17-kDa staphylococcal heparin-binding protein with broad specificity. J Med Microbiol. 50(6):547-57.

Fauquenoy S, Morelle W, Hovasse A, Bednarczyk A, Slomianny C, Schaeffer C, Van Dorsselaer A, & Tomavo S. 2008. Proteomics and glycomics analyses of N-glycosylated structures involved in *Toxoplasma gondii*-host cell interactions. Mol Cell Proteomics. 7(5):891-910.

Garate M, Marchant J, Cubillos I, Cao Z, Khan NA, & Panjwani N. 2006. In vitro pathogenicity of Acanthamoeba is associated with the expression of the mannose-binding protein. Invest Ophthalmol Vis Sci. 47(3):1056-62.

Gardiner PR, Nene V, Barry MM, Thatthi R, Burleigh B, & Clarke MW. 1996. Characterization of a small variable surface glycoprotein from *Trypanosoma vivax*. Mol Biochem Parasitol. 82(1):1-11.

Geier G, Banaj HJ, Heid H, Bini L, Pallini V, & Zwilling R. 1999. Aspartyl proteases in *Caenorhabditis elegans*. Isolation, identification and characterization by a combined use of affinity chromatography, two-dimensional gel electrophoresis, microsequencing and databank analysis. Eur J Biochem. 264(3):872-9.

Gerold P, Striepen B, Reitter B, Geyer H, Geyer R, Reinwald E, Risse HJ, & Schwarz RT. 1996. Glycosyl-phosphatidylinositols of *Trypanosoma congolense*: two common precursors but a new protein-anchor. J Mol Biol. 261(2):181-94.

Gonzalez MS, Silva LC, Albuquerque-Cunha JM, Nogueira NF, Mattos DP, Castro DP, Azambuja P, & Garcia ES. 2011. Involvement of sulfated glycosaminoglycans on the development and attachment of *Trypanosoma cruzi* to the luminal midgut surface in the vector, *Rhodnius prolixus*. Parasitology. 9:1-8.

Govrin E, & Levine A. 1999. Purification of active cysteine proteases by affinity chromatography with attached E-64 inhibitor. Protein Expr Purif. 15(3):247-50.

Guedes HL, Rezende JM, Fonseca MA, Salles CM, Rossi-Bergmann B, & De-Simone SG. 2007. Identification of serine proteases from *Leishmania braziliensis*. Z Naturforsch C. 62(5-6):373-81.

Guha-Niyogi A, Sullivan DR, &Turco SJ. 2001. Glycoconjugate structures of parasitic protozoa. Glycobiology. 11: 45R-59R.

Hayashida A, Bartlett AH, Foster TJ, & Park PW. 2009. *Staphylococcus aureus* beta-toxin induces lung injury through syndecan-1. Am J Pathol. 174(2):509-18.

Hellström U, Dillner ML, Hammarström S, & Perlmann P. 1976. Fractionation of human T lymphocytes on wheat germ agglutinin-sepharose. J Exp Med. 144(5):1381-5.

Hilgard P, & Stockert R. 2000. Heparan sulfate proteoglycans initiate dengue virus infection of hepatocytes. Hepatology. 32(5):1069-77.

Hoegl L, Korting HC, & Klebe G. 1999. Inhibitors of aspartic proteases in human diseases: molecular modeling comes of age. Pharmazie. 54(5):319-29.

Jacobs DB, & Poretz RD. 1980. Lectin induction of lymphokines in cultured murine leukocytes. Cell Immunol. 51(2):424-9.

Jacquet A, Coulon L, De Nève J, Daminet V, Haumont M, Garcia L, Bollen A, Jurado M, & Biemans R. 2001. The surface antigen SAG3 mediates the attachment of *Toxoplasma gondii* to cell-surface proteoglycans. Mol Biochem Parasitol. 116(1):35-4

Kabir S. 1998. Jacalin: a jackfruit (*Artocarpus heterophyllus*) seed-derived lectin of versatile applications in immunobiological research. J Immunol Methods. 212(2):193-211.

Klemba M, & Goldberg DE. 2002. Biological roles of proteases in parasitic protozoa. Annu Rev Biochem. 71:275-305

Kobayashi K, Kato K, Sugi T, Takemae H, Pandey K , Gong H, Tohya Y, & Akashi H. 2010. *Plasmodium falciparum* BAEBL Binds to Heparan Sulfate Proteoglycans on the Human Erythrocyte Surface. J Biol Chem. 285(3): 1716–1725.

Kong, H H, Kim, T H & Chung, D-Ii. 2000. Purification And Characterization of A Secretory serine protease of *Acanthamoeba Healyi* isolated from gae. Journal of Parasitology. 86, 12–17.

Li F, Patra KP, Yowell CA, Dame JB, Chin K, & Vinetz JM. 2010. Apical surface expression of aspartic protease plasmepsin 4, A Potential. J Biol Chem. 285(11):8076-83.

Liang OD, Ascencio F, Fransson LA, & Wadström T. 1992. Binding of heparan sulfate to *Staphylococcus aureus*. Infect Immun. 60(3):899-906.

Lord MJ, Jolliffe NA, Marsden CJ, Pateman CS, Smith DC, Spooner RA, Watson PD, & Roberts LM. 2003. Ricin. Mechanisms of cytotoxicity. Toxicol Rev. 22(1):53-64.

Love DC, Esko JD, & Mosser DM. 1993. A heparin-binding activity on Leishmania amastigotes which mediates adhesion to cellular proteoglycans. J Cell Biol. 123(3):759-66.

Ly M, Laremore TN, & Linhardt RJ. 2010. Proteoglycomics: recent progress and future challenges. OMICS. 14(4):389-99.

Maia, A.A.M. & Guimarães, M.P. Distribuição Sazonal De Larvas De Dermatobia Hominis (Linnaeus Jr., 1781) (Diptera: Cuterebridae) Em Bovinos De Corte Da Região De Governador Valadares - Minas Gerais. Arquivo Brasileiro De Medicina Veterinária E Zootecnia, Brasil. , 1985. V. 37, N. 5, P. 469-475.

Mansour MH. 1996. Purification and characterization of SM 37: a fucosyllactose determinant-bearing glycoprotein probed by a *Biomphalaria alexandrina* lectin on adult male shistosomes. J Parasitol. 82(4):586-93.

Matsumoto K, Yamamoto D, Ohishi H, Tomoo K, Ishida T, Inoue M, Sadatome T, Kitamura K, & Mizuno H. 1989. Mode of binding of E-64-c, a potent thiol protease inhibitor, to papain as determined by X-ray crystal analysis of the complex. FEBS. Lett 245, 177–180.

McKerrow JH, Caffrey C, Kelly B, Loke P, & Sajid M. 2006. Proteases in parasitic diseases. Annu Rev Pathol. 1:497-536.

Morgado-Díaz JA, Silva-Lopez RE, Alves CR, Soares MJ, Corte-Real S, & De Simone SG. 2005. An intracellular serine protease of 68 kDa in *Leishmania (Leishmania) amazonensis* promastigotes. Mem Inst Oswaldo Cruz. 100(4):377-83

Morris SR, & Sakanari JA. 1994. Characterization of The serine protease and serine protease inhibitor from the tissue-penetrating nematode *Anisakis Simplex*. J Biol Chem. 269(44):27650-6.

Morrison CJ, Hurst SF, Bragg SL, Kuykendall RJ, Diaz H, McLaughlin DW, & Reiss E. 1993. Purification and characterization of the extracellular aspartyl proteinase of *Candida albicans*: removal of extraneous proteins and cell wall mannoprotein and evidence for lack of glycosylation. J Gen Microbiol. 139 Pt 6:1177-86.

Muharsini S, Sukarsih, Riding G, Partoutomo S, Hamilton S, Willadsen P, & Wijffels G. 2000. Identification and characterisation of the excreted/secreted serine proteases of larvae of the old world screwworm fly, *Chrysomya Bezziana*. Int J Parasitol. 30(6):705-14.

Naguleswaran A, Alaeddine F, Guionaud C, Vonlaufen N, Sonda S, Jenoe P, Mevissen M, & Hemphill A. 2005. *Neospora caninum* protein disulfide isomerase is involved in tachyzoite-host cell interaction. Int J Parasitol. 35(13):1459-72.

Nolan DP, Jackson DG, Windle HJ, Pays A, Geuskens M, Michel A, Voorheis HP, & Pays E. 1997. Characterization of a novel, stage-specific, invariant surface protein in *Trypanosoma brucei* containing an internal, serine-rich, repetitive motif. J Biol Chem. 272(46):29212-21.

O'Donnell CD, & Shukla D. 2008. The importance of separan sulfate in herpesvirus infection. Virol Sin. 23(6):383-393.

Oliveira FO Jr, Alves CR, Calvet CM, Toma L, Bouças RI, Nader HB, Côrtes LMC, Krieger MA, Meirelles MN, & Souza Pereira MC. 2008. *Trypanosoma cruzi* heparin-binding proteins and the nature of the host cell heparan sulfate-binding domain. Microb Pathog. 44(4):329-38

Ortega-Barria E, & Pereira EA. 1991. A novel T. cruzi heparin-binding protein promotes fibroblast adhesion and penetration of engineered bacteria and trypanosomes into mammalian cells. Cell. 67: 411-421.

Park PW, Pier GB, Hinkes MT, & Bernfield M. 2001. Exploitation of syndecan-1 shedding by *Pseudomonas aeruginosa* enhances virulence. Nature. 411(6833):98-102.

Pereira BA, Silva FS, Rebello KM, Marín-Villa M, Traub-Cseko YM, Andrade TC, Bertho ÁL, Caffarena ER, & Alves CR. 2011. In silico predicted epitopes from the COOH-terminal extension of cysteine proteinase B inducing distinct immune responses during *Leishmania (Leishmania) amazonensis* experimental murine infection. BMC Immunol. 2011,12: 44.

Pinho RT, Beltramini LM, Alves CR, & De-Simone SG. 2009. *Trypanosoma cruzi*: isolation and characterization of aspartyl proteases. Exp Parasitol. 122(2):128-33.

Rebello KM, Côrtes LM, Pereira BA, Pascarelli BM, Côrte-Real S, Finkelstein LC, Pinho RT, d'Avila-Levy CM, & Alves CR. 2009. Cysteine proteases from promastigotes of *Leishmania (Viannia) braziliensis*. Parasitol Res. 106(1):95-104.

Redmond DL, Geldhof P, & Knox DP. 2004. Evaluation of *Caenorhabditis elegans* glycoproteins as protective immunogens against *Haemonchus contortus* challenge in sheep. Int J Parasitol. 34(12):1347-53.

Reinwald E, Rautenberg P, & Risse HJ. 1981. Purification of the variant antigens of *Trypanosoma congolense*: a new approach to the isolation of glycoproteins. Biochim Biophys Acta. 668(1):119-31.

Rüdiger H, & Gabius HJ. 2001. Plant lectins: occurrence, biochemistry, functions and applications. Glycoconj J.18(8):589-613.

Ruiz-Bustos E, Ochoa JL, Wadström T, & Ascencio F. 2001. Isolation and characterisation of putative adhesins from Helicobacter pylori with affinity for heparan sulphate proteoglycan. J Med Microbiol. 50(3):215-22.

Shenai BR, Sijwali PS, Singh A, & Rosenthal PJ. 2000 Characterization of native and recombinant falcipain-2, a principal trophozoite cysteine protease and essential hemoglobinase of *Plasmodium falciparum*. J Biol Chem. 275(37):29000-10.

Silva-Lopez RE, & Giovanni-De-Simone S. 2004. *Leishmania (Leishmania) amazonensis:* Purification and characterization of a promastigote serine protease. Exp Parasitol. 107(3-4):173-82.

Silva-Lopez RE, Coelho MG, & De Simone SG. 2005. Characterization of an extracellular serine protease of *Leishmania (Leishmania) amazonensis*. Parasitology. 131(Pt 1):85-96.

Silva-López RE, Dos Santos TR, Morgado-Díaz JA, Tanaka MN, & De Simone SG. 2010. Serine protease activities in *Leishmania (Leishmania) chagasi* promastigotes. Parasitol Res. 107(5):1151-62

Silva-Lopez RE, Morgado-Díaz JA, Alves CR, Côrte-Real S, & Giovanni-De-Simone S. 2004. Subcellular localization if an extracellular serine protease in *Leishmania Leishmania) amazonensis*. Parasitol Res. 93(4):328-31

Silva-Lopez RE, Morgado-Díaz JA, Dos Santos PT, & Giovanni-De-Simone S. 2008. Purification and subcellular localization of a secreted 75 KDa *Trypanosoma cruzi* serine oligopeptidase. Acta Trop. 07(2):159-67.

Smith WD, Newlands GF, Smith SK, Pettit D, & Skuce PJ. 2003. Metalloendopeptidases from the intestinal brush border of *Haemonchus contortus* as protective antigens for sheep. Parasite Immunol. 25(6):313-23.

Smith WD, Smith SK, Pettit D, Newlands GF, & Skuce PJ. 2000. Relative protective properties of three membrane glycoprotein fractions from *Haemonchus contortus*. Parasite Immunol. 22(2):63-71.

Soong G, Martin FJ, Chun J, Cohen TS, Ahn DS, & Prince A. 2011. *Staphylococcus aureus* Protein A mediates invasion across airway epithelial Cells through activation of RhoA GTPase signaling and proteolytic activity. J Biol Chem. 286(41):35891-8.

Stephens RS, Koshiyama K, Lewis E, & Kubo A. 2001. Heparin-binding outer membrane protein of chlamydiae. Mol Microbiol. 40(3):691-9.

Todorova VK, & Stoyanov DI. 2000. Partial Characterization of Serine proteinases secreted By Adult *Trichinella Spiralis*. Parasitol Res. 86(8):684-7.

Turner DG, Wildblood LA, Inglis NF, & Jones DG. 2008. Characterization of a galectin-like activity from the parasitic nematode, *Haemonchus contortus*, which modulates ovine eosinophil migration in vitro. Vet Immunol Immunopathol. 122(1-2):138-45.

Valdivieso E, Bermudez H, Hoebeke J, Noya O, & Cesari IM. 2003. Immunological similarity between Schistosoma and bovine cathepsin D. Immunol Lett. 89(1):81-8.

Wlodawer A, & Vondrasek J. 1998. Inhibitors of HIV-1 protease: a major success of structure-assisted drug design. Annu Rev Biophys Biomol Struct.27:249-84.

Wilson ME, & Pearson RD. 1984. Infect Immun. Stage-specific variations in lectin binding to *Leishmania donovani*. 46(1):128-34.

Yan Y, Silvennoinen-Kassinen S, Leinonen M, & Saikku P. 2006. Inhibitory effect of heparin 41 sulfate-like glycosaminoglycans on the infectivity of Chlamydia pneumoniae in HL 42 cells varies between strains. Microbes Infect. 8:866-872